# Lecture Notes in Computer Science 12392

More information about this series at http://www.springer.com/series/7409

Sven Hartmann · Josef Küng ·
Gabriele Kotsis · A Min Tjoa ·
Ismail Khalil (Eds.)

# Database and Expert Systems Applications

31st International Conference, DEXA 2020
Bratislava, Slovakia, September 14–17, 2020
Proceedings, Part II

 Springer

*Editors*
Sven Hartmann
Clausthal University of Technology
Clausthal-Zellerfeld, Germany

Josef Küng
Johannes Kepler University of Linz
Linz, Austria

Gabriele Kotsis
Johannes Kepler University of Linz
Linz, Austria

A Min Tjoa
IFS
Vienna University of Technology
Vienna, Wien, Austria

Ismail Khalil
Johannes Kepler University of Linz
Linz, Austria

ISSN 0302-9743          ISSN 1611-3349   (electronic)
Lecture Notes in Computer Science
ISBN 978-3-030-59050-5          ISBN 978-3-030-59051-2   (eBook)
https://doi.org/10.1007/978-3-030-59051-2

LNCS Sublibrary: SL3 – Information Systems and Applications, incl. Internet/Web, and HCI

This Springer imprint is published by the registered company Springer Nature Switzerland AG
The registered company address is: Gewerbestrasse 11, 6330 Cham, Switzerland

# Preface

This volume contains the papers presented at the 31st International Conference on Database and Expert Systems Applications (DEXA 2020). This year, DEXA was held as a virtual conference during September 14–17, 2020, instead of as it was originally planned to be held in Bratislava, Slovakia.

On behalf of the Program Committee we commend these papers to you and hope you find them useful.

Database, information, and knowledge systems have always been a core subject of computer science. The ever increasing need to distribute, exchange, and integrate data, information, and knowledge has added further importance to this subject. Advances in the field will help facilitate new avenues of communication, to proliferate interdisciplinary discovery, and to drive innovation and commercial opportunity.

DEXA is an international conference series which showcases state-of-the-art research activities in database, information, and knowledge systems. The conference and its associated workshops provide a premier annual forum to present original research results and to examine advanced applications in the field. The goal is to bring together developers, scientists, and users to extensively discuss requirements, challenges, and solutions in database, information, and knowledge systems.

DEXA 2020 solicited original contributions dealing with all aspects of database, information, and knowledge systems. Suggested topics included, but were not limited to:

- Acquisition, Modeling, Management and Processing of Knowledge
- Authenticity, Privacy, Security, and Trust
- Availability, Reliability and Fault Tolerance
- Big Data Management and Analytics
- Consistency, Integrity, Quality of Data
- Constraint Modeling and Processing
- Cloud Computing and Database-as-a-Service
- Database Federation and Integration, Interoperability, Multi-Databases
- Data and Information Networks
- Data and Information Semantics
- Data Integration, Metadata Management, and Interoperability
- Data Structures and Data Management Algorithms
- Database and Information System Architecture and Performance
- Data Streams and Sensor Data
- Data Warehousing
- Decision Support Systems and Their Applications
- Dependability, Reliability, and Fault Tolerance
- Digital Libraries and Multimedia Databases
- Distributed, Parallel, P2P, Grid, and Cloud Databases
- Graph Databases

- Incomplete and Uncertain Data
- Information Retrieval
- Information and Database Systems and Their Applications
- Mobile, Pervasive, and Ubiquitous Data
- Modeling, Automation, and Optimization of Processes
- NoSQL and NewSQL Databases
- Object, Object-Relational, and Deductive Databases
- Provenance of Data and Information
- Semantic Web and Ontologies
- Social Networks, Social Web, Graph, and Personal Information Management
- Statistical and Scientific Databases
- Temporal, Spatial, and High Dimensional Databases
- Query Processing and Transaction Management
- User Interfaces to Databases and Information Systems
- Visual Data Analytics, Data Mining, and Knowledge Discovery
- WWW, Databases and Web Services
- Workflow Management and Databases
- XML and Semi-structured Data

Following the call for papers which yielded 190 submissions, there was a rigorous refereeing process that saw each submission reviewed by three to five international experts. The 38 submissions judged best by the Program Committee were accepted as full research papers, yielding an acceptance rate of 20%. A further 20 submissions were accepted as short research papers.

As is the tradition of DEXA, all accepted papers are published by Springer. Authors of selected papers presented at the conference were invited to submit substantially extended versions of their conference papers for publication in special issues of international journals. The submitted extended versions underwent a further review process.

We wish to thank all authors who submitted papers and all conference participants for the fruitful discussions.

This year we have five keynote talks addressing emerging trends in the database and artificial intelligence community:

- "Knowledge Graphs for Drug Discovery" by Prof. Ying Ding (The University of Texas at Austin, USA)
- "Incremental Learning and Learning with Drift" by Prof. Barbara Hammer (CITEC Centre of Excellence, Bielefeld University, Germany)
- "From Sensors to Dempster-Shafer Theory and Back: the Axiom of Ambiguous Sensor Correctness and its Applications" by Prof. Dirk Draheim (Tallinn University of Technology, Estonia)
- "Knowledge Availability and Information Literacies" by Dr. Gerald Weber (The University of Auckland, New Zealand)
- "Explainable Fact Checking for Statistical and Property Claims" by Paolo Papotti (EURECOM, France)

In addition, we had a panel discussion on "The Age of Science-making Machines" led by Prof. Stéphane Bressan (National University of Singapore, Singapore).

This edition of DEXA features three international workshops covering a variety of specialized topics:

- BIOKDD 2020: 11th International Workshop on Biological Knowledge Discovery from Data
- IWCFS 2020: 4th International Workshop on Cyber-Security and Functional Safety in Cyber-Physical Systems
- MLKgraphs 2020: Second International Workshop on Machine Learning and Knowledge Graphs

The success of DEXA 2020 is a result of collegial teamwork from many individuals. We like to thank the members of the Program Committee and the external referees for their timely expertise in carefully reviewing the submissions.

Warm thanks to Ismail Khalil and the conference organizers as well as all workshop organizers.

We would also like to express our thanks to all institutions actively supporting this event, namely:

- Comenius University Bratislava (who was prepared to host the conference)
- Institute of Telekoopertion, Johannes Kepler University Linz (JKU)
- Software Competence Center Hagenberg (SCCH)
- International Organization for Information Integration and Web based Applications and Services (@WAS)

We hope you enjoyed the conference program.

September 2020                                                        Sven Hartmann
                                                                     Josef Küng

---

The original version of the book was revised: The paper "Bounded Pattern Matching Using Views" was accidentally published twice in the book "Database and Expert Systems Applications, Part II", and the paper "View Selection for Graph Pattern Matching"was omitted. This has been corrected. Each paper is now included once. The correction to the book is available at https://doi.org/10.1007/978-3-030-59051-2_29

# Organization

## Program Committee Chairs

| | |
|---|---|
| Sven Hartmann | Clausthal University of Technology, Germany |
| Josef Küng | Johannes Kepler University Linz, Austria |

## Steering Committee

| | |
|---|---|
| Gabriele Kotsis | Johannes Kepler University Linz, Austria |
| A Min Tjoa | Vienna University of Technology, Austria |
| Ismail Khalil | Johannes Kepler University Linz, Austria |

## Program Committee and Reviewers

| | |
|---|---|
| Javier Nieves Acedo | Azterlan, Spain |
| Sonali Agarwal | IIIT, India |
| Riccardo Albertoni | CNR-IMATI, Italy |
| Idir Amine Amarouche | USTHB, Algeria |
| Rachid Anane | Coventry University, UK |
| Mustafa Atay | Winston-Salem State University, USA |
| Faten Atigui | CEDRIC, CNAM, France |
| Ladjel Bellatreche | LIAS, ENSMA, France |
| Nadia Bennani | LIRIS, INSA de Lyon, France |
| Karim Benouaret | Université de Lyon, France |
| Djamal Benslimane | Université de Lyon, France |
| Morad Benyoucef | University of Ottawa, Canada |
| Catherine Berrut | LIG, Université Joseph Fourier Grenoble I, France |
| Vasudha Bhatnagar | University of Delhi, India |
| Athman Bouguettaya | The University of Sydney, Australia |
| Omar Boussai | ERIC Laboratory, France |
| Stephane Bressan | National University of Singapore, Singapore |
| Sharma Chakravarthy | The University of Texas at Arlington, USA |
| Cindy Chen | University of Massachusetts Lowell, USA |
| Gang Chen | Victoria University of Wellington, New Zealand |
| Max Chevalier | IRIT, France |
| Soon Ae Chun | City University of New York, USA |
| Alfredo Cuzzocrea | University of Calabria, Italy |
| Debora Dahl | Conversational Technologies, USA |
| Jérôme Darmont | Université de Lyon, France |
| Soumyava Das | Teradata Labs, USA |
| Vincenzo Deufemia | University of Salerno, Italy |
| Dejing Dou | University of Oregon, USA |

| | |
|---|---|
| Ismini Vasileiou | De Montfort University, UK |
| Krishnamurthy Vidyasankar | Memorial University, Canada |
| Marco Vieira | University of Coimbra, Portugal |
| Piotr Wisniewski | Nicolaus Copernicus University, Poland |
| Ming Hour Yang | Chung Yuan Chritian University, Taiwan |
| Xiaochun Yang | Northeastern University, China |
| Haruo Yokota | Tokyo Institute of Technology, Japan |
| Qiang Zhu | University of Michigan, USA |
| Yan Zhu | Southwest Jiaotong University, China |
| Marcin Zimniak | Leipzig University, Germany |
| Ester Zumpano | University of Calabria, Italy |

## Organizers

Institute for
Telecooperation

International Organization for

www.iiwas.org

Information Integration and
Web-based Applications & Services

# Abstracts of Keynote Talks

# Knowledge Graph for Drug Discovery

Ying Ding

The University of Texas at Austin, USA

**Abstract.** A critical barrier in current drug discovery is the inability to utilize public datasets in an integrated fashion to fully understand the actions of drugs and chemical compounds on biological systems. There is a need to intelligently integrate heterogeneous datasets pertaining to compounds, drugs, targets, genes, diseases, and drug side effects now available to enable effective network data mining algorithms to extract important biological relationships. In this talk, we demonstrate the semantic integration of 25 different databases and showcase the cutting-edge machine learning and deep learning algorithms to mine knowledge graphs for deep insights, especially the latest graph embedding algorithm that outperforms baseline methods for drug and protein binding predictions.

# Incremental Learning and Learning with Drift

Barbara Hammer

CITEC Centre of Excellence, Bielefeld University, Germany

**Abstract.** Neural networks have revolutionized domains such as computer vision or language processing, and learning technology is included in everyday consumer products. Yet, practical problems often render learning surprisingly difficult, since some of the fundamental assumptions of the success of deep learning are violated. As an example, only few data might be available for tasks such as model personalization, hence few shot learning is required. Learning might take place in non-stationary environments such that models face the stability-plasticity dilemma. In such cases, applicants might be tempted to use models for settings they are not intended for, such that invalid results are unavoidable.

Within the talk, I will address three challenges of machine learning when dealing with incremental learning tasks, addressing the questions: how to learn reliably given few examples only, how to learn incrementally in non-stationary environments where drift might occur, and how to enhance machine learning models by an explicit reject option, such that they can abstain from classification if the decision is unclear

# From Sensors to Dempster-Shafer Theory and Back: The Axiom of Ambiguous Sensor Correctness and Its Applications

Dirk Draheim

Tallinn University of Technology, Estonia
dirk.draheim@taltech.ee

**Abstract.** Since its introduction in the 1960s, Dempster-Shafer theory became one of the leading strands of research in artificial intelligence with a wide range of applications in business, finance, engineering, and medical diagnosis. In this paper, we aim to grasp the essence of Dempster-Shafer theory by distinguishing between ambiguous-and-questionable and ambiguous-but-correct perceptions. Throughout the paper, we reflect our analysis in terms of signals and sensors as a natural field of application. We model ambiguous-and-questionable perceptions as a probability space with a quantity random variable and an additional perception random variable (Dempster model). We introduce a correctness property for perceptions. We use this property as an axiom for ambiguous-but-correct perceptions. In our axiomatization, Dempster's lower and upper probabilities do not have to be postulated: they are consequences of the perception correctness property. Furthermore, we outline how Dempster's lower and upper probabilities can be understood as best possible estimates of quantity probabilities. Finally, we define a natural knowledge fusion operator for perceptions and compare it with Dempster's rule of combination.

# Knowledge Availability and Information Literacies

Gerald Weber

The University of Auckland, New Zealand

**Abstract.** At least since Tim Berners-Lee's call for 'Raw Data Now' in 2009, which he combined with a push for linked data as well, the question has been raised how to make the wealth of data and knowledge available to the citizens of the world. We will set out to explore the many facets and multiple layers of this problem, leading up to the question of how we as users will access and utilize the knowledge that should be available to us.

# Explainable Fact Checking for Statistical and Property Claims

Paolo Papotti

EURECOM, France

**Abstract.** Misinformation is an important problem but fact checkers are overwhelmed by the amount of false content that is produced online every day. To support fact checkers in their efforts, we are creating data-driven verification methods that use structured datasets to assess claims and explain their decisions. For statistical claims, we translate text claims into SQL queries on relational databases. We exploit text classifiers to propose validation queries to the users and rely on tentative execution of query candidates to narrow down the set of alternatives. The verification process is controlled by a cost-based optimizer that considers expected verification overheads and the expected claim utility as training samples. For property claims, we use the rich semantics in knowledge graphs (KGs) to verify claims and produce explanations. As information in a KG is inevitably incomplete, we rely on rule discovery and on text mining to gather the evidence to assess claims. Uncertain rules and facts are turned into logical programs and the checking task is modeled as a probabilistic inference problem. Experiments show that both methods enable the efficient and effective labeling of claims with interpretable explanations, both in simulations and in real world user studies with 50% decrease in verification time. Our algorithms are demonstrated in a fact checking website (https://coronacheck.eurecom.fr), which has been used by more than twelve thousand users to verify claims related to the coronavirus disease (COVID-19) spread and effect.

# "How Many Apples?" or the Age of Science-Making Machines (Panel)

Stéphane Bressan (Panel Chair)

National University of Singapore, Singapore

**Abstract.** Isaac Newton most likely did not spend much time observing apples falling from trees. Galileo Galilei probably never threw anything from the tower of Pisa. They conducted thought experiments.

What can big data, data science, and artificial intelligence contribute to the creation of scientific knowledge? How can advances in computing, communication, and control improve or positively disrupt the scientific method?

Richard Feynman once explained the scientific method as follows. "In general, we look for a new law by the following process. First, we guess it; don't laugh that is really true. Then we compute the consequences of the guess to see what, if this is right, if this law that we guessed is right, we see what it would imply. And then we compare those computation results to nature, or, we say, compare to experiment or experience, compare directly with observations to see if it works. If it disagrees with experiment, it's wrong and that simple statement is the key to science." He added euphemistically that "It is therefore not unscientific to take a guess."

Can machines help create science?

The numerous advances of the many omics constitute an undeniable body of evidence that computing, communication, and control technologies, in the form of high-performance computing hardware, programming frameworks, algorithms, communication networks, as well as storage, sensing, and actuating devices, help scientists and make the scientific process significantly more efficient and more effective. Everyone acknowledges the unmatched ability of machines to streamline measurements and to process large volumes of results, to facilitate complex modeling, and to run complex computations and extensive simulations. The only remaining question seems to be the extent of their unexplored potential.

Furthermore, the media routinely report new spectacular successes of big data analytics and artificial intelligence that suggest new opportunities. Scientists are discussing physics-inspired machine learning. We are even contemplating the prospect of breaking combinatorial barriers with quantum computers. However, except, possibly for the latter, one way or another, it all seems about heavy-duty muscle-flexing without much subtlety nor finesse.

Can machines take a guess?

Although the thought processes leading to the guesses from which theories are built are laden with ontological, epistemological, and antecedent theoretical assumptions, and the very formulation of the guesses assumes certain conceptual views, scientists seem to have been able to break through those glass ceilings again and again and invent entirely new concepts. Surely parallel computing, optimization algorithms, reinforcement learning, or genetic algorithms can assist in the exploration of the space of combinatorial compositions of existing concepts. In the words of Feynman again: "We set up a machine, a great computing machine, which has a random wheel and if it makes a succession of guesses and each time it guesses a hypothesis about how nature should work computes immediately the consequences and makes a comparison to a list of experimental results that it has at the other hand. In other words, guessing is a dumb man's job. Actually, it is quite the opposite and I will try to explain why." He continues: "The problem is not to make, to change or to say that something might be wrong but to replace it with something and that is not so easy."

Can machines create new concepts?

The panelists are asked to share illustrative professional experiences, anecdotes, and thoughts, as well as their enthusiasm and concerns, regarding the actuality and potential of advances in computing, communication, and control in improving and positively disrupting the scientific process.

# Contents – Part II

## Knowledge Discovery

## Machine Learning

## Semantic Web and Ontologies

**Stream Data Processing**

**Temporal, Spatial, and High Dimensional Databases**

# Contents – Part I

## Prediction and Decision Support

# Authenticity, Privacy, Security and Trust

Authenticity, Integrity, and Trust

# A Framework for Factory-Trained Virtual Sensor Models Based on Censored Production Data

Sabrina Luftensteiner[✉] and Michael Zwick[✉]

Software Competence Center Hagenberg, 4232 Hagenberg im Muehlkreis, Austria
{sabrina.luftensteiner,michael.zwick}@scch.at
https://www.scch.at

**Abstract.** In Industrial Manufacturing, the process variable is often not directly observable using a physical measuring device. For this reason, Virtual Sensors are used, which are surrogate models for a physical sensor trained on previously collected process data. In order to continuously improve such virtual sensor models, it is desirable to make use of data gathered during production, e.g. to adapt the model to client-specific tools not included in the basic training data. In many real-world scenarios, it is only possible to gather data from the production process to a limited extent, as feedback is not always available. One example of such a situation is the production of a workpiece within required quality bounds. In case the finished workpiece meets the quality criteria or else is irreversibly damaged in the process, there is no feedback regarding the model error. Only in the case where the operator is able to specify a correction to reach the desired target value in a second run, the correction error can be used as an approximation for the model error.

To make use of this additional censored data, we developed a framework for preprocessing such data prior to the training of a virtual sensor model. The approach used is independent from the chosen sensor model as well as the method used for correction of the censored data. We tested our approach in combination with three different correction methods for censored data (Tobit regression, constrained convex optimization, OptNet neural network optimization layer) using data gathered in a real-world industrial manufacturing setting. Our results show that including the data can approximate the uncensored data up to a censorship percentage of 60% while at the same time improving the performance of the virtual sensor model up to 30%.

**Keywords:** Censored regression · Virtual sensor · Industrial manufacturing

The research reported in this paper has been funded by the Federal Ministry for Climate Action, Environment, Energy, Mobility, Innovation and Technology (BMK), the Federal Ministry for Digital and Economic Affairs (BMDW), and the Province of Upper Austria in the frame of the COMET - Competence Centers for Excellent Technologies Programme managed by Austrian Research Promotion Agency FFG.

S. Hartmann et al. (Eds.): DEXA 2020, LNCS 12392, pp. 3–16, 2020.
https://doi.org/10.1007/978-3-030-59051-2_1

# 1   Introduction

When cyclically operating on workpieces in industrial manufacturing, one or more process parameters are constantly monitored and used to control the production process. In many cases, the parameter of interest is not directly observable using a physical measuring device. This can be due to the high cost of installing a physical sensor (e.g. for the initial purchase of the sensor, the ongoing maintenance fees or the long setup-times after changing tools), as well as the mechanical construction of a machine, making it impossible to mount a sensor in a useful way, even if such a sensor would be available. In these settings, virtual sensors are a viable alternative for controlling the production process and reach the required process targets.

A virtual sensor [13] is a surrogate model for a physical sensor trained on previously collected process data. Such a model takes as input process information that is actually observable, and outputs a prediction for the target parameter which is required to control the production process, e.g. to determine when a desired state of a workpiece is reached.

When virtual sensors are trained offline, there is the underlying assumption that the input data used for prediction during production follows the same distribution as the data collected beforehand for training (usually done in a controlled environment). It is therefore desirable to make use of data gathered during production, e.g. to re-calibrate the virtual sensor or to customize the virtual sensor to special process conditions (e.g. adapt the model to client-specific tools not included in the training data).

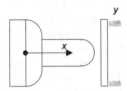

**Fig. 1.** A tool acting with feature $x$ on a workpiece causing a target feature $y$ (following [22]). The target is cyclically predicted using a previously trained model (a virtual sensor) based on data collected in a controlled environment prior to deploying the system. Once the desired target value is reached, the process is stopped.

Data gathered in a real-world scenarios, while the production machine is online, exhibits a much lower quality than training data gathered in a controlled experimental environment prior to training the model. This is especially true when input from a human operator is required, e.g. when manual recording of model errors is necessary. This is often the case because automatic measurements from physical sensors are not available, which leads to the very reason virtual sensors are needed in the first place (see above). In many real-world scenarios, it is at least possible to gather some data from the production process, which has to be recorded anyway in order for normal production to proceed. This puts no additional burden on the human operator and can be used for improving the virtual sensors prediction model.

Consider a tool (e.g. a drilling head) acting with feature $x$ on a workpiece causing a target feature $y$ (see Fig. 1). The goal is to reach a certain target value $y_t$ within lower and upper bounds $[y_l, y_u]$ to ensure a desired quality level (Fig. 2). Using a virtual sensor, the process is stopped once the virtual sensor predicts that $y_t$ is reached. The difference from the prediction to the actual target value $\Delta y = \hat{y} - y_t$ represents the model error. This setup is asymmetric in the sense that in case the virtual sensor makes predictions which result in a target value within or above the quality boundaries ($y \geq y_l$), there is no feedback regarding the model error because either the workpiece meets the quality criteria and is forwarded to the next production step, or has to be discarded due to the irreversibly of the deformation. Only in the case where the actual value lies below the quality boundaries ($y < y_l$), the operator is able to specify a correction using $\Delta y$ to reach the desired target value $y_t$ after the same workpiece underwent a correction run with $y_t' = y_t + \Delta y$ as the new target value. We want to use $\Delta y$ as an approximation to the model error and use it to further refine the production model currently in use for the specific context the data was collected (e.g. for a machine model or tool setting).

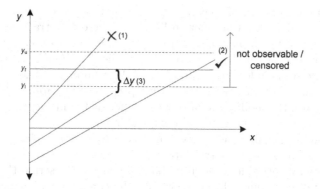

**Fig. 2.** Possible outcomes after the production of a workpiece using the prediction from a virtual sensor for controlling the process: (1) The resulting target value $y$ lies above the upper bound $y_u$, resulting in a deficient product that needs to be discarded; (2) The target value lies within the acceptable production limits $[y_l, y_u]$; (3) The resulting target value $y$ lies below the lower bound $y_l$, in this case the operator is able to apply a correction $\Delta y$ to the process which is also assumed to be an approximation for prediction error of the virtual sensor. Note that in cases (1) and (2), there is no feedback regarding the prediction error available.

What follows is the formal description of the problem setup. We are faced with a *censored regression model* $h_\theta$ [5,8,9,19] with model parameters $\theta$, where the observable target $y_i$ is censored above the threshold value $y_l$, $y_i^*$ representing the true, unobserved dependent variable and $i \in \{1, ..N\}$ the indices of observations (see Eqs. 1 and 2).

$$y_i^* = h_\theta(x_i) + \varepsilon_i \tag{1}$$

$$y_i = \begin{cases} y_l, & \text{if } y_i^* \geq y_l \\ y_i^*, & \text{if } y_i^* < y_l \end{cases} \tag{2}$$

The model assumes a homoscedastic error term $\varepsilon \sim \mathcal{N}(0, \sigma^2)$. Fitting censored regression models to data can be described using the (Log-)Maximum Likelihood formulation (Eq. 3, see [10]), where $\phi(.)$ and $\Phi(.)$ represent the probability density function and the cumulative distribution function of the standard normal distribution, and $I_i^{y_l}$ an indicator function (Eq. 4). The loss/error corresponding to the ML is given by $E = -\log L$.

$$\log L = \sum_{i=1}^{N} \left[ I_i^{y_l} \log \Phi \left( \frac{h_\theta(x_i) - y_l}{\sigma} \right) \right.$$

$$\left. + (1 - I_i^{y_l}) \left( \log \phi \left( \frac{y_i - h_\theta(x_i)}{\sigma} \right) - \log \sigma \right) \right] \tag{3}$$

$$I_i^{y_l} = \begin{cases} 1, & \text{if } y_i = y_l \\ 0, & \text{if } y_i < y_l \end{cases} \tag{4}$$

The main contribution of this paper is a model agnostic approach for integrating censored data from production lines into the offline training of (virtual sensor) models. Such an approach is especially useful in industrial environments where it is not feasible for the operator to label the data. Our results show that it is reasonable to include censored data into the training process compared to omitting it. Especially in combination with preprocessing approaches tailored for censored data, the models are improving, whereas the improvement starts to decline if the amount of censored data exceeds 60%.

This paper is organized as follows: in Sect. 2 we discuss relevant literature regarding the handling of censored data, which is mainly based on survival data analysis. Section 3 introduces our framework, where we describe the structure and workflow of our approach. Furthermore, the three model-independent correction methods we evaluated to adapt censored data prior to training are presented (see Sect. 3.2). Section 4 presents our experimental setting, the data used and the results of the experiments, where we tested our approach by evaluating the correction methods. Section 5 concludes our work and gives a short overview on possible future work.

## 2    Related Work

This section contains a brief overview into current and traditional machine learning literature regarding censored data models and applications. Approaches to decrease the negative influence of censored data in supervised learning are mentioned. However, they are often tailored to specific model architectures, not model-agnostic as presented in this paper. Furthermore, some industrial applications are mentioned, as previous approaches are mainly targeted towards survival data analysis.

## 2.1 Model Architectures for Censored Data

Regarding Support Vector Machines (SVM), [18] introduces an interesting approach. They developed a special type of SVM called SVCR, which is especially tailored for regression on censored target data. Their approach considers both, right- and left-censored data as it penalizes values outside of a pre-defined interval, which defines the boundaries for uncensored data. The approach includes all available information and also works for quadratic problems. The intervals have to be defined previously, which complicates adaptability, modularity and reusability of models. Furthermore, the approach was only tested on left-censored data.

In [17], an approach for regression trees in combination with right-censored data is proposed. The author uses the Tarone-Ware and Harrington-Fleming classes of two-sample statistics [20] to define the splitting rules of a tree.

Many popular survival models use a combination of random forest models and the Cox proportional hazard method [3] to handle right censored data. Another approach regarding random forest models is the random survival forest model (RSF) with space extensions proposed in [21]. The RSF model is adapted to right censored time-to-event data by maximizing the survival differences between subtree nodes and the use of space extensions.

A neural network approach using the Cox's proportional hazard model [3] is presented in [6]. The approach uses the input-output relationship of a simple feed-forward network as a base for a non-linear Cox's proportional hazard model.

In [11], an approach for lasso models is presented. The author suggests to use a minimizer of a convex loss function to calibrate an estimator in an one-step imputation.

## 2.2 Censored Data in Industrial Applications

In [12], an approach for monitoring right-censored failure time data in industrial manufacturing is presented, where the authors make use of one-sample non-parametric rank tests.

The approach presented in [23] is based on censored data in display advertising. The authors describe a click-through rate (CTR) estimation based on censored auction data, which is based on a survival model, and furthermore present a gradient descent method to achieve unbiased learning.

## 3 Methods

This section covers the structure of the framework, its possibilities for configuration and its different modules. Additionally, the methods available for the correction respectively adaptation of censored data and subsequent model training in combination with uncensored data are presented to give a brief insight into their functionality, including advantages and disadvantages as well as their fields of application.

## 3.1  Framework

The framework was originally created to facilitate comparisons between diverse virtual sensor models, but has been further developed to cover censored data. Since the beginning it has been kept modular to enable extensibility regarding both the preprocessing methods and virtual sensor model training. The framework can be split into 2 main components, the preprocessing of (censored) input data and the subsequent training of a virtual sensor model (see Fig. 3).

To enable fast and easy adaptations regarding the preprocessing as well as training steps, the framework offers a configuration file with various configuration options. Feature and target columns are able to be set as well as the training model and preprocessing method. Due to its extensibility, the framework can be adapted to a wide field of applications.

Figure 3 represents the workflow of the framework. The input data may contain censored data, which can be adapted by various preprocessing approaches in a first step. In case a correction, of censored data is configured, the data is sorted due to method specifications and passed to the chosen method, where it is processed and adapted. The adapted values are returned and replace the censored data, whereas the uncensored data stays untouched. After the preprocessing step is completed, the data is passed on to the training of the virtual sensor model. The virtual sensor model is defined in the configuration file, where it is also possible to define multiple models, which are iteratively trained and deployed. The available models can be easily extended as the models just have to follow some specifications, e.g. uniform method invocations. The basic model architectures we use for evaluation consist of linear regression, ridge regression, lasso, elastic net and simple random forest and feed-forward neural network models (see Sect. 4).

## 3.2  Correction Methods

Due to their independence with respect to the virtual sensor model, the following correction methods for censored data were chosen as preprocessing steps (see Fig. 3).

*Convex Optimization Approach.* Convex optimization covers a wide area regarding its application [2], e.g. machine learning and signal processing. As the adaptation of (non-)linear censored data may be performed using convex optimization, we chose to use the CVXPY library [4], which provides a domain-specific language to define convex optimization problems. To define such a problem in CVXPY, it is necessary to know the constraints, e.g. where the lower bound is (see Fig. 2). The approach is applicable for both left and right censored data as well as a mixture of censored data, e.g. due to limited records [14].

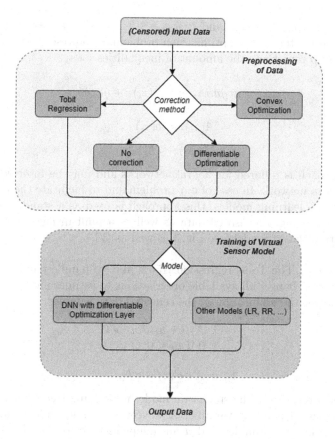

**Fig. 3.** Workflow of the Framework, consisting of two main parts: the preprocessing, including the correction of censored data, and the model training.

*Differentiable Optimization Approach.* The Differentiable Optimization (Opt-Net, see [1]) approach is based on a neural network architecture that integrates optimization problems, whereas it focuses on quadratic problems as individual layers in deep neural networks. The individual layers encode constraints and complex dependencies between hidden states. Traditional neural networks with fully-connected layers are often not able to capture such dependencies often due to simple computations. Neural networks using Differentiable Optimization are able to capture much richer behaviour and therefore expressing complex operations, enabling a shallower neural network while preserving the richness of the representation.

The approach presented in [1] uses the $n+1$st layer of a neural network as a solution for a constrained quadratic optimization problem based upon previous layers. The authors consider layers defined by a quadratic program in the form represented in Eq. 5. $z \in \mathbb{R}^n$ represents the optimization variable, $Q \in \mathbb{R}^{n*n}$ and $q \in \mathbb{R}^n$ are parameters of the optimization problem. $n$ is the amount of hidden layers in the network. In Eqs. 6 and 7, $A \in \mathbb{R}^{m*n}$, $b \in \mathbb{R}^m$, $G \in \mathbb{R}^{p x n}$ and $h \in \mathbb{R}^p$ are further parameters of the optimization problem. $m$ represents the amount of equations and $p$ represents the amount of inequalities.

$$z_{i+1} = argmin_z \frac{1}{2} z^T Q_i(z_i) z + q(z_i)^T z \tag{5}$$

$$A(z_i)z = b(z_i) \tag{6}$$

$$G(z_i)z \leq h(z_i) \tag{7}$$

This approach is tailored for neural networks and may be inserted as additional layer in a network. In case of our problem and to facilitate the usage with various machine learning models, this approach is used as a stand-alone model for the preprocessing of censored data as well as a built-in layer in the neural network virtual sensor model (see Fig. 3 as well as [14]).

*Tobit Regression.* The Tobit regression [15] is able to handle truncated as well as censored data. It uses all available observations to estimate a regression line, which is used for the adaptation of the censored data samples.

$$y_t = \begin{cases} X_t \beta + u_t \text{ if } u_t > 0 \\ 0 \text{ if } u_t \leq 0 \end{cases} \tag{8}$$

$$E_y = X\beta F(z) + \sigma f(z) \tag{9}$$

Equation 8 represents the stochastic model underlying the Tobit regression. $y$ represents the dependent variable, $X$ a vector of independent variables, $\beta$ a vector of unknown variables and $u$ an independent error term, which has a normal distribution with zero mean and a constant variance $\sigma^2$. The calculation of the adapted value is described by Eq. 9, where $z$ is $X\beta/\sigma$, $f(z)$ the unit normal density and $F(z)$ the cumulative normal distribution function. The adapted value for observations above a given limit is the error of the truncated normal error term $\sigma f(z)/F(z)$ in addition to $X\beta$.

In contrast to the Convex Optimization Approach and the OptNet Approach, the Tobit regression is only reliable for linear problems and is not recommended for quadratic problems, which also becomes obvious when looking at the experimental results (see Sect. 4). Improved Tobit regression models are able to handle left censored data as well as right censored data [14].

## 4   Experiments and Discussion

In this section we describe the experimental data used, as well as the setup for simulating censored data. Furthermore, the results of our experiments are presented including a comparison of the tested approaches to adapt the censored values prior to model training.

## 4.1   Data and Simulated Censoring

The data used for our experiments was gathered in a real industrial environment similar to Fig. 1 and is was used to train a virtual sensor (see Sect. 1). It contains observable features as inputs (i.e. machine settings, workpiece geometry and dynamic process variables), and a numerical target parameter, which is used to determine the state of the workpiece.

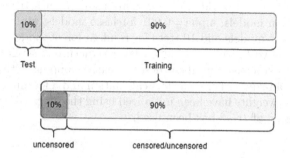

**Fig. 4.** Data partition. 10% of the available data is used as testing data, the other 90% as training data. The training data is used to simulate censored data, where 10% are used as uncensored baseline training data and the remaining data is censored according to a given percentage or limit. The cross-validation in these experiments respects both partitions.

To reflect censored data added by customers, e.g. for a recalibration of the virtual sensor, we simulate the censoring of available data, see Fig. 4. For this purpose, we divide the training data into ten parts, nine parts are censored and one part always remains uncensored, the one part representing data provided by the manufacturer. The remaining data is ordered according to the target value and the highest $n\%$ are set as a boundary value. To cover a wide range of censorship possibilities, $n$ ranges from 0 to 90% censored data with a step size of 10%.

This scenario fits the use case of virtual sensors as a pre-trained model is often provided for customers. The customers are producing further data over time, where some or all of the data is censored. This censored data has a negative influence on further model training and to avoid such biased models, adaptations to the data have to be made.

## 4.2   Model

We tested the approaches for the adaptation of censored data in combination with 6 different model types to show their model-independence:

– Linear Regression
– Ridge Regression

- Elastic Net
- Lasso
- Feed-Forward Neural Network
- Random Forest

The models have been chosen as they are often used in industry to train prediction models, see [7, 16]. The setting of the hyper-parameters has been taken from a prior grid-search conducted for training the initial offline (virtual sensor) models, to reflect a real-world use case. This resulted in setting alpha = 0.01 for ridge regression models, alpha = 0.001 for lasso models, alpha=0.002 and l1 ratio for elastic net models and 10 trees for random forest models. For the feed-forward neural network, we use a hidden layer structure of 35-25-15-7-10-20 in combination with stochastic gradient descent, early-stopping with a patience of 25 iterations, mean-squared error as loss calculation and a learning rate of 0.001. Furthermore, the weights have been initialized using the Xavier-Uniform method and a weight decay of 1e−3 has been used.

### 4.3    Results

We used ten-fold cross-validation in our experiments and selected the following metrics for the presentation of our results: the root mean squared error (RMSE) to represent the overall performance and the standard deviation $\sigma$ to represent the robustness of models during cross-validation. To enable the comparison of our approaches regarding the adaptation to censored data, we added two further experiments: completely omitting censored data (CDNA) and using the censored data in its unmodified (CDNC) form during the training of the virtual sensor model.

Figure 5 shows the results of a feed-forward neural network in combination with different amounts of censored data and correction approaches. The visualization illustrates the disparity between the different approaches. In our case, it is clearly visible that adding censored data improves the performance of the virtual sensor model, even if it is not corrected, see the results of CDNC Train and Test compared to CDNA. Further improvements are reached by using convex and differentiable optimization correction approaches, in this case the results are very similar. The differentiable optimization approach displays a slight unstable behaviour compared to the convex optimization approach, which may be due to its ability to also learn the constraints without additional input. The usage of the Tobit regression leads to worse results as the underlying problem of the data is not linear. Due to this reason, the Tobit regression was excluded from the visualization in Fig. 5. However, the results are still presented in Table 1.

Table 1 contains the testing results of the conducted experiments. It is organised according to the models used and percentages of censoring (vertically) as well as the different correction approaches (horizontally), where one entry represents a single experiment. For each experiment the root-mean squared error (RMSE) and the standard deviation ($\sigma$) are calculated. Furthermore, the column *perc.* represents the improvement of an approach compared to the baseline value in the same censored percentage. As the baseline, we chose the column CDNC

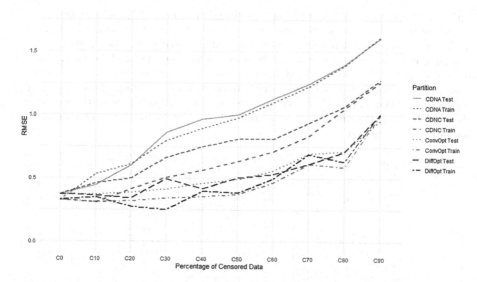

**Fig. 5.** RMSE trend of a feed-forward virtual sensor model compared over the growing amount of censored data. Test and training results are presented together (see legend). ConvOpt and DiffOpt represent the convex and differentiable optimization approaches, CDNA (censored data not added) represents training without censored data and CDNC (censored data not corrected) the unmodified usage of censored data during the training.

(censored data not corrected), which contains the results of the according models trained with unmodified censored data. As an additional baseline, we trained the models without the censored data (CDNA, censored data not added). The CDNA approach leads to less training data, as the censored data is removed from the dataset and we only use the uncensored data for training.

The results disclose that every model is negatively influenced by the censored data as can be seen by viewing the RMSE and $\sigma$ values. The errors are rising in combination with a higher amount of censored data. The idea of training the model with a smaller but uncensored dataset (CDNA) may seem easy and comfortable at first glance, but still the missing data contains information, which is needed for a more successful and proper model. This leads to the result that even additional unmodified censored data improves a model slightly. Nearly all of the tested models show this behaviour in a similar way, with exception of the lasso and elastic net model.

The results of the Tobit regression provide the worst results in our experiments, leading to an extreme degradation of the performance. The bad results are caused by the data, as the Tobit regression has been created for linear data and our data is non-linear. The convex and differentiable optimization approach seem very promising. They are able to adapt censored values in the direction of the original values and therefore lead to stronger improvements regarding the model results in comparison to models using the other approaches. Regarding the experiments from our use-case, the results seem promising in cases where the

**Table 1.** RMSE and StdDev ($\sigma$) results on the testing data. The results represent the average of the values over the different cross-validation folds. The left column covers the used virtual censor models and the percentage of censored data. The column CDNA (censored data not added) represents the results where the censored data is not used during training. CDNC (censored data not corrected) represents the results of models trained with a specific amount of censored data without correction (baseline). Tobit regression, convex and differentiable optimization approaches represent the results of the models using the given preprocessing method. The column *perc.* illustrates the improvement compared to the baseline.

| | CDNA RMSE | $\sigma$ | perc. | CDNC (Baseline) RMSE | $\sigma$ | Tobit Regression RMSE | $\sigma$ | perc. | Convex Optimization RMSE | $\sigma$ | perc. | Differentiable Optimization RMSE | $\sigma$ | perc. |
|---|---|---|---|---|---|---|---|---|---|---|---|---|---|---|
| **Linear Regression** 0% | 0.367 | 0.383 | 100.0% | 0.367 | 0.383 | 0.367 | 0.383 | 100.0% | 0.367 | 0.383 | 100.0% | 0.367 | 0.383 | 100.0% |
| 10% | 0.402 | 0.434 | 106.13% | 0.379 | 0.396 | 6.085 | 6.272 | 1606.81% | 0.373 | 0.382 | 98.59% | 0.363 | 0.406 | 95.89% |
| 20% | 0.442 | 0.48 | 103.58% | 0.427 | 0.464 | 6.411 | 6.609 | 1501.64% | 0.358 | 0.367 | 83.96% | 0.333 | 0.339 | 77.88% |
| 30% | 0.526 | 0.54 | 105.5% | 0.498 | 0.516 | 6.502 | 6.701 | 1305.33% | 0.357 | 0.373 | 71.57% | 0.455 | 0.443 | 91.3% |
| 40% | 0.577 | 0.589 | 103.32% | 0.559 | 0.56 | 6.666 | 6.865 | 1193.21% | 0.363 | 0.36 | 64.9% | 0.367 | 0.374 | 65.73% |
| 50% | 0.66 | 0.648 | 103.99% | 0.635 | 0.621 | 5.554 | 5.767 | 874.57% | 0.399 | 0.423 | 68.85% | 0.438 | 0.486 | 68.94% |
| 60% | 0.705 | 0.705 | 99.27% | 0.701 | 0.71 | 4.961 | 5.192 | 698.65% | 0.451 | 0.473 | 63.51% | 0.501 | 0.409 | 70.6% |
| 70% | 0.862 | 0.839 | 102.27% | 0.842 | 0.851 | 5.982 | 6.215 | 710.05% | 0.58 | 0.632 | 68.85% | 0.594 | 0.643 | 70.48% |
| 80% | 1.137 | 1.101 | 110.04% | 1.033 | 1.083 | 5.906 | 6.094 | 571.67% | 0.537 | 0.537 | 51.93% | 0.604 | 0.582 | 58.48% |
| 90% | 1.223 | 1.16 | 95.7% | 1.278 | 1.371 | 6.585 | 6.388 | 515.4% | 1.007 | 0.81 | 78.84% | 0.907 | 0.792 | 71.0% |
| **Ridge Regression** 0% | 0.358 | 0.379 | 100.0% | 0.358 | 0.379 | 0.358 | 0.379 | 100.0% | 0.358 | 0.379 | 100.0% | 0.358 | 0.379 | 100.0% |
| 10% | 0.384 | 0.41 | 104.68% | 0.367 | 0.387 | 6.083 | 6.249 | 1657.84% | 0.366 | 0.375 | 99.73% | 0.356 | 0.399 | 96.94% |
| 20% | 0.422 | 0.456 | 102.38% | 0.412 | 0.441 | 6.41 | 6.586 | 1554.83% | 0.351 | 0.371 | 85.03% | 0.325 | 0.343 | 78.73% |
| 30% | 0.504 | 0.512 | 104.3% | 0.483 | 0.494 | 6.501 | 6.684 | 1345.82% | 0.353 | 0.373 | 73.13% | 0.361 | 0.361 | 93.48% |
| 40% | 0.56 | 0.571 | 102.89% | 0.541 | 0.541 | 6.665 | 6.851 | 1223.33% | 0.356 | 0.365 | 65.18% | 0.361 | 0.373 | 66.93% |
| 50% | 0.648 | 0.641 | 104.26% | 0.622 | 0.609 | 5.553 | 5.752 | 893.23% | 0.393 | 0.406 | 63.22% | 0.431 | 0.468 | 69.4% |
| 60% | 0.706 | 0.707 | 101.25% | 0.698 | 0.684 | 4.959 | 5.168 | 710.97% | 0.451 | 0.49 | 64.72% | 0.502 | 0.425 | 71.94% |
| 70% | 0.866 | 0.85 | 104.26% | 0.831 | 0.843 | 5.98 | 6.187 | 719.71% | 0.582 | 0.623 | 69.99% | 0.595 | 0.634 | 71.64% |
| 80% | 1.102 | 1.072 | 107.75% | 1.023 | 1.072 | 5.905 | 6.073 | 577.16% | 0.544 | 0.546 | 53.13% | 0.611 | 0.59 | 59.74% |
| 90% | 1.229 | 1.131 | 96.85% | 1.269 | 1.367 | 6.583 | 6.361 | 518.61% | 0.996 | 0.803 | 78.5% | 0.896 | 0.784 | 70.61% |
| **Lasso** 0% | 0.358 | 0.38 | 100.0% | 0.358 | 0.38 | 0.358 | 0.38 | 100.0% | 0.358 | 0.38 | 100.0% | 0.358 | 0.38 | 100.0% |
| 10% | 0.374 | 0.397 | 102.53% | 0.365 | 0.381 | 6.083 | 6.241 | 1666.7% | 0.366 | 0.373 | 100.28% | 0.356 | 0.397 | 97.48% |
| 20% | 0.403 | 0.441 | 98.6% | 0.409 | 0.447 | 6.409 | 6.677 | 1567.69% | 0.351 | 0.371 | 85.79% | 0.325 | 0.343 | 79.44% |
| 30% | 0.473 | 0.506 | 98.85% | 0.481 | 0.501 | 6.5 | 6.677 | 1350.71% | 0.359 | 0.37 | 74.51% | 0.457 | 0.44 | 94.93% |
| 40% | 0.525 | 0.549 | 96.39% | 0.544 | 0.548 | 6.665 | 6.852 | 1224.7% | 0.356 | 0.365 | 65.41% | 0.361 | 0.379 | 66.26% |
| 50% | 0.6 | 0.605 | 96.89% | 0.623 | 0.617 | 5.553 | 5.748 | 891.29% | 0.388 | 0.401 | 62.22% | 0.426 | 0.463 | 68.4% |
| 60% | 0.636 | 0.638 | 90.91% | 0.7 | 0.693 | 4.957 | 5.171 | 708.26% | 0.452 | 0.49 | 64.62% | 0.503 | 0.425 | 71.81% |
| 70% | 0.774 | 0.771 | 92.8% | 0.834 | 0.846 | 5.979 | 6.18 | 716.53% | 0.579 | 0.627 | 69.36% | 0.593 | 0.639 | 71.01% |
| 80% | 1.112 | 1.083 | 108.16% | 1.028 | 1.077 | 5.904 | 6.057 | 574.48% | 0.556 | 0.568 | 54.11% | 0.624 | 0.613 | 60.63% |
| 90% | 1.218 | 1.149 | 95.3% | 1.274 | 1.368 | 6.585 | 6.341 | 516.83% | 0.998 | 0.804 | 78.32% | 0.898 | 0.785 | 70.47% |
| **Elastic Net** 0% | 0.357 | 0.37 | 100.0% | 0.357 | 0.37 | 0.357 | 0.37 | 100.0% | 0.357 | 0.37 | 100.0% | 0.357 | 0.37 | 100.0% |
| 10% | 0.365 | 0.375 | 101.06% | 0.361 | 0.375 | 6.082 | 6.218 | 1686.11% | 0.37 | 0.366 | 102.47% | 0.359 | 0.39 | 99.64% |
| 20% | 0.385 | 0.421 | 96.33% | 0.4 | 0.422 | 6.409 | 6.56 | 1602.98% | 0.354 | 0.359 | 88.46% | 0.328 | 0.331 | 81.97% |
| 30% | 0.451 | 0.469 | 95.84% | 0.471 | 0.472 | 6.5 | 6.669 | 1380.51% | 0.367 | 0.359 | 77.94% | 0.465 | 0.428 | 98.88% |
| 40% | 0.488 | 0.514 | 91.5% | 0.533 | 0.515 | 6.664 | 6.845 | 1250.37% | 0.362 | 0.355 | 67.83% | 0.366 | 0.369 | 68.71% |
| 50% | 0.561 | 0.558 | 91.57% | 0.613 | 0.599 | 5.552 | 5.736 | 906.2% | 0.39 | 0.392 | 63.65% | 0.428 | 0.455 | 69.98% |
| 60% | 0.603 | 0.619 | 91.19% | 0.69 | 0.675 | 4.957 | 5.134 | 718.33% | 0.46 | 0.491 | 66.6% | 0.51 | 0.426 | 73.9% |
| 70% | 0.753 | 0.772 | 91.17% | 0.826 | 0.837 | 5.978 | 6.159 | 724.16% | 0.585 | 0.654 | 70.83% | 0.598 | 0.665 | 72.49% |
| 80% | 1.123 | 1.066 | 110.1% | 1.02 | 1.068 | 5.904 | 6.036 | 578.75% | 0.564 | 0.584 | 55.3% | 0.632 | 0.628 | 61.93% |
| 90% | 1.564 | 1.434 | 123.36% | 1.268 | 1.367 | 6.584 | 6.31 | 519.26% | 0.992 | 0.81 | 78.31% | 0.892 | 0.791 | 70.31% |
| **Random Forest** 0% | 0.242 | 0.363 | 100.0% | 0.242 | 0.363 | 0.242 | 0.363 | 100.0% | 0.242 | 0.363 | 100.0% | 0.242 | 0.363 | 100.0% |
| 10% | 0.434 | 0.422 | 123.37% | 0.352 | 0.381 | 6.082 | 6.261 | 1730.39% | 0.278 | 0.351 | 79.0% | 0.267 | 0.375 | 76.09% |
| 20% | 0.521 | 0.499 | 123.5% | 0.422 | 0.422 | 6.405 | 6.601 | 1518.44% | 0.324 | 0.34 | 76.81% | 0.298 | 0.312 | 70.65% |
| 30% | 0.621 | 0.537 | 122.81% | 0.506 | 0.464 | 6.491 | 6.708 | 1283.37% | 0.37 | 0.398 | 73.17% | 0.468 | 0.468 | 98.6% |
| 40% | 0.705 | 0.642 | 122.81% | 0.574 | 0.517 | 6.659 | 6.869 | 1159.5% | 0.346 | 0.385 | 60.35% | 0.351 | 0.399 | 61.06% |
| 50% | 0.791 | 0.761 | 121.78% | 0.65 | 0.575 | 5.547 | 5.757 | 853.0% | 0.393 | 0.405 | 60.55% | 0.432 | 0.467 | 66.47% |
| 60% | 0.869 | 0.891 | 119.83% | 0.725 | 0.662 | 4.945 | 5.156 | 682.01% | 0.438 | 0.464 | 60.36% | 0.488 | 0.399 | 67.31% |
| 70% | 1.033 | 1.071 | 121.45% | 0.851 | 0.844 | 5.967 | 6.171 | 701.43% | 0.579 | 0.603 | 68.05% | 0.593 | 0.615 | 69.67% |
| 80% | 1.319 | 1.389 | 126.98% | 1.039 | 1.095 | 6.435 | 6.065 | 565.45% | 0.542 | 0.548 | 52.15% | 0.609 | 0.592 | 58.66% |
| 90% | 1.57 | 1.744 | 123.0% | 1.277 | 1.387 | 6.465 | 6.352 | 506.36% | 0.873 | 0.804 | 68.35% | 0.773 | 0.786 | 60.51% |
| **Feed-Forward NN** 0% | 0.328 | 0.325 | 100.0% | 0.328 | 0.325 | 0.328 | 0.325 | 100.0% | 0.328 | 0.325 | 100.0% | 0.328 | 0.325 | 100.0% |
| 10% | 0.529 | 0.454 | 171.58% | 0.308 | 0.299 | 5.987 | 6.105 | 1940.94% | 0.31 | 0.283 | 100.43% | 0.291 | 0.266 | 70.85% |
| 20% | 0.61 | 0.55 | 148.49% | 0.411 | 0.354 | 6.309 | 6.508 | 1534.96% | 0.317 | 0.294 | 77.17% | 0.291 | 0.266 | 70.88% |
| 30% | 0.797 | 0.689 | 157.77% | 0.505 | 0.409 | 6.398 | 6.582 | 1267.04% | 0.343 | 0.326 | 67.86% | 0.441 | 0.399 | 87.32% |
| 40% | 0.891 | 0.837 | 159.03% | 0.56 | 0.443 | 6.559 | 6.766 | 1170.4% | 0.351 | 0.323 | 68.7% | 0.356 | 0.337 | 63.53% |
| 50% | 0.976 | 0.94 | 153.75% | 0.635 | 0.502 | 5.467 | 5.686 | 861.44% | 0.375 | 0.343 | 59.04% | 0.413 | 0.405 | 65.1% |
| 60% | 1.1 | 1.063 | 154.22% | 0.713 | 0.574 | 4.886 | 5.097 | 685.15% | 0.465 | 0.448 | 65.83% | 0.516 | 0.384 | 72.29% |
| 70% | 1.224 | 1.222 | 146.91% | 0.833 | 0.734 | 5.89 | 6.131 | 706.72% | 0.613 | 0.561 | 73.53% | 0.627 | 0.573 | 75.18% |
| 80% | 1.389 | 1.454 | 132.43% | 1.049 | 1.03 | 5.81 | 5.965 | 553.86% | 0.59 | 0.556 | 56.21% | 0.657 | 0.6 | 62.66% |
| 90% | 1.616 | 1.722 | 128.36% | 1.259 | 1.305 | 6.433 | 6.225 | 511.19% | 0.995 | 0.753 | 79.07% | 0.895 | 0.734 | 71.12% |

amount of censored data is below 60%. If a higher amount of data is censored, the approaches are still effective but are losing the ability to correct well. The two approaches produce similar results, whereas the differentiable optimization approach has a higher variance. The fluctuation indicates that the approach is more unstable, which is likely provoked by the automatic derivation of the constraints (i.e. censor boundaries) in the OptNet neural network layer.

Although it leads to such bad results, the Tobit regression is still included in the framework to have a fast and simple solution for linear problems. The convex and differentiable optimization approach both produce good results, whereas the computation time using the differentiable optimization approach is higher compared to the convex optimization approach due to its underlying neural network.

## 4.4 Further Discussion

The training of networks using differentiable optimization takes longer compared to the training of the convex optimization approach but still has slightly better results. The main reason for the slow training of differentiable optimization networks is the training of a whole network and its additional background actions to discover the constraints by itself, given only the number of (in-)equalities. Unfortunately, the Python library for differentiable optimization has not been updated for a few years, although the approach is very interesting, and may become deprecated within time. Due to these developments, we recommend using the framework in combination with the convex optimization approach for convex problems or the Tobit regression for linear problems, as it has its advantage in speed compared to the convex optimization approach.

## 5 Conclusion and Future Work

In this paper, we have presented a framework which implements a model-agnostic approach for integrating censored data from production lines into the offline training of (virtual sensor) models. The framework is especially useful in environments, where human operators are not able to label the data, but it is still possible to indirectly infer the labels. Furthermore, we presented three approaches to adapt censored data in the direction of their original values, where one is used for linear problems and two may be used for non-linear problems.

Our results show that including the data can approximate the uncensored data up to a censorship percentage of 60%. At the same time, the performance of the virtual sensor models can be improved up to 30% compared to the baseline (only use uncensored data), depending on the used correction method. Generally, the results seem convincing regarding the usage of censored data in model training, especially in combination with a preprocessing correction method, given the method fits the data (linear vs. non-linear). The convex and differentiable optimization of censored data produces promising results in combination with our industrial use-case, the Tobit regression fails due to the non-linear nature of our dataset.

In the future we plan to apply the framework in a Multi-Task and Online-Learning setting, e.g. to adapt the model to client-specific tools not included in the training data. In [14], a similar framework already supports multi-task and online-learning. We plan to adapt the work in this paper for the framework in [14] and run the experimental setup presented in this paper in a multi-task setting, in order to show the influence of censored data in an online/multi-task environment.

## References

1. Amos, B., Kolter, J.Z.: OptNet: differentiable optimization as a layer in neural networks. In: Proceedings of the 34th International Conference on Machine Learning, vol. 70, pp. 136–145. JMLR. org (2017)

2. Boyd, S., Vandenberghe, L.: Convex Optimization. Cambridge University Press, Cambridge (2004)
3. Cox, D.R.: Analysis of Survival Data. Chapman and Hall/CRC (2018)
4. Diamond, S., Boyd, S.: CVXPY: a Python-embedded modeling language for convex optimization. J. Mach. Learn. Res. (2016, to appear). http://stanford.edu/~boyd/papers/pdf/cvxpy_paper.pdf
5. Efron, B.: The two sample problem with censored data. In: Proceedings of the Fifth Berkeley Symposium on Mathematical Statistics and Probability, vol. 4, pp. 831–853 (1967)
6. Faraggi, D., Simon, R.: A neural network model for survival data. Stat. Med. **14**(1), 73–82 (1995)
7. Ge, Z., Song, Z., Ding, S.X., Huang, B.: Data mining and analytics in the process industry: the role of machine learning. IEEE Access **5**, 20590–20616 (2017)
8. Ghahramani, Z., Jordan, M.I.: Learning from incomplete data (1995)
9. Greene, W.H.: Censored data and truncated distributions. Available at SSRN 825845 (2005)
10. Henningsen, A.: censReg: Maximum Likelihood estimation of censored regression (Tobit) models with cross-sectional and panel data, R package version 0.5-30 (2019)
11. Johnson, B.A., et al.: On lasso for censored data. Electron. J. Stat. **3**, 485–506 (2009)
12. Li, Z., Kong, Z.: A generalized procedure for monitoring right-censored failure time data. Qual. Reliab. Eng. Int. **31**(4), 695–705 (2015)
13. Liu, L., Kuo, S.M., Zhou, M.: Virtual sensing techniques and their applications. In: 2009 International Conference on Networking, Sensing and Control, pp. 31–36 (2009). https://doi.org/10.1109/ICNSC.2009.4919241
14. Luftensteiner, S.M.: Improving offline deep learning models by censored online data and quadratic programming. Master's thesis, University of Applied Sciences Upper Austria, Campus Hagenberg (2019)
15. McDonald, J.F., Moffitt, R.A.: The uses of Tobit analysis. Rev. Econ. Stat. 318–321 (1980)
16. Monostori, L., Márkus, A., Van Brussel, H., Westkämpfer, E.: Machine learning approaches to manufacturing. CIRP Ann. **45**(2), 675–712 (1996)
17. Segal, M.R.: Regression trees for censored data. Biometrics, 35–47 (1988)
18. Shivaswamy, P.K., Chu, W., Jansche, M.: A support vector approach to censored targets. In: Seventh IEEE International Conference on Data Mining (ICDM 2007), pp. 655–660. IEEE (2007)
19. Štajduhar, I., Dalbelo-Bašić, B.: Uncensoring censored data for machine learning: a likelihood-based approach. Expert Syst. Appl. **39**(8), 7226–7234 (2012)
20. Tarone, R.E., Ware, J.: On distribution-free tests for equality of survival distributions. Biometrika **64**(1), 156–160 (1977)
21. Wang, H., Zhou, L.: Random survival forest with space extensions for censored data. Artif. Intell. Med. **79**, 52–61 (2017)
22. Zellinger, W., et al.: Multi-source transfer learning of time series in cyclical manufacturing. J. Intell. Manuf. **31**(3), 777–787 (2019). https://doi.org/10.1007/s10845-019-01499-4
23. Zhang, W., Zhou, T., Wang, J., Xu, J.: Bid-aware gradient descent for unbiased learning with censored data in display advertising. In: Proceedings of the 22nd ACM SIGKDD International Conference on Knowledge Discovery and Data Mining, pp. 665–674. ACM (2016)

# Blockchain-Based Access Control for IoT in Smart Home Systems

Bacem Mbarek[✉], Mouzhi Ge, and Tomas Pitner

Faculty of Informatics, Masaryk University, Brno, Czech Republic
bacem.mbarek@mail.muni.cz, mouzhi.ge@muni.cz, tomp@fi.muni.cz

**Abstract.** Smart home systems are featured by a variety of connected smart household devices, where Internet of Things (IoT) is one of the critical enablers in the smart home environment. Since these smart home IoT devices are working collaboratively, the access control among the IoT devices becomes more and more important because non-authorised access can result in resource misuse, home breach threats or private information disclosure. Thus, an effective access control in smart home systems is essential to prevent from unauthorized use of the available resources. However, most of the access control schemes in smart home systems are still lack of decentralized peer trust and hard to control the security and credibility of the smart home IoT network. This paper therefore proposes a Blockchain-based Access Control (BAC) solution by integrating the Blockchain technique to IoT networks, where the agent-based policy is proposed to improve the efficiency of the Blockchain management. In order to validate the BAC solution, we demonstrate the implementation process of the proposed BAC in the parental control scenario and also evaluate performance and feasibility in a simulated smart home.

**Keywords:** Blockchain · Smart home systems · IoT · Access Control

## 1  Introduction

Nowadays, smart home systems are developed to enable home automation and increase the quality of life [18]. These systems are usually featured by using a smart phone application to interconnect and control different home devices such as lights, power plugins, cooking devices, temperature and humidity sensors as well as security systems [25]. In order to implement the home device interconnections, Internet of Things (IoT) is widely used as one typical wireless communication technique [17]. IoT is a concept of interrelated physical objects or devices that have been used in different application domains such as in smart cities, vehicular networks, military, and healthcare [11]. In the smart home context, the main goal of IoT is to connect different kinds of objects such as refrigerator, washing machines, TV, laptop, or cars [33] to facilitate people's daily life.

While the smart home system brings comfort and convenience to the home environment, it is also found to be vulnerable and exposed to different non-authorised access threats [12]. For example, if an attacker targets the functionality of the access control system by modifying the authentication headers in the

© Springer Nature Switzerland AG 2020
S. Hartmann et al. (Eds.): DEXA 2020, LNCS 12392, pp. 17–32, 2020.
https://doi.org/10.1007/978-3-030-59051-2_2

packets, this attack can create a fake identity and access without legal authorization to the IoT devices deployed in a home environment. In order to prevent such threats, access-control mechanisms are being investigated as policy enforcement components or main gateways to secure communications between IoT devices [30]. However, current access-control mechanisms used in IoT are with limited flexibility and inefficient to secure the resource constrained smart home devices [22]. These smart home devices are commonly prone to malicious attacks and require mutual authentication to guarantee the confidentiality and security of data shared in the IoT network [15]. As a typical IoT network, smart home systems urge to be a highly secure, independent and distributed platform [11]. In order to tackle this issue, the emergent Blockchain technology can be used to transform the way that data will be shared in IoT networks [2].

In this paper, we therefore propose a Blockchain-based Access Control solution, named as *BAC*, which controls the access to smart home devices and ensures secure communications between the IoT devices and the householders. In particular, we address the drawbacks of regular Blockchain access control protocols in terms of the time cost involved in the necessity of contact with the owner of the resource for each new access, by using agent-based software as an authenticated key for access control management. The contribution in this paper is twofold. First, we highlight the problem of the time cost involved in getting an access permission with the standard Blockchain platform. Second, we present an agent-based method to monitor inappropriate activities of users in the Blockchain. While we consider that Blockchain is capable of controlling the secure access and efficient data sharing in smart home systems, the *BAC* can be also applicable for diverse scaled IoT applications. Further, the collaborations of smart home devices can benefit from Blockchain to enhance the processes of authentication and coordination of data collection as well as optimizing the usage of the available resources [14,35]. In order to validate the proposed BAC solution, we demonstrated its applicability in the parental control scenario in smart home systems.

The remainder of the paper is organized as follows. Section 2 presents an overview of Blockchain with common definitions. Section 3 proposes BAC solution by integrating the Blockchain into smart home. To evaluate the proposed solution, Sect. 4 describes a typical smart home scenario and implement BAC in smart home systems. Section 5 concludes the paper and outlines future directions for this work.

## 2 Related Work

While the increasing use of IoT devices in smart home allows for real-time, low-cost data acquisition, the smart home IoT network is exposed to challenging security and access threats such as spoofing user identity, cloning access control cards, or submission of false information [18]. In order to prevent such threats, access-control mechanisms are being investigated to secure communication among the IoT devices. However, most current mechanisms used in computing systems can be inappropriate for resource constrained environments including IoT-based home automation systems [1]. Although the smart home is a specific environment, the overall nature of security threats is similar to other domains. We describe three types of threats and their consequences in smart home systems as follows.

**Confidentiality Leaking:** An attacker may try to access the sensitive information in the smart home. For example, the confidentiality threats could lead to house stealing by hacking information about air conditioning system parameters or light system operation to determine whether a house is occupied or not. Thus, the confidentiality should be considered to ensure the safety of IoT systems [36].

**Access Phishing:** An attacker can confuse the home system by letting the system agree that there is an using emergency alerts and opening doors and windows to allow an emergency exit. Therefore, without appropriate access control schemes, the smart home can be in danger [30].

**Unauthorized Access:** An attacker can use unauthorized devices in the IoT network by for example cloning some access card. Since many smart home devices can be physically attacked, an attacker may start an energy depletion attack, a form of denial of service. Thus, using simple access control system through password and key management is not enough. Dynamic key updating scheme is necessary to protect smart home network [16].

A variety of solutions have been proposed to solve the access control problems in smart home by identifying the malicious activities and exploring secure access control management mechanisms among house members [8,9,30]. In [19], the authors proposed a platform, named FairAccess, to secure the IoT network based on the token method, where Fairaccess only supports the Blockchain authorization based on access tokens using smart contracts. However, their proposal has some issues, for example, the time involved in obtaining access licenses takes long, the time cost is mainly in the necessity of contacting with the owner of the resource for each new access or for each token expiration. In [8], the authors also employed Blockchain in access control. They proposed a Blockchain-based smart home framework that is used for handling all the internal communications inside home and external communications from or to the home. In [30], authors proposed a private Blockchain as central intelligent home administrator. Their proposed model addresses the problems related to DoS attacks by controlling all incoming and outgoing packets to smart home devices. In [9], authors proposed a managing IoT device using Blockchain platform that is manipulated by Ethereum Blockchain with smart contracts for tracking meter and setting policies. It can be used to turn on and off air conditioner and light bulbs in order to save energy.

From the Blockchain perspective, authors in [7] proposed a combination of private and public Blockchains. The private Blockchain was employed to handle data flow in SHS, whereas the public Blockchain was to manage data flow over cloud storage. Their approach is composed of three layers: smart home systems (SHS), overlay network and cloud storage. To improve further the security between IoT devices and the Blockchain, different researchers proposed the integration of cloud computing in Blockchain platforms to manage a large number of IoT devices. In [21], the authors implemented a cloud server as a centralized and decentralized combined architecture and an intermediate between IoT devices and the Blockchain. In [37], the authors combined Blockchain and off-Blockchain storage to construct a personal data management platform focused on data privacy. Similarly, in [20], the authors proposed a secure service providing mechanism for IoT by using Blockchain and cloud. However, the unstable connection between cloud servers and IoT devices could cause communication delay as well as prevent the proposed model from achieving optimal performance and security scores. In [9], authors have demonstrated how Blockchain can be used to store IoT data with smart contracts. In [24], authors have proposed a new framework in which Blockchain is used to support the distributed storage of data in IoT applications. In [5], authors have proposed various protocols that enable the synchronization of IoT endpoints to the Blockchain. To this end, the authors have used several security levels and communication requirements. They have also analyzed the traffic and the power consumption created by the synchronization protocols via numerical simulations. Extended reviews of using Blockchain to store IoT data with a focus on open problems of anonymity, integrity, and adaptability as well as to architect Blockchain-based IoT applications are available in [4]. Therefore, the development of a smart home based on Blockchain system with scalable communication network is required for ensuring the safety of users as well as the transmission of aggregated IoT data.

## 3   BAC: Blockchain-Based Access Control Model

In order to provide an effective access control solution for smart home systems, we integrate Hyperledger Fabric-based Blockchain [10] to IoT network to secure the peer and trust and use agent-based policy model to improve the Blockchain efficiency. The agent-based policy aims to reduce or eliminate human interventions in the Blockchain, as every agent in the IoT device has the capacity to autonomously process Blockchain transactions. In addition, the Blockchain can organize the user behaviors by using a mobile agent which to detect inappropriate user activities with the used IoT device.

### 3.1   System Setting and Roles

In the system, the Blockchain is composed of three roles: 1) endorsing peers, 2) the ordering service, and 3) committing peers. Each householder is a transmitter/receiver that sends requests to a manager controller in the channel through the Blockchain platform. Inside the Blockchain, the smart contract is a code fragment that executes the terms of a contract [32]. A channel can be defined as a sub-network for peers communication. In the smart home context, it can be communication between IoT devices, and family members. Using different channels can divide transactions according to different boundaries according to some service logic. In the smart home systems, the transactions can be considered as an activity request or a collaboration order.

- **Endorsing peers** are the manager members in the Blockchain channel. The endorsing peers will endorse a transaction proposal, for example, the endorsement can be carried out based on specified family policies. When enough endorsing peers support a transaction proposal, it can be submitted to the ordering service to be added as a block. During the commissioning and configuration of the Blockchain network, the smart home system should firstly select the endorsing peers.
- **Ordering service** collects transactions for a channel into proposed blocks for distribution to peers. Blocks can be delivered for each defined channel. The smart home ordering service is to gather all the endorsed transactions, perform the ordering in a block, and then send the ordered blocks to the committing peers.
- **Committing peers** includes all the members of the Blockhain. The committing peers run the validation and update their copy of the Blockchain and status of transactions. Each peer receives the block, as a committing peer, can now attach the new block to its copy of the transactions. Committing peers have also the responsibility of updating the shared ledger with the list of transactions.

When a householder agrees to enter a transaction, he determines the parameters of this transaction by specifying its request, its location, authentication key. Each transaction is stored in a smart contract and transmitted to the house managers as playing the rules of endorsing peers of the Blockchain platform. The received transaction is verified by checking their smart contract. Then, the endorsing peers verifies the received transaction by executing the static agent policies. After, the verification of the received transaction by the house managers and the creation of the Block by the ordering server, the endorsing peers will submit a mobile agent to the requested user of IoT sensors to detect the inappropriate activities during the utilization of connected device.

## 3.2   Agent-Based Policy for the Blockchain Management

In the smart home systems, there can be different frequent activity requests such as turn on or off the light. For those frequently occurred requests, it is inconvenient to manually approve each request for the endorsing peers. In order to increase Blockchain efficiency for smart home IoT network, the Blockchain platform will create two kinds of agents: (1) a static agent, and (2) a mobile agent. Both the static and mobile agents are dedicated to every new transaction. The endorsing peers can monitor the selected node by using those agents. While the static agent is used to verify the received transaction, the mobile agent is to check the user activities. The mobile agent also can control the connected devices in real time, for example, if activities from some IoT device are inappropriate, the mobile agent can turn off this IoT device. The positions of these agents and their functions are depicted on Fig. 1.

**Static Agent** is a static agent that will be implemented in each peer device of the Blockchain. Once the peer device receives a transaction such as a activity request, The peer device will use the static agent to verify and check the received transaction. Those Agents intend to replace manual verification. Then, the static agent will carry on each received transaction. The static agent compares the received transaction $T$ to its predefined policies $P_s$. Then the static agent compares $P_s$ with $T$. If $P_s \neq T \pm \varepsilon$, then the transaction will be rejected. If $P_s = T \pm \varepsilon$, then the transaction will be accepted.

$$T = \begin{cases} 1, & \text{accepted} \\ 0, & \text{if } P_s \neq T \pm \varepsilon \end{cases} \qquad (1)$$

**Mobile Agent** is the agent that is created by the householder managers to control access to their household devices. The mobile agent is a standalone software entity that can start various tasks when visiting different computing nodes: such as collecting data, contorting access management tools, as well as monitoring and detecting inappropriate using of household devices. According to the request of the endorsing peers, a mobile agent can migrate to the selected house, transfer their code and data, allow them to efficiently perform computations, gather information and accomplish tasks. The agent can be encrypted by asymmetric authentication (by the public key of the selected peer).

The Blockchain platform will create a mobile agent dedicated to every requested IoT device. The mobile agent will migrate to the requested IoT device. Then, the mobile agent will execute its code in the IoT device controller. The mobile agent collects the behaviours of user $B$ that is detected by each user and reported by the sensors. Then the mobile agent compares $B$ with $P$, where $P$ is the declared policies given by Blockchain platform. If $B \neq P \pm \varepsilon$, then a potential inappropriate activities $(P)$ is suspected. The mobile agent also checks the user time connection $T$ to the IoT device, and compares it to the policy declared time $T_B$. If $T < T_B$, then a potential inappropriate activity can be reported. The mobile agent detects the inappropriate activities as depicted in Eq. (2), where $\varepsilon$ represents some measurement error threshold.

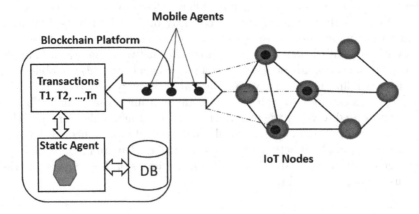

**Fig. 1.** Architecture of blockchain agents-based model

$$P = \begin{cases} 1, & \text{if } B \neq P \pm \varepsilon \text{ or } T < T_B \pm \varepsilon \\ 0, & \text{otherwise} \end{cases} \tag{2}$$

Figure 1 shows the overall architecture of integration between agent-based Blockchain and IoT networks. In this architecture, although Database (DB) is place, the analysis engine is out of the scope of this paper. It can be seen that the role of the static agent is used to manage access control for a transaction. After allowing and giving a permission for a transaction, a mobile agent will be created by the Blockchain platform and will migrate to the requested IoT node to control user behaviours with the connected device.

### 3.3 System Algorithm Design for Smart Home

Algorithm 1 presents the BAC solution by using Blockchain components and software agents as a method for enabling and access control in smart home. To

---

**Algorithm 1.** Hyperledger Fabric Blockchain-based Access Control (BAC)

EP: endorsing peers
Tr: Transaction
SA: Static agent
MA: Mobile agent
BC: Blockchain platform
C: Children
OS: Ordering Service:
DB: Data Base
P: Declared policies given by BC
B: Collected behaviours of user
T: User connection time
$T_B$: Policy declared time
res: proposal response
CE: Collected Endorsements
**while** Tr received **do**
   Verify(T, EP(SA)) {EP verifies T by executing SA}
   Send(res, EP, C) {EP sends res to C}
   Request(EP, OS) {EP requests block creation form OS}
   Verify(CE, OS) {OS verifies CE}
   Send(Block, OS, BC){OS sends the created Block to BC}
   Add(Block, BC(DB)) {BC adds the Block into its DB}
   Send(MA, EP, C) {EP sends MA to C}
   **while** MA received **do**
     **if** $B \neq P \pm \varepsilon$ **then**
       inappropriate activity is detected
     **else if** $T < T_B$, **then**
       inappropriate activity is detected
     **else**
       appropriate activity
     **end if**
   **end while**
**end while**

---

adapt Blockchain to the specific needs of IoT, we integrated 2 software agents: the static and mobile agents. First, the static agent is responsible of transactions verification. Second, the mobile agent is able to detect inappropriate activities. Once on the peer (children), the mobile agent will be acting as a Local Intrusion Detector (LID) as well as a Delegated Authenticator (DA) on behalf of endorsing peers (Parents). To this end, it will use the local processing resources of the peer to investigate its behavior. The mobile agent compares its declared policies $P$ with the user behaviour $B \neq P$, then inappropriate activity is detected. The mobile agent also compares its policy declared time $T_B$ with the user connection time $T$.

## 4   Validation of BAC in Smart Home Systems

In order to validate our solution, we demonstrate the application of BAC in the parental control scenario, which is a typical and important case in smart home systems [27]. We use the smart home setting for demonstration and evaluation purposes, while BAC is application agnostic and well-suited for diverse IoT applications and networks. Our demonstration is show how to design and implement a secure and efficient model that can manage access control between children and their parents based on the Blockchain.

Considering that in the smart home, sensors are coupled to a number of household devices to turn on or off the digital devices, which can be controlled remotely and give access to the children. Those control sensors are coupled to household devices in the smart home to monitor and control home appliances. Control sensor consists of activating or deactivating household devices and monitoring the user behavior about how appliances are used. For example, Children need request for permission before using any connected household devices. Parents will receive the transaction and react to it by allowing or rejecting usage by executing access control policies. The parents can control remotely the connected home appliances through their mobile phone or tablet or a wireless remote. Two agent-based control will be used in our system: static agent and mobile agent.

The *static agent* will be implemented in each parent manger device to verify received transactions that their children request. After executing each policies code, the static agent is also responsible for accepting or declining the received transaction. After giving access to children, the *mobile agent* will migrate to the target devices. The mobile agent is able to control the activities of children and detect e.g. the inappropriate behaviour. The mobile agent can visit each of the connected home appliances and execute customized code on each control sensor. Therefore, the mobile agent can analyse user behaviour through the connected home appliances. Apart from monitoring user behaviour, when a malicious activities are detected, the mobile agent can also turn off the device and denied further access.

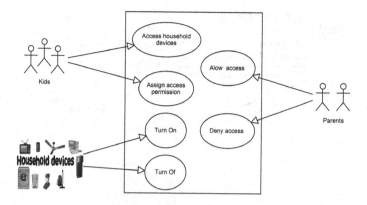

**Fig. 2.** Roles in blockchain-based model: use case diagram

## 4.1   System Implementation Process

While the BAC can enable the parents to set digital rules remotely with their devices to control child's activity on the house, digital parental access control helps to keep children safe when they're using household devices and to track their activity when using the connected devices. As shown in Fig. 2, devices at home such as television, computer and video games are connected to a private Blockchain channel and accessible only when the parents give access to their child's. Moreover, parents have a remote controllers (e.g. mobile phone, hand clock, computer) which are connected to the private Blockchain channel.

When a children enters into a transaction, he determines the parameters of this transaction by specifying its location, name of household items that he will interact with it, usage time, etc. Each transaction is stored in a smart contract and transmitted to the endorsing peers of the Blockchain platform. As shown in Fig. 3, the following steps occur during access request time.

1. The children store the endorsement requests in the smart contract and submits it to their endorsing peers, in the smart home context, the endorsing peers are usually the parents.
2. The static agents located in the parents devices can approve or reject the status after checking for consistency by aligning the smart contract.
3. The static agents send the proposal response to the IoT device applicant. Each execution captures the set of read and written data (also called the RW set), which is flowing in the Hyperledger fabric. Moreover, the Endorsement System Chaincode (ESCC) signs the proposal response on the endorsing peer. The RW sets are signed by each endorser. Every set will include the number of the version of each record.
4. The children send the proposal response to the ordering service, which is located on the Blockchain platform as a clustering code to create Blocks).
5. Afterwards, the ordering service verifies the collected endorsements and creates the Block of transactions.

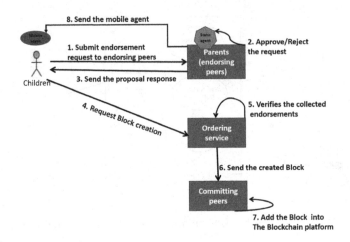

**Fig. 3.** Order-execute architecture to storage data in the Hyperledger Fabric Blockchain platform

6. The ordering service sends the Block to the committing peers which can include the parents or other family members who will be notified for validation.
7. If the validity is fulfilled, the Blockchain platform adds the checked Block into the data base of the Blockchain. Then, the Blockchain emits the notification to all peers.
8. After the validation of transaction, one of the parents static agent (in a chosen sequential manner) creates a mobile agent and sends it to the child who made the request and track his activities.

We have implemented a system prototype of the proposed BAC based on the open source Blockchain platform: Hyperledger Fabric [10], which has been reported as one of the most suitable platforms for the IoT applications [31]. In the smart home environment, all nodes such as the IoT endpoint for parents and children in the network use the SHA-1 hash algorithm and 2MB as data block sizes. Also, the IoT home devices are linked by the hyperledger in Blockchain, the foundation of the Blockchain in smart home is designed to control children activities with the household devices. To implement this in practice, first, house-holders should be Blockchain channel members. Besides that, there are two ways for endorsing peers to verify and check the authenticity of the children's requests and transactions: by their smart contract and by the mobile agent report. We created a smart contract program that is used by the children to send its transaction to the endorsing peers, which in our case the peers are the IoT endpoint from parents. A transaction invokes a smart contract called chaincode, which defines an application running on the Blockchain [3]. When the transition is issued, the mobile agent is used to monitor the behavior of the IoT device usage. For example, a child can request to watch TV for half an hour by smart contract. After the parents agree this transaction by endorsement, the mobile agent is used to monitor if watching TV in this transaction is more than half an hour or not, in case longer than half an hour, the mobile agent can turn off the TV based the defined policy.

We deployed the initial prototype to test the feasibility of the solution, and planned tests by running simulations of different scenarios of tracking children's activities in smart home. We described the implementation process of the Blockchain components and their characteristics as follows. Each peer in the Blockchain channel runs as an image in the docker container and contains the smart contract modeled as JavaScript Object Notation (JSON). The smart contract is designed and implemented by using the Hyperledger Composer, which is an extensive, open development tool set and framework to facilitate the implementation of Blockchain applications. The peer processes and orders transactions on a first-come-first-serve basis across the network. The notification generated from the blockchain network is emitted to the client using WebSockets [28]. Each Block contains a key and the key of the previous block. Therefore, the database collects Blocks that are cryptographically linked together to form a sequence of chains. The connector between Blocks helps user to trace back the different stored transactions and to know the history changes that happened in the state database. The ordering node is employed with the practical byzantine fault tolerance (PBFT) algorithm [6] to ensure the consistency of every copy of the leger.

## 4.2   Performance Evaluation

To analyze the performance of our proposed BAC solution, we used Hyperledger Caliper software [26], which is a benchmark tool developed within the Hyperledger project. This tool can produce reports containing various performance indicators, such as transactions per second, transaction latency and resource utilization. For performance evaluation, this study is mainly focused on the task execution time. In the simulations, we varied two kinds of variables to observe the transaction execution time. One variable is the number of IoT devices in the smart home, and the other one is the number of transactions with a fixed number of smart home IoT devices.

In the setting of varying the number of IoT devices in the smart home, we configured the system with a range of 50 to 250 IoT devices with the interval of 50 devices. This range is chosen because of real-world applicability. For smart buildings, we estimate that the total number of IoT devices is usually less than 200, and the IoT devices or sensors in smart home in most cases are less than 50 pieces. Considering Hyperledger Fabric nodes setup, Those IoT devices contain 2 endorsing peers that are both parental controller devices, 4 committing peers including endorsing peers and others are smart IoT devices or sensors at home. We run our simulations for 60 s. The execution time is mainly spent on processing transactions, peer communications and updating the Blockchain. The transaction is the exchange of messages between sender and receiver where a peer (e.g. children) access request, asking the manager peers (e.g. parents) for the access of the household device (e.g. TV). In our simulation, to read one transaction in the

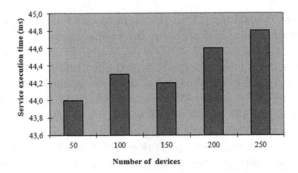

**Fig. 4.** Average execution time of each transaction by varying the number of IoT devices

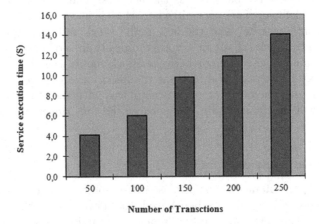

**Fig. 5.** Execution time of processing all the transactions by varying the number of transactions

Blockchain, the average time is around 39 ms. To compute the time execution of each transaction, we are using the simulation tool Hyperledger Caliper [26] to record the average time. We observed that for the one transaction the average execution time is 44 ms.

Figure 4 shows the evaluation result of the execution time for accessing usage to IoT devices. We can obtain the average execution time of one transaction by varying the number of IoT devices directly from the simulator. The main observation is that when increasing the number of IoT devices, the execution time is between (43 ms) and (45 ms), indicating that the execution time for transaction is very reasonable and implementable. Also, it can be seen that the variation of IoT device quantity has limited effects on our solution and thus the proposed BAC can potentially support even large IoT networks. One of the reasons is that in smart homes each transaction may only involve a small set of peers. Thus enlarging the network may not enlarge the scale of the transaction. Moreover, the implemented static agent helps to verify fast the received transactions.

The agent-based method in our solution can provide more efficient transaction execution.

In the setting of varying the number of transactions in the smart home, we configured the system with a fixed number of 50 IoT devices because we estimate most smart homes have less than 50 devices. Thus our performance testing can be set to be a larger scale case. Same as the previous setting, 2 IoT devices with parents as endorsing and committing peers and 2 more IoT devices as committing peers. The rest of the IoT devices are the IoT devices or sensors available in the house. We reported the observations in Fig. 5.

Figure 5 shows the total time that is used to successfully complete all the designed transactions in the Blockchain. In this setting, all the householder members who submit transactions have a designated target smart contract function. As we expected, more transactions will take more execution time. However, the average execution time per transaction is decreasing as the number of the transactions increase. This may be explained by the fact that Blockchain provides a way to execute many transactions at the same time in an efficient way [23]. For instance, when the number of transactions is equal to 50, the total time to process all the transactions is 4 s and thus each transaction needs 0.08 s on average, while for 250 transactions the total execution time is 14 s, where each transaction needs 0.056 s. It can be seen that each transaction needs less time when more transactions are carried out. One possible reason is that executing smart contract for many transactions simultaneously spend less execution time than executing smart contract program separately for each transaction. Thus, when the density of the transaction is high, instead of processing the full cycle for each transaction, the smart contract takes the full list of transactions as input values. Likewise, each receiver peer may process the list of multiple sequential transactions into one Block by using the smart contract executable program. Therefore, with the increase number of transactions, the time that is used to execute each transaction will decrease.

During the simulation, we also found that collaborations among the smart home IoT devices can achieve results that exceed their individual capabilities. In smart homes, different IoT devices may need to work collaboratively to finish one task [35]. In this context, because of the commonly limited processing, storage, and communication capabilities of these devices as well as the dynamic, open, and unpredictable environments in which they operate, implementing effective collaboration mechanisms remain a challenging concern [34]. For example, in order to finish one cooking task, the kettle, the cooking machine and the oven may need to work collaboratively, interactively or in a sequence. Potential solutions can be inspired from the extensive studies to enable the collaboration between distributed autonomous entities in the context of Multi-Agent Systems (MAS). MAS have proven flexibility, autonomy, and intelligence to solve complex problems within highly dynamic, constrained, and uncertain environments [13,29]. In our proposed BAC solution, agent-based design is used to automate the approval, monitoring and controlling activities. We believe that the agent-based design can be further developed to coordinate the communication among IoT devices and enhance the efficient automation in smart home systems.

## 5   Conclusion

In this paper, we have proposed a Blockchain-based access control (BAC) mechanism for IoT in smart home systems. The BAC solution is mainly featured by integrating the Blockchain to IoT networks with agent embedded systems. This solution is designed for the smart home IoT devices. In the BAC solution, we have modelled the agent-based policy for the Blockchain management and designed the algorithms for smart home access control. In order to validate our solution, we have implemented the BAC for a typical parental control scenario in smart home, and demonstrated the detailed processes and interactions, across from request, to endorsement and verification as well as monitoring activities in parental control. Based on the performance evaluation, it can be seen that the execution time for accessing the service in IoT devices is reasonable and the system interactions such as transaction processing are efficient. This also indicates that the BAC solution can be implemented in the real-world smart home systems. Furthermore, although the BAC solution is proposed in the context of smart home, it can be potentially used in the other scaled IoT networks with access control concerns and requirements.

As future works, we first plan to conduct more experiments with various real-world IoT applications. In most smart home systems, since the number of IoT devices is only scaled to a small limitation, we believe our solution can be effectively implemented. Further, a user study will be conducted in a real-world smart home environment, and we will also intend to investigate the interoperability and the cost of implementing our solution compared to other solutions.

## References

1. Al-Shaboti, M., Welch, I., Chen, A., Mahmood, M.A.: Towards secure smart home IoT: manufacturer and user network access control framework. In: 32nd International Conference on Advanced Information Networking and Applications, Krakow, Poland, pp. 892–899 (2018)
2. Ali, G., Ahmad, N., Cao, Y., Asif, M., Cruickshank, H.S., Ali, Q.E.: Blockchain based permission delegation and access control in Internet of Things (BACI). Comput. Secur. **86**, 318–334 (2019)
3. Brandenburger, M., Cachin, C., Kapitza, R., Sorniotti, A.: Blockchain and trusted computing: problems, pitfalls, and a solution for hyperledger fabric. arXiv:1805.08541 (2018)
4. Conoscenti, M., Vetro, A., De Martin, J.C.: Blockchain for the Internet of Things: a systematic literature review. In: 2016 IEEE/ACS 13th International Conference of Computer Systems and Applications (AICCSA), pp. 1–6. IEEE (2016)
5. Danzi, P., Kalor, A.E., Stefanovic, C., Popovski, P.: Analysis of the communication traffic for blockchain synchronization of IoT devices. In: 2018 IEEE International Conference on Communications (ICC), pp. 1–7. IEEE (2018)

6. De Angelis, S., Aniello, L., Baldoni, R., Lombardi, F., Margheri, A., Sassone, V.: PBFT vs proof-of-authority: applying the cap theorem to permissioned blockchain (2018)
7. Dorri, A., Kanhere, S.S., Jurdak, R.: Towards an optimized blockchain for IoT. In: Proceedings of the Second International Conference on Internet-of-Things Design and Implementation, pp. 173–178. ACM (2017)
8. Dorri, A., Kanhere, S.S., Jurdak, R., Gauravaram, P.: Blockchain for IoT security and privacy: the case study of a smart home. In: 2017 IEEE International Conference on Pervasive Computing and Communications Workshops (PerCom Workshops), pp. 618–623. IEEE (2017)
9. Huh, S., Cho, S., Kim, S.: Managing IoT devices using blockchain platform. In: 19th International Conference on Advanced Communication Technology, pp. 464–467. IEEE (2017)
10. Hyperledger: Hyperledger fabric (2019). https://github.com/hyperledger/fabric
11. Johnsen, F.T., et al.: Application of IoT in military operations in a smart city. In: 2018 International Conference on Military Communications and Information Systems (ICMCIS), pp. 1–8. IEEE (2018)
12. Kavallieratos, G., Chowdhury, N., Katsikas, S.K., Gkioulos, V., Wolthusen, S.D.: Threat analysis for smart homes. Future Internet 11(10), 207 (2019)
13. Kong, Y., Zhang, M., Ye, D.: A belief propagation-based method for task allocation in open and dynamic cloud environments. Knowl.-Based Syst. 115, 123–132 (2017)
14. Kum, S.W., Kang, M., Park, J.: IoT delegate: smart home framework for heterogeneous IoT service collaboration. TIIS 10(8), 3958–3971 (2016)
15. Lyu, Q., Zheng, N., Liu, H., Gao, C., Chen, S., Liu, J.: Remotely access "my" smart home in private: an anti-tracking authentication and key agreement scheme. IEEE Access 7, 41835–41851 (2019)
16. Mbarek, B., Ge, M., Pitner, T.: Self-adaptive RFID authentication for Internet of Things. In: 33rd International Conference on Advanced Information Networking and Applications, Matsue, Japan, pp. 1094–1105 (2019)
17. Mbarek, B., Ge, M., Pitner, T.: An efficient mutual authentication scheme for Internet of Things. Internet Things 9, 100160 (2020)
18. Mocrii, D., Chen, Y., Musilek, P.: IoT-based smart homes: a review of system architecture, software, communications, privacy and security. Internet of Things 1, 81–98 (2018)
19. Ouaddah, A., Abou Elkalam, A., Ait Ouahman, A.: Fairaccess: a new blockchain-based access control framework for the Internet of Things. Secur. Commun. Netw. 9(18), 5943–5964 (2016)
20. Rehman, M., Javaid, N., Awais, M., Imran, M., Naseer, N.: Cloud based secure service providing for IoTs using blockchain. In: IEEE Global Communications Conference (2019)
21. Rifi, N., Rachkidi, E., Agoulmine, N., Taher, N.C.: Towards using blockchain technology for IoT data access protection. In: 2017 IEEE 17th International Conference on Ubiquitous Wireless Broadband (ICUWB), pp. 1–5. IEEE (2017)
22. de Rivera, D.S., Bordel, B., Alcarria, R., Robles, T.: Enabling efficient communications with resource constrained information endpoints in smart homes. Sensors 19(8), 1779 (2019)
23. Selimi, M., Kabbinale, A.R., Ali, A., Navarro, L., Sathiaseelan, A.: Towards blockchain-enabled wireless mesh networks. In: Proceedings of the 1st Workshop on Cryptocurrencies and Blockchains for Distributed Systems, pp. 13–18 (2018)

24. Shafagh, H., Burkhalter, L., Hithnawi, A., Duquennoy, S.: Towards blockchain-based auditable storage and sharing of IoT data. In: Proceedings of the 2017 on Cloud Computing Security Workshop, pp. 45–50. ACM (2017)

25. Stojkoska, B.L.R., Trivodaliev, K.V.: A review of Internet of Things for smart home: challenges and solutions. J. Clean. Prod. **140**, 1454–1464 (2017)

26. Sukhwani, H., Wang, N., Trivedi, K.S., Rindos, A.: Performance modeling of hyperledger fabric (permissioned blockchain network). In: 2018 IEEE 17th International Symposium on Network Computing and Applications (NCA), pp. 1–8. IEEE (2018)

27. Vilas, A.F., Redondo, R.P.D., Rodríguez, S.S.: IPTV parental control: a collaborative model for the social web. Inf. Syst. Front. **17**(5), 1161–1176 (2015)

28. Wörner, D., von Bomhard, T.: When your sensor earns money: exchanging data for cash with bitcoin. In: Proceedings of the 2014 ACM International Joint Conference on Pervasive and Ubiquitous Computing: Adjunct Publication, pp. 295–298. ACM (2014)

29. Wray, K., Thompson, B.: An application of multiagent learning in highly dynamic environments. In: AAAI Workshop on Multiagent Interaction Without Prior Coordination (2014)

30. Xue, J., Xu, C., Zhang, Y.: Private blockchain-based secure access control for smart home systems. KSII Trans. Internet Inf. Syst. **12**(12) (2018)

31. Yu, Y., Guo, Y., Min, W., Zeng, F.: Trusted transactions in micro-grid based on blockchain. Energies **12**(10), 1952 (2019)

32. Yuan, Y., Wang, F.Y.: Towards blockchain-based intelligent transportation systems. In: IEEE 19th International Conference on Intelligent Transportation Systems, pp. 2663–2668 (2016)

33. Zaidan, A.A., Zaidan, B.B.: A review on intelligent process for smart home applications based on IoT: coherent taxonomy, motivation, open challenges, and recommendations. Artif. Intell. Rev. **53**(1), 141–165 (2020)

34. Zaidan, A.A., et al.: A survey on communication components for IoT-based technologies in smart homes. Telecommun. Syst. **69**(1), 1–25 (2018)

35. Zhang, Y., Tian, G., Zhang, S., Li, C.: A knowledge-based approach for multiagent collaboration in smart home: from activity recognition to guidance service. IEEE Trans Instrum. Measure. **69**(2), 317–329 (2020)

36. Zhang, Y., He, Q., Xiang, Y., Zhang, L.Y., Liu, B., Chen, J., Xie, Y.: Low-cost and confidentiality-preserving data acquisition for internet of multimedia things. IEEE Internet Things J. **5**(5), 3442–3451 (2018)

37. Zyskind, G., Nathan, O., et al.: Decentralizing privacy: using blockchain to protect personal data. In: 2015 IEEE Security and Privacy Workshops, pp. 180–184. IEEE (2015)

# Online Attacks on Picture Owner Privacy

Bizhan Alipour Pijani$^{(\boxtimes)}$, Abdessamad Imine$^{(\boxtimes)}$, and Michaël Rusinowitch$^{(\boxtimes)}$

Lorraine University, Cnrs, Inria, 54506 Vandœuvre-lès-Nancy, France
{bizhan.alipourpijani,abdessamad.imine,rusi}@loria.fr

**Abstract.** We present an online attribute inference attack by leveraging Facebook picture metadata (i) alt-text generated by Facebook to describe picture contents, and (ii) comments containing words and emojis posted by other Facebook users. Specifically, we study the correlation of the picture's owner with Facebook generated alt-text and comments used by commenters when reacting to the image. We concentrate on gender attribute that is highly relevant for targeted advertising or privacy breaking. We explore how to launch an online gender inference attack on any Facebook user by handling online newly discovered vocabulary using the retrofitting process to enrich a core vocabulary built during offline training. Our experiments show that even when the user hides most public data (e.g., friend list, attribute, page, group), an attacker can detect user gender with $AUC$ (area under the $ROC$ curve) from *87%* to *92%*, depending on the picture metadata availability. Moreover, we can detect with high accuracy sequences of words leading to gender disclosure, and accordingly, enable users to derive countermeasures and configure their privacy settings safely.

**Keywords:** Social network · Attribute privacy · Online inference attack

## 1 Introduction

Facebook is the dominant platform for users to interact, share pictures, and stay connected to their friends. A Facebook user has a profile, a list of friends, and a digital record of behaviors. For instance, a user's behavioral data comprise the list of pages liked by the user. Attribute inference attacks are emerging threats to user privacy in social media such as Facebook. Previous attribute inference attacks focus on inferring a user's private attributes (e.g., location, gender, sexual orientation, and/or political view) by leveraging either social friendship structures [14] or user behaviors [1]. Attribute inference attacks can be significantly improved by natural language processing techniques (*NLP*) since most Facebook data is textual or can be represented as a text. The recent development of deep learning for *NLP* allows one to capture semantic relations between

This work is supported by DIGITRUST (http://lue.univ-lorraine.fr/fr/article/digitrust/).

words from their vectorial representations, lead to an efficient attribute inference attack [22]. The approach also applies to non-textual communication modes by using *Emoji2vec* [10].

The problem of inferring user's attributes becomes qualitatively different if social structures and user behaviors are not available, which is often the case in the real scenario. Unlike previous studies, we show how to detect Facebook user's gender *online* through his/her shared images along with Facebook generated alt-text and received comments underneath those pictures. In online attacks: (i) the attacker relies on offline analysis knowledge to predict new target user attributes, and (ii) input data are collected online by an attacker crawling the targeted profile. The attacker constructs the offline analysis knowledge by collecting profiles with known attributes (in our case gender) and employs sophisticated techniques (e.g., *NLP*) to capture patterns and structures from collected data. Moreover, Facebook users can build their offline analysis knowledge and consider themselves as online targeted users to check their vulnerability to attribute inference attacks.

Machine learning classifiers, and/or vector representation techniques accurately infer private attributes from users' public data (e.g., page likes). Inferred attributes can be employed to deliver personalized advertisements [5] or privacy breaking [4]. In [11], the authors investigate Facebook users' privacy awareness and show one-half of the *479k* examined Facebook users hide their gender. Facebook users prefer to hide their gender for two reasons. First, they want camouflage against sexual harassment and stalking. The Facebook search bar lets users track down pictures of their female friends, but not the male ones [16]. Second, they want to reduce discrimination. The American Civil Liberties Union *(ACLU)*[1] accused Facebook of enabling employers to use targeting technology that excludes women from receiving job ads for some positions.

While many Facebook users hide their sensitive attributes (e.g., gender, age, political view), pictures are still available to public. A social media sharing analysis conducted by *The New York Times* revealed that *68%* of their respondents share images to give people a better sense of *who they are* and *what they care about*. Users in social media share pictures to receive feedback for their activities, especially from friends, and acquaintances, provide a great sense of connectedness. However, they lose privacy control on their posted pictures due to extra information (i.e., meta-data) added by third-party during the publication process. For any uploaded photos, Facebook implemented an object detection system to provide automatically a set of tags, called alt-text, that describe pictures content. They propose this technique to generate a description that can be used by a text reader for blind users. An attacker can use these tags to relax image processing tasks. Furthermore, when observing a picture on Facebook, people write instinctive comments to express their feeling. Automatically generated alt-text and comments (picture metadata) contain potentially sensitive information available to an attacker.

---

[1] https://www.aclu.org/blog/womens-rights/womens-rights-workplace/facebook-settles-civil-rights-cases-making-sweeping.

***Problem Statement.*** We propose a method to infer online the target user gender by using non-user generated data. This method even applies to Facebook users who are cautious about their privacy and hide any type of available information (e.g., friend list, liked pages, groups, and attributes) on their profile. Since our training dataset only contains *25,456* unique words, the input data of an online attack may contain words that do not occur in this training dataset. The new words are called *out of vocabulary words* (*OOV*). To circumvent this problem, we rely on the pre-trained vectors of an advanced *NLP* model, namely *word2vec* [18], and its version dedicated to emojis *emoji2vec* [10]. *Word2vec* and *emoji2vec*, abbreviated by *WE2V*, are trained on large datasets (e.g., Wikipedia for *word2vec*) with specific writing structure or usage. Therefore their pre-trained vectors should be adapted when we aim to apply them to a specific domain such as Facebook. *Retrofitting* technique [12] is called for adjusting the *WE2V* pre-trained vectors by combining external knowledge (*WE2V* dataset), and internal knowledge (offline collected words/emojis co-occurrence). A simple and direct approach to handle an out of vocabulary word would be to replace it by a synonym. However, this approach fails for our gender inference problem, as the word and its synonym can orient to different genders. An example taken from our dataset illustrates this point: male-posted pictures receive more comments containing the word *gorgeous*, while a synonym of this word, namely *beautiful*, is used more frequently for commenting female posted pictures. To that end, we use cosine similarity score [18] to compute similarities among words/emojis, including *OOV* words. Our approach assumes the following hypotheses:

1. The commenter's gender is hidden. As a consequence, standard homophily-based methods do not apply.
2. The target user is careful enough to conceal gender information in his/her posted comments. Therefore it is needless to process the comments written by the target user.
3. The user profile name does not disclose gender information as Facebook users often use pseudos.

***Contributions.*** Our contributions and improvements over previous works are:

1. A new online attribute inference attack based on non-user generated data composed of alt-text created by Facebook to describe the content of the picture, and comments containing words and emojis posted by friends, friends of friends, or regular users.
2. A strategy for adapting pre-trained vectors for Facebook by exploiting offline scraped comments.
3. A privacy-enhancing system that pinpoints received comments or posted pictures leading to a gender inference attack.

***Outline.*** The paper is organized as follows: we review related work in Sect. 2. In Sect. 3, we overview the system architecture. Section 4 defines the gender inference attack. Section 5 presents in detail the offline attack steps. Section 6 presents the online attack steps. Section 7 shows experimental results. In Sect. 8, we discuss the attack process, and we conclude in Sect. 9.

## 2   Related Work

Profiling has gained great attention in the past decade. Deriving user gender, for instance, is important for recommendation systems. Recently, researchers have investigated social media platforms in order to distinguish males and females from content sharing [8] and behavior [17]. Prior works claim that gender prediction is possible from the writing style [13], word usage [25] of the target user. Gender inference from the target user name can be performed across major social networks [15]. However, the performance of this type of attack is biased towards countries of origin [24]. The authors in [7] propose user gender identification through user shared images in Fotolog and Flickr, two image-oriented networks. They perform image processing on each crawled image (in offline mode), which is not feasible with online attacks. The diversity and global usage of emojis lead researchers to analyse emoji usage according to gender. The authors in [6] collect the data with *Kika Keyboard* and investigate user preferences. This work has two drawbacks that degrade the performance: (i) opposite-gender friends interaction may affect user emoji usage [20], and (ii) user cautiousness in choosing the emojis. Our work is different in two senses. First, we skip the target user emoji usage and rather rely on other Facebook users' emotional responses to solve the above limitations. Second, we exploit the idea that the picture content has a powerful impact on individuals' emotional responses.

To sum up, in contrast with previous works, we study gender inference attacks on Facebook by considering words, and emojis preferences of other Facebook users (e.g., friends) when commenting pictures published by the target user. We do not explore the user network, which has two advantages: (i) makes the attack feasible even when target personal data and his/her ego-network is unavailable, and (ii) makes the attack suitable for online mode. We showed the benefit of non-user generated data analysis to infer the picture owner's gender [2,3]. This work is different from our previous works. First, our attack is not limited to textual language, or emojis as we combine words, and emojis. Second, we propose an online gender inference attack. Third, we leverage sophisticated words, and emojis vector representation, *word2vec*, and *emoji2vec*, to handle out of vocabulary words, and emojis.

## 3   Architecture

Figures 1 and 2 depict the overall architecture of our system. First, we overview the offline training components, and next, we present our online attack ingredients.

***Offline Training.*** This procedure combines domain specific and external knowledge in the following way (see Fig. 1). *Data Crawling* collects Facebook users' data in an offline mode for training gender classifiers. Then *Data Preprocessing* prunes, cleanses and normalizes the collected data. *Feature Extraction and Feature Selection* derive a set of features that contribute the most to gender inference from an initial set obtained by n-grams and correlation of alt-text and comments. *Retrofitting* is the process of adjusting an existing word/emoji vector representation using extra knowledge (in our case, offline collected words/emojis co-occurrence). It allows us to fit *WE2V* word vector representations to our specific domain, namely Facebook. *Machine Learning* aims to select the best gender classifier among the one that we have trained, using standard evaluation metrics. We discuss in detail all the steps in Sect. 5.

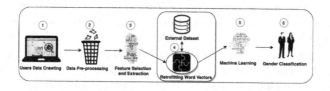

**Fig. 1.** Offline training

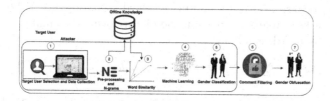

**Fig. 2.** Online attack and gender obfuscation

***Online Attack and Gender Obfuscation.*** After training the machine learning classifiers the following operations are performed online: (i) gender inference attack by following Steps *1* to *5*, and (ii) user protection from gender inference attack by applying Steps *1* to *7* (see Fig. 2). *Target User Selection and Data Collection* selects a specific user and collects his/her data in an online mode. *Pre-processing and N-grams* prunes raw data and extracts features compose of words/emojis or sequence of words/emojis. *Word/Emoji Similarity* finds similarity between word/emoji vectors in the vector space representation. *Gender Classification* applies to the target user based on extracted features. *Comment Filtering* proposes to hide received comments or published pictures that considered to be sensitive to gender. Finally, after hiding the suggested comments, *Gender Obfuscation* re-runs the gender inference attack steps to check whether the gender information secretes. These steps are detailed in Sect. 6.

## 4   Attack Description

In this work, we consider an attacker who intends to infer a picture owner gender $g$ by observing a set of pictures $P$. Each published picture $p$, where $p \in P$ contains metadata (a set of comments $c_p$, and generated alt-text $a_p$). The attacker can be anyone who can crawl data from Facebook. To accurately infer the target user's gender, the attacker has to cover three possible scenarios. In the first scenario, pictures receive no comments, or comments are unavailable due to privacy reasons (the target user conceals all comments underneath the picture). The attacker can still learn female and male preferences in picture sharing style, and infer target gender from tags or alt-texts. In the second scenario, Facebook is unable to generate automatic alt-texts due to the quality of the posted picture. The attacker learns words/emojis usage from comments posted by other Facebook users to infer the target gender. In the third scenario, both comments and alt-text are publicly available. The attacker can leverage both Facebook users' words/emojis usage and generated alt-text (tags) to infer the target gender. The gender inference attack is based on computing *Mutual Information (MI)* on all users labeled by gender $g$ given Facebook generated alt-text $a$, and/or other Facebook users posted comments $c$ for each picture $p$. Let $X$ be a random variable that takes value $1$ when the posted photo contains $a$ in the first scenario (resp., $c$ in the second scenario, resp., both $a$ and $c$ in the third scenario) and $0$ when the published picture does not contain $a$ in the first scenario (resp., $c$ in the second scenario, resp., both $a$ and $c$ in the third scenario). Let $Y$ be a random variable that takes value $1$ (resp., $0$) if the picture owner is female (resp., male). Then, we compute MI as follows:

$$MI(X;Y)_{ac} = \sum_{x\in\{0,1\}} \sum_{y\in\{0,1\}} P(X=x, Y=y)_{ac} log_2 \frac{P(X=x, Y=y)_{ac}}{P(X=x)_{ac} P(Y=y)_{ac}} \qquad (1)$$

where $P(X=x)_{ac}$ and $P(Y=y)_{ac}$ are the marginal probabilities of $x$ and $y$, and $P(X=x, Y=y)_{ac}$ is their joint probability. Based on $MI$, the most likely target gender for a set of pictures $P$ is:

$$\underset{g\in\{0,1\}}{\arg\max} MI(X;Y)_{ac} \qquad (2)$$

which is the core concept behind our inference attack. Let $x_u^i$ be the number of occurrence of feature $i$ in user pictures metadata (a or c). $MI$ measures the mutual dependence between picture owner gender and generated alt-text (resp., received comments, resp., alt-text and received comments) in the first scenario, (resp., second scenario, resp., third scenario). Each user $u$ with set of pictures $P$, and gender $g$ can be represented by a feature set $x_u = \{x_u^1, x_u^2, .., x_u^n\}$, and the label (or class) $y_u$, where $y_u \in \{0,1\}$. In the offline mode, the attacker trains machine learning algorithms with samples $(x_u, y_u)$, for all $u \in U_{training}$ as inputs, where $x_u = \{x_u^1, x_u^2, .., x_u^n\}$ and $U_{training}$ is a set of users. In the online mode, the attacker carries out the attack on a chosen target user $u_{new}$, by using the features obtained from trained algorithms. We discuss feature selection and extraction techniques in Subsection 5.3.

## 5   Offline Training

In this section, we introduce and discuss the offline components that we implement to train our machine learning algorithms.

## 5.1   Data Crawling from User Context

Let $U$ be the set of Facebook users, where $u_i$ is the $ith$ user with a set of posted pictures $P_i = \{p_i^1, p_i^2, .., p_i^n\}$. For user $u_i$, we extract $\langle g_i, A_i, C_i \rangle$ where $g_i$ is the user gender, $A_i$ is a set of alt-text presented by Facebook and $C_i$ is the set of comments posted by other Facebook users for $P_i$.

## 5.2   Data Pre-processing

Difficulty in analyzing data from social media raises from the presence of different kinds of textual errors, such as misspellings and grammatical errors. In addition to pre-processing steps detailed in [2,3], we perform spelling correction to correct word's spelling and/or intentionally repeated characters (as in *soooooooo beautiful*). However, not all misspelled words can be handled by *NLP* spell correction techniques. For example, *love u* contains a deformation letter *u* that refers to pronoun *you* that can be considered as a misspelled letter *a*, or *luv* is an abbreviation form of *love you*. We re-formulate misspelled words that cannot be corrected by spell correction techniques.

## 5.3   Features Selection and Extraction

Feature selection is the process of identifying and selecting relevant features correlated to variables of interest (gender, in our case). The purpose of feature selection is three-fold: promoting the model prediction performance, providing faster and efficient classifiers, and reducing the data dimensionality to decrease the model complexity. We select features in two different ways:

*1. Contiguous Sequence of Word/Emoji.* We compute n-grams to capture the occurrences of words/emojis in comments, and tags in alt-text in a given window size ($n$) for each gender. Our experiments revealed that *4-grams*, and *5-grams* are best suited for comments, and alt-text, respectively [2]. By retaining only $n$-grams that appear more than *50* times in total, we collect *2797* features. Table 1(a), and (b) show the discriminative alt-texts generated for female and male-owned pictures, and Facebook users' words/emojis preferences while commenting them, respectively. Additionally, we compute the probability of a person being male or female, given the picture generated alt-text, or words/emojis. For example, $p(female| \text{😊})$ is the probability that the user is female if she receives comments with 😊 emoji from other Facebook users.

*2. Correlation of Alt-text and Comments.* We construct a co-occurrence matrix to find the correlation between gender, alt-text, and received comments to distinguish females from males. The matrix records co-occurrences of words, emojis, and tags in the same picture, not necessarily in a given window size $n$. We drop rare co-occurrence pairs that appear less than *50* times in total. In that way, we collect *2103* features from all the possible combinations of words/emojis,

**Table 1.** MI result: (a) alt-text, (b) words/emojis, (c) correlation of alt-text and words, (d) correlation of alt-text and emoji.

(a)

| alt_text | MI | p(female—alt_text) | p(male—alt_text) |
|---|---|---|---|
| closeup | 0.020 | 0.79 | 0.21 |
| smiling | 0.012 | 0.73 | 0.27 |
| people smiling | 0.012 | 0.71 | 0.29 |
| 1 person | 0.008 | 0.68 | 0.32 |
| smiling closeup | 0.007 | 0.87 | 0.13 |
| person smiling | 0.006 | 0.77 | 0.23 |
| person closeup | 0.005 | 0.84 | 0.16 |
| beard | 0.004 | 0.23 | 0.77 |
| car | 0.004 | 0.25 | 0.75 |
| selfie | 0.003 | 0.81 | 0.19 |

(b)

| word/emoji | MI | p(female—word/emoji) | p(male—word/emoji) |
|---|---|---|---|
| [emoji] | 0.043 | 0.84 | 0.16 |
| [emoji] | 0.025 | 0.82 | 0.18 |
| [emoji] | 0.014 | 0.87 | 0.13 |
| [emoji] | 0.007 | 0.79 | 0.21 |
| [emoji] | 0.007 | 0.83 | 0.17 |
| beautiful | 0.006 | 0.90 | 0.10 |
| [emoji] | 0.003 | 0.36 | 0.64 |
| bro | 0.003 | 0.05 | 0.95 |
| [emoji] | 0.002 | 0.85 | 0.15 |
| gorgeous | 0.002 | 0.32 | 0.68 |

(c)

| alt_text + word | MI | p(female—alt_text + word) | p(male—alt_text + word) |
|---|---|---|---|
| 1 person, beautiful | 0.042 | 0.94 | 0.06 |
| closeup, beautiful | 0.030 | 0.92 | 0.08 |
| 1 person gorgeous | 0.018 | 0.39 | 0.61 |
| smiling, beautiful | 0.017 | 0.87 | 0.12 |
| 1 person, pretty | 0.016 | 0.89 | 0.11 |
| closeup, gorgeous | 0.011 | 0.42 | 0.58 |
| closeup, pretty | 0.011 | 0.90 | 0.10 |
| smiling, pretty | 0.007 | 0.89 | 0.11 |
| selfie, beautiful | 0.006 | 0.85 | 0.15 |
| 1 person, cute | 0.005 | 0.80 | 0.20 |

(d)

| alt_text + emoji | MI | p(female—alt_text + emoji) | p(male—alt_text + emoji) |
|---|---|---|---|
| 1 person, [emoji] | 0.037 | 0.80 | 0.20 |
| 1 person, [emoji] | 0.032 | 0.81 | 0.17 |
| closeup, [emoji] | 0.022 | 0.88 | 0.12 |
| closeup, [emoji] | 0.018 | 0.83 | 0.17 |
| 1 person, [emoji] | 0.015 | 0.89 | 0.11 |
| 1 person, [emoji] | 0.013 | 0.80 | 0.20 |
| 1 person, [emoji] | 0.010 | 0.90 | 0.10 |
| smiling, [emoji] | 0.009 | 0.87 | 0.13 |
| beard, [emoji] | 0.005 | 0.13 | 0.87 |
| beard, [emoji] | 0.001 | 0.30 | 0.70 |

and alt-text in our data set. Table 1(c), and (d) take into account the correlation of generated alt-text with received words and emojis, respectively. We also compute the probability for a person to be male or female, given picture alt-text, and received words/emojis. In total, we select *4900* features from the above categories. After choosing these features, we apply feature extraction algorithms to downsample the features and retain only the ones that contribute the most to gender prediction. We evaluated individual and combined feature extraction methods to derive the best features set [2]. We apply these methods to find the best feature set $W_{best}$.

### 5.4   Retrofitting Words/Emojis Vectors

After selecting the best feature set, we compute vector representations of these features to evaluate the similarity of the online collected words from the target profile (which may contain new words or sequences of words) to our best feature set. To that end we use *word2vec* and *emoji2vec*. Our goal is to create a set of embeddings that accounts for both our offline collected dataset, *OCD*, and original word/emoji representations learned from *WE2V*. *Retrofitting* [12] is a process that adjusts an original word vector separately using a knowledge graph (e.g. *WordNet* [19]), in our case *OCD* instead. Retrofitting has advantages of being (i) a post-processing operation that does not require to browse the corpus again, (ii) applicable to any vector model, and (iii) simple and fast to implement. Retrofitting computes a new vector $v_i$ for the word/emoji $w_i \in W_{best}$, with the objective of being close to $w_i$'s original vector $v_i'$, when it exists, and also to vectors $v_j$ representing $w_j$ that are the $w_i$'s nearest overlapped words/emojis in *WE2V* or *OCD*. For that, we try to minimize this objective function:

$$\sum_{i=1}^{n} \left[ \alpha_i \|v_i - v_i'\|^2 + \sum_{j:w_j \in WE2V} \gamma_{ij} \|v_i - v_j\|^2 + \sum_{j:w_j \in OCD} \beta_{ij} \|v_i - v_j\|^2 \right] \quad (3)$$

We set $\alpha_i = 1$ when $w_i \in WE2V$ and $0$ otherwise. An *overlapped* word/emoji belongs by definition to both $OCD$ and $WE2V$ datasets. The distance between a pair of vectors is defined to be the Euclidean distance. For $w_i$ in $WE2V \backslash OCD$ we take $\beta_{ij} = 0$ and $\gamma_{ij}$ is the *Cosine Similarity* score between $v_i$ and nearest overlapped words/emojis vectors $v_j$ in $WE2V$ dataset. Cosine similarity is widely used to measure the similarity between two non-zero vectors by measuring the cosine of angle between them. For $w_i$ in $OCD$, we take $\gamma_{ij} = 0$ and $\beta_{ij}$ is the *Pointwise Mutual Information* (*PMI*) score [9] between $w_i$, and overlapped co-occurring words $w_j$. *PMI* has been extensively used in the field of *NLP* to measure words closeness based on their co-occurrence probability. *PMI* is formulated as:

$$PMI(w_i, w_j) = \log \frac{p(w_i, w_j)}{p(w_i)p(w_j)}$$

where $p(w_i)$, and $p(w_j)$ represent the probabilities that a comment contains $w_i$, or $w_j$, and $p(w_i, w_j)$ represents the probability that a comment contains both $w_i$ and $w_j$). Therefore, we calculate the vector $v_i$ by taking the average of the nearest overlapped words/emojis vectors $v_j$, considering their cosine similarity $\gamma_{ij}$, or pointwise mutual information $\beta_{ij}$ score according to the cases as follows:

$$v_i = \frac{\sum_{j:w_j \in WE2V} \gamma_{ij} v_j + \sum_{j:w_j \in OCD} \beta_{ij} v_j + \alpha_i v_i'}{\sum_{j:w_j \in WE2V} \gamma_{ij} + \sum_{j:w_j \in OCD} \beta_{ij} + \alpha_i} \tag{4}$$

The advantage of adjusting the pre-trained words/emojis vector by using offline extracted data co-occurrences is two-fold: (i) handling non-overlapped words/emojis easily, and (ii) using sophisticated distributional embeddings (*WE2V*) that make the retrofitted vectors robust and suitable for gender inference attack process. For the feasibility of the computation, we truncate each sum in Eq. 4 by summing only the *10* most significant terms (corresponding to the closest words to $w_i$). In the case of having a sequence of words as the best feature, we first retrofit each word vector separately. Then we take the average of the vectors associated with the words in the sequence. [21]. For example, consider *beautiful lady* as a best feature, we first retrofit *beautiful* and *lady* separately. Next, we get a vector for *beautiful lady* by averaging the retrofitted vectors of *beautiful* and *lady*. Figure 3 illustrates the separate word retrofitting, where the blue dots are *word2vec* vectors, and orange dots are the retrofitted vectors.

# 6    Online Attack and Gender Obfuscation

Using our offline knowledge, the online phase consists of classifying a target user with unknown gender to the male or female category. For the demonstration purpose, we assume the target user is given.

## 6.1    Pre-processing and N-Grams

This step follows by reformulating the words to their normal form, as detailed in Subsection 5.2. Next, as $W_{best}$ contains both single and sequences of words, n-grams (up to 3-grams) permit to find new words or sequences of words in the online extracted comments.

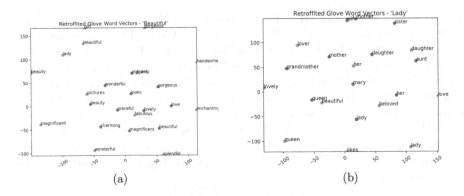

**Fig. 3.** Retrofitting: (a) Beautiful, (b) Lady (Color figure online)

## 6.2  Word/Emoji Similarity

We count the online occurrences of words belonging to $W_{best}$ and set the value of $w_{best}$ feature, where $w_{best} \in W_{best}$ to its number of occurrence. Otherwise, we find the closest words to $w_{best}$ by using their vector representation and set the $w_{best}$ value to the number of occurrence of close words. For each new word $w_{new}$, we compute the closest $w_{best}$ as follows:

$$\underset{w_{new}, w_{best}}{\arg\max} \; cosine(w_{new}, w_{best})$$

We ignore the result if the cosine similarity value is less than *50%*. For sequence of words in $w_{new}$ we proceed as in Subsection 5.4. This mechanism of replacing a new word by its closest word in the best feature set allows one to handle out of vocabulary words.

## 6.3  Gender Classification

In this step, the attacker has to evaluate the trained machine learning algorithm in an online mode. Given a target user, the algorithm outputs (i) *female*, (ii) *male*, or (iii) *unknown*. The output depends on the prediction probability threshold. We set this threshold to be *0.70*. For example, the output is *female* if the algorithm gives prediction probability of *0.70* to *female*, and *0.30* to *male*. Moreover, the output is *unknown* if the algorithm prediction probability for *female* is *0.65*, and *0.55* for *male*. In our experiments (see Sect. 7), we present the result of *700* users as they are labelled *female*, or *male*. Although the threshold empirically derived from our dataset, it is an arbitrary choice to be adapted to other datasets. It helps to prevent inaccurate attacks due to a lack of input information. For example, if the user has only one picture with an alt-text.

## 6.4   Comment/Picture Filtering

We now describe the filtering option proposed by our system for protecting user privacy. First, we discuss comment filtering, and later we define the picture filtering option.

*Comments Filtering Option.* It suggests users hide comments that contain sensitive words, or sequence of words that disclose picture owner gender. Facebook offers two comments filtering options. First, the users can set up a list of words, phrases, or emojis that they do not want to receive from commenters. Facebook hides matching comments entirely from the published photos. Second, the users can manually select the comments and make them invisible from photos. The advantage of hiding comment is that it is still visible to the commenter and his/her friends, which reduces tension between the commenter and the picture owner.

*Pictures Filtering Option.* As a suggestion, the user can hide the picture if the generated alt-text or combination of alt-text and comments leak the picture owner gender. Facebook settings allow users to restrict the picture visibility. We use *LIME* [23] to alert the user about comments, or pictures that reveal owner gender and require to be hidden. *LIME* computes an explanation as a list of features (with weights) that either contribute to the prediction (in green) or are evidence against it (in red). Figure 4(a) presents the most contributing features, while, Fig. 4(b) shows the least contributing features for a user, labelled by *female*.

| Contribution? | Feature | Value | | Contribution? | Feature | Value |
|---|---|---|---|---|---|---|
| +2.155 | hair | 1.000 | | -0.130 | dad | 1.000 |
| +0.161 | sobeautiful | 1.000 | | -0.132 | boy | 1.000 |
| +0.144 | absolutely | 1.000 | | -0.152 | beautifulpicture | 0.000 |
| +0.087 | stunning | 1.000 | | -0.204 | sopretty | 0.000 |
| +0.050 | beauty | 1.000 | | -2.348 | <BIAS> | 1.000 |
| (a) | | | | (b) | | |

**Fig. 4.** Lime output: (a) Most contributed features (b) Least contributed features (Color figure online)

As for gender obfuscation, the user can select some most contributing features as Fig. 4(a), according to his/her desires, and then hides the comments, or pictures containing those features by following the above steps. Next, the user can re-run our online system component to check the gender inference attack vulnerability. The user can repeat the process until the output is *unknown*.

## 7   Experiments

In this section, we evaluate our approach for all three scenarios and demonstrate offline and online experiments.

***Offline Experiments.*** Using a Python crawler, we have randomly collected *627,776* pictures and their *1,332,219* comments. Facebook was unable to generate alt-text for *24833* pictures. We have kept those pictures for our second attack scenario, where we rely only on words/emojis usage for commenting pictures. The experiments are achieved by applying the classifiers from Python library *scikit-learn*. For result robustness, we apply several supervised machine learning algorithms such as *Logistic Regression, Random Forests, K-Nearest Neighbors, Naive Bayes* and *Decision Tree*. To evaluate the classifier, we select the same number of males and females to prevent bias classification. Train-test splitting was preferable in this study as it runs k-times faster than k-fold. We choose the train-test size to be *70–30*, which gives the best accuracy. To address the problem of fairly estimating the performance of each classifier, we set aside a validation dataset. We train and adjust the hyper-parameters to optimize the performance of classifiers by using this dataset. Eventually, we evaluate the classifiers on the test dataset. Considering the extracted gender as the ground-truth, to evaluate our attack, we compute the *AUC-ROC* curve. The *AUC-ROC* curve is a performance measurement for classification problems at various threshold settings. In Fig. 5, we show the *AUC-ROC* results for all three scenarios. In the first scenario, we rely on alt-text to infer the picture owner's gender. Figure 5(a) displays trained algorithms results on the extracted alt-texts features. Based on that, our trained algorithms can infer the target user gender with an *AUC* of *87%*. In the second scenario, we conduct an inference attack by training algorithms only on commenters' comments. As illustrated in Fig. 5(b), the performance increases to *90% AUC*. In the third scenario, we train the classifiers by using the co-occurrence of alt-texts and words/emojis. Based on the Fig. 5(c), *Logistic Regression* model which had *87% AUC* in the first scenario, and *90%* in the second scenario gets *5%*, and *2% AUC* boost in this scenario, respectively, which is a fairly substantial gain in performance.

To conclude, *Logistic Regression* performs the best in all scenarios. It is a discriminative model that is appropriate when the dependent variable is binary (i.e., has two classes). The results confirm our hypothesis that gender and picture contents have an impact on Facebook users' emotional responses. As a result, an attacker can train standard classifiers by using pictures metadata contained (i) other Facebook users' words/emojis preferences, and (ii) generated alt-text to infer the picture owner's gender. Note, as we rely solely on non-target generated data, the results cannot be compared to previous works that exploit data published by the target.

***Online Experiments.*** We have applied our online experiment to *700* users with their *21,713* pictures and their *64,940* corresponding comments. We have evaluated the performance of each classifier with *AUC-ROC*. As illustrated in Fig. 5 (d, e, and f), *Logistic Regression* outperforms other classifiers in all three scenarios. Notably, the combination of alt-text and words/emojis boosts the performance of the classifiers in comparison to the other scenarios. To sum up, *Logistic Regression* is a suitable classifier for this task that can be trained by an attacker to perform a gender inference attack in online mode.

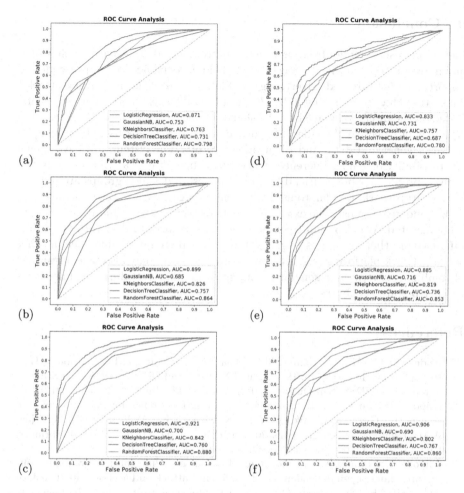

**Fig. 5.** AUC result of logistic regression trained on: (a) only alt-text (b) only commenters reactions (c) both alt-text and commenters reactions (d) removing alt-text features (e) removing commenters reactions features (f) removing alt-text and commenters reactions features.

## 8  Discussion

The best scenario for the attacker is, as expected, the third scenario when he has access to Facebook generated alt-text and commenters' comments. The attack may work even when the target publishes nothing else than pictures. By applying our system to his published pictures, a Facebook user can check if he/she is vulnerable to gender inference attack of the above type. To counter the aforementioned privacy violations, we offer two countermeasures, namely: (i) hide some comments, or (ii) hide some pictures when they strongly contribute to the attack as explained in Subsection 6.4.

# 9   Conclusion

Identifying users' gender from their online activities and data sharing behavior is an important topic in the growing research field of social networks. It provides an opportunity for targeted advertising, profile customization, or privacy attacks. This study has investigated *627,776* pictures and their *1,332,219* comments. Based on the intensive analysis of the shared images, this work has demonstrated (i) a new perspective of gender inference attack on Facebook users by relying on non-user generated data, and (ii) a privacy protection system. We have shown the possibility of gender inference attack even when all user attributes/activities such as profile attributes, friend list, liked pages, and joined groups are hidden.

As future work, we plan to use sophisticated algorithms to take the output of *LIME* and give the user the exact comments or pictures to be hidden in order to be obfuscated. We also plan to extend the current work to deal with the online inference of other attributes (e.g., age) and to explore online inference in other social network platforms (e.g. Twitter, Instagram). We may also take advantage of combining several user-generated content from different online social networks to infer private attributes.

# References

1. Abdelberi, C., Ács, G., Kâafar, M.A.: You are what you like! Information leakage through users' interests. In: 19th Annual Network and Distributed System Security Symposium, NDSS, San Diego, California, USA. The Internet Society (2012)
2. Alipour, B., Imine, A., Rusinowitch, M.: Gender inference for Facebook picture owners. In: Gritzalis, S., Weippl, E.R., Katsikas, S.K., Anderst-Kotsis, G., Tjoa, A.M., Khalil, I. (eds.) TrustBus 2019. LNCS, vol. 11711, pp. 145–160. Springer, Cham (2019). https://doi.org/10.1007/978-3-030-27813-7_10
3. Alipour, B., Imine, A., Rusinowitch, M.: You are what emojis say about your pictures: language-independent gender inference attack on Facebook. In: The 35th ACM/SIGAPP Symposium On Applied Computing, Brno, Czech Republic, pp. 1826–1834. ACM (2020)
4. Belinic, T.: Personality profile of social media users how to get maximum from it, April 2009. https://medium.com/krakensystems-blog/personality-profile-of-social-media-users-how-to-get-maximum-from-it-5e8b803efb30
5. Cadwalladr, C., Graham Harrison, E.: How Cambridge analytica turned Facebook 'likes' into a lucrative political tool, May 2018. https://www.theguardian.com/technology/2018/mar/17/facebook-cambridge-analytica-kogan-data-algorithm
6. Chen, Z., Lu, X., Ai, W., Li, H., Mei, Q., Liu, X.: Through a gender lens: learning usage patterns of emojis from large-scale Android users. In: Proceedings of the 2018 World Wide Web Conference on World Wide Web, WWW, Lyon, France, pp. 763–772. ACM (2018)
7. Cheung, M., She, J.: An analytic system for user gender identification through user shared images. TOMCCAP **13**(3), 1–20 (2017). Article no. 30
8. De Choudhury, M., Sharma, S.S., Logar, T., Eekhout, W., Nielsen, R.C.: Gender and cross-cultural differences in social media disclosures of mental illness. In: Proceedings of the Conference on Computer Supported Cooperative Work and Social Computing, CSCW, Portland, OR, USA, pp. 353–369. ACM (2017)

9. Church, K.W., Hanks, P.: Word association norms, mutual information, and lexicography. Comput. Linguist. **16**(1), 22–29 (1990)
10. Eisner, B., Rocktäschel, T., Augenstein, I., Bošnjak, M., Riedel, S.: emoji2vec: learning emoji representations from their description. arXiv preprint arXiv:1609.08359 (2016)
11. Farahbakhsh, R., Han, X., Cuevas, Á., Crespi, N.: Analysis of publicly disclosed information in Facebook profiles. CoRR, abs/1705.00515 (2017)
12. Faruqui, M., Dodge, J., Jauhar, S.K., Dyer, C., Hovy, E., Smith, N.A.: Retrofitting word vectors to semantic lexicons. arXiv preprint arXiv:1411.4166 (2014)
13. Flekova, L., Carpenter, J., Giorgi, S., Ungar, L.H., Preotiuc-Pietro, D.: Analyzing biases in human perception of user age and gender from text. In: Proceedings of the 54th Annual Meeting of the Association for Computational, Berlin, Germany, pp. 843–854. ACL (2016)
14. Gong, N.Z., Liu, B.: You are who you know and how you behave: attribute inference attacks via users' social friends and behaviors. In: 25th Security Symposium, Austin, TX, USA, pp. 979–995. USENIX (2016)
15. Karimi, F., Wagner, C., Lemmerich, F., Jadidi, M., Strohmaier, M.: Inferring gender from names on the web: a comparative evaluation of gender detection methods. In: Proceedings of the 25th International Conference on World Wide Web, WWW, Montreal, Canada, pp. 53–54. ACM (2016)
16. Lenton, A.: Facebook wants you to search photos of your female friends at the beach, but not your male mates, February 2019. https://www.whimn.com.au/talk/people/facebook-wants-you-to-search-photos-of-your-female-friends-at-the-beach-but-not-your-male-mates/news-story/bbc21ee6883bd07bfbbbe76a0c8ca54c
17. Ludu, P.S.: Inferring gender of a Twitter user using celebrities it follows. CoRR, abs/1405.6667 (2014)
18. Mikolov, T., Chen, K., Corrado, G., Dean, J.: Efficient estimation of word representations in vector space. arXiv preprint arXiv:1301.3781 (2013)
19. Miller, G.A.: WordNet: a lexical database for English. Commun. ACM **38**(11), 39–41 (1995)
20. Nguyen, D., et al.: Why gender and age prediction from tweets is hard: lessons from a crowdsourcing experiment. In: Proceedings of the 25th International Conference on Computational Linguistics, COLING, Dublin, Ireland, pp. 1950–1961. ACL (2014)
21. Pagliardini, M., Gupta, P., Jaggi, M.: Unsupervised learning of sentence embeddings using compositional n-gram features. arXiv preprint arXiv:1703.02507 (2017)
22. Perozzi, B., Al-Rfou, R., Skiena, S.: DeepWalk: online learning of social representations. In: Proceedings of the 20th ACM SIGKDD International Conference on Knowledge Discovery and Data Mining, pp. 701–710 (2014)
23. Ribeiro, M.T., Singh, S., Guestrin, C.: "Why should i trust you?" Explaining the predictions of any classifier. In: Proceedings of the 22nd ACM SIGKDD International Conference on Knowledge Discovery and Data Mining, pp. 1135–1144 (2016)
24. Santamaría, L., Mihaljevic, H.: Comparison and benchmark of name-to-gender inference services. PeerJ Comput. Sci. **4**, e156 (2018)
25. Sap, M., et al.: Developing age and gender predictive lexica over social media. In: Proceedings of the 2014 Conference on Empirical Methods in Natural Language Processing, EMNLP, Doha, Qatar, pp. 1146–1151. ACL (2014)

# Cloud Databases and Workflows

# Distributed Caching of Scientific Workflows in Multisite Cloud

Gaëtan Heidsieck[1]([⊠]), Daniel de Oliveira[4], Esther Pacitti[1],
Christophe Pradal[1,2], François Tardieu[3], and Patrick Valduriez[1]

[1] Inria & LIRMM, Univ. Montpellier, Montpellier, France
gaetan.heidsieck@inria.fr
[2] CIRAD & AGAP, Univ. Montpellier, Montpellier, France
[3] INRAE & LEPSE, Univ. Montpellier, Montpellier, France
[4] UFF, Niteroi, Brazil

**Abstract.** Many scientific experiments are performed using scientific workflows, which are becoming more and more data-intensive. We consider the efficient execution of such workflows in the cloud, leveraging the heterogeneous resources available at multiple cloud sites (geo-distributed data centers). Since it is common for workflow users to reuse code or data from other workflows, a promising approach for efficient workflow execution is to cache intermediate data in order to avoid re-executing entire workflows. In this paper, we propose a solution for distributed caching of scientific workflows in a multisite cloud. We implemented our solution in the OpenAlea workflow system, together with cache-aware distributed scheduling algorithms. Our experimental evaluation on a three-site cloud with a data-intensive application in plant phenotyping shows that our solution can yield major performance gains, reducing total time up to 42% with 60% of same input data for each new execution.

**Keywords:** Multisite cloud · Distributed caching · Scientific workflow · Workflow system · Workflow scheduling

## 1 Introduction

In many scientific domains, *e.g.*, bio-science [7], complex numerical experiments typically require many processing or analysis steps over huge datasets. They can be represented as scientific workflows, or workflows, for short, which facilitate the modeling, management and execution of computational activities linked by data dependencies. As the size of the data processed and the complexity of the computation keep increasing, these workflows become data-intensive [7], thus requiring high-performance computing resources.

The cloud is a convenient infrastructure for handling workflows, as it allows leasing resources at a very large scale and relatively low cost. In this paper, we consider the execution of a large workflow in a multisite cloud, *i.e.*, a cloud with geo-distributed cloud data centers (sites). Note that a multisite option is now

© Springer Nature Switzerland AG 2020
S. Hartmann et al. (Eds.): DEXA 2020, LNCS 12392, pp. 51–65, 2020.
https://doi.org/10.1007/978-3-030-59051-2_4

well supported by all popular public clouds, *e.g.*, Microsoft Azure, Amazon EC2, and Google Cloud, which provide the capability of using multiple sites with a single cloud account, thus avoiding the burden of using multiple accounts. The main reasons for using multiple cloud sites for data-intensive workflows is that they often exceed the capabilities of a single site, either because the site imposes usage limits for fairness and security, or simply because the datasets are too large. In scientific applications, there can be much heterogeneity in the storage and computing capabilities of the different sites, *e.g.*, on premise servers, HPC platforms from research organizations or federated cloud sites at the national level [4]. As an example in plant phenotyping, greenhouse platforms generate terabytes of raw data from plants, which are typically stored at data centers geographically close to the greenhouse to minimize data transfers. However, the computation power of those data centers may be limited and fail to scale when the analyses become more complex, such as in plant modeling or 3D reconstruction. Other computation sites are then required.

Most Scientific Workflow Management Systems (workflow systems) can execute workflows in the cloud [12]. Some examples are Swift/T, Pegasus, SciCumulus, Kepler and OpenAlea [9]. Our work is based on OpenAlea [14], which is widely used in plant science for simulation and analysis. Most existing systems use naive or user-based approaches to distribute the tasks across sites. The problem of scheduling a workflow execution over a multisite cloud has started to be addressed in [11], using performance models to predict the execution time on different resources. In [10], we proposed a solution based on multi-objective scheduling and a single site virtual machine provisioning approach, assuming homogeneous sites, as in public cloud.

Since it is common for workflow users to reuse code or data from other workflows [5], a promising approach for efficient workflow execution is to cache intermediate data in order to avoid re-executing entire workflows. Furthermore, a user may need to re-execute a workflow many times with different sets of parameters and input data depending on the previous results generated. Fragments of the workflow, *i.e.* a subset of the workflow activities and dependencies, can often be reused. Another important benefit of caching intermediate data is to make it easy to share with other research teams, thus fostering new analyses at low cost.

Caching has been supported by some workflow systems, *e.g.*, Kepler, VisTrails and OpenAlea. Kepler [1] provides a persistent cache on the cloud, but at a single site, and does not support multisite. VisTrails [3] provides a persistent cache, but only for local execution on a personal desktop. In [6], we proposed an adaptive caching method for OpenAlea that automatically determines the most suited intermediate data to cache, taking into account workflow fragments, but only in the case of a single cloud site. Another interesting single site method, also exploiting workflow fragments, is to compute the ratio between re-computation cost and storage cost to determine what intermediate data should be stored [16]. All these methods are single site (centralized). The only distributed caching method for workflow execution in a multisite cloud we are aware of is restricted to hot metadata (frequently accessed metadata) [8], ignoring intermediate data.

Caching data in a multisite cloud with heterogeneous sites is much more complex. In addition to the trade-off between re-computation and storage cost at single sites, there is the problem of site selection for placing cached data. The problem is more difficult than data allocation in distributed databases [13], which deals only with well-defined base data, not intermediate data produced by tasks. Furthermore, the scheduling of workflow executions must be cache-aware, *i.e.*, exploit the knowledge of cached data to decide between reusing and transferring cached data versus re-executing the workflow fragments.

In this paper, we propose a distributed solution for caching of scientific workflows in a multisite cloud. Based on a distributed and parallel architecture [13], composed of heterogeneous sites (including on premise servers and shared-nothing clusters), we propose algorithms for adaptive caching, cache site selection and dynamic workflow scheduling. We implemented our caching solution in OpenAlea, together with a multisite scheduling algorithm. Based on a real data-intensive application in plant phenotyping, we provide an extensive experimental evaluation using a cloud with three heterogeneous sites.

This paper is organized as follows. Section 2 presents our real use case in plant phenotyping. Section 3 introduces our workflow system architecture in multisite cloud. Section 4 describes our cache management solution. Section 5 gives our experimental evaluation. Finally, Sect. 6 concludes.

## 2    Use Case in Plant Phenotyping

In this section, we introduce a real use case in plant phenotyping that will serve as motivation for the work and basis for the experimental evaluation. In the last decade, high-throughput phenotyping platforms have emerged to allow for the acquisition of quantitative data on thousands of plants in well-controlled environmental conditions. For instance, the seven facilities of the French Phenome project[1] produce each year 200 Terabytes of data, which are various (images, environmental conditions and sensor outputs), multiscale and originate from different sites. Analyzing such massive datasets is an open, yet important, problem for biologists [15].

The Phenomenal workflow [2] has been developed in OpenAlea to analyze and reconstruct the geometry and topology of thousands of plants through time in various conditions. It is composed of nine fragments such as image binarization, 3D volume reconstruction, organ segmentation or intercepted light simulation. Different users can conduct different biological analyses by reusing some workflow fragments on the same dataset to test different hypotheses [6]. To save both time and resources, they want to reuse the intermediate results that have already been computed rather than recompute them from scratch.

The raw data comes from the Phenoarch platform, which has a capacity of 1,680 plants within a controlled environment (*e.g.*, temperature, humidity, irrigation) and automatic imaging through time. The total size of the raw image

---

[1] https://www.phenome-emphasis.fr/phenome_eng/.

dataset for one experiment is 11 Terabytes. To limit data movement, the raw
data is stored at a server near to the experimental platform, with both data
storage and computing resources. However, these computing resources are not
enough to process a full experiment in a relatively short time. Thus, scientists
who need to do a full experiment will execute the Phenomenal workflow at a
more powerful site by transferring the raw data for each new analysis.

In this Phenomenal use case, the cloud is composed of heterogeneous sites,
with both on premise servers close to the experimental platform and other more
powerful cloud sites. The on premise server has high storage capacity and hosts
the raw data. Other sites are used to computational intensive executions, with
high-performance computing resources. On premise servers are used locally to
execute some Phenomenal fragments that do not require powerful resources. In
this case, one has to choose between transferring the raw data or some intermediate data to a powerful site or re-executing some fragments locally before
transferring intermediate data. The trade-off between data re-computation and
data transfer is complex in a multisite cloud with much heterogeneity. In particular, one needs to pay attention to cached data placement, so as to avoid
bottlenecks on the most used intermediate data.

## 3   Multisite Cloud Workflow System Architecture

In this section, we present our workflow system architecture that integrates
caching and reuse of intermediate data in a multisite cloud. We motivate our
design decisions and describe our architecture in terms of nodes and components
(see Fig. 1), which are involved in the processing of workflows.

Our architecture capitalizes on the latest advances in distributed and parallel data management to offer performance and scalability [13]. We consider a
distributed cloud architecture with on premise servers, where raw data is produced, e.g., by a phenotyping experimental platform in our use case, and remote
sites, where the workflow is executed. The remote sites (data centers) are shared-
nothing clusters, i.e., clusters of server machines, each with processor, memory
and disk. We adopt shared-nothing as it is the most scalable and cost-effective
architecture for big data analysis.

In the cloud, metadata management has a critical impact on the efficiency of
workflow scheduling as it provides a global view of data location, e.g., at which
nodes some raw data is stored, and enables task tracking during execution [8].
We organize the metadata in three repositories: catalog, provenance database
and cache index. The catalog contains all information about users (access rights,
etc.), raw data location and workflows (code libraries, application code). The
provenance database captures all information about workflow execution. The
cache index contains information about tasks and cache data produced, as well as
the location of files that store the cache data. Thus, the cache index itself is small
(only file references) and the cached data can be managed using the underlying
file system. A good solution for implementing these metadata repositories is a

key-value store, such as Cassandra[2], which provides efficient key-based access, scalability and fault-tolerance through replication in a shared-nothing cluster.

The raw data (files) are initially produced and stored at some cloud sites, *e.g.*, in our use case, at the phenotyping platform. During workflow execution, the intermediate data is generated and consumed at one site's node in memory. It gets written to disk when it must be transferred to another node (potentially at the same site), or when explicitly added to the cache. The cached data (files) can later be replicated at other sites to minimize data transfers.

We extend the workflow system architecture proposed in [9] for single site. It is composed of six modules: workflow manager, global scheduler, local scheduler, task manager, data manager and metadata manager, to support both execution and intermediate data caching in a multisite cloud. The workflow manager provides a user interface for workflow definition and processing. Before workflow execution, the user selects a number of virtual machines (VMs), given a set of possible instance formats, *i.e.*, the technical characteristics of the VMs, deployed on each site's nodes. When a workflow execution is started, the workflow manager simplifies the workflow by removing some workflow fragments and partitions depending on the raw input data and the cached data (see Sect. 4). The global scheduler uses the metadata (catalog, provenance database, and cache index) to schedule the workflow fragments of the simplified workflow. The VMs on each site are then initialized, *i.e.*, the programs required for the execution of the tasks are installed and all parameters are configured. The local scheduler schedules the workflow fragments received on its VMs.

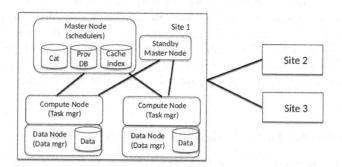

**Fig. 1.** Multisite workflow system architecture

The data manager module handles data transfers between sites during execution (for both newly generated intermediate data and cached data) and manages cache storage and replication. At a single site, data storage is distributed between nodes. Finally, the task manager (on each VM) manages the execution of fragments on the VMs at each site. It exploits the provenance metadata to decide whether or not the task's output data should be placed in the cache,

---

[2] https://cassandra.apache.org.

based on the cache provisioning algorithm described in Sect. 4. Local scheduling and execution can be performed as in [6].

Figure 1 shows how these components are involved in workflow processing, using the traditional master-worker model. In this architecture, we consider two types of cloud sites, *i.e.*, coordinator and participant. The relationship between the site is also based on the master-worker model, the coordinator site, managing the participant sites. The workflow manager and the global scheduler modules are implemented on the coordinator site. The remaining modules are implemented on all sites.

At each site, there are three kinds of nodes: master, compute and data nodes, which are mapped to cluster nodes at configuration time, *e.g.* using a cluster manager like Yarn (http://hadoop.apache.org). There is one active master node per site. There is also a standby node to deal with master node failure. The master nodes are the only ones to communicate across sites. The local scheduler and metadata management modules are implemented on the master node, which manages communication, metadata and scheduling. The master nodes are responsible for transferring data between sites during execution.

## 4   Multisite Cache-Aware Workflow Execution

In this section, we present in more details how the global scheduler performs multisite cache-aware workflow execution. In particular, the global scheduler must decide which data to cache (cache data selection) and where (cache site selection), and where to execute workflow fragments (execution site selection). Since these decisions are not independent, we propose a cost function to make a global decision, based on the cost components for individual decisions. We start by giving an overview of distributed workflow execution. Then, we present the methods and cost functions for cache data selection, cache site selection and execution site selection. Finally, we introduce our cost function for the global decision.

### 4.1   Distributed Workflow Execution Overview

We consider a multisite cloud with a set of sites $S = \{s_1, ..., s_n\}$. A workflow $W(A, D)$ is a a directed acyclic graph (DAG) of computational activities $A$ and their data dependencies $D$. A task $t$ is the instantiation of an activity during execution with specific associated input data. A fragment $f$ of an instantiated workflow is a subset of tasks and their dependencies.

The execution of a workflow $W(A, D)$ in $S$ starts at a coordinator site $s_c$ and proceeds in three main steps:

1. The global scheduler at $s_c$ simplifies and partitions the workflow into fragments. Simplification uses metadata to decide whether a task can be replaced by corresponding cached data references. Partitioning uses the dependencies in $D$ to produce fragments.

2. For each fragment, the global scheduler at $s_c$ computes a cost function to make a global decision on which data to cache where, and on which site to execute. Then, it triggers fragment execution and cache placement at the selected sites.
3. At each selected site, the local scheduler performs the execution of its received fragments using its task manager (to execute tasks) and data manager (to transfer the required input data). It also applies the decision of the global scheduler on storing new intermediate data into the cache.

We introduce basic cost functions to reflect data transfer and distributed execution. The time to transfer some data $d$ from site $s_i$ to site $s_j$, noted $T_{tr}(d, s_i, s_j)$, is defined by

$$T_{tr}(d, s_i, s_j) = \frac{Size(d)}{TrRate(s_i, s_j)} \tag{1}$$

where $TrRate(s_i, s_j)$ is the transfer rate between $s_i$ and $s_j$.

The time to transfer input and cached data, $In(f)$ and $Cached(f)$ respectively, to execute a fragment $f$ at site $s_i$ is $T_{input}(f, s_i)$:

$$T_{input}(f, s_i) = \sum_{s_j}^{S} (T_{tr}(In(f), s_j, s_i) + T_{tr}(Cached(f), s_j, s_i)) \tag{2}$$

The time to compute a fragment $f$ at site $s$, noted $T_{compute}(f, s)$, can be estimated using Amdahl's law [17]:

$$T_{compute}(f, s) = \frac{(\frac{\alpha}{n} + (1 - \alpha)) * W(f)}{CPU_{perf}(s)} \tag{3}$$

where $W(f)$ is the workload for the execution of $f$, $CPU_{perf}(s)$ is the average computing performance of the CPUs at site $s$ and $n$ is the number of CPUs at site $s$. We suppose that the local scheduler may parallelize task executions. Therefore, $\alpha$ represents the percentage of the workload that can be executed in parallel.

The expected waiting time to be able to execute a fragment at site $s$ is noted $T_{wait}(s)$, which is the minimum expected time for $s$ to finish executing the fragments in its queue.

The time to transfer the intermediate data generated by fragment $f$ at site $s_i$ to site $s_j$, noted $T_{write}(Output(f), s_i, s_j)$, is defined by:

$$T_{write}(Output(f), s_i, s_j) = T_{tr}(Output(f), s_i, s_j) \tag{4}$$

where $Output(f)$ is the data generated by the execution of $f$.

## 4.2   Cache Data Selection

To determine what new intermediate data to cache, we consider two different methods: greedy and adaptive. Greedy data selection simply adds all new data to the cache. Adaptive data selection extends our method proposed in [6] to multisite cloud. It achieves a good trade-off between the cost saved by reusing cached data and the cost incurred to feed the cache.

To determine if it is worth adding some intermediate data $Output(f)$ at site $s_j$, we consider the trade-off between the cost of adding this data to the cache and the potential benefit if this data was reused. The cost of adding the data to site $s_j$ is the time to transfer the data from the site where it was generated. The potential benefit is the time saved from loading the data from $s_j$ to the site of computation instead of re-executing the fragment. We model this trade-off with the ratio between the cost and benefit of the cache, noted $p(f, s_i, s_j)$, which can be computed from Eqs. 2, 3 and 4,

$$p(f, s_i, s_j) = \frac{T_{write}(Output(f), s_i, s_j)}{T_{input}(f, s_i) + T_{compute}(f, s_i) - T_{tr}(Output(f), s_j, s_i)} \quad (5)$$

In the case of multiple users, the probability that $Output(f)$ will be reused or the number of times fragment $f$ will be re-executed is not known when the workflow is executed. Thus, we introduce a threshold $Threshold$ (computed by the user) as the limit value to decide whether a fragment output will be added to the cache. The decision on whether $Output(f)$ generated at site $s_i$ is stored at site $s_j$ can be expressed by

$$\epsilon_{i,j} = \begin{cases} 1, & \text{if } p(f, s_i, s_j) < Threshold. \\ 0, & \text{otherwise.} \end{cases} \quad (6)$$

### 4.3   Cache Site Selection

Cache site selection must take into account the data transfer cost and the heterogeneity of computing and storage resources. We propose two methods to balance either storage load ($bStorage$) or computation load ($bCompute$) between sites. The $bStorage$ method allows preventing bottlenecks when loading cached data. To assess this method at any site $s$, we use a load indicator, noted $L_{bStorage}(s)$, which represents the relative storage load as the ratio between the storage used for the cached data ($Storage_{used}(s)$) and the total storage ($Storage_{total}(s)$).

$$L_{bStorage}(s) = \frac{Storage_{used}(s)}{Storage_{total}(s)} \quad (7)$$

The $bCompute$ method balances the cached data between the most powerful sites, i.e., with more CPUs, to prevent computing bottlenecks during execution. Using the knowledge on the sites' computing resources and usage, we use a load indicator for each site $s$, noted $L_{bCompute}(s)$, based on CPUs idleness ($CPU_{idle}(s)$) versus total CPU capacity ($CPU_{total}(s)$).

$$L_{bCompute}(s) = \frac{1 - CPU_{idle}(s)}{CPU_{total}(s)} \quad (8)$$

The load of a site $s$, depending on the method used, is represented by $L(s)$, ranging between 0 (empty load) and 1 (full). Given a fragment $f$ executed at site $s_i$, and a set of sites $s_j$ with enough storage for $Output(f)$, the best site $s^*$ to add $Output(f)$ to its cache can be obtained using Eq. 1 (to include transfer time) and Eq. 6 (to consider multiple users),

$$s^*(f)_{s_i} = \operatorname*{argmax}_{s_j}(\epsilon_{i,j} * \frac{(1 - L(s_j))}{T_{write}(Output(f), s_i, s_j)}) \tag{9}$$

## 4.4   Execution Site Selection

To select an execution site $s$ for a fragment $f$, we need to estimate the execution time for $f$ as well as the time to feed the cache with the result of $f$. The execution time $f$ at site $s$ ($T_{execute}(f, s)$) is the sum of the time to transfer input and cached data to $s$, the time to get computing resources and the time to compute the fragment. It is obtained using Eqs. 2 and 3.

$$T_{execute}(f, s) = T_{input}(f, s) + T_{compute}(f, s) + T_{wait}(s) \tag{10}$$

Given a fragment $f$ executed at site $s_i$ and its intermediate data $Output(f)$, the time to write $Output(f)$ to the cache ($T_{feed\_cache}(f, s_i, s_j)$) can be defined as:

$$T_{feed\_cache}(f, s_i, s_j, \epsilon_{i,j}) = \epsilon_{i,j} * T_{write}(Output(f), s_i, s_j) \tag{11}$$

where $s_j$ is given by Eq. 9.

## 4.5   Global Decision

At Step 2 of workflow execution, for each fragment $f$, the global scheduler must decide on the best combination of individual decisions regarding cache data, cache site, and execution site. These individual decisions depend on each other. The decision on cache data depends on the site where the data is generated and the site where it will be stored. The decision on cache site depends on the site where the data is generated and the decision of whether or not the data will be cached. Finally, the decision on execution site depends on what data will be added to the cache and at which site. Using Eqs. 10 and 11, we can estimate the total time ($T_{total}$) for executing a fragment $f$ at site $s_i$ and adding its intermediate data to the cache at another site $s_j$:

$$T_{total}(f, s_i, s_j, \epsilon_{i,j}) = T_{execute}(f, s_i) + T_{feed\_cache}(f, s_i, s_j, \epsilon_{i,j}) \tag{12}$$

Then, the global decision for cache data ($\epsilon(f)$), cache site ($s^*_{cache}$) and execution site ($s^*_{exec}$) is based on minimizing the following equation for the $n^2$ pairs of sites $s_i$ and $s_j$

$$(s^*_{exec}, s^*_{cache}, \epsilon(f)) = \operatorname*{argmin}_{s_i, s_j}(T_{total}(f, s_i, s_j, \epsilon_{i,j})) \tag{13}$$

This decision is done by the coordinator site at before each fragment execution and only takes into account the cloud site's status at that time. Note that $s^*_{exec}$, $s^*_{cache}$ can be the coordinator site and can be the same site.

## 5   Experimental Evaluation

In this section, we first present our experimental setup, which features a heterogeneous multisite cloud with multiple users who re-execute part of the workflow. Then, we compare the performance of our multisite cache scheduling method against two baseline methods. We end the section with concluding remarks.

### 5.1   Experimental Setup

Our experimental setup includes a multisite cloud, with three sites in France, a workflow implementation and an experimental dataset. *Site 1* in Montpellier is a server close to the Phenoarch phenotyping platform. It has the smallest number of CPUs and largest amount of storage among the sites. The raw data is stored at this site. *Site 2* is the coordinator site, located in Lille. *Site 3*, located in Lyon, has the largest number of CPUs and the smallest amount of storage.

To model site heterogeneity in terms of storage and CPU resources, we use heterogeneity factor $H$ in three configurations: $H = 0$, $H = 0.3$ and $H = 0.7$. For the three sites altogether, the total number of CPUs is 96 and the total storage on disk for intermediate data is 180 GB (The raw data is stored on an additional node at Site 1). On each site, several nodes are instantiated for the executions, they have a determined number of CPUs from 1, 2, 4, 8 or 16 CPUs. The available disk size for each node is limited by implementation. With $H = 0$ (homogeneous configuration), each site has 32 CPUs (two 16 CPUs nodes) and 60 GB (30 GB each). With $H = 0.3$, we have 22 CPUs and 83 GB for Site 1, 30 CPUs and 57 GB for Site 2 and 44 CPUs and 40 GB for Site 3. With $H = 0.7$ (most heterogeneous configuration), we have 6 CPUs and 135 GB for Site 1, 23 CPUs and 35 GB for Site 2 and 67 CPUs and 10 GB for Site 3.

The input dataset for the Phenomenal workflow is produced by the Phenoarch platform (see Sect. 2). Each execution of the workflow is performed on a subset of the input dataset, *i.e.* 200 GB of raw data, which represents the execution of 15,000 tasks. For each user, 60% of the raw data is reused from previous executions. Thus each execution requires only 40% of new raw data. For the first execution, no data is available in the cache.

We implemented our cache-aware scheduling method, which we call *cacheA*, in OpenAlea and deployed it at each site using the Conda multi-OS package manager. The metadata distributed database is implemented using Cassandra. Communication between the sites is done using the protocol library ZeroMQ. Data transfer between sites is done through SSH. We have also implemented two baseline methods, *Sgreedy* and *Agreedy*, based on the *SiteGreedy* and *Act-Greedy* methods described in [10], respectively. The *Sgreedy* method extends *SiteGreedy*, which schedules each workflow fragment at a site that is available for execution, with our cache data/site selection methods. Similarly, the *Agreedy* method extends *ActGreedy*, which schedules each workflow fragment at a site that minimizes a cost function based on execution time and input data transfer time, with our cache data/site selection methods. These baseline methods perform execution site selection followed by cache data/site selection while *CacheA* makes a global decision.

## 5.2 Experiments

We compare *CacheA* with the two baseline methods in terms of execution time and amount of data transferred. We define total time as execution time plus transfer time. In experiment 1, we consider a workflow execution with caching or without. In Experiment 2, we consider multiple users who execute the same workflow on similar input data, where 60% of the data is the same. In Experiment 3, we consider different heterogeneous configurations for one workflow execution.

**Experiment 1: With Caching.** In this basic experiment, we compare two workflow executions: with caching, using *CacheA* and *bStorage*; and without caching, using *ActGreedy*. We consider one re-execution of the workflow on different input datasets, from 0% to 60% of same reused data.

    *CacheA* outperforms *ActGreedy* from 20% of reused data. Below 20%, the overhead of caching outweighs its benefit. For instance, with no reuse (0%), the total time with *CacheA* is 16% higher than with *ActGreedy*. But with 30%, it is 11% lower, and with 60%, it is 42% lower.

**Experiment 2: Multiple Users.** Figure 2 shows the total time of the workflow for the three scheduling methods, four users, $H = 0.7$ and our two cache site selection methods: (a) *bStorage*, and (b) *bCompute*.

    Let us first analyze the results in Figure 2a (*bStorage* method). For the first user execution, *CacheA* outperforms *Sgreedy* in terms of execution time by 8% and in terms of data and intermediate data transfer times by 51% and 63%, respectively. The reason *Sgreedy* is slower is that it schedules some compute-intensive fragments at Site 1, which has the lowest computing resources. Furthermore, it does not consider data placement and transfer time when scheduling fragments.

    Again for the first user execution, *CacheA* outperforms *Agreedy* in terms of total time by 24%, when considering data transfer time to the cache. However, *CacheA* execution time is a bit slower (by 9%). The reason that *Agreedy* is slower in terms of total time is that it does not take into account the placement of the cached data, which leads to larger amounts (by 67%) of cache data to transfer. For other users' executions (when cached data exists), *CacheA* outperforms *Sgreedy* in terms of execution time by 29%, and for the fourth user execution, by 20%. This is because *CacheA* better selects the cache site in order to reduce the execution time of the future re-executions. In addition, *CacheA* balances the cached data and computations. It outperforms *Sgreedy* and *Agreedy* in terms of intermediate data transfer times (by 59% and 15%, respectively), and cache data transfer times (by 82% and 74%, respectively).

    Overall, *CacheA* outperforms *Sgreedy* and *Agreedy* in terms of total times by 61% and 43%, respectively. The workflow fragments are not necessarily scheduled

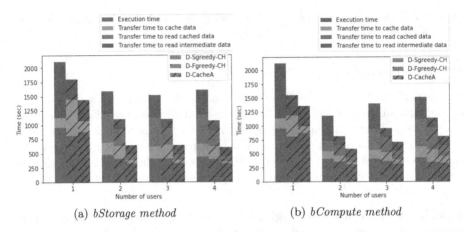

(a) *bStorage method*    (b) *bCompute method*

**Fig. 2.** Total times for multiple users (60% of same raw data per user) for three scheduling methods (*Sgreedy, Agreedy* and *CacheA*).

to the site with shortest execution time, but to the site that minimizes overall total time. Considering the multiuser perspective, *CacheA* outperforms baseline methods, reducing the total time for each new user (up to 6% faster for the fourth user compared to the second).

Let us now consider Fig. 2b (*bCompute* method). For the first user execution, *CacheA* outperforms *Sgreedy* and *Agreedy* in terms of total time by 36% and 10% respectively. *bCompute* stores the cache data on the site with most idle CPUs, which is often the site with the most CPUs. This leads the cached data to be stored close to where it is generated, thus reducing data transfers when adding data to the cache. For the second user, *CacheA* outperforms *Sgreedy* and *Agreedy* in terms of total time by 46% and 21% respectively. The cached data generated by the first user is stored on the sites with more available CPUs, which minimizes the intermediate and reused cached data transfers. From the third user, the storage at some site gets full, *i.e.* for the third user's execution, Site 3 storage is full and from the fourth user's execution, Site 2 storage is full. Thus, the performance of the three scheduling methods decreases due to higher cache data transfer times. Yet, *CacheA* still outperforms *Sgreedy* and *Agreedy* in terms of total time by 49% and 25% respectively.

**Experiment 3: Cloud Site Heterogeneity.** We now compare the three methods in the case of heterogeneous sites by considering the amount of data transferred and execution time. In this experiment (see Fig. 3), we consider only one user who executes the workflow and that previous executions with 60% of the same raw data have generated some cached data. We use the *bStorage* method for cache site selection.

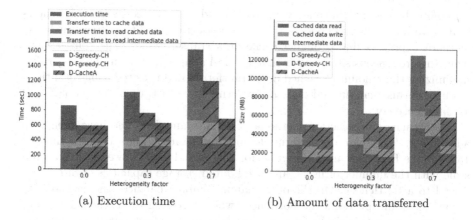

(a) Execution time    (b) Amount of data transferred

**Fig. 3.** Execution times and amounts of data transferred for one user (60% of same raw data used), on heterogeneous sites with three scheduling methods (*Sgreedy*, *Agreedy* and *CacheA*).

Figure 3 shows the execution times and the amount of data transferred using the three scheduling methods in case of heterogeneous sites. With homogeneous sites ($H = 0$), the three methods have almost the same execution time. *CacheA* outperforms *Sgreedy* in terms of amount of intermediate data transferred and total time by 44% and 26%, respectively. *CacheA* has execution time similar to *Agreedy* (3.1% longer). The cached data is balanced as the three sites have same storage capacities. Thus, the total times of *CacheA* and *Agreedy* are almost the same.

With heterogeneous sites ($H > 0$), the sites with more CPUs have less available storage but can execute more tasks, which leads to a larger amount of intermediate and cached data transferred between the sites. For $H = 0.3$, *CacheA* outperforms *Sgreedy* and *Agreedy* in terms of total time (by 40% and 18%, respectively) and amount of data transferred (by 47% and 21%, respectively).

With $H = 0.7$, *CacheA* outperforms *Sgreedy* and *Agreedy* in terms of total time (by 58% and 42%, respectively) and in terms of amount of data transferred (by 55% and 31%, respectively). *CacheA* is faster because its scheduling leads to a smaller amount of cached data transferred when reused (48% smaller than *Agreedy*) and added to the cache (62% smaller than *Agreedy*).

## 5.3   Concluding Remarks

Our cache-aware scheduling method *CacheA* always outperforms the two baseline methods (which also benefit from our cache/data selection method), both in the case of multiple users and heterogeneous sites.

The first experiment (with caching) shows that storing and reusing cached data becomes beneficial when 20% or more of the input data is reused. The second experiment (multiple users) shows that *CacheA* outperforms *Sgreedy* and

*Agreedy* in terms of total time by up to 61% and 43%, respectively. It also shows that, with increasing numbers of users, the performance of the three scheduling methods decreases due to higher cache data transfer times. The third experiment (heterogeneous sites) shows that *CacheA* adapts well to site heterogeneity, minimizing the amount of cached data transferred and thus reducing total time. It outperforms *Sgreedy* and *Agreedy* in terms of total time by up to 58% and 42% respectively.

Both cache site selection methods *bCompute* and *bStorage* have their own advantages. *bCompute* outperforms *bStorage* in terms of data transfer time by 13% for the first user and up to 17% for the second user. However, it does not scale with the number of users, and the limited storage capacities of Site 2 and 3 lead to a bottleneck. On the other hand, *bStorage* balances the cached data among sites and prevents the bottleneck when accessing the cached data, thus reducing re-execution times. In summary, *bCompute* is best suited for compute-intensive workflows that generate smaller intermediate datasets while *bStorage* is best suited for data-intensive workflows where executions can be performed at the site where the data is stored.

## 6   Conclusion

In this paper, we proposed a solution for distributed caching of scientific workflows in a cloud with heterogeneous sites (including on premise servers and shared-nothing clusters). Based on a distributed and parallel architecture, we proposed algorithms for adaptive caching, cache site selection and dynamic workflow scheduling. We implemented our solution in OpenAlea, together with a multisite scheduling algorithm. Using a real data-intensive application in plant phenotyping (Phenomenal), our extensive experimental evaluation using a cloud with three heterogeneous sites shows that our solution can yield major performance gains. In particular, it reduces much execution times and data transfers, compared to two baseline scheduling methods (which also use our cache/data selection method).

**Acknowledgments.** This work was supported by the #DigitAg French initiative (www.hdigitag.fr), the SciDISC and HPDaSc Inria associated teams with Brazil, the Phenome-Emphasis project (ANR-11-INBS-0012) and IFB (ANR-11-INBS-0013) from the Agence Nationale de la Recherche and the France Grille Scientific Interest Group.

## References

1. Altintas, I., Barney, O., Jaeger-Frank, E.: Provenance collection support in the Kepler scientific workflow system. In: Moreau, L., Foster, I. (eds.) IPAW 2006. LNCS, vol. 4145, pp. 118–132. Springer, Heidelberg (2006). https://doi.org/10.1007/11890850_14
2. Artzet, S., Brichet, N., Chopard, J., Mielewczik, M., Fournier, C., Pradal, C.: OpenAlea.phenomenal: a workflow for plant phenotyping, September 2018. https://doi.org/10.5281/zenodo.1436634

3. Callahan, S.P., Freire, J., Santos, E., Scheidegger, C.E., Silva, C.T., Vo, H.T.: VisTrails: visualization meets data management. In: ACM SIGMOD International Conference on Management of Data (SIGMOD), pp. 745–747 (2006)
4. Crago, S., et al.: Heterogeneous cloud computing. In: 2011 IEEE International Conference on Cluster Computing, pp. 378–385. IEEE (2011)
5. Garijo, D., Alper, P., Belhajjame, K., Corcho, O., Gil, Y., Goble, C.: Common motifs in scientific workflows: an empirical analysis. Future Gener. Comput. Syst. (FGCS) **36**, 338–351 (2014)
6. Heidsieck, G., de Oliveira, D., Pacitti, E., Pradal, C., Tardieu, F., Valduriez, P.: Adaptive caching for data-intensive scientific workflows in the cloud. In: Hartmann, S., Küng, J., Chakravarthy, S., Anderst-Kotsis, G., Tjoa, A.M., Khalil, I. (eds.) DEXA 2019. LNCS, vol. 11707, pp. 452–466. Springer, Cham (2019). https://doi.org/10.1007/978-3-030-27618-8_33
7. Kelling, S., et al.: Data-intensive science: a new paradigm for biodiversity studies. Bioscience **59**(7), 613–620 (2009)
8. Liu, J., et al.: Efficient scheduling of scientific workflows using hot metadata in a multisite cloud. IEEE Trans. Knowl. Data Eng. **31**(10), 1–20 (2018)
9. Liu, J., Pacitti, E., Valduriez, P., Mattoso, M.: A survey of data-intensive scientific workflow management. J. Grid Comput. **13**(4), 457–493 (2015). https://doi.org/10.1007/s10723-015-9329-8
10. Liu, J., Pacitti, E., Valduriez, P., de Oliveira, D., Mattoso, M.: Multi-objective scheduling of scientific workflows in multisite clouds. Future Gener. Comput. Syst. (FGCS) **63**, 76–95 (2016)
11. Maheshwari, K., Jung, E., Meng, J., Vishwanath, V., Kettimuthu, R.: Improving multisite workflow performance using model-based scheduling. In: IEEE International Conference on Parallel Processing (ICPP), pp. 131–140 (2014)
12. de Oliveira, D., Baião, F.A., Mattoso, M.: Towards a taxonomy for cloud computing from an e-science perspective. In: Antonopoulos, N., Gillam, L. (eds.) Cloud Computing. CCN, pp. 47–62. Springer, London (2010). https://doi.org/10.1007/978-1-84996-241-4_3
13. Özsu, M.T., Valduriez, P.: Principles of Distributed Database Systems, 4th edn. Springer, Cham (2020). https://doi.org/10.1007/978-3-030-26253-2
14. Pradal, C., Fournier, C., Valduriez, P., Cohen-Boulakia, S.: OpenAlea: scientific workflows combining data analysis and simulation. In: International Conference on Scientific and Statistical Database Management (SSDBM), pp. 11:1–11:6 (2015)
15. Tardieu, F., Cabrera-Bosquet, L., Pridmore, T., Bennett, M.: Plant phenomics, from sensors to knowledge. Curr. Biol. **27**(15), R770–R783 (2017)
16. Yuan, D., et al.: A highly practical approach toward achieving minimum data sets storage cost in the cloud. IEEE Trans. Parallel Distrib. Syst. **24**(6), 1234–1244 (2013)
17. Zhang, J., Luo, J., Dong, F.: Scheduling of scientific workflow in non-dedicated heterogeneous multicluster platform. J. Syst. Softw. **86**(7), 1806–1818 (2013)

# Challenges in Resource Provisioning for the Execution of Data Wrangling Workflows on the Cloud: A Case Study

Abdullah Khalid A. Almasaud, Agresh Bharadwaj, Sandra Sampaio[✉],
and Rizos Sakellariou

Department of Computer Science, The University of Manchester, Manchester, UK
s.sampaio@manchester.ac.uk

**Abstract.** Data Wrangling (DW) is an essential component of any big data analytics job, encompassing a large variety of complex operations to transform, integrate and clean sets of unrefined data. The inherent complexity and execution cost associated with DW workflows make the provisioning of resources from a cloud provider a sensible solution for executing these workflows in a reasonable amount of time. However, the lack of detailed profiles of the input data and the operations composing these workflows makes the selection of resources to run these workflows on the cloud a hard task due to the large search space to select appropriate resources, their interactions, dependencies, trade-offs and prices that need to be considered. In this paper, we investigate the complex problem of provisioning cloud resources to DW workflows, by carrying out a case study on a specific Traffic DW workflow from the Smart Cities domain. We carry out a number of simulations where we change resource provisioning, focusing on what may impact the execution of the DW workflow most. The insights obtained from our results suggest that fine-grained cloud resource provisioning based on workflow execution profile and input data properties has the potential to improve resource utilization and prevent significant over- and under-provisioning.

## 1 Introduction

Data Wrangling (DW) is the most widely used term to refer to the process of transforming data, from the format in which it was originally collected into a desired format suitable for analysis [4]. The reason for the transformation of "raw" data is the need to turn unrefined data into a valuable asset, from which intelligence is to be obtained and used to benefit business and science. A DW workflow has similarities with traditional Extraction-Transformation-Loading (ETL) processes commonly used in data warehousing [16]. The main similarity consists in that DW encompasses operations for preparing data to be integrated before disseminated, including operations for data cleaning, format transformation, summarization and integration [11]. However, DW and ETL differ in their level of reusability with regards to the number of applications whose

S. Hartmann et al. (Eds.): DEXA 2020, LNCS 12392, pp. 66–75, 2020.
https://doi.org/10.1007/978-3-030-59051-2_5

requirements each is able to fulfil; while ETL activities are carefully designed to fulfil the requirements of multiple use cases, DW workflows are tailored for the purposes of a single data analysis task, making their reusability a challenge.

The fact that each DW workflow is tailored to a specific task causes such workflows to significantly differ in their level of complexity, depending on a number of factors, e.g., number of inputs, format of inputs and need for complex transformations [11], availability of metadata [8], need for metadata reconciliation [15], data quality [3], size of inputs, to name a few. While some DW workflows are simple, involving a couple of nodes, other workflows may involve dozens of nodes encapsulating complex processing, such as the transformation of a JSON formatted file into a CSV one, aggregation of values in multiple columns, the joining of two files, etc. In addition, complex DW workflows are usually expensive in terms of the resources they consume and can incur long execution times, leading data wranglers to resort to cloud services for executing their workflows in a reasonable amount of time. As a result, research on resource allocation for complex DW workflows is timely, particularly in domains such as Smart Cities, where typical road traffic analysis workflows present a high-level of complexity, as they perform operations that go beyond simple data manipulations over a single dataset. Thus, the main challenge in this scenario is how to best provision resources to fulfil the performance requirements of this type of workflow, while seeking to find a balance between the conflicting interests of cloud service users, i.e., minimization of financial cost, against those of cloud service providers, i.e., the maximization of resource utilization.

Cloud resource provisioning encompasses all activities that lead to the selection and use of all resources (e.g., CPU, storage and network) needed for the execution of a job submitted by a cloud service user, considering Quality of Service (QoS) requirements and Service Level Agreement (SLA) [14]. Provisioning of resources can be done 'on-demand', whereby resources are promptly provided to urgent jobs, or by long-term reservation, where resources are reserved for later use. While each approach presents advantages, on-demand provisioning often causes too many jobs to simultaneously use the same resource, leading to interference and performance degradation. On the other hand, long-term reservation often causes many resources to be in an idle state [1].

This paper considers the problem of cloud resource provisioning for complex and data intensive DW workflows, by providing an investigation into the impact of varying levels of cloud resource provisioning on the performance of these workflows. The main aim is to provide insights that can be used in the development of solutions that answer the following questions:

1. What is a 'good' amount of resources to choose for the execution of complex DW workflows, aiming at avoiding significant over- and under-provisioning (i.e., preventing more or less resources than the amount actually needed to be allocated [9])?
2. How can the execution profile of workflows, size of input datasets and intermediate results be used in the development of criteria for resource provisioning?

3. How can the information in (2) above be effectively and efficiently used in resource provisioning?

To obtain useful insights, we show results for a number of simulations exploring the performance behaviour of complex DW workflows under varying levels of resource provisioning, considering the resources that have most impact on these workflows. As a use case, we take a typical data analysis workflow in the Smart Cities domain. Our simulations are performed using a widely used Cloud Workflow Simulator, WorkflowSim [2], as well as real-world data. The main contributions of this paper can be summarized as follows:

- Identification of properties and profile information of complex DW workflows that can be used in cloud resource provisioning.
- Identification of the resources that most impact on the performance of these workflows and their level of impact.
- Design and rationale of a number of simulations for revealing the level of impact of cloud resources on the performance of DW workflows.
- Discussion of results, derived insights, and challenges to be addressed in future research.

The rest of the paper is organized as follows: Sect. 2 describes some related work. Section 3 provides background information on the workflows used in our investigation. Section 4 describes the rationale behind each performed simulation. Section 5 provides a description of the simulations, the obtained results and discussion, and Sect. 6 concludes and describes further work.

## 2    Related Work

Early work on cloud resource provisioning mostly focused on the development of general techniques for static and dynamic provisioning, as the survey by Guruprasad et al. [1] indicates, where approaches more susceptible to resource over- or under-provisioning are described. Greater concern about performance and other SLA requirements led to the development of QoS based techniques, an example being the work by Singh et al. [12]; more specifically, this work suggests that identification, analysis and classification of cloud workloads, taking into account QoS metrics, should be performed before scheduling, to avoid violation of SLA. A survey by the same authors, in [13], classifies various works in cloud resource provisioning according to different types of provisioning mechanisms, and focuses on typical cloud workload types, such as Web sites, online transaction processing, e-commerce, financial, and internet applications as well as mobile computing, which account for the bulk of cloud workloads.

Recent Software Engineering trends towards self-management, minimization of energy consumption as well as the impact of machine learning, complex data preparation and analysis on the success of both business and science have significantly influenced research in cloud resource provisioning. Examples of work addressing self-management include Gill et al. [6], which addresses limitations in resource management by proposing an autonomic resource management technique focused on self-healing and self-configuration; and Gill and Buyya [5], which addresses self-management of cloud resources for execution of clustered workloads. An example of cloud resource provisioning work considering energy consumption is Gill et al. [7], which proposes a technique for resource scheduling that minimizes energy consumption considering a multitude of resources, in order to better balance the conflicting requirements of high reliability/availability and minimization of the number of active servers. Exploring different types of workloads, the work by Pietri et al. [10] proposes a cloud resource provisioning approach to handle large and complex scientific workflows, where an algorithm for efficiently exploring the search space of alternative CPU frequency configurations returns Pareto-efficient solutions for cost and execution trade-offs.

Similarly to the work by Pietri et al. [10], the work in this paper focuses on cloud resource provisioning for workloads resulting from the execution of complex workflows. Pietri et al. focus on scientific workflows, while we focus on data-intensive DW workflows, which share similarities with subsets of activities found in many scientific workflows, in that these also require Data Wrangling (DW). Traffic DW workflows, in particular, are highly complex because of the presence of functions that go beyond simple data manipulations over a single data unit. Rather, examples of such functions include spatio-temporal join operations using time, latitude and longitude proximities to integrate files, functions to iterate over a number of rows in a file to remove redundant data, etc. Considering that DW workflows are data-intensive and require complex analysis, identification of the resources that may mostly impact on the performance of this type of workload (e.g., via job profiling), and the level of impact of these resources (via experimentation or cloud simulation) can provide awareness of the challenges that need to be addressed.

## 3   A Traffic DW Workflow

The work in this paper makes use of a DW Workflow from the Smart Cities domain [11]. In particular, this workflow (illustrated in Fig. 1) answers the following traffic-related question: *What is the typical Friday Journey Time (JT) for the fragment of Chester Road stretching from Poplar Road to the Hulme area between 17:00 and 18:00?* Note that input *File 1* and *File 2* are "raw" traffic data files, each of size 1 GB, describing data from two collection sites on Chester Road (in the city of Manchester, UK). *File 3* holds information about distances between data collection sites across the city, with less than 100KB. Each of the main files is reduced and prepared for integration by having extraneous columns and rows that do not match the specified week day and day time removed, as well as some single columns split into two. *Files 1* and *2* are then merged vertically using the union operation before being horizontally merged with *File 3*. Note that, as *File 1* and *File 2* are significantly reduced at this point, the merge with *File 3* does no incur high execution costs as it generally would if the reduction of these files had not taken place before this merge. The information is then grouped by ID of collection site before the data is summarised and journey time, calculated. In total, there are 13 operations preceded by an operation for uploading the files onto the environment used.

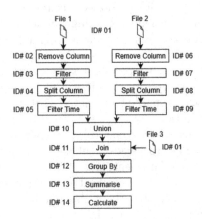

**Fig. 1.** An illustration of the Workflow used in the simulations.

## 4    Methodology

Profiling of the workflow described in Sect. 3[1] has revealed that traffic DW contains operations that are I/O or CPU intensive, or a combination of both, depending on functionality and input/output size. For example, the *ID1-Read* operation is I/O intensive, however, when consuming *File 3*, it incurs a much lower cost than *Files 1* and *2*, due to file size. On the other hand, operations *ID13-Summarise* and *ID14-Calculate* are mostly CPU intensive. To observe over- and under-provisioning, investigation into the levels of performance improvement or degradation, as variations on the amount of the most impacting resources are made, is required. To fulfil this purpose, three sets of simulations were performed using WorkflowSim [2], in which the execution of the workflow was simulated. The first set encompasses simulations where variations of CPU *Million Instructions Per Second* (MIPS) are made while all other simulation parameters remain fixed, to observe how variations in the availability of CPU resources in isolation impact on the execution time of the workflow. The second set encompasses simulations where parameters that define maximum available bandwidth are varied while other parameters remain fixed, including CPU MIPS, to observe how variations in bandwidth in isolation affect the execution time of the workflow. The third set encompasses simulations where the number of VMs are varied while all other parameters remain fixed, to observe how the different types of parallelism, inter- and intra-operator, can be explored and performance gains obtained, while

---

[1] Using *workflow-profiler* (https://github.com/intel/workflow-profiler).

mimicking a cloud environment where multiple nodes are available for the execution of a task. Note that WorkflowSim, by default, performs task clustering by allocating a single Virtual Machine (VM) per branch of a workflow. Also note that the choice of parameters used in the simulations was made by performing additional simulations (outside the main scope of this paper) with each parameter in WorkflowSim, and selecting the ones that had the most impact on execution time. We believe the three sets of simulations we describe serve to help identify a 'good' amount of resources for the execution of complex DW workflows, by revealing the number of and which operations are CPU or I/O bound, the extent to which specific resources should be increased or decreased to obtain performance gains and what correlations exist between parameters (such as input size, CPU/IO-bound classification and resource availability), potentially resulting in the development of models that can be used to avoid significant over- and under-provisioning. The results of the three sets of simulations are presented in the next section.

# 5   Simulation, Results and Discussion

Simulations were carried out by installing and running the WorkflowSim simulator on a Windows 10 computer with an Intel i7-7500U processor and 16 GB memory using Java version 1.8.0_191. A number of runs were repeated up to five times to validate that the simulator gave consistent results every time. The results of the three sets of experiments are presented and analyzed next.

*Simulation Set 1 (CPU MIPS):* For this set of simulations, a single-CPU VM that operates a range of MIPS values equally distributed between a minimum and maximum, $Mmin = 1000$ and $Mmax = 5000$, with 1000 MIPS steps is assumed, with all other simulation variables remaining fixed. Figure 2 shows how the total execution time of the complete CPU differs as the value of provisioned MIPS increases. On the level of individual operations of the workflow, the performance improvement varies as shown in Fig. 3; we observe that the reduction in execution time is more significant for operations that are CPU intensive. As expected, the main observation is a clear linear inverse relation between workflow execution time and MIPS in this set of simulations, as detailed in Fig. 2. The increase in MIPS in this simulation proved beneficial in reducing the execution time of all DW operations in the workflow, although the degree of reduction of the execution time needs to be taken into account as most cloud service providers will charge higher fees for more powerful CPUs.

*Simulation Set 2 (Bandwidth):* I/O bound tasks are usually affected by the bandwidth and transfer rates available in the execution environment. To explore potential execution time reductions, a number of three different bandwidth values are used in this set of simulations. The values used are 1, 15 and 225 MB/s, with all other parameters remaining fixed. It can be observed that the impact of the variation in bandwidth on performance may not be as profound as it was

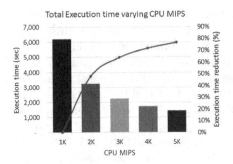

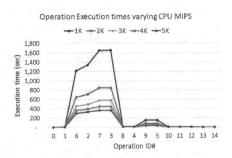

**Fig. 2.** Workflow execution time with varying CPU MIPS values.

**Fig. 3.** Execution times of individual operations varying CPU MIPS.

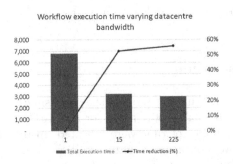

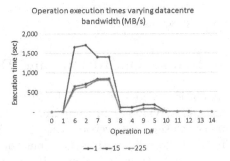

**Fig. 4.** Workflow execution time with varying max available bandwidth.

**Fig. 5.** Execution times of individual operations for each bandwidth.

in the case of CPU MIPS, as shown in Fig. 4. Considering the execution time of individual operations in the workflow, it is also observed that execution time improvements are more significant for certain operations, as shown in Fig. 5, specifically, those that were profiled as both I/O and CPU bound and that also process large inputs. Even though, execution time improvements are not equally significant for all operations, increase in bandwidth is still beneficial for reducing the execution time of the whole workflow. It is worth pointing out that, increasing bandwidth from 15 MB/s to 225 MB/s, leads to very small savings in execution time. This observation raises the question of whether this increase is worth paying for.

*Simulation Set 3 (Number of Virtual Machines):* In this set of simulations, the number of VMs used in the execution is varied while other parameters remain fixed. First, the variation is performed on the "original" two-branched workflow, to a maximum of the number of branches in the workflow, exploring *inter-operator parallelism*. Next, the input data files are partitioned so that the same workflow operations are performed on fragments of the original files, increasing the level of parallelism by exploring *intra-operator parallelism.* When explor-

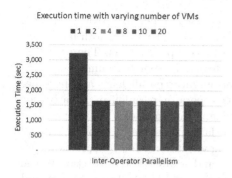

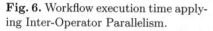

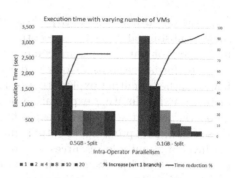

**Fig. 6.** Workflow execution time applying Inter-Operator Parallelism.

**Fig. 7.** Workflow execution time applying Intra-Operator Parallelism.

ing **inter-operator parallelism**, it is observed that the increase in number of VMs does not have any impact on the timings of individual operations, but it reduces the total execution time of the complete workflow, by assigning operations located in different branches of the workflow to run on different VMs. Figure 6 shows how the increase in VMs is beneficial, up to the number of branches in the workflow. To explore **intra-operator parallelism**, the input data is partitioned by a factor of 2 and 10, combining both types of parallelism, as shown in Fig. 7. It is observed that the total workflow execution time on a single VM is not significantly affected. However, data partitioning allows an increase in the degree of parallelism, by increasing the number of VMs used in the execution of the workflow. This results in a reduction of the total execution time of the whole workflow (as shown in Fig. 7), as partitions of the same file are simultaneously input, processed and output on different VMs. Further increases in the number of VMs without further data partitioning can lead to no performance gains, as can be seen from the *0.1 GB-split* case in Fig. 7, where an increase to 20 VMs is ideal for obtaining performance gains, but if further increases are desired without further data partitioning, i.e., to more than 20 VMs, no further performance gains are obtained.

*Discussion:* Three main observations are derived from the results presented in the previous section, discussed in the following: **(1)** CPU MIPS is the parameter that mostly impacts execution time and one of the most costly cloud resources. However, a balanced combination of CPU MIPS provisioning with the provisioning of other impacting resources can result in financially viable parameter configurations, while still providing similar performance gains. **(2)** Bandwidth has a more modest impact on execution time, showing less significant performance gains with increases in availability, as the size of intermediate results gradually decreases due to the application of data reduction operations, rendering no more than 56% improvement at best. Presence of data reduction operations early in the workflow execution can potentially lessen the benefit of higher bandwidths, generating opportunities for resource release before the workflow execution is over. (3) Variations in the number of provisioned VMs show a substantial impact on execution time, particularly when both inter- and intra-operator parallelisms are combined to speed up execution. The extent to which performance gains are observed depends mainly upon the number of workflow branches, limiting exploitation of inter-operator parallelism, and intermediate data size, limiting exploitation of intra-operator parallelism. Clearly, not all operations in the workflow benefit from higher numbers of VMs, particularly those that input, process and output smaller data sizes. Finally, resource balancing involving multiple resources for obtaining execution time reductions incurs different cost and performance implications, and so an effective solution to the problem of finding a 'good' amount of resources to balance financial cost and performance benefits, avoiding under- and over-provisioning, probably involves combining not only the Pareto-efficient set of configurations that finds the best cost-benefit balance involving multiple resources for the execution of one job, but also for several jobs that may be waiting to be executed simultaneously.

## 6   Conclusions and Further Work

To help finding answers to the research questions that motivate this paper (Sect. 1), we performed a number of simulations using a representative DW workflow. The results have shown that, depending on the execution profile of the DW workflow, more than one resource can have a significant impact on execution performance; and that job execution profiles, if considered when provisioning cloud resources, have the potential to improve decision making and avoid over- and under-provisioning. While choices regarding which subset of resources to focus on and their provisioning levels have an impact on performance and financial costs, these decisions are, to a large extent, job-dependent. Therefore, we believe that models to find the configurations that return the best cost-performance trade-off for a job, such as the work by Pietri et al. [10], should be extended to consider multiple resources as well as multiple jobs, based on the execution profile of the individual jobs and their performance requirements. Numerous challenges in developing such a solution need to be faced, such as: (i) how to efficiently obtain profiles of the jobs, which may require a large number of experiments or simulations; (ii) how to accurately identify the most relevant profile metadata to be used for cloud resource provisioning; (iii) how to devise an efficient and effective mechanism for making use of the profile information at the time of cloud resource provisioning. We intend to investigate these challenges in the future, experimenting also with a variety of different DW workflows. One direction to address these challenges may be that job profiles are generated at run time and Machine Learning or related training techniques become important components of an effective solution.

**Acknowledgement.** Partial support from the H2020 I-BiDaaS project (grant agreement No. 780787) is gratefully acknowledged.

# References

1. Bhavani, B.H., Guruprasad, H.S.: Resource provisioning techniques in cloud computing environment: a survey. Int. J. Res. Comput. Commun. Technol. **3**, 395–401 (2014)
2. Chen, W., Deelman, E.: Workflowsim: a toolkit for simulating scientific workflows in distributed environments. IEEE 8th International Conference on E-Science, pp. 1–8 (2012)
3. Sampaio, S.D.F.M., Dong, C., Sampaio, P.: $DQ^2S$ - a framework for data quality-aware information management. Expert Syst. Appl. **42**(21), 8304–8326 (2015)
4. Furche, T., Gottlob, G., Libkin, L., Orsi, G., Paton, N.W.: Data wrangling for big data: challenges and opportunities. In: Proceedings of the 19th International Conference on Extending Database Technology, EDBT 2016, Bordeaux, France, 15–16 March 2016, Bordeaux, France, pp. 473–478 (2016)
5. Gill, S.S., Buyya, R.: Resource provisioning based scheduling framework for execution of heterogeneous and clustered workloads in clouds: from fundamental to autonomic offering. J. Grid Comput. **17**(3), 385–417 (2019)
6. Gill, S.S., Chana, I., Singh, M., Buyya, R.: RADAR: self-configuring and self-healing in resource management for enhancing quality of cloud services. Concurrency and Computation: Practice and Experience **31**(1), (2019)
7. Gill, S.S., et al.: Holistic resource management for sustainable and reliable cloud computing: an innovative solution to global challenge. J. Syst. Softw. **155**, 104–129 (2019)
8. Hellerstein, J.M., et al.: Ground: a data context service. In: CIDR 2017, 8th Biennial Conference on Innovative Data Systems Research, Online Proceedings, Chaminade, CA, USA, 8–11 January 2017 (2017)
9. Nahrstedt, K.: To overprovision or to share via QoS-aware resource management? In: Proceedings of the Eighth International Symposium on High Performance Distributed Computing (Cat. No. 99TH8469), Redondo Beach, CA, USA, 6 August, pp. 205–212 (1999)
10. Pietri, I., Sakellariou, R.: A Pareto-based approach for CPU provisioning of scientific workflows on clouds. Future Gener. Comput. Syst. **94**, 479–487 (2019)
11. Sampaio, S., Aljubairah, M., Permana, H.A., Sampaio, P.: A conceptual approach for supporting traffic data wrangling tasks. Comput. J. **62**(3), 461–480 (2019)
12. Singh, S., Chana, I.: Q-aware: quality of service based cloud resource provisioning. Comput. Electr. Eng. **47**, 138–160 (2015)
13. Singh, S., Chana, I.: Cloud resource provisioning: survey, status and future research directions. Knowl. Inf. Syst. **49**(3), 1005–1069 (2016). https://doi.org/10.1007/s10115-016-0922-3
14. Singh, S., Chana, I.: A survey on resource scheduling in cloud computing: issues and challenges. J. Grid Comput. **14**(2), 217–264 (2016)
15. Stonebraker, M., Ilyas, I.F.: Data integration: the current status and the way forward. IEEE Data Eng. Bull. **41**(2), 3–9 (2018)
16. Vassiliadis, P.: A survey of extract-transform-load technology. Int. J. Data Warehouse. Min. **5**, 1–27 (2009)

# Genetic Programming Based Hyper Heuristic Approach for Dynamic Workflow Scheduling in the Cloud

Kirita-Rose Escott$^{(\boxtimes)}$ (D), Hui Ma (D), and Gang Chen

Victoria University of Wellington, Wellington, New Zealand
{kirita.escott,hui.ma,aaron.chen}@ecs.vuw.ac.nz

**Abstract.** Workflow scheduling in the cloud is the process of allocating tasks to limited cloud resources to maximise resource utilization and minimise makespan. This is often achieved by adopting an effective scheduling heuristic. Most existing heuristics rely on a small number of features when making scheduling decisions, ignoring many impacting factors that are important to workflow scheduling. For example, the MINMIN algorithm only considers the size of the tasks when making scheduling decisions. Meanwhile, many existing works focused on scheduling a static set of workflow tasks, neglecting the dynamic nature of cloud computing. In this paper, we introduce a new and more realistic workflow scheduling problem that considers different kinds of workflows, cloud resources, and impacting features. We propose a Dynamic Workflow Scheduling Genetic Programming (DSGP) algorithm to automatically design scheduling heuristics for workflow scheduling to minimise the overall makespan of executing a long sequence of dynamically arriving workflows. Our proposed DSGP algorithm can work consistently well regardless of the size of workflows, the number of available resources, or the pattern of workflows. It is evaluated on a well-known benchmark dataset by using the popular WorkflowSim simulator. Our experiments show that scheduling heuristics designed by DSGP can significantly outperform several manually designed and widely used workflow scheduling heuristics.

**Keywords:** Cloud computing · Dynamic workflow scheduling · Genetic programming

## 1 Introduction

Cloud computing is a distributed computing paradigm, which enables the delivery of IT resources over the Internet and follows the pay-as-you-go billing method [13]. These resources are provided by cloud providers for cloud users to execute their application-specific *workflows*. Workflows are widely used models for modelling and managing complex distributed computations [16]. To maximise revenue, cloud users aim to minimise their operation costs by effectively using

© Springer Nature Switzerland AG 2020
S. Hartmann et al. (Eds.): DEXA 2020, LNCS 12392, pp. 76–90, 2020.
https://doi.org/10.1007/978-3-030-59051-2_6

cloud resources to process their workflows and, in the meantime, achieve the best possible performance in workflow execution.

Workflow scheduling aims to allocate workflow tasks to cloud resources so that the overall performance, e.g., the makespan involved in executing a long sequence of dynamically arriving workflows, can be minimised. The heterogeneity of the workflow tasks and the wide range of available cloud resources make this an NP-hard optimisation problem [18]. No algorithms have ever been developed to satisfactorily solve this problem in an efficient and scalable manner. Therefore, heuristics are needed to find near-optimal solutions efficiently.

While designing any new heuristics, it is important to distinguish static workflow scheduling from its dynamic counterpart. In particular, static working scheduling requires the scheduling decision to be made *a priori* over all workflows pending for execution. On the other hand, dynamic working scheduling requires cloud resources to be allocated to ready tasks at runtime [1]. This is because new requests for workflow execution arrive at the cloud progressively making runtime resource allocation inevitable.

Several heuristics have been designed to address the dynamic workflow scheduling problem, such as *Heterogeneous Earliest Finish Time (HEFT), MIN-MIN* [3] and *MAXMIN* [4]. All these heuristics have demonstrated effectiveness in some related studies [12]. However, they only rely on a very small number of features, such as task size, to make scheduling decisions. Hence important information that is vital for workflow scheduling, e.g., virtual machine speed, the waiting time, and the expected completion time of every pending task, have been completely ignored. Due to this reason, these heuristics only work well on specific scheduling problem instances and cannot be reliably applied to a variety of different workflows and cloud settings. It is hence highly desirable to design new heuristics with consistently good performance across many workflow scheduling scenarios.

When designing scheduling heuristics for the dynamic workflow scheduling problem, several properties must be jointly considered, including the size of tasks, the dependencies of tasks, the amount of available resources and the specifications of those resources [14]. Numerous combined use of these properties is possible in practice to construct the scheduling heuristic. It is hence difficult to manually design an effective heuristic that is capable of solving many complex scheduling problems consistently well in the cloud.

Genetic Programming (GP) is a prominent Evolutionary Computation (EC) technology for designing heuristics for various difficult problems [17]. Designing scheduling heuristics manually in the cloud is time-consuming and requires substantial domain knowledge. Hence, GP possesses a clear advantage in designing these heuristics automatically. Its effectiveness has already been demonstrated in many combinatorial optimization problems, including job shop scheduling and other resource allocation problems [9,10,15].

In this paper, we have the goal to develop a new GP-based algorithm for automatically designing workflow scheduling heuristics in the cloud. Driven by this goal, this paper has three main research objectives:

1. Develop a new *Dynamic Workflow Scheduling Genetic Programming* (DSGP) algorithm to automatically design scheduling heuristics for the dynamic workflow scheduling problem.
2. Conduct an extensive experimental evaluation of DSGP by using the Workflowsim simulator and a popular benchmark dataset with 12 different workflow applications. Scheduling heuristics designed by DSGP is further compared to several existing scheduling heuristics, including HEFT, RR, and MCT [7].
3. Analyse the experiment results to investigate essential properties of workflows and cloud resources with a substantial impact on the performance of designed scheduling heuristics.

It is infeasible to conduct a real-world evaluation of any scheduling heuristic designed by DSGP due to restrictions on hardware and computation cost. To cope with this challenge, this paper adopts a simulated approach based on the combined use of multiple simulation technologies, including GridSim [5], CloudSim [6], and WorkflowSim [7]. With the help of these simulators, we can accurately and quickly evaluate any scheduling heuristic, ensuring the efficient and effective operation of DSGP [7].

The rest of the paper is organised as follows: Sect. 2 presents previous work related to workflow scheduling in the cloud environment. Section 3 gives an overview of the workflow scheduling problem. Section 4 outlines the proposed DSGP approach. Section 5 presents the experimental evaluation results and analysis. The conclusions and future work are outlined in Sect. 6.

## 2   Related Work

Scheduling cloud resources in the application layer to serve the cloud user is a major topic in cloud resource management and scheduling research [19]. Thus there are many existing works on this topic.

Some research studies workflow scheduling as a static problem, which assumes that all the workflows are known at the time of scheduling. Some meta-heuristic approaches propose to make the given workflow to resources (VMs) in clouds.

Greedy algorithms have previously been proposed to solve the workflow scheduling problem in the cloud. Typically, these algorithms generate a singular heuristic based on the current state of the tasks and resources. Mahmood et al. [8] propose an adaptive GA approach that focuses on minimising processing and communication costs of tasks with hard deadlines, as well as a greedy algorithm to be used for comparison. Greedy algorithms are known as simple and relatively quick, however, they do not always produce the best solution [8].

There are many existing workflow scheduling algorithms that consider different constraints with the aim to optimise cloud objectives such as makespan, cost, and application performance. Arabnejad et al. [2] propose a heuristic scheduling algorithm, Budget Deadline Aware Scheduling (BDAS), that addresses eScience workflow scheduling under budget and deadline constraints in Infrastructure as a Service (IaaS). Yu et al. [18] propose a Flexible Scheduling Genetic Programming (FSGP) that employs GP to generate scheduling heuristics for the workflow scheduling problem. While these algorithms produce promising results, they are only able to solve static workflow scheduling problems. In the cloud services can be dynamically added or removed at any given time. This means that services can be added on the go and while working with workflows a number of resources can be defined during runtime [11]. This means that the aforementioned algorithms may not perform well, if at all, in a dynamic cloud environment.

As far as we know, GP has never been applied to the dynamic workflow scheduling problem. However, due to the success of GP being used in similar problems, we are confident that GP can evolve heuristics capable of producing a near-optimal resource allocation. Our proposed DSGP will generate scheduling heuristics for the diverse collection of workflows in a cloud with the aim to minimise the overall makespan.

## 3 Preliminaries

### 3.1 Problem Overview

In the workflow scheduling problem, a workflow is commonly depicted as a directed acyclic graphs $DAG([n], [e])$, as demonstrated in Fig. 1, where $n$ is a set of nodes representing $n$ dependent tasks, $[e]$ indicates the data flow dependencies between tasks. Each workflow has a task with no predecessors named *entry* task and a task with no successors names *exit* task. Assuming workflows with different patterns, and tasks arriving from time to time, *dynamic workflow scheduling* allocates tasks to virtual machines with the objective of minimizing the makespan of the execution.

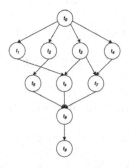

**Fig. 1.** A DAG with 10 tasks in the workflow

In the process of allocation, the following constraints must be satisfied.

- A task can only be allocated once all its predecessors have been executed.
- Each task can be allocated to any virtual machine, which will process the task.
- Each virtual machine can only process at most one task at any given time.

## 3.2   Formal Definition

The following equations present the basic properties of tasks and virtual machines that will be used to formulate the dynamic workflow scheduling problem.

The execution time of task $t_i$ on virtual machine $v_k$ is given by Eq. (1).

$$ET_{ik} = \frac{s_i}{m_k} \tag{1}$$

The size $s_i$ of task $t_i$ divided by the speed $m_k$ of virtual machine $v_k$.

Each task $t_i$ can be allocated to a virtual machine $v_k$ once all of the parents $aParent(t_i)$ have completed processing. The allocation time, denoted by $AT_i$, is given by Eq. (2).

$$AT_i = max_{p \in aParent(t_i)} FT_p \tag{2}$$

The actual start time of each task $t_i$, denoted by $ST_i$, is decided either by the time that virtual machine $v_k$ becomes idle, or when all the parent tasks of $t_i$ are completed, whichever is later. Virtual machine $v_k$ becomes idle when the previous task $t_{prev}$ being executed on $v_k$ completes processing. The actual start time is given by Eq. (3).

$$ST_i = max\{FT_{t_{prev}}, AT_i\} \tag{3}$$

The waiting time of task $t_i$, denoted $WT_i$, is given by Eq. (4), as the difference between the arrival time and the start time of a task.

$$WT_i = ST_i - AT_i \tag{4}$$

The relative finish time of task $t_i$ is given by Eq. (5).

$$RFT_i = WT_i + ET_{ik} \tag{5}$$

The expected completion time $ECT_i$ of task $t_i$ and it's children $aChild(t_i)$ is the maximum expected completion time possible and is given by Eq. (6).

$$ECT_i = total_{c \in aChild(t_i)} ET_{ik} \tag{6}$$

The makespan of a workflow $MS_T$ with tasks is the time taken to process all tasks in workflow $T$. The time taken is calculated by subtracting the earliest start time $ST$ from the latest finish time $FT$, as shown in Eq. (7).

$$MS_T = max\, FT - min\, ST \tag{7}$$

The total number of tasks $TNS$ defined in Eq. (8) sums the number of tasks $N$ in workflow $j$ of all $Num$ workflows. $TMS$ is the sum of makespan $MS_T$, of workflow $j$ of all $Num$ workflows, as shown in Eq. (9).

$$TNS = \sum_{j=1}^{Num} Nj \qquad (8)$$

$$TMS = \sum_{j=1}^{Num} MS_T j \qquad (9)$$

The overall average makespan of all test workflows $AMS$ is obtained by dividing $TMS$ by $TNS$, as presented in Eq. (10).

$$AMS = \frac{TMS}{TNS} \qquad (10)$$

The objective of the dynamic workflow scheduling problem is to minimize the overall average makespan of all the workflows. i.e., min $AMS$. We aim to find a rule $r$, which produces a schedule with minimal AMS. The fitness of a rule $r$ is evaluated by the $AMS$, of all workflows scheduled using the rule. Fitness is defined in Eq. (11).

$$fitness(r) = AMS(r) \qquad (11)$$

Based on the above, the goal is to find $r^*$ such that it minimizes $AMS(r^*)$ over all possible rules.

## 4   Design Scheduling Heuristic Through Genetic Programming

In this section, we present our new DSGP algorithm to solve the workflow scheduling problem. We will first describe the outline of the algorithm. We then describe the representation of the heuristics generated by DSGP, as well as the function set and the terminal set to be used by DSGP. Finally, we introduce the genetic operators to evolve new scheduling heuristics, including *crossover*, *mutation*, and *reproduction*.

### 4.1   Outline of DSGP

DSGP aims to evolve scheduling heuristics to help the Workflow Scheduler to schedule cloud resources for workflow execution. As mentioned in the introduction, we rely on the Workflowsim simulator to simulate the processing of dynamically arriving workflows over an extensive period of time [7]. In Workflowsim, the Workflow Scheduler obtains all *ready tasks* regularly from the Workflow engine. A task is ready if and only if all of its parent tasks in a workflow have been processed by virtual machines in the cloud. Given this, we decide to prioritise those tasks with a higher number of children in the corresponding workflows. This is because, once they are completed, a larger number of child tasks will become ready for processing by the Workflow Scheduler, ensuring the cloud to make continued progress in executing these workflows. In comparison, processing a large (or time-consuming) task with no children can often cause delays in the execution of all pending workflow, therefore increasing the overall makespan [18].

Being a hyper-heuristic approach, DSGP consists of a *training phase* and a *testing phase*. During the training phase, DSGP is designed to evolve a scheduling heuristic, driven by the goal to minimize the total makespan involved in executing a group of workflows by using a fixed collection of cloud resources (i.e., VMs in cloud data centers). The best heuristic discovered by DSGP during training is subsequently utilized in the testing phase to schedule various workflows.

As discussed, DSGP is designed to search for effective scheduling heuristics to solve the dynamic workflow scheduling problem. The pseudo-code in ALGO-RITHM 1 details the algorithmic design of DSGP.

In our experimental studies, the training set for DSGP consists of 12 different workflow types and a fixed collection of cloud resources. DSGP follows a standard GP framework with an embedded simulation as the evaluation process. The output of the training stage is the best scheduling heuristic/rule discovered by DSGP. This heuristic will be further evaluated in the testing stage to determine its true usefulness for scheduling various workflows in simulated cloud environments.

## 4.2   Representation

We represent a scheduling heuristic in the form of a GP tree. Example GP trees are illustrated in Fig. 2 and 3. The terminal nodes of the GP tree capture a range of properties concerning workflows and cloud resources. Our study of relevant research shows that those are the most important properties to be considered by a scheduling heuristic to minimize the makespan of workflow execution. The intermediate nodes of the GP tree are arithmetic functions, as summarized in Table 1.

## 4.3   Terminal Set and Functional Set

Demonstrated in Table 1, the terminals used to represent properties of workflows and cloud resources, in this case, virtual machines, that have effects on the total makespan of a workflow. Properties such as the size of a task, the execution time of a task and, the speed of a virtual machine. The function set, also described in Table 1, denotes the operators used; addition, subtraction, multiplication, and protected division. Protected division means that it will return 1 if divided by 0.

**Algorithm 1.** DSGP Algorithm

**Input:** A set of workflows
**Output:** Heuristic Rule
Initialize a population of rules $R$;
**while** *does not meet stop criteria* **do**
    **for** *a rule r in R* **do**
        $TMS = 0$;
        **for** *a workflow in a set of workflows* **do**
            Initialise the simulator;
            Prioritise tasks in set of ready tasks;
            **for** *a task in a set of ready tasks* **do**
                **vm = vmSelection(vms,** r**)**;
                **if** *vm is idle* **then**
                    allocate(task, vm);
                **end**
            **end**
            $TMS$ += calculateMakespan(workflow);
        **end**
        calculate $AMS$
        fit = fitness($AMS$)
    **end**
    TournamentSelection;
    Crossover;
    Mutation;
    Reproduction;
**end**
return the best rule

## 4.4  Crossover, Mutation, and Reproduction

Crossover, mutation, and reproduction are commonly used techniques in GP. We aim to use these techniques to generate scheduling heuristics which will be then evaluated using a determined fitness function, discussed in this section.

*Crossover* refers to the genetic operator which recombines the genetic information of two parents to create a new offspring. This is demonstrated in Fig. 2, where the left branch of Parent A is combined with the right branch of Parent B to become offspring A'. Likewise, the left branch of Parent B is combined with the right branch of Parent A to become offspring B'.

*Mutation* refers to the genetic operator which changes the sub-tree of a tree to maintain genetic diversity from one generation to the next. This is demonstrated in Fig. 3, where the sub-tree of the parent is replaced by a new sub-tree in the child. Mutation is used to prevent instances from becoming too similar to each other, which, can lead to poor performance [18].

*Reproduction* refers to the genetic operator which selects individuals from the current generation and passes the best of those individuals to the next generation as offspring.

**Table 1.** The terminal and function set for the proposed FSGP approach

| Terminal name | Definition |
|---|---|
| $TS$ | The total size of a task $t_i$ |
| $VS$ | The speed of a virtual machine $v_j$ |
| $ET$ | Execution time of a task $t_i$ |
| $WT$ | Waiting time of a task $t_i$ |
| $RFT$ | Relative finish time of a task $t_i$ |
| $ECT$ | Expected completion time of a task $t_i$ |
| Function name | Definition |
| Add, Subtract, Multiply | Basic arithmetic operations $(+, -, \times)$ |
| Protected Division | Protected division, return 1 if the denominator is 0 (%) |

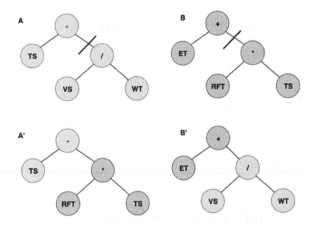

**Fig. 2.** Crossover example

## 5 Experimental Evaluation

In this section, we describe the experiments we conducted to evaluate the DSGP algorithm. We can perform this evaluation by using the WorkflowSim simulator to simulate a real cloud environment and execute each of the workflows using the generated heuristic in that environment. As our approach aims to generate scheduling heuristics which minimise the average total makespan, the fitness of a generated heuristic is determined by the normalised average total makespan of all the workflows scheduled by the heuristic.

### 5.1 Dataset

Four benchmark workflow patterns are used in our experiments. Fig. 4 demonstrates visualisations of the CyberShake, Inspiral, Montage and SIPHT workflows. We used three different sizes for each of the patterns, medium, large and,

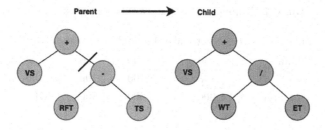

**Fig. 3.** Mutation example

extra-large, described in Table 2. Thus, we tested 12 workflows in total, with the number of tasks ranging between 50 and 1,000.

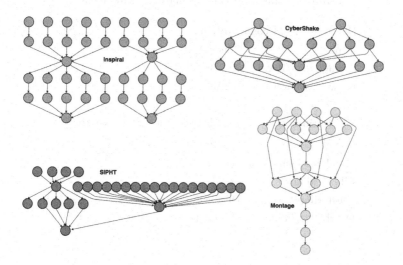

**Fig. 4.** Visual representation of workflow applications

## 5.2  Simulator

We used the WorkflowSim simulator in our experiments to evaluate the effectiveness of the generated heuristics. The configuration settings are described in Table 3. CondorVM extends the Cloudsim VM, it has a local storage system and can indicate whether it is busy or not [7]. In WorkflowSim, VMs are created with the same parameters. We have VMs of small, medium, large and extra large size. We also use three different sets of VMs in our experiments. We have a small set of 4, which has 1 each of the different sized VMs, a set of 16, which has 4 of each VM and a set of 32, which has 8 of each VM. We have 12 workflows, and 3 sets of VMs to test each workflow, therefore we have 36 test scenarios.

The storage size is 10 GB, the RAM is 512 MB, Network Bandwidth is 10Mbit/s and we use one CPU. We use the local file system, as opposed to

**Table 2.** Number of tasks in workflow applications

| Application size | CyberShake | Inspiral | Montage | SIPHT |
|---|---|---|---|---|
| Medium | 50 | 50 | 50 | 60 |
| Large | 100 | 100 | 100 | 100 |
| Extra large | 1,000 | 1,000 | 1,000 | 1,000 |

the shared file system which has only one storage for the data center. Work-flowSim also has a Clustering Engine which merges tasks into Jobs which are processed together, in our experiments we disable this feature.

**Table 3.** Configuration of simulation environment

| Datacenter/VM configuration | |
|---|---|
| Type | CondorVM |
| Storage Size/RAM | 10 GB/512 MB |
| Network Bandwidth | 10 Mbit/s |
| # CPUs | 1 |
| VM Ratio Size | {Small, Medium, Large, X-Large} |
| # VMs | {8, 16, 32} |
| Simulator configuration | |
| File System | Local |
| Clustering | Disabled |
| Overheads/Failures | Disabled |

### 5.3 Baseline Algorithms

WorkflowSim has dynamic implementations of traditional workflow scheduling algorithms. We examined those implementations and give a summary of each below [1,3,4,7].

- HEFT always selects the fastest available virtual machine to schedule a task to. We implemented our own version of HEFT in WorkflowSim for comparison
- MINMIN takes the task with the minimum size and allocates it to the first idle virtual machine
- MAXMIN takes the task with the largest size and allocates it to the first idle virtual machine
- FCFS allocates the first ready task to the first available virtual machine
- MCT allocates a task to the virtual machine with the minimum expected completion time for that task
- RR selects the virtual machine a task is to be allocated to in a circular order

## 5.4   Parameter Settings

In our experiments we set the population size to 512, and the number of generations to 51. The crossover, mutation and reproduction rates are 85%, 10% and 5% respectively [9]. The tournament size for tournament selection is 7 and the maximum depth of a tree is set to 17. Tournament selection is used to encourage the survival of effective heuristics in the population. We run the our experiments 30 times to verify our results.

## 5.5   Results

The results of our experiments show that DSGP outperforms competing algorithms. Table 4 shows the average makespan of each of the algorithms for all workflow application scenarios. The smaller the value, the better the performance of the algorithm. We see that FCFS has the worst performance, and DSGP has the best performance. We calculate the average makespan for DSGP as the mean average makespan over 30 independent runs.

Table 5 compares the results of all approaches for all of the testing scenarios. We can see that DSGP achieves the best results on more of the test scenarios than the competing approaches. We see that the RR and FCFS approaches consistently performed the worst.

Of all the approaches, HEFT performed secondary to our DSGP approach. In both the number of test scenarios that it achieved the best results and in the overall average makespan of all test scenarios. MINMIN, MAXMIN, and MCT, achieved the best results for 6 of the 36 test scenarios between them. Whereas, RR and FCFS did not achieve the best results for any of the test scenarios.

**Table 4.** Average makespan of all workflow applications

| | Scheduling Heuristic | | | | | | |
|---|---|---|---|---|---|---|---|
| Workflow Application | DSGP | HEFT | MINMIN | MAXMIN | RR | FCFS | MCT |
| All Workflow Sizes {Medium, Large, X Large} | **36.90** ± 14.55 | 37.41 ± 13.48 | 57.84 ± 24.34 | 40.49 ± 13.75 | 207.01 ± 118.37 | 207.52 ± 118.36 | 48.97 ± 19.97 |

## 5.6   Discussion

We analysed the terminals of the best heuristic rules generated in each run to determine which terminals affected the performance of the heuristics generated by DSGP. We found that there were some purposeless combinations such as "$(ECT - ECT)$", which have no impact on the trees and were removed from the analysis. From here, we counted the frequency of terminal appearances, shown in Table 6.

We can observe from Table 6 that the most frequently occurring terminals are expected completion time ($ECT$) and task size ($TS$). In the following, we show one of the heuristic rules generated by our DSGP to analyse the relationship between the terminals and the heuristics rules.

**Table 5.** Average makespan of individual workflow application scenarios

| Workflow | Application Size | Num VMs | DSGP | HEFT | MINMIN | MAXMIN | RR | FCFS | MCT |
|---|---|---|---|---|---|---|---|---|---|
| Cybershake | Medium | Small | 15.9989 | **14.1131** | 23.211 | 15.3275 | 15.4486 | 16.7487 | 18.0751 |
| | | Medium | **14.0997** | 14.2128 | 16.3827 | 14.2584 | 15.5911 | 15.3292 | 14.1279 |
| | | Large | **6.0264** | 9.4706 | 9.4792 | 9.441 | 9.5293 | 9.4523 | 9.4436 |
| | Large | Small | 13.9852 | **7.9588** | 25.4003 | 18.4432 | 87.3345 | 87.3345 | 13.9744 |
| | | Medium | 8.0987 | 5.707 | 7.7164 | 7.0453 | 29.0844 | 29.0844 | 9.6625 |
| | | Large | 6.5287 | **2.8697** | 3.4996 | 2.9647 | 4.8269 | 4.9214 | 3.4289 |
| | X Large | Small | 9.135 | 8.0487 | 6.1173 | **5.6271** | 237.518 | 237.518 | 5.9525 |
| | | Medium | **3.0255** | 4.8862 | 4.8939 | 5.9262 | 89.8922 | 89.8922 | 4.8771 |
| | | Large | 2.5018 | **1.4865** | 2.1634 | 3.5278 | 11.4675 | 11.4675 | 2.4932 |
| Inspiral | Medium | Small | **103.4092** | 112.5431 | 162.6154 | 136.9769 | 129.1874 | 120.7618 | 118.9919 |
| | | Medium | **52.6484** | 91.7377 | 102.0764 | 106.6661 | 88.306 | 90.5894 | 89.8665 |
| | | Large | **31.882** | 94.566 | 94.3443 | 94.4015 | 96.2813 | 94.5207 | 95.4195 |
| | Large | Small | 85.9817 | 70.4122 | **56.6749** | 69.0277 | 332.6658 | 332.6658 | 61.3014 |
| | | Medium | 60.6875 | **41.4328** | 112.5836 | 63.0693 | 166.9795 | 193.6208 | 89.5669 |
| | | Large | **21.7495** | 28.4321 | 36.2024 | 32.1388 | 33.4726 | 33.5923 | 32.7614 |
| | X Large | Small | 91.244 | 77.2779 | 32.2015 | 33.1887 | 1063.3361 | 1063.3361 | **31.882** |
| | | Medium | 28.584 | 40.2824 | 29.4118 | 22.712 | 743.3518 | 743.3518 | **22.9864** |
| | | Large | 30.5142 | **14.6363** | 67.3942 | 53.9798 | 58.3216 | 55.1827 | 68.513 |
| Montage | Medium | Small | **4.5023** | 4.7461 | 4.9686 | 4.8121 | 6.3935 | 6.3795 | 4.8058 |
| | | Medium | **3.9144** | 4.8263 | 5.7214 | 5.1894 | 6.5424 | 6.5337 | 5.1818 |
| | | Large | **1.9999** | 4.9277 | 4.9664 | 4.9572 | 6.2118 | 6.2125 | 4.9568 |
| | Large | Small | 5.183 | **2.3159** | 5.1986 | 5.0767 | 15.8117 | 15.8117 | 5.1466 |
| | | Medium | 5.0404 | **2.0527** | 5.2787 | 5.5415 | 8.9132 | 8.9131 | 5.1955 |
| | | Large | 4.7851 | **1.7169** | 2.0262 | 1.8659 | 3.788 | 3.7858 | 1.7486 |
| | X Large | Small | 4.8342 | **2.0076** | 2.7838 | 2.6907 | 44.1525 | 44.1525 | 2.7691 |
| | | Medium | 1.3619 | 1.4293 | 7.2686 | 7.2607 | 23.2355 | 23.2355 | 7.1617 |
| | | Large | 1.8457 | **1.0318** | 4.9664 | 1.8851 | 3.7543 | 3.7543 | 1.8539 |
| SIPHT | Medium | Small | 204.9747 | **132.0703** | 362.3385 | 131.4414 | 301.155 | 301.155 | 298.8751 |
| | | Medium | **84.9519** | 98.2268 | 165.3731 | 109.2362 | 180.0472 | 180.0423 | 127.7043 |
| | | Large | 84.3457 | **75.4759** | 84.6456 | 75.8628 | 76.365 | 77.6026 | 76.7125 |
| | Large | Small | 114.3366 | 106.0498 | 178.111 | **78.6599** | 345.7508 | 345.7508 | 120.9216 |
| | | Medium | 59.5024 | 61.0234 | 138.9487 | **45.6752** | 224.5103 | 224.5103 | 62.1764 |
| | | Large | 48.0613 | **23.4687** | 47.6233 | 43.0831 | 46.4149 | 50.6955 | 37.3639 |
| | X Large | Small | **72.3466** | 107.3811 | 93.8539 | 79.6473 | 1541.926 | 1541.926 | 144.3428 |
| | | Medium | **24.4861** | 61.7895 | 90.9403 | 46.2486 | 1220.3527 | 1220.352 | 82.0233 |
| | | Large | **16.1024** | 16.1154 | 84.7539 | 113.6452 | 184.5964 | 180.3763 | 80.6106 |

**Table 6.** Frequency of Each Terminal

| Terminal | $VS$ | $ET$ | $ECT$ | $RFT$ | $WT$ | $TS$ |
|---|---|---|---|---|---|---|
| Frequency | 364 | 487 | 544 | 486 | 479 | 507 |

$$Heuristic_1 = (((WT - TS) + (ECT * RFT)) - ((ET - VS) - (ET * WT)))$$

In this paper, we schedule tasks according to the value calculated by the heuristic. The smaller the value is, the faster it will be executed. From the $Heuristic_1$, we observe that smaller ECT and a higher VS can lead to a smaller value calculated by this heuristic. As RFT and ET are the third and fourth most frequently occurring terminals, we know they have also played a role in many of the heuristic rules generated by DSGP. This means that a heuristic rule containing RFT will have good performance. Unlike other heuristics, the heuristics generated by DSGP contain more features that affect the makespan. This can be confirmed by observing other heuristics generated by our approach.

$$Heuristic_2 = (((RFT * WT) - (ET - ECT)) + ((VS + VS) + (ET + ET)))$$

We can see that $Heuristic_2$ supports the claim that a smaller ECT and a higher VS can lead to a smaller value calculated by the heuristic, therefore, a quicker execution for a task.

## 6  Conclusion

In this paper, we present a genetic programming hyper-heuristic approach for dynamic workflow scheduling in the cloud. The novelty of our paper is that our approach addresses the scheduling of workflows in a dynamic cloud environment where tasks arrive periodically. The experimental results show that our DSGP approach addresses the workflow scheduling problem for multiple workflow patterns with varied types of resources in a dynamic environment. Our DSGP approach outperforms competing algorithms in a majority of test scenarios and in terms of the overall average makespan for all test scenarios. In the future, our proposed algorithm can be extended to consider multiple factors, such as cost, security levels, and deadline as well as makespan.

## References

1. Abdelkader, D.M., Omara, F.: Dynamic task scheduling algorithm with load balancing for heterogeneous computing system. Egypt. Inform. J. **13**(2), 135–145 (2012)
2. Arabnejad, V., Bubendorfer, K., Ng, B.: Budget and deadline aware e-science workflow scheduling in clouds. IEEE Trans. Parallel Distrib. Syst. **30**(1), 29–44 (2018)
3. Blythe, J., et al.: Task scheduling strategies for workflow-based applications in grids. In: 2005 IEEE International Symposium on Cluster Computing and the Grid, CCGrid 2005, vol. 2, pp. 759–767. IEEE (2005)
4. Braun, T.D., et al.: A comparison of eleven static heuristics for mapping a class of independent tasks onto heterogeneous distributed computing systems. J. Parallel Distrib. Comput. **61**(6), 810–837 (2001)
5. Buyya, R., Murshed, M.: GridSim: a toolkit for the modeling and simulation of distributed resource management and scheduling for grid computing. Concur. Comput.: Pract. Exp. **14**(13–15), 1175–1220 (2002)
6. Calheiros, R.N., Ranjan, R., Beloglazov, A., De Rose, C.A., Buyya, R.: CloudSim: a toolkit for modeling and simulation of cloud computing environments and evaluation of resource provisioning algorithms. Softw.: Pract. Exp. **41**(1), 23–50 (2011)
7. Chen, W., Deelman, E.: WorkflowSim: a toolkit for simulating scientific workflows in distributed environments. In: 2012 IEEE 8th International Conference on E-Science, pp. 1–8. IEEE (2012)
8. Mahmood, A., Khan, S.A., Bahlool, R.A.: Hard real-time task scheduling in cloud computing using an adaptive genetic algorithm. Computers **6**(2), 15 (2017)

9. Masood, A., Mei, Y., Chen, G., Zhang, M.: Many-objective genetic programming for job-shop scheduling. In: 2016 IEEE Congress on Evolutionary Computation (CEC), pp. 209–216. IEEE (2016)
10. Nguyen, S., Zhang, M., Johnston, M., Tan, K.C.: A computational study of representations in genetic programming to evolve dispatching rules for the job shop scheduling problem. IEEE Trans. Evol. Comput. **17**(5), 621–639 (2012)
11. Raghavan, S., Sarwesh, P., Marimuthu, C., Chandrasekaran, K.: Bat algorithm for scheduling workflow applications in cloud. In: 2015 International Conference on Electronic Design, Computer Networks & Automated Verification (EDCAV), pp. 139–144. IEEE (2015)
12. Rahman, M., Hassan, R., Ranjan, R., Buyya, R.: Adaptive workflow scheduling for dynamic grid and cloud computing environment. Concurr. Comput.: Pract. Exp. **25**(13), 1816–1842 (2013)
13. Sahni, J., Vidyarthi, D.P.: A cost-effective deadline-constrained dynamic scheduling algorithm for scientific workflows in a cloud environment. IEEE Trans. Cloud Comput. **6**(1), 2–18 (2015)
14. Sonmez, O., Yigitbasi, N., Abrishami, S., Iosup, A., Epema, D.: Performance analysis of dynamic workflow scheduling in multicluster grids. In: Proceedings of the 19th ACM International Symposium on High Performance Distributed Computing, pp. 49–60 (2010)
15. Tay, J.C., Ho, N.B.: Evolving dispatching rules using genetic programming for solving multi-objective flexible job-shop problems. Comput. Ind. Eng. **54**(3), 453–473 (2008)
16. Taylor, I.J., Deelman, E., Gannon, D.B., Shields, M., et al.: Workflows for e-Science: Scientific Workflows for Grids, vol. 1. Springer, London (2007). https://doi.org/10.1007/978-1-84628-757-2
17. Xie, J., Mei, Y., Ernst, A.T., Li, X., Song, A.: A genetic programming-based hyper-heuristic approach for storage location assignment problem. In: 2014 IEEE congress on evolutionary computation (CEC), pp. 3000–3007. IEEE (2014)
18. Yu, Y., Feng, Y., Ma, H., Chen, A., Wang, C.: Achieving flexible scheduling of heterogeneous workflows in cloud through a genetic programming based approach. In: 2019 IEEE Congress on Evolutionary Computation (CEC), pp. 3102–3109. IEEE (2019)
19. Zhan, Z.H., Liu, X.F., Gong, Y.J., Zhang, J., Chung, H.S.H., Li, Y.: Cloud computing resource scheduling and a survey of its evolutionary approaches. ACM Comput. Surv. (CSUR) **47**(4), 63 (2015)

# Data and Information Processing

# View Selection for Graph Pattern Matching

Xin Wang[1]([✉]), Xiufeng Liu[2], Yuxiang Chen[1], Xueyan Zhong[1],
and Ping Cheng[3]

[1] Southwest Petroleum University, Chengdu, China
xinwang.ed@gmail.com, cyx@swpu.edu.cn, zhongxueyan@sohu.com
[2] Technical University of Denmark, Kongens Lyngby, Denmark
xiuli@dtu.dk
[3] Sichuan ChangHong Electric Co. Ltd., Mianyang, China
ping.cheng@changhong.com

**Abstract.** View-based techniques have been investigated on relational data, XML and graphs and proven effective for querying big data. While the pivot of using materialized views for query answering is view selection. Though explored for several years, the view selection problem for graph pattern matching has not been investigated. To this end, we investigate the problem in this paper. We first formalize the problem and show its intractability feature. We next propose an appropriate cost model and develop an effective algorithm to identify a set of view definitions under resource constraints, *e.g.*, space storage and query processing cost, from a query workload. We finally verify the performance of the algorithms, and show that our view selection algorithm can identify a set of views that not only have low cost, *e.g.*, storage space (below specified budget) and query processing cost but also can answer queries•efficiently.

## 1 Introduction

Graph pattern matching (GPM) has been widely used in *e.g.*, social analysis. The GPM problem can be formalized as following: given a pattern query Q, and a data graph $G$, it is to find all the matches of Q in $G$. Here, the matching semantic is typically defined in terms of subgraph isomorphism, which makes the problem intractable. In practice, social graphs are often very large, when matching is performed on big social data, users are often unable to afford such high computational costs. To make query answering feasible in big graphs, one may apply view-based techniques [1,2], which increase by orders of magnitude the speed of queries by using pre-computed caches. However, the problem of materialized view selection, which plays an important role in profiting from the above notable achievements of answering pattern matching using views, is not well discussed. In this work, we study the problem of selecting views to materialize.

© Springer Nature Switzerland AG 2020
S. Hartmann et al. (Eds.): DEXA 2020, LNCS 12392, pp. 93–110, 2020.
https://doi.org/10.1007/978-3-030-59051-2_7

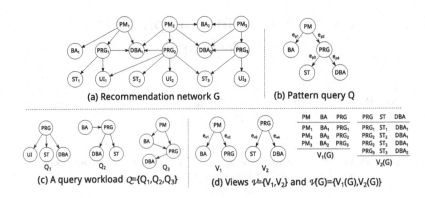

**Fig. 1.** Graph $G$, pattern Q, query workload $\mathcal{Q}$ and views $\mathcal{V}$, $\mathcal{V}(G)$

*Example 1.* A fraction of a recommendation network is depicted as a graph $G$ in Fig. 1(a), where each node denotes a person with job title (*e.g.,* project manager (PM), business analyst (BA), database administrator (DBA), programmer (PRG), user interface designer (UI) and software tester (ST)); and each edge indicates collaboration, *e.g.,* $(PM_1, PRG_1)$ indicates that $PRG_1$ worked well with $PM_1$ on a project led by $PM_1$. Since there is only a single edge type, we hence do not explicitly mark edge types in the figure.

To build a team for software development, one issues a pattern query Q depicted in Fig. 1(b), to impose search conditions. The team members need to satisfy the following requirements: (1) with expertises: PM, BA, PRG, DBA and ST; and (2) meeting the following collaborative experience: (a) BA and PRG worked well under the project manager PM; and (b) PRG supervised ST and DBA. Then it would be a daunting task to perform graph pattern matching via subgraph isomorphism [3] on social graphs, since it takes $O(|G|!|G|)$ time to identify all the isomorphic matches of Q in $G$ [3]. Here $|G| = |V| + |E|$ indicates the size of $G$ and is often particularly large, *e.g.,* with billions of nodes and edges.

While one can do better by leveraging a set of cached views. Suppose a query workload $\mathcal{Q} = \{Q_1, Q_2, Q_3\}$ (Fig. 1 (c)), is given. Then one can identify a set of views $\mathcal{V} = \{V_1, V_2\}$, that can answer most pattern queries in $\mathcal{Q}$, and obtain materialize views $\mathcal{V}(G) = \{V_1(G), V_2(G)\}$ (often with cost constraints), as shown in Fig. 1 (d). Then pattern query Q can be efficiently computed by "merging" materialized views in $\mathcal{V}(G)$ with techniques, *e.g.,* [2]. In doing so, one key issue has to be settled: how to identify a set of view definitions for caching?   □

**Contributions.** This paper investigates the aforementioned problem: view selection for graph pattern matching. We focus on matching with *subgraph isomorphism* [3].

(1) To make practical application of pattern matching using views, it is critical to identify an appropriate set of views for materialization. In light of this, we study the *view selection problem*, denoted as VSP. In Sect. 3, we

first formalize the problem and show intractability of the problem via static analysis.

(2) We introduce a cost model by combining constraints of storage space and query processing cost (Sect. 3.1). We next provide an algorithm which performs exhaustive search to find the optimal solution (Sect. 3.2). To strike a balance between result quality and efficiency, we still develop an efficient algorithm, that is able to identify high quality results and in the meanwhile has a function of $|\mathcal{Q}|$ and $||\mathcal{Q}||$, *i.e.*, the total size of pattern queries and cardinality of a query workload $\mathcal{Q}$, as its computational complexity (Sect. 3.3). Since $|\mathcal{Q}|$ and $||\mathcal{Q}||$ are often small and view selection can be processed offline, our algorithm hence fills the critical void for VSP (Sect. 3.3).

(3) Using real-life and synthetic data, we experimentally verify the performances of our algorithms and view-based pattern matching (Sect. 4). We find the following. (1) Our view selection algorithm performs well: it spends less than 65 seconds to find a set of views $\mathcal{V}$ under resource constraint from a query workload $\mathcal{Q}$ with size $|\mathcal{Q}| = 120$, such that each pattern query in $\mathcal{Q}$ is contained by $\mathcal{V}$. (2) View-based matching method is almost 11 times faster than conventional matching method on Youtube, a recommendation network with 1.4 million nodes and 3 million edges, and scales well with the data size.

In a summary of the scientific contributions, this work explored view selection problem for graph pattern matching and gives a full treatment for the problem. Taking these together, the results fill one critical void for view-based pattern matching and yield a promising approach to querying big graphs.

**Related Work.** We next categorize related work as follows.

*Query Answering Using Views.* Answering queries using views has been well studied for relational data (see [4–6] for surveys), XML data [7–9], and graph data [1,2,10]. This paper adopts subgraph isomorphism, as the semantic of pattern matching, instead of graph simulation [11] and bounded simulation [12], that are applied by [1] and [10], and serves as one of fundamental techniques for view-based matching techniques.

Query containment checking plays a vital role in view-based query answering. The containment problem for conjunctive queries is NP-complete, and is undecidable for relational algebra [13]. In [14], the containment checking of simple XPath queries is shown coNP-complete. When disjunction, DTDs and variables are taken into account, the problem ranges from coNP-complete to EXPTIME-complete to undecidable for various XPath classes [8]. When graph queries is considered, the containment checking becomes polynomial time solvable [10] and intractable [2] for matching via simulation and subgraph isomorphism, respectively.

*View Selection.* The view selection problem has been investigated for relational, XML and RDF data. (i) For relational data, the problem has been well studied (see [15] for a survey). Techniques for centralized scenario have been proposed by [16–18], for example, [16] introduced techniques to find exact and inexact answers; [17] introduced a lattice framework to select views; [19] provided competitive heuristics to optimize query processing time, especial for some important special cases of the general data warehouse scenario; and an evolutionary game theory-based method to materialized view selection is studied in [18]. In contrast to prior work which considers view selection in a centralized context, [20,21] explored the problem in a distributive scenario, and proposes solutions under resource constraints. (ii) A host of techniques are developed for XML data. [22,23] studied the problem with storage space constraints. [24] provides techniques to identify views of XPath queries. (iii) On RDF data, [25] introduced techniques to select a set of indexes as materialized views based on workload of SPARQL queries. [26] developed methods for RDF view selection, towards minimizing a combination of query processing, view storage, and view maintenance costs. In a dynamic circumstance, the view selection is based on a dynamic query workload, that is built incrementally and changes over time. As an representative, [27] proposed to materialize the most frequently accessed tuples of the view rather than materializing all tuples of the view. Unlike [27], [28] explores partial view selection over an evolving graph. To sum up, to our best knowledge, there is yet any prior work on view selection for graph pattern matching.

## 2   Preliminaries

In this section, we first review data graphs, pattern queries and graph pattern matching. We next revisit pattern containment.

### 2.1   Graphs, Patterns and Graph Pattern Matching

We start with basic notations of graph pattern matching.

**Data Graphs.** A *data graph* is a node-labeled, directed graph $G = (V, E, L_v, L_e)$, where (1) $V$ is a finite set of data nodes; (2) $E \subseteq V \times V$, where $(v, v') \in E$ denotes a *directed* edge from node $v$ to $v'$; and (3) $L_v(\cdot)$ and $L_e(\cdot)$ are functions such that for each node $v$ in $V$, $L_v(v)$ is a label from an alphabet $\Sigma$ and for each edge $e$ in $E$, $L_e(e)$ is a label from the same alphabet. Intuitively, $L_v(\cdot)$ specifies *e.g.*, job titles, social roles, ratings, etc on nodes and $L_e(\cdot)$ designates relationship between a pair of nodes [29].

**Pattern Queries.** A *pattern query* (or shortened as pattern) is a directed graph $Q = (V_p, E_p, f_v, f_e)$, where (1) $V_p$ is the set of *pattern nodes*, (2) $E_p$ is the set of *pattern edges*, and (3) $f_v(\cdot)$ and $f_e(\cdot)$ are functions defined on $V_p$ and $E_p$, respectively, such that for each $u \in V_p$, $f_v(u)$ is a label in $\Sigma$ and for each $e_p$ in $E_p$, $f_e(e_p)$ is also a label from $\Sigma$.

**Graph Pattern Matching** [3]. A *match* of $Q$ in $G$ via *subgraph isomorphism* is a subgraph $G_s$ of $G$ that is isomorphic to $Q$, *i.e.*, there exists a *bijective function* $h$ from $V_p$ to the node set $V_s$ of $G_s$ such that (1) for each node $u \in V_p$, $f_v(u) = L_v(v)$ $(v \in V_s)$; and (2) $(u, u')$ is an edge in $Q$ if and only if $(h(u), h(u'))$ is an edge in $G_s$ and in the meanwhile $f_e(u, u') = L_e(h(u), h(u'))$. Then the match result of Q in $G$, denoted as $Q(G)$, is a set consisting of all the matches of Q in $G$.

Indeed, each match $G_s$ in $Q(G)$ corresponds to a distinct mapping $h$. Then for a pattern edge $e_p = (u, u')$, we derive a set $S(e_p)$ from $Q(G)$ by letting $S(e_p) = \{(v, v')|v = h(u), v' = h(u'), (v, v') \in G_s, G_s \in Q(G)\}$, and denote $S(e_p)$ as the match set of $e_p$.

*Example 2.* Consider pattern query Q shown in Fig. 1 (b), where each node carries a search condition (job title), and each edge indicates a collaboration relationship. When Q is posed on the network $G$ of Fig. 1 (a), the match set $Q(G)$ includes following matches: $\{(\mathsf{PM}_1, \mathsf{BA}_1, \mathsf{PRG}_1, \mathsf{ST}_1, \mathsf{DBA}_1), (\mathsf{PM}_2, \mathsf{BA}_2, \mathsf{PRG}_2, \mathsf{ST}_2, \mathsf{DBA}_1), (\mathsf{PM}_2, \mathsf{BA}_2, \mathsf{PRG}_2, \mathsf{ST}_3, \mathsf{DBA}_1), (\mathsf{PM}_3, \mathsf{BA}_2, \mathsf{PRG}_3, \mathsf{ST}_3, \mathsf{DBA}_2)\}$. □

## 2.2 View-Based Pattern Matching

We now review views, view extensions and pattern containment [2]. To make the paper self-contained, we cite them as follows (rephrased).

**Views and View Extensions.** A *view* (*a.k.a.* view definition) V is also a pattern query. Its match result $V(G)$ in a data graph $G$ is denoted as *view extension*, or *extension* when it is clear from the context [30]. As shown in Fig. 1(c), a set of views $\mathcal{V} = \{V_1, V_2\}$ are defined, with extensions $\mathcal{V}(G) = \{V_1(G), V_2(G)\}$ on $G$ cached.

**Pattern Containment.** A pattern $Q = (V_p, E_p, f_v)$ is *contained* in a set of views $\mathcal{V} = \{V_1, \cdots, V_n\}$, denoted by $Q \sqsubseteq \mathcal{V}$, if there exists a mapping $\lambda$ from $E_p$ to powerset $\mathcal{P}(\bigcup_{i\in[1,n]} E_{V_i})$, such that for any graph $G$, $S(e_p) \subseteq \bigcup_{e'_p \in \lambda(e_p)} S(e'_p)$ for any edge $e_p \in E_p$.

*Example 3.* Recall Example 1, where a set of views $\mathcal{V} = \{V_1, V_2\}$ are defined. One may verify that for pattern edge $(\mathsf{PM}, \mathsf{BA})$ in $V_1$, $S = \{(\mathsf{PM}_1, \mathsf{BA}_1), (\mathsf{PM}_2, \mathsf{BA}_2), (\mathsf{PM}_3, \mathsf{BA}_2)\}$, similarly for other pattern edges. Furthermore, it can be easily verified that $S(e_p) \subseteq \bigcup_{e'_p \in \lambda(e_p)} S(e'_p)$ for any edge $e_p$ in Q (Fig. 1(b)), where pattern edges $e_{p_i}$ of Q are mapped via $\lambda$ to edges $e_{v_i}$ in either $V_1$ or $V_2$, respectively ($i \in [1, 4]$). Thus Q is contained in $\mathcal{V} = \{V_1, V_2\}$.

Intuitively, mapping $\lambda$ maps $E_p$ of Q to the edge set of each $V_i$ in $\mathcal{V}$, and bridges the gap between maches of Q in $G$ and view extensions $\mathcal{V}(G)$, for any $G$. Indeed, *pattern containment* determines whether a pattern query Q can be answered by using a set of views $\mathcal{V}$. Proposition 1 makes *pattern containment* checking feasible.

**Proposition 1:** *For a set of view definitions $\mathcal{V}$ and a pattern query* $Q = (V_p, E_p, f_v)$, $Q \sqsubseteq \mathcal{V}$ *if and only if* $E_p = \bigcup_{V_i \in \mathcal{V}} H^Q_{V_i}$. □

Here, $H^Q_{V_i}$ refers to the *shadow* from $V_i$ to Q, and is defined as the union of edge sets of matches of $V_i$ in Q (Q and $V_i$ are viewed as data graph and pattern

**Table 1.** A summary of notations

| Symbols | Notations |
|---|---|
| $G = (V, E, L_v, L_e)$ | A data graph |
| $Q = (V_p, E_p, f_v, f_e)$ | A pattern query or pattern for short |
| $Q(G)$ | Match result of $Q$ in $G$ |
| $S(e_p)$ | Match set of a pattern edge $e_p$ in $G$ |
| $\mathcal{V} = \{V_1, \ldots, V_n\}$ | A set of view definitions $V_i = (V_{V_i}, E_{V_i}, f_{V_i})$ |
| $\mathcal{V}(G) = \{V_1(G), \ldots, V_n(G)\}$ | A set of view extensions $V_i(G)$ |
| $Q \sqsubseteq \mathcal{V}$ | $Q$ is contained in $\mathcal{V}$. |
| $H_V^Q$ | The shadow of a view $V$ in $Q$ |
| $\mathcal{Q} = \{Q_1, \ldots, Q_n\}$ | A query workload consisting of a set of patterns $Q_i = (V_{p_i}, E_{p_i}, f_{v_i})$ |
| $\mathcal{Q} \sqsubseteq \mathcal{V}$ | $Q_i \sqsubseteq \mathcal{V}$ for each $Q_i \in \mathcal{Q}$ |
| $\gamma(V)$ | The estimated size of $V(G)$ |
| $\gamma(\mathcal{V})$ | The sum of $\gamma(V)$ of all view extensions in $\mathcal{V}(G)$ |
| $|Q|$ (resp. $|V|$) | Size (total number of nodes and edges) of a pattern $Q$ (resp. view definition $V$) |
| $|Q(G)|$ (resp. $|V(G)|$) | Total size of matches of $Q$ (resp. $V$) in $G$ |
| $|\mathcal{V}|$ (resp. $|\mathcal{Q}|$) | Total size of view definitions in $\mathcal{V}$ (resp. pattern queries in $\mathcal{Q}$) |
| $\|\mathcal{V}\|$ (resp. $\|\mathcal{Q}\|$) | The number of view definitions in $\mathcal{V}$ (resp. pattern queries in $\mathcal{Q}$) |
| $|\mathcal{V}(G)|$ | Total size of matches in $\mathcal{V}(G)$ |

query, respectively). Intuitively, when treating $V_i$ as data graph, $Q$ as pattern query, a shadow $H_{V_i}^Q$ from $V_i$ to $Q$ then represents a part of $Q$ such that each edge in $H_{V_i}^Q$ must be appeared in a match of $V_i$ in $Q$.

*Example 4.* Consider a set of view definitions $\mathcal{V} = \{V_1, V_2\}$ shown in Fig. 1 (d). One may verify that the shadow from $V_1$ (resp. $V_2$) to $Q$ is $H_{V_1}^Q = \{(\text{PM}, \text{PRG}), (\text{PM}, \text{BA})\}$ (resp. $H_{V_2}^Q = \{(\text{PRG}, \text{ST}), (\text{PRG}, \text{DBA})\}$), and $Q \sqsubseteq \mathcal{V}$ since $E_p = H_{V_1}^Q \cup H_{V_2}^Q$. □

We also use the following notations. We denote (1) $|V_p| + |E_p|$ as the size $|Q|$ of $Q$ and $|V| + |E|$ as the size $|G|$ of $G$, (2) $Q(G)$ (resp. $V(G)$) as match result of $Q$ (resp. $V$) in $G$, and (3) $\mathcal{V}(G) = \{V_1(G), \ldots, V_n(G)\}$ as a set of view extensions $V_i(G)$. Abusing notation of containment [2], we use $\mathcal{Q} \sqsubseteq \mathcal{V}$ to denote that $\mathcal{Q}$ is contained in $\mathcal{V}$, *i.e.*, each $Q_i$ in $\mathcal{Q}$ is contained in $\mathcal{V}$.

The notations of the paper are summarized in Table 1.

# 3   View Selection

In this section, we study the view selection problem to profit from *answering graph pattern matching using views*. Informally, the problem can be stated as following: let $\mathcal{D}$ be a database, and $\mathcal{Q}$ be a query workload including a set of queries, it is to find a set of views $\mathcal{V}$ to materialize, such that $\mathcal{Q}$ can be answered by using $\mathcal{V}$ and its extension $\mathcal{V}(\mathcal{D})$, and in the meanwhile, the costs, *e.g.*, "space cost", "query cost" and "maintenance cost" can be minimized [15].

Along the same line, we formalize the *view selection problem* (abbreviated as VSP), for graph pattern matching as below.

**View Selection Problem.** Given a query workload $\mathcal{Q}$ consisting of a set of pattern queries $\{Q_1, \cdots, Q_n\}$ over a graph $G$, where each $Q_i$ in $\mathcal{Q}$ is associated with a non-negative weight $\psi(Q_i)$, representing *e.g.*, query frequency. VSP is to find a set of views $\mathcal{V}$ for materialization, such that $Q \sqsubseteq \mathcal{V}$ for each $Q$ in $\mathcal{Q}$, the storage space is under a budget $\mathcal{B}$ and the total cost is minimized.

The problem, however, is nontrivial: it is NP-hard, as indicated by Theorem 2 (1).

**Theorem 2.** *(1) The view selection problem is NP-hard (decision problem). (2) Given a query workload* $\mathcal{Q} = \{Q_1, \cdots, Q_n\}$, *and a bound* $\mathcal{B}$, *it is in* $O(|\mathcal{Q}|(|\mathcal{Q}|^3|\mathcal{Q}|! + \|\mathcal{Q}\|))$ *time to find a set of views* $\mathcal{V}$ *to materialize such that* $\mathcal{Q} \sqsubseteq \mathcal{V}$ *and the estimated storage space of* $\mathcal{V}(G)$ *is no more than* $\mathcal{B}$.

*Proof.* We next prove Theorem 2(1). We defer the proof of Theorem 2(2) in Sect. 3.3 by providing an algorithm as a constructive proof.

We prove NP-hardness of VSP by showing intractability of its special case, which enforces that each pattern $Q_i$ in $\mathcal{Q}$ takes a weight $\psi(Q_i) = 1$ as its frequency and another weight $c(Q_i)$ as its cost. Note that there exist different metrics to measure the cost, a formal cost model for VSP will be given shortly. To simplify discussion, we let $c(Q_i)$ represent query processing cost of $Q_i$ using $\mathcal{V}$, and enforce that $c(Q_i)$ equals to the number of views that can be used to answer $Q_i$; moreover we estimate $|\mathcal{V}(G)|$ as total number of edges of $\mathcal{V}$. Then the decision problem of VSP is to decide, given a workload $\mathcal{Q} = \{Q_1, \cdots, Q_n\}$, a bound $\mathcal{B}$ as budget of storage space and a bound $\mathcal{C}$ to limit total cost, whether there is a set of views $\mathcal{V}$ defined on $G$ such that $\mathcal{Q} \sqsubseteq \mathcal{V}$, $\Sigma_{Q \in \mathcal{Q}} c(Q) \leq \mathcal{C}$ and $|\mathcal{V}(G)| \leq \mathcal{B}$. We prove the NP-hardness by reduction from the NP-complete *set basis problem* (SBP) [31].

Given a set $U$, a collection of subsets of $U$, $\mathcal{S} = \{S_1, \cdots, S_n\}$, and an integer $L \leq n$, SBP is to decide whether exists a collection $\mathcal{T}$ of subsets of $U$ with $|\mathcal{T}| = L$, such that for each $S_i \in \mathcal{S}$, there exists a subcollection of $\mathcal{T}$ whose union is exactly $S_i$. We show that there exists a collection $\mathcal{T}$ as a set basis of $U$ w.r.t. $\mathcal{S}$, if and only if there exists a set of views $\mathcal{V}$ with $\mathcal{Q} \sqsubseteq \mathcal{V}$, $\Sigma_{Q \in \mathcal{Q}} c(Q) \leq \mathcal{C}$ and $|\mathcal{V}(G)| \leq \mathcal{B}$.

Given such an instance of SBP, we construct an instance of VSP as follows: (a) for each element $x_j \in U$, we create a unique edge $e_{x_j}$; (b) for each $S_i$ in $\mathcal{S}$, we construct a pattern query $Q_i$, which consists of edges $e_{x_j}$ that correspond to elements $x_j$ of $S_i$; (c) we define a query workload $\mathcal{Q} = \{Q_1, \cdots, Q_n\}$, where each $Q_i$ takes a weight $\psi(Q_i) = 1$ and a weight $c(Q_i) = |T'|$, where $T' = \{T_j | S_i \cap T_j \neq \emptyset, S_i \in \mathcal{S}, T_j \in \mathcal{T}\}$; and (d) we set $\mathcal{B} = |U| * L$ and $\mathcal{C} = n * L$.

The construction is obviously in PTIME. We next verify correctness of the reduction.

(1) Assume that there exists such a collection $\mathcal{T}$. Let $\mathcal{V}$ be the set of view definitions, where each $V_i$ in $\mathcal{V}$ takes edges $e_{x_j}$ that corresponds to elements $x_j$ of $T_i \in \mathcal{T}$ and is associated with weight $|E_{p_i}|$, as $\gamma(V_i)$. One can then verify that $Q_i \sqsubseteq \mathcal{V}$ for each $Q_i$ in $\mathcal{Q}$, and in the meanwhile, $|\mathcal{V}(G)| \leq \mathcal{B} = |U| * L$, $\Sigma_{Q \in \mathcal{Q}} c(Q) \leq \mathcal{C} = n * L$.

(2) Conversely, if there is a set of views $\mathcal{V}$ with $\mathcal{Q} \sqsubseteq \mathcal{V}$, $|\mathcal{V}(G)| \leq \mathcal{B}$ and $\Sigma_{Q \in \mathcal{Q}} c(Q) \leq \mathcal{C}$, then it is easy to see that the corresponding set $\mathcal{T}$ is a set basis of $U$ w.r.t. $\mathcal{S}$.

As SBP is NP-complete, so is NP-hardness of VSP.                    □

**Remark.** Note that the storage space budget $\mathcal{B}$ should not set too small, since otherwise even materializing a single view $V$ on a graph $G$ may trivially exceed the budget.

## 3.1   Cost Model

In literatures, *e.g.*, [15], the cost often refers to query processing cost, view maintenance cost or their combination. To simplify discussion, we consider query processing cost only, and define the cost for evaluating a pattern $Q_i$ with views as follows:

$$Cost(Q_i, \mathcal{V}) = \psi(Q_i) * Qc(Q_i, \mathcal{V}),$$

where $Qc(Q_i, \mathcal{V})$ indicates the query processing cost of pattern $Q_i$ with views $\mathcal{V}$, $\mathcal{V}(G)$.

In practice, before we actually materialize the chosen views, it is hard to know their storage space, *i.e.*, the exact size of view extensions on a graph $G$, not to mention query processing cost. Thus, one may only be able to estimate these costs.

To compute the estimated "storage space" (denoted as $\gamma(V)$), one can adopt the approach of sampling [22], *i.e.*, running a view $V$ on a representative portion of the graph $G$ and multiplied by a scaling factor $\eta$, to estimate $|V(G)|$.

Recall that the view-based query processing cost of a pattern $Q_i$ is dominated by the pattern size $|Q_i|$ and the total size of view extensions $|\mathcal{V}(G)|$ [2]. More precisely, to answer a pattern $Q_i$, only those views $V$ whose shadow $H_V^{Q_i}$ is not empty can contribute to the matches of $Q_i$. As a result, we define a function to estimate the query processing cost of a pattern $Q_i$ as follows.

$$Qc(Q_i, \mathcal{V}) \approx \tau * \sum_{V \in \mathcal{V}'} \gamma(V),$$

where $\mathcal{V}' = \{V|H_V^{Q_i} \neq \emptyset, V \in \mathcal{V}\}$ and $\tau$ is a parameter, that is used to adjust $e.g.$, joining cost among various view extensions and can set by users via learning and sampling.

Though simplified, our problem can be generalized by using more complex cost models. While study on the problem enables us to tackle the issue more conveniently.

Abusing notations, we refer to $\gamma(\mathcal{V})$ as the sum of $\gamma(V)$, $i.e.$, total size of all view extensions in $\mathcal{V}(G)$, and denote $Cost(\mathcal{Q}, \mathcal{V})$ as the total query processing cost for patterns in $\mathcal{Q}$ with views $\mathcal{V}$ and $\mathcal{V}(G)$, $i.e.$, $Cost(\mathcal{Q}, \mathcal{V}) = \Sigma_{Q_i \in \mathcal{Q}} Cost(Q_i, \mathcal{V})$.

## 3.2  A Baseline Algorithm

Typically, approaches to VSP consist of two main tasks. The first task identifies candidate views which are promising for materialization based on the query workload provided. Using candidate views, the second task selects the set of views to materialize under the resource constraints. Along the same line, we next introduce an algorithm, denoted as Naive (not shown), to select views for graph pattern matching.

*Candidates Generation.* For a pattern query $Q_i$, a view (along with its extension) is referred to as a *candidate view* if it can provide partial or complete answers to answer $Q_i$. Thus, a view $V_j$, whose shadow $H_{V_j}^{Q_i}$ is not an empty set, can contribute partial or complete answers and be chosen as a candidate view. Then, the complete set of candidate views of a pattern $Q_i$ is a set $C_i = \{V_j|H_{V_j}^{Q_i} \neq \emptyset\}$. Similarly, for a given query workload $\mathcal{Q} = \{Q_1, \cdots, Q_n\}$, the complete set of candidate views is a set $C = \bigcup_{i \in [1,n]} C_i$.

*View Selection.* Given a set of candidate views $C$ generated as above, a bound $B$ indicating space constraint and a query workload $\mathcal{Q}$, one may apply exhaustive search to find an optimal solution. In a nutshell, Naive enumerates every possible subset of $C$ as view set $\mathcal{V}$ and identifies a $\mathcal{V}_{OPT}$ such that $\gamma(\mathcal{V}_{OPT}) \leq B$ and the cost $Cost(\mathcal{Q}, \mathcal{V}_{OPT})$ is minimized.

Theoretically, it takes $\binom{||C||}{1} + \binom{||C||}{2} + \cdots + \binom{||C||}{||C||}$ rounds comparisons to obtain the final result, where $||C||$ already equals to $\binom{n}{1} + \binom{n}{2} + \binom{n}{3} + \cdots + \binom{n}{n} = 2^n - 1$, which is in exponential of $|\mathcal{Q}|$ (note that $n$ is the total number of distinct edges in $\mathcal{Q}$ and is bounded by $|\mathcal{Q}|$). Thus, Naive is in $O(2^{2^n})$ time, which is often not affordable.

## 3.3  An Efficient Heuristic

Since Naive is prohibitively expensive and not applicable in practice, we hence need more efficient solutions. Below we present a heuristic algorithm (VScHeu). As will be verified, VScHeu is not only far more efficient than Naive, but also can identify a set of views $\mathcal{V}$ with high quality, $i.e.$, the total cost $Cost(\mathcal{Q}, C)$ is close to the optimal solution, and in the meanwhile, $\gamma(\mathcal{V})$ is no more than the storage space budget.

The core idea of the algorithm is that it iteratively identifies the "best" view until a termination condition is satisfied. This calls for a "goodness" metric for view selection.

*Goodness Metric.* The "goodness" of a view $\mathcal{V}$ lies in two parts: influence of query processing cost and increment of storage space when introducing $\mathcal{V}$. As influence of query processing cost caused by $\mathcal{V}$ is a little tricky, to simplify discussion, we define a function $g(\mathsf{V})$ for measuring "goodness" of V as follows. While, we remark that one can incorporate cost of each pattern to develop a more sophisticated measurement.

$$g(\mathsf{V}) = \frac{\delta(\mathsf{V})}{\gamma(\mathsf{V})},$$

$$\delta(\mathsf{V}) = \sum_{Q_i \in \mathcal{Q}} |H_{\mathsf{V}}^{Q_i} \cap I_u(Q_i)| * \psi(Q_i),$$

where $I_u(Q_i)$ is a set that contains all the *uncovered* edges of the pattern $Q_i$ in $\mathcal{Q}$ and $E_{p_i}$ is the edge set of pattern $Q_i$. Here, *uncovered* edges of $Q_i$ indicates edges that do not appear in any shadow $H_{\mathsf{V}}^{Q_i}$ for each V in $\mathcal{V}$.

Intuitively, $g(\mathsf{V})$ is a function of $\delta(\mathsf{V})$ and $\gamma(\mathsf{V})$. In particular, $\delta(\mathsf{V})$ measures the effectiveness of V. The rational behind that is if a view V can cover more uncovered edges of $Q_i$, then V is a good choice. The higher $g(\mathsf{V})$ is, the better V is.

**Algorithm.** The algorithm, denoted as VScHeu, is shown in Fig. 2. Given a query workload $\mathcal{Q} = \{Q_1, \cdots, Q_n\}$ and a bound $\mathcal{B}$, it applies a greedy strategy to find a set of view definitions $\mathcal{V}$ to answer $\mathcal{Q}$ such that $\gamma(\mathcal{V}) \leq \mathcal{B}$ and $Cost(\mathcal{Q}, \mathcal{V})$ is small as well.

Before illustrating the algorithm, we first introduce data structures used by VScHeu. (1) An index $\mathcal{I}_u$, which maintains *uncovered* edges of each pattern $Q_i$ in $\mathcal{Q}$, is used to infer *coverage rate* of each candidate view definition. (2) We use another index $\mathcal{I}_e$ to maintain "goodness" value, *i.e.*, $g(\mathsf{V})$ for each view V. For fast fetch, the indexes can be implemented with map structures, then given a key, its value can be fetched efficiently.

Algorithm VScHeu first initializes maps $\mathcal{I}_u$, $\mathcal{I}_e$, a set $\mathcal{V}$ and an integer uncvd as counter (line 1). More specifically, for each edge $e$ that is appeared in certain $Q_i$ in $\mathcal{Q}$, VScHeu constructs a view definition $\mathsf{V}_e$ with $e$ only, computes $g(\mathsf{V}_e)$, and sets $\mathcal{I}_e(\mathsf{V}_e)$ as $g(\mathsf{V}_e)$, in the meanwhile, it also initializes $\mathcal{V}$ with a view $\mathsf{V}_e$ with the largest $g(\mathsf{V}_e)$; for each $Q_i$ in $\mathcal{Q}$, VScHeu sets $\mathcal{I}_u(Q_i)$ as $E_{p_i}$; VScHeu also sets uncvd as $\Sigma_{Q_i \in \mathcal{Q}} |E_{p_i}|$. After initialization, VScHeu iteratively updates $\mathcal{V}$ by either including a single-edge view $\mathsf{V}_e$ with highest "goodness" value $g(\mathsf{V}_e)$, or replacing an existing view $\mathsf{V}_j$ with an extended one, until termination condition is encountered (lines 2–12). During above process, it is crucial to generate a view with highest "goodness", VScHeu conducts following for the issue. For each view $\mathsf{V}_j$ in $\mathcal{V}$, it generates a new view $\mathsf{V}_j'$ by extending $\mathsf{V}_j$ with a single-edge view $\mathsf{V}_k$ in $\mathcal{I}_e$ (line 5), computes increased "goodness" $\Delta g(\mathsf{V}_j')$ if $\mathsf{V}_j'$ is introduced (line 6), and updates max and $\mathsf{V}_m$ if $\mathsf{V}_j'$ is the best candidate so far (line 7). After **for** loop, the view with largest "goodness" is determined, VScHeu then compares

---

*Input:* Workload $\mathcal{Q} = \{Q_1, \cdots, Q_n\}$ with weight on each $Q_i$,
    and a bound $\mathcal{B}$.
*Output:* A set of view definitions $\mathcal{V}$.

1.  initialize $\mathcal{I}_u, \mathcal{I}_e, \mathcal{V}$ and uncvd; set $V_m := \emptyset$; max:=0;
2.  **while** uncvd $> 0$ **do**
3.      pick a view $V_e$ from $\mathcal{I}_e$ with largest $g(V_e)$; max:=0;
4.      **for each** $V_j$ in $\mathcal{V}$ **do**
5.          generate $V_j'$ by enlarging $V_j$ with $V_k$ in $\mathcal{I}_e$;
6.          compute $g(V_j'), \Delta g(V_j') = g(V_j') - g(V_j)$;
7.          **if** max $< \Delta g(V_j)$ **then** max $:= g(V_j); V_m := V_j$ ;
8.      **if** $g(V_e) >$ max **then**
9.          $\mathcal{V} := \mathcal{V} \cup \{V_e\}$; update $\mathcal{I}_u$ and uncvd;
10.     **else**
11.         update $\mathcal{V}$ with $V_m$; update $\mathcal{I}_u, \mathcal{I}_e$ and uncvd;
12.     **if** $\gamma(\mathcal{V}) > \mathcal{B}$ **then** restore $\mathcal{V}$; **break** ;
13. **return** $\mathcal{V}$;

---

**Fig. 2.** Algorithm VScHeu

"goodness" of $V_e$ and $V_m$, and updates $\mathcal{V}, \mathcal{I}_u$ and uncvd (lines 8–11). Note that if $\mathcal{V}$ is updated with an enlarged view $V_m$, then *uncovered* edges of each $Q_i \in \mathcal{Q}$ that are affected by $V_m$ have to be recomputed, which leads to the update of $\mathcal{I}_e$. The construction of $\mathcal{V}$ finishes when either $\gamma(\mathcal{V})$ exceeds the bound $\mathcal{B}$ (line 12), or counter uncvd reaches 0, which indicates $\mathcal{Q} \sqsubseteq \mathcal{V}$ (line 2). Finally, VScHeu returns set $\mathcal{V}$ as result (line 13).

*Example 5.* Suppose we are given a query workload $\mathcal{Q} = \{Q_1, Q_2, Q_3\}$ that are shown in Fig. 3(a), and a large bound $\mathcal{B}$. To simplify illustration, we set $\tau = 1$ and let $\psi(Q_i) = 1$ for each pattern $Q_i$ in $\mathcal{Q}$; we also assume that all $V_e$ take the same $\gamma(\cdot)$ value, for each $e$ appeared in $\mathcal{Q}$. VScHeu works as follows. (I) VScHeu initializes $\mathcal{I}_e(V_{e_i})$ for each $e_i$ ($i \in [1,6]$) appeared in $\mathcal{Q}$, i.e., $g(V_{e_3}) = \frac{3}{\gamma(V_{e_3})}, g(V_{e_2}) = \frac{2}{\gamma(V_{e_2})}, g(V_{e_4}) = \frac{2}{\gamma(V_{e_4})}, g(V_{e_5}) = \frac{1}{\gamma(V_{e_5})}, g(V_{e_6}) = \frac{1}{\gamma(V_{e_6})}, g(V_{e_1}) = \frac{1}{\gamma(V_{e_1})}$. (ranked in descending order of $g()$ values). (II) VScHeu repeatedly updates $\mathcal{V}$ as follows. (i) It includes $V_{e_3}$ in $\mathcal{V}$ since $g(V_{e_3})$ is maximum. (ii) It generates $V_2$ by extending $V_{e_3}$ with $V_{e_2}$, as $\Delta g(V_2) = \frac{4}{\gamma(V_1)} - \frac{2}{\gamma(V_{e_3})}$ is assumed larger than $g(V_{e_2})$. (iii) It first includes $V_{e_4}$ in $\mathcal{V}$ and then generates $V_2$ by merging $V_{e_4}$ with $V_{e_3}$. (iv) It generates $V_4$ and $V_{e_1}$, along the same line as above. (III) VScHeu found uncvd = 0, and returns $\mathcal{V} = \{V_{e_1}, V_2, V_3, V_4\}$ (shown in Fig. 3(b)), as final result. □

*Correctness.* It suffices to show that (1) VScHeu always terminates, and when it terminates, $\gamma(\mathcal{V}) \leq \mathcal{B}$; and (2) VScHeu always selects the best view definition in each round iteration.

(1) The **while** loop of VScHeu terminates when either condition (a) (line 2) or condition (b) (line 12) is encountered. Observe that counter uncvd records

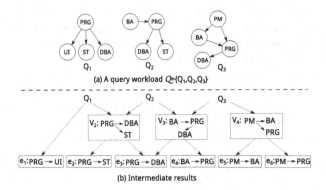

(a) A query workload $Q=\{Q_1,Q_2,Q_3\}$

(b) Intermediate results

**Fig. 3.** Procedure of VScHeu

sum of all the *uncovered* edges, and initially equals to $\Sigma_{Q_i \in Q}|E_{p_i}|$. During iteration, after a view definition $V_e$ (resp. $V_m$) is chosen, VScHeu calculates how many extra edges are *covered* by $V_e$ (resp. $V_m$), and updates uncvd by subtracting this number. When uncvd turns to 0, *i.e.*, each query in $Q$ is *covered* by $V$, condition (a) is then satisfied. On the other hand, condition (b) guarantees that when VScHeu terminates $\gamma(V)$ can not exceed bound $B$.

(2) Indeed, VScHeu applies a greedy strategy to choose view definitions. At each round iteration, VScHeu selects a view with the highest "effectiveness" value, to update $V$. The strategy guides VScHeu to collect a set of views $V$ that not only contain as many pattern queries in $Q$ as possible, but also with small $\gamma(V)$.

*Complexity.* The initialization for $\mathcal{I}_e$ and $\mathcal{I}_u$ is in $O(\Sigma_{Q_i \in Q}|E_{p_i}| + ||Q||)$ time, since there exists $\Sigma_{Q_i \in Q}|E_{p_i}|$ edges and $||Q||$ pattern queries in $Q$. Algorithm VScHeu then iteratively selects view definitions. The **while** loop repeats at most $\Sigma_{Q_i \in Q}|E_{p_i}|$ times, since at each iteration, at least one edge in $Q$ should be *covered*. For a single iteration, it takes VScHeu (a) $\Sigma_{Q_i \in Q}|E_{p_i}|$ time to pick a single-edge view with maximum "effectiveness" (line 3); (b) $O((\Sigma_{Q_i \in Q}|E_{p_i}|)^2 * \Sigma_{Q_i \in Q}|Q_i|!|Q_i|)$ time to identify the best view for extension, as at most $\Sigma_{Q_i \in Q}|E_{p_i}|$ views in $V$ and $\mathcal{I}_e$, respectively, and "effectiveness" computation takes $O(\Sigma_{Q_i \in Q}|Q_i|!|Q_i|)$ time (lines 4–6); and (c) $||Q||$ times to update $\mathcal{I}_u$ (lines 9, 11). One may verify that $\Sigma_{Q_i \in Q}|E_{p_i}|$ is bounded by $O(|Q|)$, and $O(\Sigma_{Q_i \in Q}|Q_i|!|Q_i|)$ is further bounded by $O(|Q|!|Q|)$, hence, algorithm VScHeu is in $O(|Q|(|Q|^3|Q|! + ||Q||))$ time. $\qquad\square$

*Remark.* (1) When generating view definitions, we consider nontrivial case, *i.e.*, only extend view definitions that are connected after expansion. (2) Our algorithm can be easily adapted when other cost constraints are considered. (3) Though the computational complexity is in an exponential order of $|Q|$, as verified in our experimental study, VScHeu works well over $Q$ with a reasonable size. Better still, $Q$ is often small in practice. (4) As a part of input of VScHeu, the bound $B$ should be appropriately set, since a too big or small $B$ does not make sense. Moreover, when facing limited storage space, views identified by VScHeu

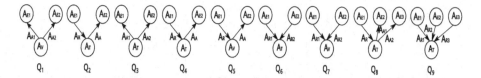

**Fig. 4.** A set of typical patterns in CHQ

might not be able to answer all the pattern queries, then approximate pattern matching using views [2] can be applied.

## 4  Experimental Evaluation

Using real-life and synthetic data, we conducted experimental studies to evaluate performances of the algorithms, *i.e.*, the efficiency, effectiveness and scalability.

**Experimental Setting.** We used the following data in our test.

*(1) Query Workload.* We used following query workloads.

(a) CHQ. We obtain a query log, that contains 97801 query sentences, from our industrial collaborator. Each query in the query log is a natural language question that searches contents about entertainment. For each query sentence, we used CRF++ to recognize explicit entities and relations, and manually extracted a set of structured semantic summaries. Here, a structured semantic summary is a structured representation of the semantic summary of a query sentence, in which each node carries an annotation from $\{A_E, A_T, A_V\}$ and each edge takes annotation from $\{A_R, A_A\}$, where $A_E, A_T, A_V, A_R, A_A$ refers to entity, type, value, relation and attribute, respectively. Interested readers may refer to [32] for more details about structured semantic summaries. We then treat each structured semantic summary as a pattern query and generate a query workload $\mathcal{Q}$ with 30 typical patterns. The frequency $\psi(Q_i)$ of each pattern $Q_i$ is the ratio of its appearance in the query log. Figure 4 (a) depicts structures of some typical pattern queries in $\mathcal{Q}$.

(b) YTB. We generated a set of 50 views for Youtube, where each node specifies videos with Boolean search conditions specified by *e.g.*, age (A), length (L), category (C), rate (R) and visits (V). We simply let $\psi(Q_i) = 1$ for each $Q_i$ in $\mathcal{Q}$.

(c) SYN. We generated a set of synthetic workloads $\mathcal{Q}$, each of which includes a set of 40 (resp. 60, 80, 100, 120) pattern queries $Q$, whose node labels are drawn from the same $\Sigma$ as that used for synthetic graph generation (see below for synthetic graph generation), and with sizes of $(2, 1)$, $(3, 2)$, $(4, 3)$ and $(4, 4)$. We randomly associate an integer to each $Q_i$ in $\mathcal{Q}$ to indicate its weight $\psi(Q_i)$.

*(2) Real-life & Synthetic Graphs.* (a) Movie, a crawled knowledge graph with $87K$ nodes and $167K$ edges. Each node has attributes such as name, genre

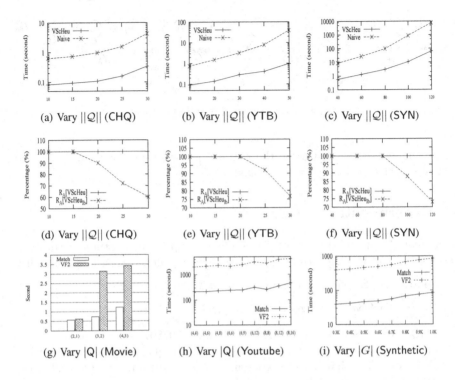

**Fig. 5.** Performance evaluation

and rating, and each edge from a person to a movie indicates that the person played in (resp. directed) the movie. (b) Youtube, a network with 1.4$M$ nodes and 3$M$ edges. Each node is a video with attributes such as category, rate, and edge edge from $v$ to $v'$ indicates that $v'$ is in the related list of $v$. (c) Synthetic, a synthetic graph generated by a generator. The generator produces random graphs, controlled by the number $|V|$ of nodes, the number $|E|$ of edges, and an alphabet $\Sigma$ for node labels.

*(3) Cost Estimation.* To estimate storage size of views, we extracted a subgraph $\overline{G_s}$ with size $|G_s| \approx 10\% * |G|$ from original $G$, for each graph used in the test, and set $\eta = 10$. When computing $\gamma(\mathsf{V})$, our view selection algorithms performance matching computation using $\mathsf{V}$ and $G_s$, and set $\gamma(\mathsf{V}) = \eta * |\mathsf{V}(G_s)|$. The time used for computing $\gamma(\mathsf{V})$ is not counted. To estimate query process cost, we simply set $\tau = 0.05$.

*(4) Implementation.* We implemented VScHeu and Naive for view selection, and a different version VScHeu$_{lb}$ of VScHeu which takes a small bound $\mathcal{B}$ as input and terminates once budget $\mathcal{B}$ is used out. Then views $\mathcal{V}$ identified by VScHeu$_{lb}$ may not be able to answer all the patterns of $\mathcal{Q}$. We also implemented Match [2] for matching with views and VF2 that performs matching on original graphs, all in Java.

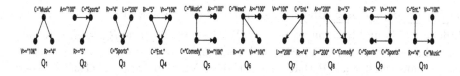

**Fig. 6.** Sample Views from YTB

All the experiments were run on a machine with an Intel Core(TM)2 Duo 3.00 GHz CPU and 4 GB memory, using Ubuntu. Each experiment was run 10 times and the average is reported.

**Experimental Results.** We next present our findings.

**Exp-1: View Selection.** We studied the efficiency and effectiveness of algorithm VScHeu vs. Naive, and then conducted tests to see the influence of bound $\mathcal{B}$, using CHQ, YTB and SYN. In this set of tests, we used a large storage space bound $\mathcal{B} = 100M$ for VScHeu, and ensures $\mathcal{Q} \sqsubseteq \mathcal{V}$ for $\mathcal{V}$ identified by VScHeu.

*Efficiency of* VScHeu. We varied the size of query workload $\|\mathcal{Q}\|$ from 10 to 30, 10 to 50 and 40 to 120, in 5, 10, 20 increments, for CHQ, YTB and SYN respectively, and evaluated the efficiency of VScHeu and Naive. As shown in Figs. 5(a), 5(b) and 5(c), (1) VScHeu is much more efficient than Naive, *e.g.*, it takes VScHeu 0.34 (resp Naive 4.21) seconds to identify a set of views from a query workload CHQ with $\|\mathcal{Q}\| = 30$; (2) VScHeu scales better than Naive with $\|\mathcal{Q}\|$, since for larger $\|\mathcal{Q}\|$, the computational cost of VScHeu grows much slower than Naive.

*Effectiveness of* VScHeu. Under the same setting as efficiency test, we compared $\gamma(\mathcal{V})$ and $Cost(\mathcal{Q}, \mathcal{V})$ of VScHeu and Naive, respectively. To do this, we defined two ratios: $R_\gamma = \frac{\gamma(\mathcal{V}_o)}{\gamma(\mathcal{V}_h)}$ and $R_C = \frac{Cost(\mathcal{Q}, \mathcal{V}_o)}{Cost(\mathcal{Q}, \mathcal{V}_h)}$, where $\mathcal{V}_h$ and $\mathcal{V}_o$ are view sets identified by VScHeu and Naive, respectively. Table 2 tells us following. (1) Views identified by VScHeu are close to that of Naive, *w.r.t.* storage space and query processing time, *e.g.*, $R_\gamma$ and $R_C$ are 94% and 95%, respectively on YTB with $\|\mathcal{Q}\| = 10$. This verifies that VScHeu not only is much more efficient than Naive, but also can achieve high quality results. (2) Naive achieves lower $R_\gamma$ and $R_C$ than VScHeu, which is as expected. (3) Both $R_\gamma$ and $R_C$ decrease with the increase of $\|\mathcal{Q}\|$. This is because for a larger input, *e.g.*, query log, exhaustive search often obtains higher performance advantages.

**Table 2.** Effectiveness comparison (vary $\|\mathcal{Q}\|$)

|     | $R_\gamma$ |     |     |     |     | $R_C$ |     |     |     |     |
|-----|------|-----|-----|-----|-----|------|-----|-----|-----|-----|
| CHQ | 10   | 15  | 20  | 25  | 30  | 10   | 15  | 20  | 25  | 30  |
|     | 92%  | 87% | 84% | 82% | 79% | 93%  | 91% | 90% | 87% | 84% |
| YTB | 10   | 20  | 30  | 40  | 50  | 10   | 20  | 30  | 40  | 50  |
|     | 94%  | 90% | 84% | 79% | 76% | 95%  | 92% | 90% | 87% | 85% |
| SYN | 40   | 60  | 80  | 100 | 120 | 40   | 60  | 80  | 100 | 120 |
|     | 86%  | 80% | 73% | 68% | 61% | 88%  | 83% | 79% | 74% | 68% |

*Influence of Bound $\mathcal{B}$.* Under the same setting as efficiency test, we analyzed the influence of bound $\mathcal{B}$ to the set of views. We first define a ratio $R_A = \frac{||Q'||}{||Q||}$ to measure the result quality of VScHeu and VScHeu$_{lb}$, *i.e.,* how many pattern queries can be answered with $\mathcal{V}$ that's identified by VScHeu (resp. VScHeu$_{lb}$), where $Q'$ indicates a subset of $Q$ that is contained by views identified by VScHeu$_{lb}$. For VScHeu$_{lb}$, we set its bounds $\mathcal{B}'$ as $30M$, $30M$ and $50M$ on CHQ, YTB and SYN, respectively. Figures 5(d), 5(e) and 5(f) tell us that (1) $R_A$ of VScHeu always remains 100%, owing to its large storage space budget; and (2) for small $||Q||$, VScHeu$_{lb}$ can find a set of views $\mathcal{V}$ with limited storage space, such that $R_A = 100\%$; while when $||Q||$ gets larger, but storage space remains the same, the set of views identified by VScHeu$_{lb}$ can not contain all the pattern queries, which leads to the decrease of $R_A$.

*Case Study.* We examined views identified by VScHeu and show a set of typical views generated from YTB in Fig. 6. As can be seen, most of the views take very simple structures, *i.e.,* tree structures, and contain no more than 3 edges. This simple structure facilitates composition of views and enables them to answer more pattern queries.

**Exp-2 View-Based Pattern Matching.** We also investigate the effectiveness of view-based pattern matching, using real-life and synthetic graphs.

*Efficiency of* Match. Figures 5(g) and 5(h) show the results on Movie and Youtube, respectively. The x-axis represents pattern size ($|V_p|, |E_p|$). On Movie, we only used patterns with size $(2, 1)$, $(3, 2)$ and $(4, 3)$, since queries are very small; moreover, we kept structure of patterns in $\mathcal{V}$ unchanged, but replaced semantic summaries with concrete values and materialized $\mathcal{V}$ on Movie. The results tell us the following. (1) Match substantially outperforms VF2, *e.g.,* taking on average 9.3% of its running time over Youtube. (2) Two algorithms spend more time on larger patterns. Nonetheless, Match is less sensitive than VF2, since it reuses previous computation cached in view extensions hence saves computational cost.

*Scalability of* Match. Using large synthetic graphs, we evaluated the scalability of Match and VF2. Fixing pattern size with $|V_p| = 4$ and $|E_p| = 6$, we varied the node number $|V|$ of data graphs from $0.3M$ to $1M$ , in $0.1M$ increments, and set $|E| = 2|V|$. As shown in Fig. 5(i), Match scales better with $|G|$ than VF2, which is consistent with the complexity analysis, and the observations in Figs. 5(g) and 5(h).

**Summary.** We obtain the following findings through our experimental evaluation. (1) Given a query workload, our view selection method can efficiently identify a set of view definitions to answer as many pattern queries, under constraints. For instance, using less than 100 megabytes storage space, our algorithm VScHeu spends only 65 seconds to find a view set $\mathcal{V}$ for a query workload with $||Q|| = 120$, such that all the pattern queries in $Q$ is contained in $\mathcal{V}$. (2) Answering graph pattern matching using views is effective in querying big graphs. For example, by using views, pattern matching via subgraph isomorphism takes only 9.3% of the time needed for computing matches directly in Youtube; and view-based matching algorithm scales much better with data size than batch algorithm.

# 5   Conclusion

We have studied *view selection* problem, for pattern queries defined in terms of subgraph isomorphism, and developed an effective algorithm with cost constraint, to identify a set of view definitions from a query workload. Our experimental results have verified the effectiveness and efficiency of our algorithms. These results extend the earlier studies of answering graph pattern matching using views.

The study of view selection is still in its infancy. One typical issue is to generalize VSP, and consider more cost constraints, *e.g.*, "maintenance cost", and develop effective estimation strategies. Another issues concerns VSP under distributed environment.

# References

1. Fan, W., Wang, X., Wu, Y.: Answering pattern queries using views. IEEE Trans. Knowl. Data Eng. **28**(2), 326–341 (2016)
2. Wang, X.: Answering graph pattern matching using views: a revisit. In: Benslimane, D., Damiani, E., Grosky, W.I., Hameurlain, A., Sheth, A., Wagner, R.R. (eds.) DEXA 2017. LNCS, vol. 10438, pp. 65–80. Springer, Cham (2017). https://doi.org/10.1007/978-3-319-64468-4_5
3. Cordella, L.P., Foggia, P., Sansone, C., Vento, M.: A (sub)graph isomorphism algorithm for matching large graphs. IEEE Trans. Pattern Anal. Mach. Intell. **26**, 1367–1372 (2004)
4. Halevy, A.Y.: Answering queries using views: a survey. VLDB J. **10**(4), 270–294 (2001)
5. Lenzerini, M.: Data integration: a theoretical perspective. In: PODS (2002)
6. Afrati, F.N., Chirkova, R.: Answering Queries Using Views. Synthesis Lectures on Data Management, 2nd edn. Morgan & Claypool Publishers, San Rafael (2019)
7. Gerome, M., Suciu, D.: Containment and equivalence for an XPath fragment. In: PODS (2002)
8. Neven, F., Schwentick, T.: XPath containment in the presence of disjunction, DTDs, and variables. In: Calvanese, D., Lenzerini, M., Motwani, R. (eds.) ICDT 2003. LNCS, vol. 2572, pp. 315–329. Springer, Heidelberg (2003). https://doi.org/10.1007/3-540-36285-1_21
9. Wang, J., Yu, J.X.: Revisiting answering tree pattern queries using views. ACM Trans. Database Syst. **37**(3), 18:1–18:34 (2012)
10. Fan, W., Wang, X., Wu, Y.: Answering graph pattern queries using views. In: ICDE, pp. 184–195 (2014)
11. Henzinge, M.R., Henzinger, T., Kopke, P.: Computing simulations on finite and infinite graphs. In: FOCS (1995)
12. Fan, W., Li, J., Ma, S., Tang, N., Wu, Y., Wu, Y.: Graph pattern matching: from intractability to polynomial time. In: PVLDB (2010)
13. Abiteboul, S., Hull, R., Vianu, V.: Foundations of Databases. Addison-Wesley, Boston (1995)
14. Miklau, G., Suciu, D.: Containment and equivalence for an XPath fragment. In: Proceedings of the Twenty-first ACM SIGACT-SIGMOD-SIGART Symposium on Principles of Database Systems, 3–5 June 2002, Madison, Wisconsin, USA, pp. 65–76 (2002)

15. Mami, I., Bellahsene, Z.: A survey of view selection methods. SIGMOD Rec. **41**(1), 20–29 (2012)
16. Caprara, A., Fischetti, M., Maio, D.: Exact and approximate algorithms for the index selection problem in physical database design. IEEE Trans. Knowl. Data Eng. **7**(6), 955–967 (1995)
17. Harinaraya, V., Rajaraman, A., Ullman, J.D.: Implementing data cubes efficiently. In: SIGMOD, pp. 205–216 (1996)
18. Sohrabi, M.K., Azgomi, H.: Evolutionary game theory approach to materialized view selection in data warehouses. Knowl. Based Syst. **163**, 558–571 (2019)
19. Gupta, H., Mumick, I.S.: Selection of views to materialize in a data warehouse. IEEE Trans. Knowl. Data Eng. **17**(1), 24–43 (2005)
20. Mami, I., Bellahsene, Z., Coletta, R.: View selection under multiple resource constraints in a distributed context. In: Liddle, S.W., Schewe, K.-D., Tjoa, A.M., Zhou, X. (eds.) DEXA 2012. LNCS, vol. 7447, pp. 281–296. Springer, Heidelberg (2012). https://doi.org/10.1007/978-3-642-32597-7_25
21. Mami, I., Bellahsene, Z., Coletta, R.: A constraint optimization method for large-scale distributed view selection. In: Hameurlain, A., Küng, J., Wagner, R. (eds.) Transactions on Large-Scale Data- and Knowledge-Centered Systems XXV. LNCS, vol. 9620, pp. 71–108. Springer, Heidelberg (2016). https://doi.org/10.1007/978-3-662-49534-6_3
22. Tang, N., Yu, J.X., Tang, H., Özsu, M.T., Boncz, P.: Materialized view selection in XML databases. In: Zhou, X., Yokota, H., Deng, K., Liu, Q. (eds.) DASFAA 2009. LNCS, vol. 5463, pp. 616–630. Springer, Heidelberg (2009). https://doi.org/10.1007/978-3-642-00887-0_55
23. Katsifodimos, A., Manolescu, I., Vassalos, V.: Materialized view selection for XQuery workloads. In: SIGMOD, pp. 565–576 (2012)
24. Mandhani, B., Suciu, D.: Query caching and view selection for XML databases. In: VLDB, pp. 469–480 (2005)
25. Castillo, R., Leser, U.: Selecting materialized views for RDF data. In: Daniel, F., Facca, F.M. (eds.) ICWE 2010. LNCS, vol. 6385, pp. 126–137. Springer, Heidelberg (2010). https://doi.org/10.1007/978-3-642-16985-4_12
26. Goasdoué, F., Karanasos, K., Leblay, J., Manolescu, I.: View selection in semantic web databases. Proc. VLDB Endow. **5**(2), 97–108 (2011)
27. Zhou, J., Larson, P., Goldstein, J., Ding, L.: Dynamic materialized views. In: Proceedings of the 23rd International Conference on Data Engineering, ICDE, pp. 526–535 (2007)
28. Koloniari, G., Pitoura, E.: Partial view selection for evolving social graphs. In: GRADES, co-located with SIGMOD/PODS 2013, p. 9 (2013)
29. Kumar, R., Novak, J., Tomkins, A.: Structure and evolution of online social networks. In: KDD (2006)
30. Halevy, A.Y.: Theory of answering queries using views. SIGMOD Rec. **29**(4)
31. Papadimitriou, C.H.: Computational Complexity. Addison-Wesley, Boston (1994)
32. Wang, X., Yang, L., Zhu, Y., Zhan, H., Jin, Y.: Querying knowledge graphs with natural languages. In: Hartmann, S., Küng, J., Chakravarthy, S., Anderst-Kotsis, G., Tjoa, A.M., Khalil, I. (eds.) DEXA 2019. LNCS, vol. 11707, pp. 30–46. Springer, Cham (2019). https://doi.org/10.1007/978-3-030-27618-8_3

# Extending Graph Pattern Matching
# with Regular Expressions

Xin Wang[1(✉)], Yang Wang[1], Yang Xu[2], Ji Zhang[3], and Xueyan Zhong[1]

[1] Southwest Petroleum University, Chengdu, China
xinwang.ed@gmail.com, wangyang@swpu.edu.cn, zhongxueyan@sohu.com
[2] Southwest Jiaotong University, Chengdu, China
xuyang@my.swjtu.edu.cn
[3] University of Southern Queensland, Toowoomba, Australia
Ji.Zhang@usq.edu.cn

**Abstract.** Graph pattern matching, which is to compute the set $M(Q, G)$ of matches of $Q$ in $G$, for the given pattern graph $Q$ and data graph $G$, has been increasingly used in emerging applications *e.g.*, social network analysis. As the matching semantic is typically defined in terms of subgraph isomorphism, two key issues are hence raised: the semantic is often too rigid to identify meaningful matches, and the problem is intractable, which calls for efficient matching methods. Motivated by these, this paper extends matching semantic with regular expressions, and investigates the top-$k$ graph pattern matching problem. (1) We introduce *regular patterns*, which revise traditional pattern graphs by incorporating regular expressions; extend traditional matching semantic by allowing edge to regular path mapping. With the extension, more meaningful matches could be captured. (2) We propose a relevance function, that is defined in terms of tightness of connectivity, for ranking matches. Based on the ranking function, we introduce the *top-$k$ graph pattern matching* problem, denoted by TopK. (3) We show that TopK is intractable. Despite hardness, we develop an algorithm with *early termination property*, *i.e.*, it finds top-$k$ matches without identifying entire match set. (4) Using real-life and synthetic data, we experimentally verify that our top-$k$ matching algorithms are effective, and outperform traditional counterparts.

## 1 Introduction

Graph pattern matching has been widely used in social data analysis [3,23], among other things. A number of algorithms have been developed for graph pattern matching that, given a pattern graph $Q$ and a data graph $G$, compute $M(Q, G)$, the set of matches of $Q$ in $G$ (*e.g.*, [10,16]). As social graphs are typically very large, with millions of nodes and billions of edges, several challenges to social data analysis with graph pattern matching are brought out. (1) Traditionally, graph pattern matching is defined in terms of subgraph isomorphism [15]. This semantic only allows edge-to-edge mapping, which is often too strict to

© Springer Nature Switzerland AG 2020
S. Hartmann et al. (Eds.): DEXA 2020, LNCS 12392, pp. 111–129, 2020.
https://doi.org/10.1007/978-3-030-59051-2_8

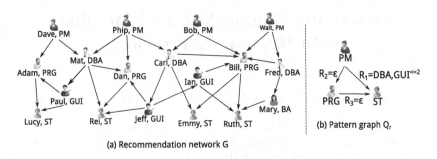

(a) Recommendation network G

(b) Pattern graph $Q_r$

**Fig. 1.** Querying big graphs

identify important and meaningful matches. (2) Existing matching algorithms often return an excessive number of results, which makes inspection a daunting task. While, in practice, users are often interested in top-ranked matches [20]. (3) The sheer size of social graphs makes matching computation costly: it is NP-complete to decide whether a match exists (cf. [21]), not to mention identifying the complete match set.

These highlight the need for extending matching semantic and exploring the *top-k graph pattern matching* problem. That is given a pattern graph $Q$ enriched with more meaningful constrains, a data graph $G$ and an integer $k$, it is to find $k$ best matches of $Q$ from $M(Q, G)$, based on certain functions measuring match quality. The benefits of identifying top-$k$ matches are twofold: (1) users only need to identify $k$ matches of $Q$ rather than a large match set $M(Q, G)$; (2) if we have an algorithm for computing top-$k$ matches with *the early termination property*, *i.e.*, it finds top-$k$ matches of $Q$ *without* computing the entire match set $M(Q, G)$, then we do not have to pay the price of full-fledged graph pattern matching.

*Example 1.* A fraction of a recommendation network is given as a graph $G$ in Fig. 1(a). Each node in $G$ denotes a person, with attributes such as *job title*, *e.g.*, project manager (PM), database administrator (DBA), programmer (PRG), business analyst (BA), user interface developer (GUI) and software tester (ST). Each directed edge indicates a recommendation, *e.g.*, (Dave, Adam) indicates that Dave recommended Adam earlier.

To build up a team for software development, one issues a pattern graph $Q_r$ depicted in Fig. 1(b) to find qualified candidates. The search intention constrains team members to satisfy following requirements: (1) with *job title*: PM, PRG and ST; (2) satisfying following recommendation relations: (i) PM recommended PRG before; (ii) ST has been recommended by PRG, and also been recommended by PM via a chain, that includes a DBA and at most two GUIs. Here, recommendation relations given above can be represented as regular expressions, *i.e.*, $R_1 = \mathsf{DBA}, \mathsf{GUI}^{\leq 2}$, $R_2 = \epsilon$ and $R_3 = \epsilon$, where $\epsilon$ indicates an empty string.

Traditionally, graph pattern matching is defined in terms of subgraph isomorphism, which only supports edge-to-edge mapping and is too rigid to find matches. To identify matches of $Q_r$, matching semantic needs to be extended.

Indeed, when edge to regular path mapping is supported, match set $M(Q_r, G)$ includes following matches: $t_{r1} = (\text{Dave}, \text{Adam}, \text{Lucy})$, $t_{r2} = (\text{Phip}, \text{Dan}, \text{Rei})$ and $t_{r3} = (\text{Bob}, \text{Bill}, \text{Ruth})$.

In practice, $M(Q_r, G)$ is possibly a large set, while users may only be interested in top-$k$ matches of $Q_r$. It is hence unnecessary and too costly to compute the entire large set $M(Q_r, G)$. In light of this, an algorithm with the *early termination property* is desired, as it identifies top-$k$ matches without inspecting entire $M(Q_r, G)$.

To measure relevance of top-$k$ matches, one may consider tightness of connectivity [4]. Observe that matches $t_{r1}$ and $t_{r2}$ are connected more tightly than $t_{r3}$ due to shorter inner distances, and are considered more relevant to the query. Hence, $\{t_{r1}, t_{r2}\}$ makes a good candidate for top-2 matches in terms of relevance.

□

This example shows that *top-k graph pattern matching* with enriched matching semantic may rectify the limitations of existing matching algorithms. While to make practical use of it, several problems have to be solved.

**Contributions.** This paper gives a full treatment for *top-k graph pattern matching*.

(1) We incorporate regular expressions in pattern graphs and introduce *regular patterns* (or *r-pattern* for short) $Q_r$. We revise the traditional notion of subgraph isomorphism by supporting edge to regular path mapping, and define graph pattern matching with the revised notion, that's given $Q_r$ and $G$, it is to compute the set $M(Q_r, G)$ of matches of $Q_r$ in $G$, where each edge $e_p$ of $Q_r$ is mapped to a path in $G$ satisfying regular expression specified by $e_p$. We also introduce a function to rank matches of $Q_r$, namely, *relevance function* $w(\cdot)$ that is defined in terms of tightness of connectivity. Based on the function, we introduce the *top-k graph pattern matching problem* (Sect. 2).

(2) We investigate the *top-k graph pattern matching problem* (Sect. 3). Given a graph $G$, a *r-pattern* $Q_r$, and a positive integer $k$, it is to find $k$ top-ranked matches of $Q_r$, based on the relevance function $w(\cdot)$. We show that the decision version of the problem is NP-hard. Despite hardness, we provide an algorithm with *early termination property* [8], that is, it stops as soon as top-$k$ matches are found, without computing entire match set.

(3) Using both real-life and synthetic data, we experimentally verify the performance of our algorithms (Sect. 4). We find the following. Our top-$k$ matching algorithms are not only effective, but also efficient. For example, on *Youtube*, our algorithm takes only 10 seconds, and inspects 38% of matches of $Q_r$, when it terminates. Moreover, they are not sensitive to the increase of $k$, and size $|G|$ of $G$.

These results yield a promising approach to querying big graphs. **All the proofs, and complexity analyses can be found in [1].**

**Related Work.** We next categorize the related work as follows.

*Pattern Matching with Regular Expressions.* Given a pattern graph $Q$ and a data graph $G$, the problem of graph pattern matching is to find all the matches in $G$ for $Q$. Typically, matching is defined in terms of subgraph isomorphism [6] and graph simulation [16]. Extended from traditional notions, a host of work [2,9,11,13,22,24] incorporate matching semantic with regular expressions and explored regular pattern matching problem. Among these, a few works focus on regular simple path queries (RSPQs), which is a special case of regular pattern matching. For example, [11] introduced a parallel algorithm to answer regular reachability queries based on partial evaluation. [24] still developed parallel algorithms via a set of operations with/without indices, for extremely large graphs; while the technique is more appropriate for regular reachability queries. [2] characterizes the frontier between tractability and intractability for (RSPQs, and summarizes the data complexity. Path indexes were used by [13] to generate efficient execution plans for RSPQs. For generalized regular pattern matching, [22] formalized regular queries as nonrecursive Datalog programs, and shows elementary tight bounds for the containment problem for regular queries. [9] incorporates regular expressions on pattern graphs, while the matching semantic is graph simulation, which is different to ours.

*Top-k Graph Pattern Matching.* Top-$k$ searching method was first developed by [7,8] for relational data, where the idea of *early termination* were introduced. Along the same line, a host of work focus on top-$k$ search on graph data. For example, [27] proposes to rank matches, *e.g.*, by the total node similarity scores, and identify $k$ matches with highest ranking scores. [15] investigates top-$k$ query evaluation for twig queries, which essentially computes isomorphism matching between rooted graphs. To provide more flexibility for top-$k$ pattern matching, [4] extends matching semantics by allowing edge-to-path mapping, and proposes to rank matches based on their compactness. To diversify search results and balance result quality and efficiency, an algorithm with approximation ratio is proposed to identify diversified top-$k$ matches [25]. Instead of matching with subgraph isomorphism, graph simulation [16] is applied as matching semantic, and pattern graph is designated with an output node, referred to as "query focus" in [12], then match result includes a set of nodes that are matches of the "query focus".

Our work differs from the prior work in the following. (1) The problem we studied has different matching semantics from earlier works. With the new semantic, more sensible matches could be captured. (2) The high computational cost of previous methods for top-$k$ matching hinders their applications. In contrast, our algorithm possesses *early termination property* and yields a practical method for querying real-life social graphs.

## 2    Preliminaries

In this section, we first review data graphs and pattern graphs. We then introduce graph pattern matching, followed by the *top-k graph pattern matching problem*.

## 2.1   Data Graphs and Pattern Graphs

We start with notions of data graphs and pattern graphs.

**Data Graphs.** A *data graph* (or simply a graph) is a directed graph $G = (V, E, L_v)$, where (1) $V$ is a finite set of nodes; (2) $E \subseteq V \times V$, in which $(v, v')$ denotes an edge from $v$ to $v'$; and (3) $L_v$ is a function defined on $V$ such that for each node $v$ in $V$, $L_v(v)$ is a tuple $(A_1 = a_1, \cdots, A_n = a_n)$, where $A_i = a_i (i \in [1, n])$, indicating that the node $v$ has a value $a_i$ for the attribute $A_i$, and denoted as $v.A_i = a_i$.

We shall use the following notations for data graphs $G$. (1) A *path* $\rho$ from $v_0$ to $v_n$, denoted as $\rho(v_0, v_n)$, in $G$ is a sequence $v_0 \xrightarrow{e_1} v_1 \xrightarrow{e_2} \cdots v_{n-1} \xrightarrow{e_n} v_n$, where (a) $v_i \in V$ for each $i \in [0, n]$, and (b) $e_j = (v_{j-1}, v_j)$ in $E$ for each $j \in [1, n]$. (2) We say a path $\rho$ is nonempty if it contains more than one edge.

**Pattern Graphs.** A *pattern graph* is a directed graph $Q_r = (V_p, E_p, f_v, f_e)$, where (1) $V_p$ is the set of *pattern nodes*; (2) $E_p$ is the set of *pattern edges*; (3) $f_v$ is a function defined on $V_p$ such that for each node $u \in V_p$; $f_v(u)$ is a predicate defined as a conjunction of atomic formulas of the form of '$A$ op $a$', such that $A$ denotes an attribute of the node $u$, $a$ is a value, and op is a comparison operator in the set $\{<, \leq, =, \neq, >, \geq\}$; and (4) $f_e$ is a function on $E_p$ such that $f_e(e_p)$ associates a regular expressions $R$ on pattern edge $e_p$. We define $R$ as below:

$$R ::= \epsilon \mid l \mid l^{\leq k} \mid l^+ \mid RR .$$

Here (1) $\epsilon$ is the empty string; (2) $l$ is a label in $\Sigma$; (3) $k$ is a positive integer, and $l^{\leq k}$ is a string $l \cdots l$, that consists of $j$ ($j \in [1, k]$) occurrence of label $l$; (4) $l^+$ indicates one or more occurrences of label $l$; and (5) $RR$ indicates concatenation.

Intuitively, the predicate $f_v(u)$ of a node $u$ specifies a search condition on labels and data contents. We say that a node $v$ in a data graph $G$ satisfies the search condition of a pattern node $u$ in $Q_r$, denoted as $v \sim u$, if for each atomic formula '$A$ op $a$' in $f_v(u)$, there exists an attribute $A$ in $L_v(v)$ such that $v.A$ op $a$. The function $f_e$ assigns a regular expression on each edge $(u, u')$ of $Q_r$. As will be seen shortly, regular expressions enable matching from edge-to-edge mapping to edge to regular path mapping. We refer to a pattern as a *normal pattern* and denote it as $Q = (V_p, E_p, f_v)$, when each edge of $Q$ takes a unit weight, *i.e.*, $R = \epsilon$.

We denote $|V_p| + |E_p|$ as $|Q|$ (the size of $Q$), and $|V| + |E|$ as $|G|$ (the size of $G$).

*Example 2.* As shown in Fig. 1(b), pattern graph $Q_r$ is a *r-pattern*, in which each node takes a predicate, *e.g.*, *job title* = "PM", to specify search conditions on nodes; and each edge carries a regular expression as edge weight, *e.g.*, $f_e(\text{PM}, \text{ST}) = (\text{DB}, \text{GUI}^{\leq 2})$. As will be seen shortly, regular expressions associated on pattern edges $e_p$ impose constrains on paths that are mapped from $e_p$.

Table 1. A summary of notations

| Symbols | Notations |
|---|---|
| $Q = (V_p, E_p, f_v)$ | A *normal pattern* |
| $Q_r = (V_p, E_p, f_v, f_e)$ | A *regular pattern* |
| $Q_T$ | A spanning tree of pattern graph $Q_r$ (resp. $Q$) |
| $G = (V, E, L_v)$ | A data graph |
| $G_q = (V_q, E_q, L'_v, L_e)$ | A twisted graph of $G$ |
| $M(Q_r, G)$ (resp. $M(Q, G)$) | A set of matches of $Q_r$ (resp. $Q$) in $G$ |
| $|V_p| + |E_p|$ | $|Q_r|$ (resp. $|Q|$), the size of $Q_r$ (resp. $Q$) |
| $|V| + |E|$ | $|G|$, the size of $G$ |
| $w(\cdot)$ | The relevance function |
| $q_R = (v, v', u_s, u_t, R)$ | A regular reachability query |
| $Q_r(e_p)$ | A *r-pattern* that contains a single edge $e_p$ |
| $Q_r(R)$ | A query automaton of regular expression $R$ |

## 2.2  Graph Pattern Matching Revised

We now introduce the graph pattern matching problem based on the revised pattern graph. Consider a data graph $G = (V, E, L_v)$ and a *r-pattern* $Q_r = (V_p, E_p, f_v, f_e)$.

**Graph Pattern Matching.** A *match* of $Q_r$ in $G$ is an $n$-ary node-tuple $t = \langle v_1, \cdots, v_n \rangle$, where $v_i \in V$ ($i \in [1, n]$, $n = |V_p|$), and there exists a *bijective function h* from $V_p$ to the nodes in $t$ such that (1) for each node $u \in V_p$, $h(u) \sim u$; and (2) $(u, u')$ is an edge in $Q_r$ if and only if there exists a path $\rho$ from $h(u)$ to $h(u')$ in $G$ that satisfies the regular expression $f_e(u, u') = R$, i.e., the path $\rho$ constitutes a string that is in the regular language defined by $R$.

Indeed, (1) for a *r-pattern* $Q_r$, the regular expression $f_e(e_p)$ of a pattern edge $e_p = (u_1, u_2)$ imposes restriction of mapping from an edge $e_p$ to a path $\rho$ in $G$: that is, a pattern edge $e_p$ can be mapped to a nonempty path $\rho = v_1 \xrightarrow{e_1} v'_1 \xrightarrow{e_2} \cdots v'_{n-1} \xrightarrow{e_n} v_2$ in a data graph $G$, such that the string of node labels $L_v(v_1.A)L_v(v_2.A) \cdots L_v(v_n.A)$ satisfies $f_e(e_p)$; and (2) a *normal pattern* enforces edge-to-edge mappings, as found in subgraph isomorphism. Traditional graph pattern matching is defined with *normal patterns* [6].

We shall use following notations for graph pattern matching.

(1) The *answer* to $Q_r$ (resp. $Q$) in $G$, denoted by $M(Q_r, G)$ (resp. $M(Q, G)$), is the set of node-tuples $t$ in $G$ that matches $Q_r$ (resp. $Q$). (2) Abusing notation of *match*, in a graph $G$, we say a pair of nodes $(v, v')$ is a *match* of a pattern edge $e_p = (u, u')$, if (a) $v$, $v'$ are the matches of $u$, $u'$, respectively, and (b) either $\rho(v, v')$ satisfies $R$ when $f_e(e_p) = R$, or $(v, v')$ is an edge when $e_p$ belongs to a *normal pattern* $Q$. (3) Given a $Q_r$, we use $Q_r(e_p)$ to denote a *r-pattern* that contains a single edge $e_p = (u_s, u_t)$ with regular expression $R$ associated, where

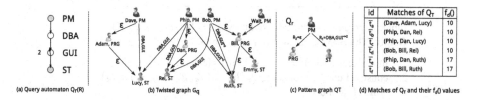

(a) Query automaton $Q_r(R)$    (b) Twisted graph $G_q$    (c) Pattern graph QT    (d) Matches of $Q_T$ and their $f_d()$ values

**Fig. 2.** Automaton $Q_r(R)$, Twisted graph $G_q$, spanning tree $Q_T$ of $Q_r$ and matches of $Q_T$

$e_p$ is in $Q_r$. Then, for a pair of nodes $(v, v')$, to determine whether it is a match of $Q_r(e_p)$ is equivalent to a regular reachability query $q_R = (v, v', u_s, u_t, R)$.

*Example 3.* Consider *r-pattern* $Q_r$ in Fig. 1(b). Each edge $e_p$ of $Q_r$ is associated with a regular expression that specifies labels on the path that $e_p$ maps to. Taking a match {Dave, Adam, Lucy} as example, it is mapped via $h$ from $V_p$, and for each pattern edge $(u, u')$, there exists a path from $h(u)$ to $h(u')$ satisfying regular expression $f_e(u, u')$, *e.g.*, (PM, ST) is mapped to a path Dave → Mat → Paul → Lucy, satisfying $R_1$ of $Q_r$.

**Remark.** General regular expressions, which consider operations *e.g.*, alternation and Kleene star [20], possess more powerful expressive abilities via higher computational cost. To strike a balance between expressive power and computational cost, we choose a subclass $R$ of general regular expressions, as $R$ already has sufficient expressive power to specify edge relationships commonly found in practice.

### 2.3 Top-$k$ Graph Pattern Matching Problem

In practice, match set could be excessively large on big graphs, while users are interested in the best $k$ matches of pattern graphs [18]. This motivates us to study the *top-k graph pattern matching problem* (TopK). To this end, we first define a function to rank matches, and then introduce TopK.

**Relevance Function.** On a match $t$ of $Q_r$ in $G$, we define the relevance function $w(\cdot)$ as following:

$$w(t) = \sqrt{\frac{|E_p|}{f_d(t)}},$$

where $f_d(t) = \Sigma_{(u,u')\in E_p, h(u), h(u')\in t}(\text{dist}(h(u), h(u')))^2$. That is, the relevance function favors those matches that are connected tightly. The more tightly the nodes are connected in a match $t$, the more relevant $t$ is to $Q_r$, as observed in study [4]. Thus, matches with high $w(\cdot)$ values are preferred.

**Top-$k$ Matching Problem.** We now state the *top-k graph pattern matching problem*. Given a graph $G$, a *r-pattern* $Q_r$, and a positive integer $k$, it is to find a set $S$ of matches of $Q_r$, such that $|S| = k$ and

---

*Input:* A graph $G$, a reachability query $q_R = (v_s, v_t, u_s, u_t, R)$.
*Output:* The Boolean answer ans to $q_R$ in $G$.

1. initialize a stack $T := (u_s, v_s)$; ans:=false;
2. construct query automaton $Q_r(R) = (V_q, E_q, L_s, L_i, u_s, u_t)$;
3. **while** $T \neq \emptyset$ **do**
4.    $(u, v) := T.\text{pop}()$;
5.    **for each** child $u'$ of $u$ **and each** child $v'$ of $v$ **do**
6.       **if** $u' = u_t$ **and** $v' \sim u'$ **then** ans:=true; **break** ;
7.       **else if** $\gamma(u') \leq L_i(u')$ **and** $v' \sim u'$ **then** push $(u', v')$ in T;
8. **return** ans;

---

**Fig. 3.** Procedure RegReach

$$w(\mathcal{S}) = \arg\max_{\mathcal{S}' \subseteq M(Q_r, G), |\mathcal{S}'| = k} \sum_{t \in \mathcal{S}'} w(t).$$

Abusing notation $w(\cdot)$, we use $w(\mathcal{S})$ to denote $\sum_{t \in \mathcal{S}} w(t)$, referred to as the *relevance* of $\mathcal{S}$ to $Q_r$. Thus, TopK is to identify a set of $k$ matches of $Q_r$, that maximizes total relevance. In other words, for all $\mathcal{S}' \subseteq M(Q_r, G)$, if $|\mathcal{S}'| = k$, then $w(\mathcal{S}) \geq w(\mathcal{S}')$.

*Example 4.* Recall graph $G$ and *r-pattern* $Q_r$ in Fig. 1. The relevance values of matches of $Q_r$ are $w(t_{r1}) = \sqrt{\frac{3}{11}}$, $w(t_{r2}) = \sqrt{\frac{3}{11}}$, $w(t_{r3}) = \sqrt{\frac{3}{18}}$, thus, $t_{r1}$, $t_{r2}$ are more relevant to $Q_r$ than $t_{r3}$.

We summarize notations used in the paper in Table 1.

## 3    Efficient Top-$k$ Graph Pattern Matching

In this section, we study the *top-k graph pattern matching problem*. The main results of the section are as follows.

**Theorem 1.** *The* TopK *(1) is* NP-*hard (decision problem); (2) can be solved by an early termination algorithm in* $O(|G|^{|Q_r|} \cdot \log(|G|^{|Q_r|}) + |G|!|G|)$ *time.*

Due to intractability, there does not exist polynomial time algorithm for TopK. To tackle the issue, one may simply develop an algorithm, that is denoted as Naive$_r$, and works as following: (1) find all the matches of $Q_r$ in $G$, and (2) pick $k$ most relevant matches from match set $M(Q_r, G)$. Though straightforward, the algorithm is quite cost-prohibitive, as it requires to (1) process mapping from an edge to a regular path instantly, (2) find out entire match set from big $G$, and (3) pick $k$ top-ranked matches from (possibly very big) match set $M(Q_r, G)$.

To rectify these, two main challenges have to be settled: (1) how to efficiently deal with edge to regular path mapping when identifying matches of $Q_r$? and (2) how to terminate earlier as soon as top-$k$ matches are identified? In the following, we provide techniques to address the issues, for TopK.

## 3.1 Query Automaton

A regular pattern $Q_r$ takes a regular expression $R$ on each edge. To find matches of $Q_r$ in a graph $G$, one needs to identify matches $(v, v')$ of $Q_r(e_p)$ in $G$, for each pattern edge $e_p = (u_s, u_t)$. This calls for techniques to evaluate regular reachability queries $q_R = (v, v', u_s, u_t, R)$. In light of this, we introduce *query automaton*, which is a variation of nondeterministic finite state automaton (NFA), to facilitate evaluation of regular reachability queries.

*Query Automaton.* A *query automaton* $Q_r(R, u_s, u_t)$ (or shortened as $Q_r(R)$ when it is clear from context) of a regular expression $R$ along with $u_s$, $u_t$ as initial and final states, is defined as a six tuple $(V_q, E_q, L_s, L_i, u_s, u_t)$, where (1) $V_q$ is a finite set of states, (2) $E_q \subseteq V_q \times V_q$ is a set of transitions between the states, (3) $L_s$ is a function that assigns each state a label in $R$, (4) $L_i$ is a function that assigns a positive integer or a symbol "*" to a state, and (5) $u_s$ and $u_t$ in $V_q$ are the initial and final states, respectively.

Abusing notations, we say that a node $v$ in $G$ is a *match* of a state $u_v$ in $Q_r(R, u_s, u_t)$, denoted as $v \sim u_v$, if and only if (a) there exists an attribute $A$ in $L_v(v)$ such that $v.A = L_s(u_v)$, and (b) there exist a path $\rho$ from $v$ to $t$ and a path $\rho'$ from $u_v$ to $u_t$, such that $\rho$ and $\rho'$ have the same label. A state $u$ is said a *child* of $u'$, if there is an edge $(u', u)$ in $E_q$, i.e., $u'$ can transit to $u$. In contrast to the transition of traditional NFA, the transitions of our query automaton follows the restriction imposed by regular expression $R$, i.e., at state $u_v$, for each edge $(v, v')$ on a path in graph $G$, a transition $u_v \to u'_v$ can be made via $(u_v, u'_v) \in E_q$ if $v \sim u_v$ and $v' \sim u'_v$.

Given a regular expression $R$ along with initial and final states, its corresponding query automaton $Q_r(R)$ can be constructed in $O(|R|\log|R|)$ time, following the conversion of [17]; moreover the function $L_i$ is defined as follows: for each entry in $R$, if it is of the form $l^{\leq k}$ (resp. $l^*$), then the corresponding state $s$ is associated with an integer $k$ (resp. label $*$) via $L_i(\cdot)$ to indicate maximum number (resp. unlimited number) of occurrence of $s$; otherwise, $L_i(s)$ is set as 1 by default.

*Example 5.* Recall regular expression $R_1 = $ DBA, GUI$^{\leq 2}$ of $Q_r$ in Fig. 1(b). Its query automaton $Q_r(R)$ is depicted in Fig. 2(a). The set $V_q$ has four states PM, DBA, GUI, ST, where the initial and final states are PM and ST, respectively. The set $E_q$ of transitions is {(PM, DBA), (DBA, GUI), (GUI, GUI), (GUI, ST)}. Observe that the edge (GUI, GUI) is associated with weight 2, indicating that transition from GUI to GUI is restricted to no more than twice. In contrast to NFA, the query automaton $Q_r(R)$ is to accept paths in *e.g.*, $G$ of Fig. 1(a), and its transitions are made by matching the labels of its states with the value of attribute *job title* of nodes on the paths.  □

**Lemma 1.** *In a graph $G$, a pair of nodes $(v_s, v_t)$ is a match of a regular reachability query $q_R = (v_s, v_t, u_s, u_t, R)$ if and only if $v_s$ is a match of $u_s$ in $Q_r$.*

---

*Input:* Data graph $G$, pattern graph $Q_r$ and integer $k$.
*Output:* A $k$-element set of matches of $Q_r$.

1. set M := $\emptyset$;
2. a *twisted graph* $G_q$:=TwistGen$(G, Q_r)$;
3. generate a spanning tree $Q_T$ of $Q_r$;
4. M := TMat$(G_q, Q_r, Q_T, k)$;
5. **return** M;

**Procedure** TMat
*Input:* $G_q, Q_r, Q_T, k$.
*Output:* A set of matches of $Q_r$.

1. a min-heap S := $\emptyset$; set L := $\emptyset$;
2. pick a node $u$ from $Q_T$; initialize can$(u)$;
3. **for each** $v$ in can$(u)$ **do**
4.     identify match set H of $Q_T$ s.t. $\bar{t}$ in H contains $v$ as a match of $u$;
5.     L := L $\cup$ H;
6. sort matches $\bar{t}$ in L in ascending order of $f_d(\bar{t})$;
7. **while** L $\neq \emptyset$ **do**
8.     pick the top-most match $\bar{t}$ in L; generate match $t$ with $\bar{t}$;
9.     S := S $\cup \{t\}$; L := L $\setminus \{\bar{t}\}$;
10.    **if** termination condition is satisfied **then**
11.        **break** ;
12. **if** $|S| > k$ **then** retain $k$ matches with least $w(t)$ in S;
13. **return** S;

---

**Fig. 4.** Algorithm TopKR

## 3.2   An Early Termination Algorithm

As is shown, the inefficiency of the algorithm Naive$_r$ is mostly caused by the search strategy it applied. To avoid exhaustive search, we develop an algorithm for TopK with *early termination property*, *i.e.*, it stops as soon as top-$k$ matches are found, without identifying entire (possibly very big) match set.

Below, we start from a notion *twisted graph* (TG for short) and its construction strategy, followed by the illustration of the *early termination* algorithm.

*Twisted Graph.* Given a graph $G$ and a *r-pattern* $Q_r$, the twisted graph $G_q$ of $Q_r$ in $G$ is a quadruple $(V_q, E_q, L'_v, L_e)$, where $V_q$ and $E_q$ are the sets of nodes and edges, respectively; $L'_v$ is defined the same as $L_v$ in $G$; and $L_e$ specifies a set of regular expressions, *i.e.*, node labels on $\rho(v, v')$ in graph $G$, for each edge $(v, v')$ of $G_q$.

Note that, each edge $(v, v')$ in $G_q$ may take multiple regular expressions, as there may exist multiple paths from $v$ to $v'$ in $G$, that satisfy different regular expressions imposed by pattern edges. All the regular expressions associated on a single edge of $G_q$ are concatenated with disjunction.

*Example 6.* Recall graph $G$ and $r$-pattern $Q_r$ in Fig. 1. The corresponding twisted graph $G_q$ is shown Fig. 2 (b). Observe that the regular expression associated with each edge in $G_q$ imposes either edge-to-edge mapping (via $\epsilon$) or edge-to-path mapping.                                                                    □

**Construction of TG.** The construction algorithm for TG, denoted as TwistGen (not shown), takes graph $G$ and pattern $Q_r$ as input, and works as follows. It first identifies node pairs $(v, v')$ in $G$ with $v \sim u$ and $v' \sim u'$, for each pattern edge $e_p = (u, u')$ of $Q_r$. It then verifies whether there exists a path from $v$ to $v'$ in $G$, that satisfies regular expression $R$ imposed by $e_p$. It next includes nodes $v$, $v'$ and an edge $(v, v')$ with regular expression $R$ in $G_q$, if there exists a path from $v$ to $v'$ satisfying $R$. Indeed, the verification of $(v, v')$ as a match of $(u, u')$ can be processed via evaluation of regular reachability queries. Below, we present a procedure RegReach, which evaluates regular reachability queries by simulating query automaton.

*Regular Reachability Queries.* The procedure, denoted as RegReach is shown in Fig. 3. It takes a graph $G$ and a regular reachability query $q_R = (v, v', u_s, u_t, R)$ as input, and returns a boolean value to indicate whether there exists a path $\rho(v, v')$ in $G$ satisfying regular expression $R$. More specifically, it first initializes a stack T with node pair $(u_s, v_s)$, and a boolean variable ans as false (line 1). It then constructs a query automaton $Q_r(R)$ following the conversion of [17] (line 2). In the following, it performs depth first search to identify whether $v_s$ can reach $v_t$ via a path satisfying $R$ (lines 3–7). RegReach first pops out the upper most element $(u, v)$ from stack T (line 4), it next verifies whether the final state of automaton is encountered, *i.e.*, the child $v'$ of $v$ satisfies search condition imposed by the child $u'$ of $u$, which is $u_t$. If so, RegReach sets ans as true, and breaks **while** loop (line 6); otherwise, it checks whether (a) $\gamma(u')$ is larger than $L_i(u')$, *i.e.*, occurrence upper bound of state $u$, where $\gamma(u')$ indicates the visiting time of $u'$ during the traversal, and (b) $v' \sim u'$, and pushes $(u', v')$ in T, if both conditions (a) and (b) are satisfied (line 7). After **while** loop terminates, RegReach returns ans as final result (line 8).

*Example 7.* Given graph $G$ in Fig. 1(a), and a regular reachability query $q_R =$ (Dave, Lucy, PM, ST, $R_1$), RegReach evaluates $q_R$ as following. It first initializes stack with (PM, Dave), and then constructs a query automaton $Q_r(R)$, as shown in Fig. 2(a). It next conducts guided search in graph $G$. The search starts from PM in $Q_r(R)$ and Dave in $G$, respectively, and visits DB, GUI, ST in $Q_r(R)$ and Mat, Paul, Lucy in $G$. As the terminal state in $Q_r(R)$ is reached, the path Dave → Mat → Paul → Lucy is accepted by the query automaton, and a truth answer is returned.                                                                    □

**Proposition 1.** *Given a list $L_c$ of matches $\bar{t}_i$ ($i \in [1, n]$) of $Q_T$ in $G_q$, that is sorted in ascending order of $f_d(\bar{t}_i)$. Do sorted access to entries of $L_c$. An entry $\bar{t}_j$ of $L_c$ can not contribute to a top-k match of $Q_r$, if (1) at least k valid matches have been identified from the first l matches of $Q_T$ (l < j) and (2) $f_d(t^m) < f_d(\bar{t}_j)$.*

Here, $Q_T$ is a spanning tree of $Q_r$, $t_i$ refers to a match of $Q_r$, that is generated from $\bar{t}_i$, $t^m$ indicates a match of $Q_r$, that's generated from $\bar{t}_l$ ($l < j$) and has the maximum $f_d(\cdot)$ value among all the matches that have been generated.

We are now ready to outline the algorithm for TopK.

**Algorithm.** The algorithm, denoted as TopKR, is shown in Fig. 4. It first initializes a set M to maintain top-$k$ matches of $Q_r$ (line 1), constructs a *twisted graph* $G_q$ with procedure TwistGen that is introduced earlier (line 2), and generates a spanning tree $Q_T$ of $Q_r$, with an algorithm introduced in [26] (line 3). TopKR then invokes TMat to identify top-$k$ matches (line 4), and returns output M of TMat as final result (line 5). In particular, TMat works in two steps: (a) it generates a set of matches of $Q_T$ (a spanning tree of $Q_r$), as candidate matches of $Q_r$; and (b) it repeatedly verifies whether these candidate matches are embedded in true matches, until termination condition is met. Below, we elaborate TMat with details.

Procedure TMat takes a twisted graph $G_q$, a pattern $Q_r$, its spanning tree $Q_T$ and an integer $k$ as input, and works as following. (1) It initializes an empty min-heap S to maintain matches of $Q_r$, and an empty set L to keep track of candidate matches (line 1). (2) It picks a node $u$ from pattern $Q_T$, and initializes a set can($u$) that includes all the candidate matches $v$ of $u$, i.e., $v \sim u$, in $G_q$ (line 2). (3) TMat generates matches $\bar{t}$ of $Q_T$ as candidate matches of $Q_r$ (lines 3–4). Specifically, TMat traverses $G_q$ starting from node $v$, following the structure of $Q_T$, and identifies matches $\bar{t}$ of $Q_T$. After traversal finished, TMat extends L with H, which contains a set of match candidates. The above process repeats |can($u$)| times. When all the candidate matches are generated, TMat ranks $\bar{t}$ in L in ascending order of their $f_d(\cdot)$ values (line 6). (4) TMat then iteratively verifies whether each of candidate matches can turn to a valid match of $Q_r$ until termination condition is encountered (lines 7–11). In each round iteration, only the top most candidate $\bar{t}$ will be used to generate true match (line 8). Note that the generation simply follows a structural expansion on $G_q$ with $\bar{t}$. After match $t$ is generated, TopKR extends S with it, and removes $\bar{t}$ from L (line 9). At this moment, TopKR verifies termination condition specified in Proposition 1, and breaks **while** loop if S already contains at least $k$ matches, and no other match with less relevance exists (lines 10–11). After loop, if S contains more than $k$

**Table 2.** Varying $|Q_r|$ and $d_w(e)$ ($k = 10$)

| | | Youtube | | Amazon | |
|---|---|---|---|---|---|
| | | $d_w(e) = 3$ | $d_w(e) = 4$ | $d_w(e) = 3$ | $d_w(e) = 4$ |
| $|Q_r|$ | (3,3) | 38% | 36% | 46% | 44% |
| | (4,6) | 46% | 37% | 50% | 48% |
| | (5,10) | 68% | 65% | 60% | 50% |
| | (6,12) | 90% | 88% | 76% | 72% |
| | (7,14) | 100% | 100% | 100% | 100% |

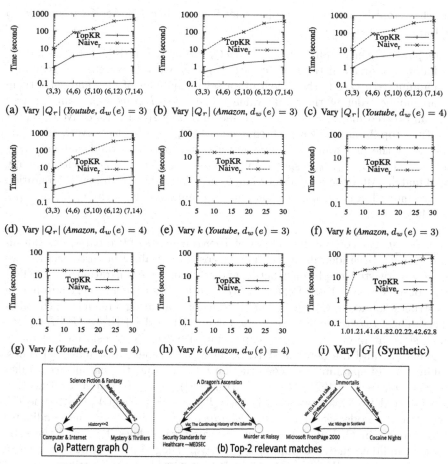

(a) Vary $|Q_r|$ (*Youtube*, $d_w(e) = 3$)  (b) Vary $|Q_r|$ (*Amazon*, $d_w(e) = 3$)  (c) Vary $|Q_r|$ (*Youtube*, $d_w(e) = 4$)

(d) Vary $|Q_r|$ (*Amazon*, $d_w(e) = 4$)  (e) Vary $k$ (*Youtube*, $d_w(e) = 3$)  (f) Vary $k$ (*Amazon*, $d_w(e) = 3$)

(g) Vary $k$ (*Youtube*, $d_w(e) = 4$)  (h) Vary $k$ (*Amazon*, $d_w(e) = 4$)  (i) Vary $|G|$ (Synthetic)

(j) Case study on *Amazon*

**Fig. 5.** Performance evaluation

**Table 3.** Varying $k$ and $d_w(e)$ ($|Q_r| = (4,6)$)

|  |  | Youtube | | Amazon | |
|---|---|---|---|---|---|
|  |  | $f_e(e) = 3$ | $f_e(e) = 4$ | $f_e(e) = 3$ | $f_e(e) = 4$ |
| $k$ | 5 | 46% | 42% | 37% | 34% |
|  | 10 | 46% | 42% | 37% | 34% |
|  | 15 | 46% | 42% | 37% | 34% |
|  | 20 | 46% | 42% | 62% | 64% |
|  | 25 | 68% | 66% | 62% | 64% |
|  | 30 | 93% | 89% | 100% | 98% |

matches, TopKR simply picks top-$k$ ones from it (line 12). TopKR returns set S as top-$k$ match set (line 13).

*Example 8.* Given graph $G$ and *r-pattern* $Q_r$ in Fig. 1, TopKR finds top-2 matches as follows. It first constructs a *twisted graph* $G_q$ with TwistGen, and generates a spanning tree $Q_T$. It then identifies a set of matches of $Q_T$, and sorts them in ascending order of their $f_d()$ values. The *twisted graph* $G_q$, spanning tree $Q_T$, and sorted match candidates are shown in Fig. 2(b-d), respectively. TopKR identifies top-2 matches with match candidates $\bar{t}_a, \cdots, \bar{t}_l$ successively. In particular, after the first four candidates are checked, only two valid matches $t_a$, $t_b$ are obtained from $\bar{t}_a$, $\bar{t}_b$, respectively, whereas $\bar{t}_c$ and $\bar{t}_d$ can not contribute any valid match. After the fourth iteration, TopKR found that termination condition is encountered, and terminates computation immediately, with $\{t_a, t_b\}$ as top-2 matches.                                                                                    □

*Complexity.* The computational cost of TopKR consists of two parts, *i.e.*, initialization and match identification. (1) The cost of initialization also includes two parts, *i.e.*, twisted graph construction and spanning tree generation. For the first part, observe that it takes RegReach (a) $O(|R|\log|R|)$ time for query automaton construction, and (b) $O(|V|(|V|+|E|))$ time for evaluating a regular reachability query. Thus, *twisted graph* construction is in $O(|V|^2(|V|+|E|)+|E_p||R|\log|R|)$ time, since at most $|E_p|$ regular expressions need to be transformed to query automatons, and the depth first traversal will be repeated at most $|V|$ times. Taking cost of spanning tree construction into account, the initialization takes TopK $O(|Q_r|+|V|^2(|V|+|E|)+|E_p||R|\log|R|)$ time. (2) Match identification is conducted by TMat, which is in $O(|V_q|!|V_q|+|V_q|^{|V_p|}\log(|V_q|^{|V_p|}))$ time ($V_q$ refers to the node set of the *twisted graph* $G_q$). Specifically, the candidates generation is in $O(|V_q|!|V_q|)$ time in the worst case (lines 3–5), and the sorting for candidate matches (line 7) is in $O(|V_q|^{|V_p|}\log(|V_q|^{|V_p|}))$ time, since there may exist at most $|V_q|^{|V_p|}$ matches in L (line 6). The match verification is in $O(|V_q|^{|V_p|}+|V_q|^{|V_p|}\log(|V_q|^{|V_p|}))$ time (lines 7-11). To see this, observe that (a) match identification takes $O(|V_q|^{|V_p|})$ time, as the **while** loop (line 7) runs at most $|V_q|^{|V_p|}$ times, and each loop takes TMat constant time to generate a match $t$ (line 8) and verify termination condition (line 10); and (b) top-$k$ match selection (line 12) takes $O(|V_q|^{|V_p|}\log(|V_q|^{|V_p|}))$ time. (3) Putting initialization and match identification together, TopKR is in $O(|Q_r|+|V|^2(|V|+|E|)+|E_p||R|\log|R|+|G|^{|Q_r|}\cdot\log(|G|^{|Q_r|})+|G|!|G|)$ time, which is bounded by $O(|G|^{|Q_r|}\cdot\log(|G|^{|Q_r|})+|G|!|G|)$, since parameters $|Q_r|$, $|V|^2(|V|+|E|)$ and $|E_p||R|\log|R|$ can be omitted from the perspective of worst-case time complexity. The analysis above completes the proof of Theorem 1(2).                                                    □

## 4    Experimental Evaluation

Using real-life and synthetic data, we conducted two sets of experiments to evaluate performance of our algorithms.

**Experimental Setting.** We used the following datasets.

*(1) Real-life graphs.* We used two real-life graphs: (a) *Amazon* [19], a product co-purchasing network with $548K$ nodes and $1.78M$ edges. Each node has attributes such as title, group and sales-rank, and an edge from product $x$ to $y$ indicates that people who buy $x$ also buy $y$. (b) *Youtube* [5], a recommendation network with $1.6M$ nodes and $4.5M$ edges. Each node is a video with attributes such as category, age and rate, and each edge from $x$ to $y$ indicates that $y$ is in the related list of $x$.

*(2) Synthetic data.* We designed a generator to produce synthetic graphs $G = (V, E, L_v)$, controlled by the number of nodes $|V|$ and edges $|E|$, where $L_v$ are assigned from a set of 15 labels. We generated synthetic graphs following the linkage generation models [14]: an edge was attached to the high degree nodes with higher probability. We use $(|V|, |E|)$ to denote the size of $G$.

*(3) Pattern generator.* We designed a generator to produce meaningful patterns. The generator uses following parameters: $|V_p|$ for the number of pattern nodes, $|E_p|$ for the number of pattern edges, label $f_v$ from an alphabet $\Sigma$ of labels taken from corresponding data graphs, and bounds $b$ and $c$ for $f_e(e_p)$ of *r-patterns*, such that one edge $e_p$ is constrained by a regular expression $l_1^{\leq b}, \cdots, l_k^{\leq b}$ ($1 \leq k \leq c$) and has weight $\Sigma_{i \in [1,k]} b$. For synthetic graphs $G$, we manually constructed a set of 10 patterns with node labels drawn from the same $\Sigma$ of $G$. We refer to $d_w(e)$ as the maximum edge weight of a pattern $Q_r$ and denote $(|V_p|, |E_p|, d_w(e))$ as the size $|Q_r|$ of $Q_r$. We use $(|V_p|, |E_p|, d_w(e))$ and $(|V_p|, |E_p|)$ to indicate the size of $Q_r$ interchangeably, when it is clear from the context.

*(4) Implementation.* We implemented the following algorithms, all in Java. (a) TopKR, for identifying top-$k$ matches of $Q_r$. (b) Naive$_r$, a naive algorithm for TopK, that performs regular reachability queries with techniques of [11] and generates a TG, identifies matches from the TG with revised VF2 [6], and selects top-$k$ matches. (c) TopK$_u$, for computing top-$k$ matches for a normal pattern $Q$. Observe that all the matches of a normal pattern $Q$ have the same relevance value. Hence, TopK$_u$ applies a different strategy to generate twisted graphs and identify top-$k$ matches. More specifically, in the initialization stage, TopK$_u$ generates a TG *w.r.t.* a normal pattern $Q$ by including those edges in $G$ that have endpoints as matches of endpoints of a pattern edge $e$ in $Q$. During matching computation, TopK$_u$ simply generates a list L of candidate matches of $Q$, and repeatedly verifies whether each of them can turn to a valid match, until $k$ matches are found. Here, no ranking for entries in L and complex verification for termination (Proposition 1) are needed. (d) Naive, a naive top-$k$ algorithm for normal patterns, that constructs twisted graphs first along the same line as TopK$_u$, finds the complete set of matches using VF2 [6], ranks and selects top-$k$ matches from the match set.

All the tests were run on a machine with an Intel Core(TM)2 Duo 3.7 GHz CPU, and 16 GB of memory, using Ubuntu. Each experiment was repeated 5 times and the average is reported.

**Experimental Results.** We next present our findings.

**Exp-1: Effectiveness.** We studied the effectiveness of the algorithms TopK vs. Naive$_r$ (resp. TopK$_u$ vs. Naive) on real-life graphs. To measure effectiveness of our *early termination* algorithms, we define a ratio IR as $\frac{|M^i(Q_r,G)|}{|M(Q_r,G)|}$, where $|M^i(Q_r, G)|$ indicates the amount of matches identified by the algorithm when it terminates. We do not report the effectiveness of Naive$_r$ and Naive, as their IR are always 100%.

*Varying $|Q_r|$.* Fixing $k=10$ and $d_w(e) = 3$ (resp. $d_w(e) = 4$) for all $e$ in $Q_r$, we varied $|Q_r|$ from $(3,3)$ to $(7,14)$, and evaluated IR over two real-life graphs. The results that are reported in Table 2, tell us the following: (1) TopKR is able to reduce excessive matches. For example, when $|Q_r| = (3,3)$ and $d_w(e) = 3$, TopKR only inspected 38% (resp. 46%) matches when top-$k$ matches are identified, on *Youtube* (resp. *Amazon*). On average, IR for TopKR is only 66.8% (resp. 64.6%) on *Youtube* (resp. *Amazon*), which indicates that TopKR benefits from its *early termination property.* (2) the effectiveness of *early terminate property* degrades when $|Q_r|$ gets larger. For example, when $|Q_r| = (7,14)$, IR = 100% on both graphs, indicating that TopKR can not terminate earlier. The main reason is that: for larger patterns, the $f_d(\cdot)$ value grows bigger for seen matches, hence termination condition can not be easily satisfied.

Under the same setting, we evaluated IR for normal patterns and found that TopK$_u$ effectively reduces excessive matches. For example, when $|Q| = (3,3)$, TopK$_u$ only inspected 9% (resp. 11%) matches when top-$k$ matches are identified, on *Youtube* (resp. *Amazon*). On average, IR for TopK$_u$ is only 36.2% (resp. 32.2%) on *Youtube* (resp. *Amazon*), which indicates that TopK$_u$ benefits from its early termination property.

*Varying $k$.* Fixing pattern size $|Q_r| = (4,6)$ and $d_w(e) = 3$ (resp. $d_w(e) = 4$), we varied $k$ from 5 to 30 in 5 increments, and reported IR for TopKR on real-life graphs. As shown in Table 3, for *regular patterns* with $d_w(e) = 3$ (resp. $d_w(e) = 4$), the ratio IR of TopKR increased from 46% to 93% (resp. 42% to 89%) and 37% to 100% (resp. 34% to 98%), on *Youtube* and *Amazon*, respectively, when $k$ increased from 5 to 30. In general, with the increase of $k$, IR gets larger either, as expected. While unlike the trend observed when varying $|Q_r|$, when $k$ changes from 5 to 15, IR keeps unchanged for all cases. This is because, a fixed number of matches of the spanning tree of $Q_r$ (as match candidates) needs to be examined before TopKR can terminate.

In the same setting as test for $Q_r$, we evaluated IR for normal patterns. We find that the ratio IR of TopK$_u$ increased from 9% to 56% and 9% to 53%, on *Youtube* and *Amazon*, respectively, when $k$ increased from 5 to 30. The increment of IR is mainly caused by that, more matches have to be examined for choosing as top-$k$ matches, for a larger $k$, which is as expected.

*Varying $d_w(e)$.* From Tables 2 and 3, we can also see the influence of the changes of $d_w(e)$. That is when $d_w(e)$ gets larger, IR drops with slight decrease as well. This shows that the *early termination property* favors patterns with large $d_w(e)$.

**Exp-2: Efficiency.** We next evaluated the efficiency of the algorithm TopKR vs. Naive$_r$ (resp. TopK$_u$ vs. Naive) in the same setting as in Exp-1.

*Varying* $|Q_r|$. As shown in Figs. 5(a)–5(d), (1) TopKR substantially outperforms Naive$_r$: it takes TopKR, on average, 3.71% (resp. 3.94%) and 2.42% (resp. 2.48%) of the time of Naive$_r$, on *Youtube* and *Amazon*, respectively, for patterns with $d_w(e) = 3$ (resp. $d_w(e) = 4$). The performance advantage is mainly caused by (a) a much more efficient strategy, applied by TopKR, to identify matches of spanning trees of $Q_r$, and (b) *early termination property* TopKR possesses; and (2) Naive$_r$ is more sensitive to $|Q_r|$ than TopKR, since the cost of match set computation, which accounts for more than 95% of entire running time of Naive$_r$, heavily depends on $Q$.

Under the same setting, we evaluated efficiency for normal patterns. The results tell us that (1) TopK$_u$ substantially outperforms Naive. For example, TopK$_u$ takes, on average, 0.68% and 1.04% of the time of Naive, on *Youtube* and *Amazon*, respectively.

*Varying* $k$. Figures 5(e)–5(h) report the efficiency results for patterns with size $|Q_r| = (4, 6)$ and $d_w(e) = 3$ (resp. $d_w(e) = 4$). We find the following: (1) TopKR and Naive$_r$ are both insensitive to $k$, since most of the computational time is spent for twisted graph construction and matching evaluation. (2) TopKR always outperforms Naive$_r$, which is consistent with the observations in Figs. 5(a)–5(d).

In the same setting, we conducted efficiency test for normal patterns and found that TopK$_u$ always outperforms Naive. For example, TopK$_u$ only takes, on average, 0.42% (resp. 1.45%) time of Naive, on *Youtube* and *Amazon*, respectively.

*Varying* $d_w(e)$. Figures 5(a)–5(h) also show the influence on the efficiency with respect to the changes of $d_w(e)$. We find that it takes longer for TopKR on patterns with larger $d_w(e)$, since patterns with larger edge weight may result in a bigger match set, thus more time is required to verify matches, and select top-$k$ ones from them. For example, TopKR spends 109% and 122% of the time when $d_w(e)$ of $Q_r$ increased from 3 to 4. on *Youtube* and *Amazon*, respectively.

**Exp-3 Scalability.** Fixing $|Q_r| = (4, 6)$ and $d_w(e) = 2$, we varied $|G|$ from $(1M, 2M)$ to $(2.8M, 5.6M)$ in $(0.2M, 0.4M)$ increments, and evaluated the scalability of the algorithms with synthetic data. The results, shown in Fig. 5(i), tell us that TopKR is less sensitive to $|G|$ than Naive$_r$, hence scale better with $|G|$ than Naive$_r$, and moreover, both algorithms are not sensitive to $d_w(e)$.

**Exp-4 Case Study.** On *Amazon*, we manually inspected top-2 matches found by TopKR for $Q_r$ of Fig. 5(j)(a). As shown in Fig. 5(j)(b), two matches found by TopKR are quite relevant to $Q_r$; while they have different relevance values, since the match on the right hand side has a longer path, which starts from Immortalis to Microsoft FrontPage 2000 via two history nodes, *i.e.*, "A Lie and a Libel", and "Vikings in Scotland".

**Summary.** Our top-$k$ matching algorithms work well. (a) They are effective: top-10 matches of patterns with size $(3, 3, 3)$ can be found by TopKR when only 38% (resp. 46%) matches on *Youtube* (resp. *Amazon*) are identified. (b) They are efficient: it only takes TopKR 0.782s (resp. 0.477s) to find top-10 matches of $Q_r$ with $|Q_r| = (3, 3, 3)$ over *Youtube* (resp. *Amazon*). (c) TopKR scales well with $|G|$.

## 5    Conclusion

We have introduced the *top-k graph pattern matching problem*. We have revised
the semantic of graph pattern matching by allowing edge to regular path mapping, and defined functions to measure match relevance. We have established the
complexity for TopK, and provided an algorithm with *early termination property* for TopK. As verified analytically and experimentally, our method remedies
the limitations of prior algorithms, via semantic extension, and substantially
improves efficiency on big social graphs.

## References

1. Full version. https://github.com/xgnaw/sun/raw/master/regularGPM.pdf
2. Bagan, G., Bonifati, A., Groz, B.: A trichotomy for regular simple path queries on
   graphs. In: PODS, pp. 261–272 (2013)
3. Brynielsson, J., Högberg, J., Kaati, L., Martenson, C., Svenson, P.: Detecting social
   positions using simulation. In: ASONAM (2010)
4. Cheng, J., Zeng, X., Yu, J.X.: Top-k graph pattern matching over large graphs. In:
   ICDE, pp. 1033–1044 (2013)
5. Cheng, X., Dale, C., Liu, J.: Youtube (2008). http://netsg.cs.sfu.ca/youtubedata/
6. Cordella, L.P., Foggia, P., Sansone, C., Vento, M.: A (sub)graph isomorphism algorithm for matching large graphs. TPAMI **26**(10), 1367–1372 (2004)
7. Fagin, R.: Combining fuzzy information from multiple systems. JCSS **58**(1), 83–99
   (1999)
8. Fagin, R., Lotem, A., Naor, M.: Optimal aggregation algorithms for middleware.
   JCSS **66**(4), 614–656 (2003)
9. Fan, W., Li, J., Ma, S., Tang, N., Wu, Y.: Adding regular expressions to graph
   reachability and pattern queries. In: ICDE, pp. 39–50 (2011)
10. Fan, W., Li, J., Ma, S., Tang, N., Wu, Y., Wu, Y.: Graph pattern matching: from
    intractable to polynomial time. PVLDB **3**(1), 264–275 (2010)
11. Fan, W., Wang, X., Wu, Y.: Performance guarantees for distributed reachability
    queries. PVLDB **5**(11), 1304–1315 (2012)
12. Fan, W., Wang, X., Wu, Y.: Diversified top-k graph pattern matching. PVLDB
    **6**(13), 1510–1521 (2013)
13. Fletcher, G.H.L., Peters, J., Poulovassilis, A.: Efficient regular path query evalua-
    tion using path indexes. In: EDBT, pp. 636–639 (2016)
14. Garg, S., Gupta, T., Carlsson, N., Mahanti, A.: Evolution of an online social aggre-
    gation network: an empirical study. In: IMC (2009)
15. Gou, G., Chirkova, R.: Efficient algorithms for exact ranked twig-pattern matching
    over graphs. In: SIGMOD (2008)
16. Henzinger, M.R., Henzinger, T.A., Kopke, P.W.: Computing simulations on finite
    and infinite graphs. In: FOCS (1995)
17. Hromkovic, J., Seibert, S., Wilke, T.: Translating regular expressions into small -
    free nondeterministic finite automata. J. Comput. Syst. Sci. **62**(4), 565–588 (2001)
18. Ilyas, I.F., Beskales, G., Soliman, M.A.: A survey of top- k query processing tech-
    niques in relational database systems. ACM Comput. Surv. **40**(4), 1–58 (2008)
19. Leskovec, J., Krevl, A.: Amazon dataset, June 2014. http://snap.stanford.edu/
    data/index.html

20. Mendelzon, A.O., Wood, P.T.: Finding regular simple paths in graph databases. In: VLDB, pp. 185–193 (1989)
21. Papadimitriou, C.H.: Computational Complexity. Addison-Wesley, Boston (1994)
22. Reutter, J.L., Romero, M., Vardi, M.Y.: Regular queries on graph databases. Theory Comput. Syst. **61**(1), 31–83 (2017). https://doi.org/10.1007/s00224-016-9676-2
23. Terveen, L.G., McDonald, D.W.: Social matching: a framework and research agenda. ACM Trans. Comput.-Hum. Interact. **12**, 401–434 (2005)
24. Wang, H., Han, J., Shao, B., Li, J.: Regular expression matching on billion-nodes graphs. CoRR, abs/1904.11653 (2019)
25. Wang, X., Zhan, H.: Approximating diversified top-$k$ graph pattern matching. In: Hartmann, S., Ma, H., Hameurlain, A., Pernul, G., Wagner, R.R. (eds.) DEXA 2018. LNCS, vol. 11029, pp. 407–423. Springer, Cham (2018). https://doi.org/10.1007/978-3-319-98809-2_25
26. Wikipedia. Prim's algorithm (2019). https://en.wikipedia.org/wiki/Prim's_algorithm
27. Zou, L., Chen, L., Lu, Y.: Top-k subgraph matching query in a large graph. In: Ph.D. Workshop in CIKM (2007)

# Construction and Random Generation of Hypergraphs with Prescribed Degree and Dimension Sequences

Naheed Anjum Arafat[1](✉), Debabrota Basu[2], Laurent Decreusefond[3], and Stéphane Bressan[1]

[1] School of Computing, National University of Singapore, Singapore, Singapore
naheed_anjum@u.nus.edu
[2] Data Science and AI Division, Chalmers University of Technology, Gothenburg, Sweden
[3] LTCI, Télécom Paris, Institut Polytechnique de Paris, Paris, France

**Abstract.** We propose algorithms for construction and random generation of hypergraphs without loops and with prescribed degree and dimension sequences. The objective is to provide a starting point for as well as an alternative to Markov chain Monte Carlo approaches. Our algorithms leverage the transposition of properties and algorithms devised for matrices constituted of zeros and ones with prescribed row- and column-sums to hypergraphs. The construction algorithm extends the applicability of Markov chain Monte Carlo approaches when the initial hypergraph is not provided. The random generation algorithm allows the development of a self-normalised importance sampling estimator for hypergraph properties such as the average clustering coefficient.

We prove the correctness of the proposed algorithms. We also prove that the random generation algorithm generates any hypergraph following the prescribed degree and dimension sequences with a non-zero probability. We empirically and comparatively evaluate the effectiveness and efficiency of the random generation algorithm. Experiments show that the random generation algorithm provides stable and accurate estimates of average clustering coefficient, and also demonstrates a better effective sample size in comparison with the Markov chain Monte Carlo approaches.

## 1 Introduction

While graphs are the prevalent mathematical models for modern applications, being natural representations of varied objects such as transportation, communication, social and biological networks [19], to mention a few, they only capture binary relationships. Hypergraphs introduce the opportunity to represent $n$-ary relationships and thus create a more general, albeit more complex and generally more computationally expensive, alternative [4, 15, 22].

Indeed many real-world systems are more naturally modelled as hypergraphs, as exemplified by the cases of multi-body systems, co-authorship networks and

© Springer Nature Switzerland AG 2020
S. Hartmann et al. (Eds.): DEXA 2020, LNCS 12392, pp. 130–145, 2020.
https://doi.org/10.1007/978-3-030-59051-2_9

parliamentary relations [3, 21, 22]. While the applications are numerous, the properties of the underlying hypergraphs are yet to be fully understood. Just as it is the case for graphs in network science [5, 12, 18], configuration modelling or the random generation of hypergraphs with prescribed degree and dimension sequences precisely allows the fabrication of suitable hypergraphs for empirical and simulation-based studies of hypergraph properties [8].

We study and propose algorithms for construction and random generation of hypergraphs with prescribed degree and dimension sequences (Sect. 4, 5). In addition, we present the necessary background on hypergraphs and $(0, 1)$-matrices in Sect. 2 and synthesise related works in Sect. 3.

Recently, Chodrow [8] proposed a Markov chain Monte Carlo (MCMC) algorithm to address this problem of generating a labelled hypergraph with a prescribed degree and dimension sequences. The limitation of the MCMC algorithm is that it requires an initial hypergraph with the prescribed degree and dimension sequences as a starting point. It is not always the case that such an initial hypergraph is available. Therefore, we present in Sect. 4 a deterministic algorithm for constructing an initial hypergraph as a starting point for the existing MCMC approach. At each iteration, our algorithm constructs the edge with the largest dimension using distinct vertices having the largest degrees.

We present in Sect. 5 a random generation algorithm for generating hypergraphs as an alternative to the existing MCMC approach. Our generation algorithm leverage properties and methods devised for $(0, 1)$-matrices [20] with row- and column-sums coinciding with the hypergraph specification. If no row or column in the matrix contains only zeros, every $(0, 1)$-matrix corresponds to the incidence matrix of a hypergraph with parallel-edges but no loop [4, Chap. 17]. The column-sums of an incidence matrix represent degrees of the vertices and the row-sums represent dimensions of the edges of a hypergraph. At each iteration, the random generation algorithm generates the edge with the largest dimension using distinct, randomly selected vertices that satisfy the characterisation theorem for $(0, 1)$ matrices (Theorem 1).

We further leverage our random generation algorithm to propose a self-normalised importance sampling (SNIS) estimator [16] for estimating hypergraph properties in Sect. 6.

We prove the correctness of both the algorithms (Theorems 2 and 3). Furthermore, we prove that the generation algorithm generates any random hypergraph having prescribed degree and dimension sequences with non-zero probability (Theorem 4). We evaluate the effectiveness (Sect. 7) of the MCMC algorithm enabled with our construction algorithm and the random generation algorithm with SNIS estimator by estimating the average clustering coefficient of the projected graphs of the family of hypergraphs having prescribed degree and dimension sequence and also computing corresponding effective samples size [16].

We conclude in Sect. 8 by summarising our findings.

## 2    Hypergraphs and $(0, 1)$-matrices

In this section, we describe selected concepts of hypergraphs and inequalities involving $(0, 1)$-matrices relevant for the transposition of properties and algorithms for $(0, 1)$-matrices to hypergraphs.

**Definition 1 (Hypergraph [4]).** *A hypergraph $H = (V, E)$ is a tuple of a vertex set $V = \{v_1, \ldots, v_n\}$ and an edge set $E = \{e_1, \ldots, e_m\}$ where each edge is a subsets of $V$. Here $V$ is a set and $E$ is a multiset.*

Unless otherwise stated, hypergraphs are *labelled*, may contain *parallel-edges* but *no self-loop*. The polyadic relations i.e. the edges connecting the vertices in a hypergraph is presentable as a $(0, 1)$-matrix, called the incidence matrix.

**Definition 2 (Incidence Matrix).** *The incidence matrix $M = [m_{ij}]_{i,j=1,1}^{m,n}$ of a labelled hypergraph $H = (V, E)$ is a $(0, 1)$-matrix with columns representing labels of vertices in $V$ and rows representing edge set $E$ where*

$$m_{i,j} = \begin{cases} 1 & \text{if } v_j \in e_i, \\ 0 & \text{otherwise.} \end{cases}$$

The incidence matrix of a hypergraph is not unique. Even if the vertices are arranged in a total-order, such as in descending order of degrees and lexicographic order of labels among the vertices with same degree, any permutation of the rows would provide another incidence matrix of the same hypergraph.

**Property 1.** Every incidence matrix of a hypergraph whose degree sequence is $(a)_n$ and dimension sequence is $(b)_m$ is contained in the set of $(0, 1)$-matrices of dimension $m \times n$ whose column-sums are $(a)_n$ and row-sums are $(b)_m$. *Thus, any algorithm that uses the characterisation of sequences $(a)_n$ and $(b)_m$ to construct an $m \times n$-dimensional $(0, 1)$-matrix with column-sum $(a)_n$ and row-sums $(b)_m$ can be leveraged to construct a hypergraph with degree-sequence $(a)_n$ and dimension sequence $(b)_m$.* This observation constitute the core of our random hypergraph generation proposed in Sect. 5.

**Property 2.** In order to design the proposed algorithms and to prove their correctness, we use the Gale-Rysers characterisation of $(0, 1)$-matrices (Theorem 1). Before discussing the theorem, we intend to remind us the notion of dominance between sequences.

**Definition 3 (Dominance [17]).** *$(a)_n$ is defined to be dominated by $(b)_m$ if the corresponding zero-padded sequences $(a^*)_l$ and $(b^*)_l$, where $l = \max\{m, n\}$ satisfy -*
*(i) sum of the first $k$ components of $(a^*)_l$ is smaller than or equal to sum of the first $k$ components of $(b^*)_l$ and*

(ii) *sum of all the components of* $(a^*)_l$ *is equal to the sum of all the components of* $(b^*)_l$. *Mathematically,*

$$(a)_n \prec (b)_m \Longleftrightarrow \begin{cases} \sum_{i=1}^{k} a_i^* \leq \sum_{i=1}^{k} b_i^*, & k = 1, 2, \dots, l-1 \\ \sum_{i=1}^{l} a_i^* = \sum_{i=1}^{l} b_i^*. \end{cases}$$

$a_i^* = a_i$ *for* $i \leq n$, $a_i^* = 0$ *for* $i > n$, $b_i^* = b_i$ *for* $i \leq m$ *and* $b_i^* = 0$ *for* $i > m$.

**Theorem 1 (Gale-Rysers [14,20] Characterisation of Matrices).** *If* $(a)_n = (a_1, a_2, \dots, a_n)$ *and* $(b)_m = (b_1, b_2, \dots, b_m)$ *are two monotonically non-increasing, non-negative integer sequences, the necessary and sufficient condition for the existence of a* $(0, 1)$-*matrix with column sums* $(a)_n$ *and row sums* $(b)_m$ *is that* $(a)_n$ *is dominated by the conjugate sequence of* $(b)_m$.

The conjugate sequence of $(b)_m$ is a sequence whose $i^{th}$ component is the number of components in $(b)_m$ that are greater than or equal to $i$. We denote the conjugate sequence of $(b)_m$ as $\bar{b}_n$.[1] A sequence-pair $((a)_n, (b)_m)$ satisfying the dominance condition in Gale-Rysers characterisation is said to be *realisable by a* $(0, 1)$-*matrix*. Conversely, such a matrix is said to *realise* $(a)_n, (b)_m$.

**Property 3.** Another observation is that if we construct a new sequence $(a')_{n-1} = (a_2, a_3, \dots, a_n)$ from a monontonically non-increasing positive integer sequence $(a)_n = (a_1, a_2, a_3, \dots, a_n)$, the conjugate sequence of $(a)_{n-1}$ can be derived from the conjugate sequence of $(a)_n$ by reducing the first $a_1$ components of $\bar{a}$ by 1.

**Lemma 1 ([17]).** *Let* $(a)_n = (a_1, a_2, \dots, a_n)$ *be a positive monotonically non-increasing. If we construct a new sequence* $(a')_{n-1} \triangleq (a_2, \dots, a_n)$, *then the conjugate sequence of* $(a')_{n-1}$ *is*

$$(\bar{a'}) = (\bar{a}_1 - 1, \dots, \bar{a}_{a_1} - 1, \bar{a}_{a_1+1}, \dots, \bar{a}_n).$$

*Example 1.* Let $(a) = (4, 2, 2, 1)$. Its conjugate sequence is $(\bar{a}) = (4, 3, 1, 1, 0, \dots)$. By removing $a_1$, we get a new sequence $(a') = (2, 2, 1)$. The conjugate sequence of $(a')$ is $(3, 2, 0, 0, \dots)$ which is exactly the sequence derived from $(4, 3, 1, 1, 0, \dots)$ by reducing first four components by 1 i.e. $(4 - 1, 3 - 1, 1 - 1, 1 - 1, 0, \dots)$.

Fulkerson and Ryser [13] state a necessary condition that preserves dominace after reducing the values of a fixed number of components by 1 in sequences $(a)_n$ and $(b)_n$ related by dominance.

---

[1] Since the number of 1's in a row of an $m \times n$-dimensional $(0, 1)$ matrix cannot exceed $n$, the length of the conjugate sequence of row sums $(b)_m$ is upper bounded by $n$.

**Lemma 2 (Fulkerson-Ryser's Lemma** [13]**).** *Let* $(a)_n$ *and* $(b)_n$ *be two monotonically non-increasing integer sequences. Let* $(u)_n$ *be sequence obtained from* $(a)_n$ *by reducing components at indices* $i_1, i_2, \ldots, i_K$ *by 1. Similarly, let* $(v)_n$ *be obtained from* $(b)$ *by reducing components at indices* $j_1, j_2, \ldots, j_K$ *by 1. If* $i_1 \leq j_1, i_2 \leq j_2, \ldots, i_K \leq j_K$, *and* $(a)_n \prec (b)_n$, *we get* $(u)_n \prec (v)_n$.

We leverage this lemma to prove correctness of our construction algorithm (Sect. 5).

## 3  Related Works

### 3.1  Graphs with a Prescribed Degree Sequence

There are two main frameworks for the generation of random graphs with a prescribed degree sequence. The first framework is *direct sampling* [12], that constructs the graph incrementally edge-by-edge. Among algorithms based on direct sampling, [6] and [2] introduced the concept of *stubs* and the procedure of stub-matching as an algorithm for counting the number of labelled graphs with a prescribed degree sequence. The stub-matching procedure may generate graphs with loops and parallel-edges, which is often undesirable. Rejecting the generated random graph until a simple graph is generated is proposed as a remedy. However, this approach is inefficient for large degree values as an exponential number of samples might get rejected [5,12]. Furthermore, there is no obvious way to extend this algorithm for graphs into an algorithm for hypergraphs [8].

As an alternative to the stub-matching algorithm, [5] proposed an algorithm that uses the Erdös-Gallai's characterisation to generate simple graphs. This algorithm generates all simple graphs following a given degree sequence with a non-zero probability. [5] also proposes an importance sampling scheme to estimate the number of simple graphs following the prescribed degree sequence. Motivated by their work on simple graphs, in this paper, we devise a self-normalised importance sampling scheme (Sect. 6) using our random generation algorithm (Sect. 5) to estimate average clustering coefficient of projected graphs of hypergraphs having a prescribed degree and dimension sequences (Sect. 7).

The second framework proposes *MCMC* algorithms [12,18,19] that iteratively switch edges of an initial graph with a given degree sequence to obtain the final graph. MCMC algorithms try and show that the intermediate hypergraphs form a Markov chain whose stationary distribution converges to the uniform distribution over the set of all graphs with the given degree sequence [12]. However, it is challenging to prove mixing-time bounds for algorithms in this family, and mixing results are known only for regular graphs [5].

### 3.2  Hypergraphs with Prescribed Degree and Dimension Sequences

Chodrow [8] proposed a hypergraph configuration modelling approach to the uniform distribution of labelled hypergraphs with prescribed degree and dimension sequence. The hypergraphs under investigation have parallel-edges but no

---

**Algorithm 1.** Constructing initial hypergraphs

---

**Input:** Degree and dimension sequences, $(a)_n$ and $(b)_m$, sorted in descending order
**Output:** Hypergraph $H = (V, E)$
1: **Initialise:** $V \leftarrow \{1, \ldots, n\}$, $E \leftarrow \phi$, $(a)^1 \leftarrow (a)_n$, $(b)^1 \leftarrow (b)_m$
2: **for** $j = 1, 2, \cdots, m$ **do**
3:     Construct edge $e_j = \{v_1, \cdots, v_{b_1^j}\}$
4:     Construct $(a)_n^{j+1}$ by reducing the first $b_1^j$ components of $(a)_n^j$ by 1.
5:     Construct $(b)^{j+1} = (b_2^j, b_3^j, \ldots, b_m^j)$
6:     $E \leftarrow E \cup \{e_j\}$
7:     Sort sequence $(a)^{j+1}$ in descending order.
8: **end for**

---

self-loop. He proposed an MCMC algorithm that, as it is done in similar algorithms for graphs, sequentially switches edges of a labelled initial hypergraph satisfying the prescribed degree and dimension sequences. As the lag at which to sample a hypergraph from the Markov chain tends to infinity, he showed that the algorithm outputs uniformly at random a hypergraph from the set of all hypergraphs having the prescribed degree and dimension sequences.

However, in practice, the initial hypergraph is not always available. Additionally, due to lack of mixing time results about the chain, there is no principled guideline for choosing lag. These observations motivated us to develop both a deterministic algorithm to construct an initial hypergraph facilitating the MCMC algorithm, as well as a random generation algorithm that does not need an initial hypergraph as an alternative to the MCMC algorithm.

## 4    Construction of an Initial Hypergraph

We leverage the properties elaborated in Sect. 2 to construct a hypergraph with prescribed degree and dimension sequences. The construction algorithm addresses the limitation of the MCMC algorithm [8] by providing a starting point for it. Our algorithm uses the methodology proposed by Ryser [20] for $(0, 1)$-matrices and by Gale [14] for flows in bipartite-graphs. We illustrate the pseudocode in Algorithm 1. *At each iteration, Algorithm 1 constructs the edge with the largest dimension using distinct vertices having the largest degrees.*

In Algorithm 1, the aim is to construct a hypergraph with $n$ vertices, $m$ edges, degree sequence $(a)_n$, and dimension sequence $(b)_m$. Algorithm 1 takes non-increasingly sorted sequences $(a)_n$ and $(b)_m$ as input. It initialises $(a)^1$ as $(a)_n$ and $(b)^1$ as $(b)_m$. At each iteration $j \in \{1, \ldots, m\}$, it constructs an edge by selecting $b_1^j$ distinct vertices with maximal non-zero degrees[2]. Then it constructs $(a)^{j+1}$ by reducing the degrees of the selected vertices in $(a)^j$ by 1 and refers to $(b_2^j, \ldots, b_m^j)$ as $(b)^{j+1}$. It proceeds to construct the next edge using $(a)^{j+1}$ and $(b)^{j+1}$, and continues until all $m$ edges are constructed.

---

[2] Here the ties are broken using the lexicographic order of the vertex-labels.

We prove that the construction of edge $e_j$ at every iteration $j$ is feasible, meaning, the residual sequences $(a)^{j+1}$ and $(b)^{j+1}$ are realisable by a hypergraph.

**Theorem 2.** *If the sequences $(a)^j$ and $(b)^j$ are realisable by a hypergraph with $m$ edges and $n$ vertices, the sequences $(a)^{j+1}$ and $(b)^{j+1}$, constructed at iteration $j$, are realisable by a hypergraph with $(m-1)$ edges and $n$ vertices.*

**Proof Sketch.** We prove the theorem by induction on $m$. If $m = 0$, the algorithm terminates with a hypergraph with empty edges ($E = \phi$), which is the only hypergraph with 0 edges and $n$ vertices.

Suppose $m > 0$. By induction hypothesis, $(a)^j, (b)^j$ are realisable by a hypergraph $H$ with $m$ edges. Taking an incidence matrix $M$ of $H$ and applying Theorem 1, we get $(a)^j \prec (\bar{b})^j$. By construction, $(a)^{j+1}$ is the same as $(a)^j$ except the first $b_1^j$ components are reduced by 1. By construction of $(b)^{j+1}$ and Lemma 1, the conjugate $(\bar{b})^{j+1}$ is the same as $(\bar{b})^j$ except the first $b_1^j$ components reduced by 1. Thus Lemma 2 implies that $(a)^{j+1} \prec (\bar{b})^{j+1}$. By Theorem 1, an $(m-1) \times n$-dimensional incidence matrix $M'$ of some hypergraph $H'$ exists that realises sequences $(a)^{j+1}, (b)^{j+1}$.

# 5    Random Generation of Hypergraphs

In this section, we propose a random generation algorithm (Algorithm 2) using the characterisation (Theorem 1) for $(0, 1)$-matrices. In Algorithm 2, we iteratively construct edges in descending order of cardinality and stochastically assign the vertices to the edges such that Theorem 1 is satisfied. Algorithm 2 leverages design methods proposed for $(0, 1)$-matrices in [7].

Three observations are central to the development of Algorithm 2.

**Observation 1.** *If there are two sequences $(b)^j = (b_1^j, b_2^j, \ldots, b_m^j)$ and $(b)^{j+1} = (b_2^j, \ldots, b_m^j)$, Lemma 1 implies that we can construct the conjugate sequence of $(b)^{j+1}$, namely $(\bar{b})^{j+1}$, from the conjugate sequence of $(b)^j$, namely $(\bar{b})^j$, by reducing first $b_1^j$ components of $(\bar{b})^j$ by 1.*

**Observation 2.** *If we randomly select $K$ non-zero components from $(a)^j$ whose indices are $i_1, \ldots, i_K$ and reduce them by 1, we obtain a residual sequence $(a)^{j+1}$. If we select those $K$ components in such a way that after reduction the dominance $(a)^{j+1} \prec (\bar{b})^{j+1}$ holds, we can construct an $(m-1) \times n$-dimensional $(0, 1)$-matrix with residual column-sums $(a)^{j+1}$ and row-sums $(b)^{j+1}$. This is direct consequence of Gale-Rysers theorem (Theorem 1). The constructed $(0, 1)$-matrix is an incidence matrix of a hypergraph with $m - 1$ edges and $n$ vertices having degree sequence $(a)^{j+1}$ and dimension sequence $(b)^{j+1}$.*

**Observation 3.** *Since our interest is to reduce $K$ non-zero components of $(a)^j$ by 1 while preserving the dominance $(a)^{j+1} \prec (\bar{b})^{j+1}$, we search for the indices*

---

**Algorithm 2.** Generating random hypergraphs

---

**Input:** Degree and dimension sequences, $(a)_n$ and $(b)_m$, sorted in descending order
**Output:** Hypergraph $H = (V, E)$
  1: **Initialise:** $V \leftarrow \{1, \ldots, n\}$, $E \leftarrow \phi$, $(a)^1 \leftarrow (a)_n$, $(b)^1 \leftarrow (b)_m$.
  2: $(\bar{b})^1 \leftarrow$ conjugate sequence of $(b)^1$
  3: **for** $j = 1, 2, \cdots, m$ **do**
  4:    Construct $(\bar{b})^{j+1}$ from $(\bar{b})^j$ by reducing first $b_1^j$ components in $(\bar{b})^j$ by 1.
  5:    Compute critical indices $\{k_1^j, k_2^j, \ldots\}$ where $(a)^j \not\prec (\bar{b})^{j+1}$ (Equation (1)).
  6:    Compute corresponding margins of violation $\{n_1^j, n_2^j, \ldots\}$ (Equation (2)).
  7:    $e_j \leftarrow \phi$, $k_0^j \leftarrow 0$
  8:    **while** $k_i^j \in \{k_1^j, k_2^j, \ldots\}$ **do**
  9:        $o_i \rightarrow$ An integer sampled from $[n_i^j - |e_j|, \min(b_1^j - |e_j|, k_i^j - k_{i-1}^j)]$ uniformly at random.
10:       $O_i \rightarrow o_i$ indices selected from $\mathcal{I}_i^j = [k_{i-1}^j + 1, k_i^j]$ uniformly at random.
11:       $e_j \leftarrow e_j \cup O_i$
12:       Reduce components in $(a)^j$ at positions $O$ by 1.
13:    **end while**
14:    $E \leftarrow E \cup \{e_j\}$.
15:    $(a)^{j+1} \leftarrow (a)^j$ sorted in descending order.
16:    Construct $(b)^{j+1} = (b_2^j, b_3^j, \ldots, b_m^j)$.
17: **end for**

---

*in* $(a)^j$ *where the violation of dominance* $(a)^j \not\prec (\bar{b})^{j+1}$ *occur. We label an index* $1 \le k < n$ *to be **critical** if*

$$\sum_{i=1}^{k} a_i^j > \sum_{i=1}^{k} \bar{b}_i^{j+1}. \tag{1}$$

$k$ *being a critical index implies that in order to preserve dominance* $(a)^{j+1} \prec (\bar{b}^{j+1})$ *within integer interval* $[1, k]$, *we need to reduce at least*

$$n \triangleq \sum_{i=1}^{k} a_i^j - \sum_{i=1}^{k} \bar{b}_i^{j+1} \tag{2}$$

*number of* 1*'s at or before index* $k$ *in* $(a)^j$. *We say* $n$ *is the **margin of violation** corresponding to the critical index* $k$. *At every iteration, we enlist all the critical indices and their corresponding margins of violation.*

    Algorithm 2 takes the degree and dimension sequences, $(a)_n$ and $(b)_m$ respectively, sorted in descending order as input. We refer to them as $(a)^1 = (a)_n$ and $(b)^1 = (b)_m$ (Line 1). Following that, it constructs the conjugate $(\bar{b})^1$ of the initial dimension sequence $(b)^1$ (Line 2).

    At each iteration $j \in \{1, \ldots, m\}$, the algorithm constructs a conjugate sequence for dimensions of $(m - j)$ edges, namely $(\bar{b})^{j+1}$, from the conjugate sequence for dimensions of $(m - j + 1)$ edges, namely $\bar{b}^j$, by reducing the first $b_1^j$ components in $(\bar{b})^j$ by 1 (Line 4). This is a consequence of Observation 1.

Following Observation 3, Algorithm 2 uses $(a)^j$ and $(\bar{b})^{j+1}$ to compute all the critical indices $\{k_1^j, k_2^j, \ldots\}$ (Line 5) and their corresponding margins of violations $\{n_1^j, n_2^j, \ldots\}$ (Line 6). The critical indices partition $\{1, \ldots, n\}$ in integer intervals $\mathcal{I}_i^j \triangleq [k_{i-1}^j + 1, k_i^j]$.

Now, we select indices from these partitions and aggregate them to resolve the critical indices. These selected indices construct a new edge $e_j$. Following Observation 2, this operation would reduce the problem of generating $(m-j+1)$ edges satisfying $(a)^j$ and $(b)^j$ to generating $(m-j)$ edges satisfying $(a)^{j+1}$ and $(b)^{j+1}$ conditioned on $e_j$.

Specifically, in Line 7, the algorithm begins the edge construction considering the edge $e_j$ to be empty. In Lines 8–13, Algorithm 2 selects *batches of vertices from integer interval $\mathcal{I}_i^j$ of indices and reduce 1 from them till all the critical vertices $k_i^j$'s are considered*. As these batches of vertices are selected, they are incrementally added to $e_j$.

Now, we elaborate selection of the batches of vertices from these intervals as executed in Lines 9–10. At the $i^{th}$ step of selecting vertices, the algorithm uniformly at random select $o_i$ indices from $\mathcal{I}_i^j$. $o_i$ is an integer uniformly sampled from the following lower and upper bounds:

- *Lower bound:* Since at least $n_i^j$ vertices have to be selected from $[1, k_i^j]$ to reinstate dominance and $|e_j|$ vertices have already been selected from $[1, k_{i-1}^j]$, the algorithm needs to select at least $n_i^j - |e_j|$ vertices from $\mathcal{I}_i^j$
- *Upper bound:* There are $(k_i^j - k_{i-1}^j)$ indices in interval $\mathcal{I}_i^j$. After selecting $|e_j|$ vertices, the algorithm can not select more than $b_1^j - |e_j|$ vertices. Thus, the maximum number of vertices selected from $\mathcal{I}_i^j$ is $\min(k_i^j - k_{i-1}^j, b_1^j - |e_j|)$.

Subsequently, the algorithm adds the $o_i$ vertices at those indices to the partially constructed edge $e_j$ (Line 11) and reduce the components at those selected indices in sequence $(a)^j$ by 1 (Line 12).

After adding the edge $e_j$ to the edge set $E$ (Line 14), the algorithm sorts $(a)^j$ in descending order to construct $(a)^{j+1}$, removes $b_1^j$ from $(b)^j$ to construct $(b)^{j+1}$ (Line 15–16). In next iteration, the algorithm focuses on generating $(m-j)$ edges satisfying $(a)^{j+1}$ and $(b)^{j+1}$ conditioned on $e_j$.

In order to prove correctness of Algorithm 2, we prove Theorems 3 and 4.

**Theorem 3.** *If the sequences $(a)^j$ and $(b)^j$ are realisable by a hypergraph with $m$ edges and $n$ vertices, the sequences $(a)^{j+1}$ and $(b)^{j+1}$ as constructed by the algorithm at iteration $j+1$ are realisable by a hypergraph with $(m-1)$ edges and $n$ vertices.*

**Proof Sketch.** This proof is similar to the proof of Theorem 2 in spirit. The only difference is in the inductive step, where we need to prove that the choice of batches of vertices leads to $(a)^{j+1}$ and $(b)^{j+1}$ such that $(a)^{j+1} \prec (\bar{b})^{j+1}$.

After reducing 1 from the selected indices in $(a)^j$, the resulting sequence $(a)^{j+1}$ follows the inequality $\sum_{i=1}^{k} a_i^{j+1} \leq \sum_{i=1}^{k} \bar{b}_i^{j+1}$ at every index $k < n$.

Following Eq. 2, if index $k$ is critical, $\sum_{i=1}^{k} a_i^{j+1} \leq \sum_{i=1}^{k} a_i^j - n_k = \sum_{i=1}^{k} \bar{b}^{j+1}$.
If $k$ is not critical, $\sum_{i=1}^{k} a_i^{j+1} < \sum_{i=1}^{k} a_i^j \leq \sum_{i=1}^{k} \bar{b}^{j+1}$ by Eq. (1). After all the
critical indices are considered, $\sum_{i=1}^{n} a_i^{j+1} = (\sum_{i=1}^{n} a_i^j) - b_1^j = (\sum_{i=1}^{n} b_i^j) - b_1^j = \sum_{i=1}^{n} b_i^{j+1}$. Consequently, $(a)^{j+1} \prec (\bar{b})^{j+1}$.

**Theorem 4.** *Algorithm 2 constructs every hypergraph realisation of $(a)_n, (b)_m$ with a non-zero probability.*

**Proof sketch.** Let us begin with an arbitrary hypergraph realisation $H_1 = (V, E_1 = \{e_1, \ldots, e_m\})$ of sequences $(a)^1 = (a)_n, (b)^1 = (b)_m$ such that $|e_1| \geq |e_2| \geq \ldots \geq |e_m|$.

At iteration 1, Algorithm 2 allocates vertices to edge $e_1$ with a probability
$\mathbb{P}(e_1) = \frac{o_1}{k_1(\min(b_1^1, k_1^1) - n_1^1 + 1)} \frac{o_2}{(k_2^1 - k_1^1)(\min(k_2^1 - k_1^1, b_1^1 - o_1) - (n_2^1 - o_1) + 1)} \cdots \mathbb{P}(e_1)$ is non-zero. Compute the conditional probabilities $\mathbb{P}(e_2|e_1), \ldots, \mathbb{P}(e_m|e_{m-1}, \ldots, e_1)$ in a similar manner. Each of the probabilities is non-zero as a consequence of Theorem 3. *The joint probability* with which the algorithm constructs the edge-sequence $E_1 \triangleq \{e_1, \ldots, e_m\}$ is $\mathbb{P}(E_1) \triangleq \mathbb{P}(e_1)\mathbb{P}(e_2|e_1)\ldots\mathbb{P}(e_m|e_{m-1}, \ldots, e_1)$. $\mathbb{P}(E_1) > 0$ as it is a product of non-zero terms. There are $c(E_1) = \frac{m!}{\prod_j mult^{E_1}(e_j)!}$ distinct permutations of $E_1$ that result in the same hypergraph as $H_1$. Here, $mult^{E_1}(e_j)$ is the multiplicity of edge $e_j$ in multiset $E_1$. Let $[E_1]$ be the set of all permutations of $E_1$. Thus, the algorithm constructs $H_1$ with probability $\mathbb{P}(H_1) \triangleq \sum_{E \in [E_1]} \mathbb{P}(E)$. $\mathbb{P}(H_1)$ being a sum of non-zero terms, is non-zero.

## 6    Self-Normalised Importance Sampling Estimator

In practice, it is desirable to apply a generation algorithm that samples hypergraphs from an uniform distribution over the population of hypergraphs $\mathcal{H}_{ab}$ having the prescribed degree and dimension sequences $(a)_n$ and $(b)_m$. Uniform generation is desirable, as uniformly generated sample hypergraphs from $\mathcal{H}_{ab}$ can be used to estimate properties of the hypergraph population $\mathcal{H}_{ab}$.

However, enumerating all hypergraphs from the population $\mathcal{H}_{ab}$ is computationally infeasible as the problem of explicit enumeration of $(0, 1)$-matrices with given row- and column-sums is #P-hard [11]. This result not only makes unbiased estimation of properties of $\mathcal{H}_{ab}$ computationally infeasible but also hardens the validation of uniformity or unbiasedness of any random generation algorithm. Testing whether a generation algorithm is uniform using state-of-the art algorithms for uniformity testing [1] for the unknown discrete space $\mathcal{H}_{ab}$ is also computationally infeasible due to the astronomically large number of samples required by the testing algorithm.

The inaccessibility of population space $\mathcal{H}_{ab}$ motivates us to design an *importance sampling based estimator*. We use this estimator to estimate properties of hypergraphs in $\mathcal{H}_{ab}$ even if the induced distribution of generation algorithm is not uniform. Importance sampling assigns weights to estimates derived from non-uniformly generated hypergraph samples using the probability at which the hypergraphs are generated. Importance sampling has been adopted to design estimators for properties of matrices [7] and graphs [5]. We adopt it to design an estimator for properties of hypergraphs.

*Stepwise Design of the Estimator.* Let the uniform distribution over hypergraphs $H \in \mathcal{H}_{ab}$ be $\mathcal{U}(H) \triangleq \frac{1}{|\mathcal{H}_{ab}|}$. We are interested in estimating expected value $\mathbb{E}_{\mathcal{U}}[f]$ of some hypergraph property $f : \mathcal{H}_{ab} \to \mathbb{R}$. For example, $f$ can be the average clustering coefficient of the projection of the hypergraphs to graphs. If we are able to access $\mathcal{U}$ and draw $N$ i.i.d samples $H'_1, \ldots, H'_N$ accordingly, the Monte Carlo estimate of $\mu(f) \triangleq \mathbb{E}_{\mathcal{U}}[f]$ is $\frac{1}{N} \sum_{i=1}^{N} f(H'_i)$. In practice, computing $\mu(f)$ is not feasible as $|\mathcal{H}_{ab}|$ is computationally infeasible to compute.

Thus, we draw $N$ independent edge-sequences $E_1, \ldots, E_N$ from the space of edge-sequences $\mathcal{E}_{ab}$ leading to the hypergraphs in $\mathcal{H}_{ab}$. Using Algorithm 2, we generate $N$ such edge-sequences $\{E_i\}_{i=1}^{N}$ with probabilities $\{\mathbb{P}(E_i)\}_{i=1}^{N}$ respectively. We denote the hypergraph associated to an edge-sequence $E_i$ as $H(E_i)$. The uniform distribution $\mathcal{U}$ over the space of hypergraphs $\mathcal{H}_{ab}$ induces a distribution denoted as $\hat{\mathcal{U}}$ over the space of edge-sequences $\mathcal{E}_{ab}$. Subsequently, we evaluate property $f$ on the generated hypergraphs $\{H(E_i)\}_{i=1}^{N}$ and apply Eq. 3 to estimate the population mean $\mu(f)$.

$$\hat{\mu}(f) = \frac{1}{N} \sum_{i=1}^{N} \frac{\hat{\mathcal{U}}(E_i)}{\mathbb{P}(E_i)} f(H(E_i)) \triangleq \sum_{i=1}^{N} w_i f(H(E_i)) \tag{3}$$

This is analogous to endowing an *importance weight* $w_i$ to a sample $H(E_i)$. The sample mean $\hat{\mu}$ is an unbiased estimator of population mean $\mu$, the proof of which we omit for brevity.

Computing $\hat{\mu}$ is infeasible, as it requires computing $\hat{\mathcal{U}}$ which again requires $|\mathcal{H}_{ab}|$ and consequently $|\mathcal{E}_{ab}|$ to be computed. Hence we adopt a self-normalised importance sampling estimator (SNIS) that uses normalised weights $w_i^{SNIS} \triangleq \frac{w_i}{\sum_i w_i}$. Although SNIS is a biased estimator, it works well in practice [5,7].

$$\tilde{\mu}(f) \triangleq \frac{\sum_{i=1}^{N} \frac{\hat{\mathcal{U}}(E_i)}{\mathbb{P}(E_i)} f(H(E_i))}{\sum_{i=1}^{N} \frac{\hat{\mathcal{U}}(E_i)}{\mathbb{P}(E_i)}} = \sum_{i=1}^{N} \frac{\frac{1}{\mathbb{P}(E_i)}}{\sum_{i=1}^{N} \frac{1}{\mathbb{P}(E_i)}} f(H(E_i)) \triangleq \sum_{i=1}^{N} w_i^{SNIS} f(H(E_i)) \tag{SNIS}$$

The effectiveness of the importance sampling estimator $\hat{\mu}$ in estimating $f$ is theoretically defined as the effective sampling size $ESS \triangleq N \frac{Var[\mu(f)]}{Var[\hat{\mu}(f)]}$ and often approximated for SNIS estimator $\tilde{\mu}$ as $\frac{1}{\sum_{i=1}^{N} (w_i^{SNIS})^2}$ [16]. ESS represents the number of i.i.d samples from $\mathcal{U}$ required to obtain a Monte Carlo estimator $\tilde{\mu}$ with the same accuracy as that of the uniform estimator $\mu$.

## 7    Performance Evaluation

In order to evaluate the effectiveness of Algorithm 2, we generate multiple random hypergraphs, project the random hypergraphs into simple unweighted graphs, and empirically estimate ($\tilde{\mu}(CC)$) the *average clustering coefficient* ($CC$) on the projected graphs. For simplicity, *we use the alias SNIS to imply the algorithmic pipeline of generating several sample hypergraphs using Algorithm 2 and then applying the estimator of Equation SNIS*.

In order to evaluate the efficiency of our generation algorithm (Algorithm 2), we measure the CPU-time to generate a certain number of random hypergraphs on different datasets.

## 7.1  Datasets

We use six graphs and two hypergraph datasets to evaluate the performance of Algorithm 2 and compare with that of the MCMC algorithm.

*Graphs.* We use the pseudo-fractal family of scale-free simple graphs [10]. Pseudo-fractal graphs are a family ($G_t$), for integer $t$, of simple graphs where every graph $G_t$ has $3^t, \ldots, 3^2, 3, 3$ vertices of degree $2, \ldots, 2^t, 2^{t+1}$ respectively. The average clustering coefficient $CC_t$ of graph $G_t$ is $\frac{4}{5}\frac{6^t+3/2}{2^t(2^t+1)}$ and approaches $4/5$ as $t$ grows [10]. We are unaware of any analytical form for the average clustering coefficient of projected random graphs[3] generated following the same degree sequence as $G_t$. However, we observe (Fig. 1) that the empirical expected value of $CC_t$ converges to ~0.27 as $t$ grows. We construct six graphs $\{G_1, \ldots, G_6\}$ from this family. $\{G_1, \ldots, G_6\}$ have degree sequences of sizes 6, 15, 42, 123, 366 and 1095 respectively, and dimension sequence of sizes 9, 27, 81, 243, 729 and 2187 respectively. The dimension sequence of each graph is a sequence of 2's.

*Hypergraphs.* We use the Enron email correspondences and legislative bills in US congress as hypergraph datasets [3](https://www.cs.cornell.edu/~arb/data/). In *Enron* dataset, the vertices are email addresses at Enron and an edge is comprised of the sender and all recipients of the email. The degree and dimension sequences are of sizes 4423 and 15653 respectively. In *congress-bills* dataset, the vertices are congresspersons and an edge is comprised of the sponsor and co-sponsors (supporters) of a legislative bill put forth in both the House of Representatives and the Senate. The degree and dimension sequences are of sizes 1718 and 260851 respectively.

## 7.2  Competing Algorithms

We compare the performance of *SNIS algorithm*, i.e. the SNIS estimator built on our random generation algorithm, with the *MCMC algorithm* [8]. We make two design choices regarding the MCMC algorithm. At first, as the choice for initial hypergraph, we use our construction Algorithm 2. Secondly, as the choice for how many iterations to run the Markov chain, we perform autocorrelation analysis on the Markov chain to select a lag value $l$. After selecting $l$, we select

---

[3] Parallel-edges are absent after projection.

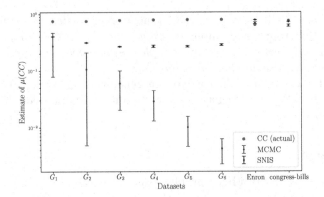

**Fig. 1.** Average clustering coefficients (in log-scale) of the projected random hypergraphs of different datasets and corresponding estimates $\mu(CC)$ using SNIS and MCMC algorithms.

random hypergraphs from the chain at every $l$-th hop until required number of hypergraphs are generated. Following standard autocorrelation analysis on Markov chain literature [9], $l$ is selected as the lag at which the autocorrelation function of average clustering coefficient estimate drops below 0.001. On datasets $G_1$, $G_2$, $G_3$, $G_4$, $G_5$, $G_6$, Enron and congress-bills, we observed and used lag values of 17, 23, 115, 129, 90, 304, 9958 and 905 respectively.

### 7.3   Effectiveness

*Comparative Analysis of Estimates of* $\mu(CC)$. On graph dataset $G_1$–$G_6$, we construct 500 random graphs (without loops) using both MCMC and Algorithm 2. On dataset *Enron* and *congress-bills*, we generate 100 and 20 random hypergraphs (without loops) respectively using both MCMC and Algorithm 2. In Fig. 1, we illustrate $CC$ estimates derived using SNIS, MCMC and the actual dataset. In Fig. 1, we observe that on average after projection the value of clustering coefficient of the multi-graph is much less than that of a simple graph. We also observe that the average clustering coefficient for the hypergraphs empirically converge to 0.27 while the average clustering coefficient of corresponding simple graphs converge to 0.8. This observation is rather expected as parallel-edges decrease the number of triadic closures that would have existed in simple graph. We also observe that, the standard deviation of the SNIS estimates are in significantly smaller than that of the MCMC estimates and closer to the $CC$ of actual data. On Enron and congress-bills hypergraphs, MCMC and SNIS yield comparable estimate for CC. *Figure 1 indicates that in practice the efficiency and stability of SNIS is either competitive or better than that of MCMC.*

*Effective Sample Sizes of Estimates of* $\mu(CC)$. Effective sample size (ESS) (Sect. 6) represents the number of i.i.d samples from an uniform sampler required to obtain a uniform Monte Carlo estimator with the same accuracy

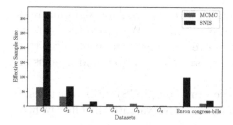

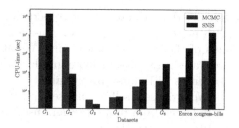

**Fig. 2.** Effective sample sizes of SNIS and MCMC algorithms on $G_1$–$G_6$, Enron and congress-bills datasets. Higher effective sample size indicates better quality of samples.

**Fig. 3.** CPU-time (in second, log-scale) to generate 500 hypergraphs for $G_1$–$G_6$, 100 hypergraphs for Enron and 20 hypergraphs for congress-bills datasets. Lower CPU-time is better.

as that of the SNIS estimator $E_p$. ESS of SNIS algorithm is approximated by $(\sum_{i=1}^{N}(w_i^{SNIS})^2)^{-1}$. The ESS of MCMC samples is defined as $\frac{N}{1+2\sum_{l=1}^{\infty}\rho(CC^l)}$ [9], where $\rho(CC^l)$ is the *autocorrelation* function at lag $l$. We consider the summation up-to the lag value for which the autocorrelation drops less than 0.001. We compute the ESS of estimate of $CC$ from both MCMC and SNIS algorithms and plot them in Fig. 2. In Fig. 2, we observe that the SNIS estimate of $CC$ exhibits higher effective sample size than the estimate using MCMC algorithm. *This observation implies that one can estimate $CC$ using much less number of SNIS samples than MCMC samples.* Although the distinction is not much when the hypergraphs are dense, as apparent from similar values of SNIS for graphs $G_4$, $G_5$ and $G_6$.

### 7.4  Efficiency

We measure the total CPU time (in seconds) taken by the MCMC and Algorithm 2 to generate 500 random graphs for the datasets $G_1$–$G_6$, 100 random hypergraphs for Enron dataset, and 20 random hypergraphs for congress-bills datasets respectively. We plot the CPU times in Fig. 3 for the datasets under consideration. In Fig. 3, we observe that the MCMC algorithm is time-efficient than Algorithm 2. In particular, it takes less CPU time in generating random hypergraphs with relatively large number of vertices and edges. However, since each run of Algorithm 2 generates hypergraphs independently from previous runs, multiple such hypergraphs can be generated in parallel for the purpose of property estimation. However, such generation is not possible using MCMC algorithm, as previously generated hypergraph are used to switch edges and generate a new hypergraph. We leave potential parallelism as a future work.

## 8  Conclusion

We present two algorithms for construction as well as random generation of hypergraphs with prescribed degree and dimension sequences. Our algorithms leverage the transposition of properties and algorithms devised for $(0, 1)$-matrices with prescribed row- and column-sums to hypergraphs. We prove the correctness of the proposed algorithms. We also prove that the generation algorithm generates any random hypergraph following prescribed degree and dimension sequences with non-zero probability.

We propose a self-normalised importance sampling (SNIS) estimator to estimate hypergraph properties and use it to empirically evaluate the effectiveness of random generation.

We compare the effectiveness of the generation algorithm by comparing the SNIS and MCMC estimates of the average clustering coefficient of the projected graphs obtained from the family of hypergraphs having prescribed degree and dimension sequences. As another measure of quality, we compare the effective sample sizes of the SNIS and MCMC estimates.

Experimental results reveal that the SNIS estimates are often more accurate and stable at estimating the average clustering coefficient and have higher effective sample sizes compared to the MCMC estimates. Although the present implementation of our generation algorithm takes longer to generate the same number of samples than the MCMC algorithm, we are currently devising a parallel version of our algorithm.

**Acknowledgement.** This work was funded by Singapore Institute for Data Science under project WATCHA, by MOE2017-T2-1-141 grant from Singapore Ministry of Education, by WASP-NTU joint grant for Adversarial Machine Learning in Big Data Era by the Knut and Alice Wallenberg Foundation, and by National Research Foundation (NRF) Singapore under its Corporate Laboratory@University Scheme, National University of Singapore, and Singapore Telecommunications Ltd.

# References

1. Batu, T., Canonne, C.L.: Generalized uniformity testing. In: 2017 IEEE 58th Annual Symposium on Foundations of Computer Science (FOCS). IEEE (2017)
2. Bender, E.A., Canfield, E.R.: The asymptotic number of labeled graphs with given degree sequences. J. Combin. Theory, Ser. A **24**(3), 296–307 (1978)
3. Benson, A.R., Abebe, R., Schaub, M.T., Jadbabaie, A., Kleinberg, J.: Simplicial closure and higher-order link prediction. Proc. Nat. Acad. Sci. **115**(48), E11221–E11230 (2018)
4. Berge, C.: Graphs and Hypergraphs. Elsevier Science Ltd. (1985)
5. Blitzstein, J., Diaconis, P.: A sequential importance sampling algorithm for generating random graphs with prescribed degrees. Internet Math. **6**(4), 489–522 (2011)
6. Bollobás, B.: A probabilistic proof of an asymptotic formula for the number of labelled regular graphs. Eur. J. Comb. **1**(4), 311–316 (1980)
7. Chen, Y., Diaconis, P., Holmes, S.P., Liu, J.S.: Sequential Monte Carlo methods for statistical analysis of tables. J. Am. Stat. Assoc. **100**(469), 109–120 (2005)
8. Chodrow, P.S.: Configuration models of random hypergraphs. arXiv preprint arXiv:1902.09302 pp. 1–20 (2019)
9. Cowles, M.K., Carlin, B.P.: Markov chain Monte Carlo convergence diagnostics: a comparative review. J. Am. Stat. Assoc. **91**(434), 883–904 (1996)

10. Dorogovtsev, S.N., Goltsev, A.V., Mendes, J.F.F.: Pseudofractal scale-free web. Phys. Rev. E **65**(6), 066122 (2002)
11. Dyer, M., Kannan, R., Mount, J.: Sampling contingency tables. Random Struct. Algorithms **10**(4), 487–506 (1997)
12. Fosdick, B.K., Larremore, D.B., Nishimura, J., Ugander, J.: Configuring random graph models with fixed degree sequences. SIAM Rev. **60**(2), 315–355 (2018)
13. Fulkerson, D.R., Ryser, H.J.: Multiplicities and minimal widths for (0, 1)-matrices. Can. J. Math. **14**, 498–508 (1962)
14. Gale, D., et al.: A theorem on flows in networks. Pac. J. Math. **7**(2), 1073–1082 (1957)
15. Klamt, S., Haus, U.U., Theis, F.: Hypergraphs and cellular networks. PLoS Comput. Biol. **5**(5), e1000385 (2009)
16. Kong, A.: A note on importance sampling using standardized weights. University of Chicago, Department of Statistics, Technical report 348 (1992)
17. Marshall, A.W., Olkin, I., Arnold, B.C.: Inequalities: Theory of Majorization and Its Applications, vol. 143. Springer, New York (1979). https://doi.org/10.1007/978-0-387-68276-1
18. Milo, R., Kashtan, N., Itzkovitz, S., Newman, M.E., Alon, U.: On the uniform generation of random graphs with prescribed degree sequences. arXiv preprint cond-mat/0312028 (2003)
19. Newman, M.: Networks: An Introduction. Oxford University Press, Oxford (2018)
20. Ryser, H.: Combinatorial properties of matrices of zeros and ones. Can. J. Math. **9**, 371–377 (1957)
21. Yang, W., Wang, G., Bhuiyan, M., Choo, K.: Hypergraph partitioning for social networks based on information entropy modularity. J. Netw. Comput. Appl. **86**, 59–71 (2016)
22. Wang, Y., Zheng, B.: Hypergraph index: an index for context-aware nearest neighbor query on social networks. Soc. Netw. Anal. Min. **3**(4), 813–828 (2013)

# Knowledge Discovery

# A Model-Agnostic Recommendation Explanation System Based on Knowledge Graph

Yuhao Chen[ID] and Jun Miyazaki[✉][ID]

Tokyo Institute of Technology, 2-12-1 Ookayama, Meguro-ku, Tokyo, Japan
chen@lsc.c.titech.ac.jp, miyazaki@c.titech.ac.jp

**Abstract.** Recommender systems have been gaining attention in recent decades for the ability to ease information overload. One of the main areas of concern is the explainability of recommender systems. In this paper, we propose a model-agnostic recommendation explanation system, which can improve the explainability of existing recommender systems. In the proposed system, a task-specialized knowledge graph is introduced, and the explanation is generated based on the paths between the recommended item and the user's history of interacted items. Finally, we implemented the proposed system using Wikidata and the Movie-Lens dataset. Through several case studies, we show that our system can provide more convincing and diverse personalized explanations for recommended items compared with existing systems.

**Keywords:** Recommender system · Knowledge graph · Model-agnostic · Explainability · Justification

## 1 Introduction

Due to the scale of the Internet and the rapid growth of information resources, it is becoming difficult for people to obtain desired information from a large amount of data. Recommender systems play an important role in all aspects of our daily lives as they are one of the main methods for addressing this information overload. In recent decades, recommender systems have been increasingly researched and significant progress has been made. High-quality personalized explanations of recommendations can boost trust, effectiveness, efficiency, and satisfaction [13]. As such, the explainability of recommender systems is one of the main areas of concern. Since the widely used machine learning-based recommender systems are lacking in explainability, the explainability of recommender systems has become more important than ever.

Explainability serves two purposes in recommender systems: transparency and justification.

- Transparency is the property that clarifies the recommendation process and enables users to understand the mechanism of the system.

S. Hartmann et al. (Eds.): DEXA 2020, LNCS 12392, pp. 149–163, 2020.
https://doi.org/10.1007/978-3-030-59051-2_10

**a. Model-agnostic**

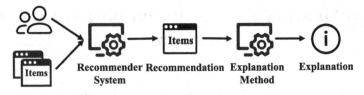

**b. Model-intrinsic**

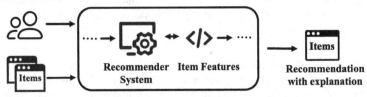

**Fig. 1.** Model-agnostic approach and Model-intrinsic approach.

– Justification is the property that provides consistent explanations independent of the recommendation module.

These two purposes correspond to two different implementations, model-agnostic and model-intrinsic approaches. The model-intrinsic method integrates the explanation process into a recommendation model, as shown in Fig. 1b. Therefore, its recommender mechanism is transparent on the some level [4]. The model-agnostic approach treats a trained recommendation module as a black-box and uses an independent method to explain the recommended item, as shown in Fig. 1a. The latter approach is particularly important for modern recommender systems because it can provide reasons for recommended items even when the recommendation module is too complex to be interpreted by uses. Specifically, in industries, most recommender systems are based on very complicated hybrid models which make it almost impossible to use model-intrinsic model to generate explanation. The model-agnostic model can also provide explainability when the system provider does not want to disclose the algorithm in the recommendation module [18]. The explainability discussed in this work is for justification purposes.

Since the model-agnostic model normally uses only information on users and items, as shown in Fig. 1a, variety of explanations is limited and the explanations lack persuasiveness. To address the aforementioned problems, we introduce a task-specialized knowledge graph to the model-agnostic approach, as shown in Fig. 2. The underlying assumption is that the information contained in datasets is not enough to generate high-quality personalized explanations. Therefore, additional *General common knowledge* is needed. In this paper, a knowledge graph is used as General common knowledge for generating high-quality personalized explanations.

**Fig. 2.** Proposed approach

## 2   Related Work

### 2.1   Recommender System

A recommender system extracts useful information for users and predicts users' preferences on their unseen items. Recommender systems are classified into collaborative filtering-based, content-based, and a hybrid of the two [1]. In particular, collaborative filtering recommender systems have been an important research direction for a long time. These systems make recommendations by capturing user-item relationships. Classic collaborative filtering algorithms, such as ItemKNN [15] and SAR [2], are all important milestones for information retrieval. Recently, deep-learning-based recommender systems, such as NCF [9], Wide and Deep Model [7], CNN+attention [8], have gained significant attention due to their ability to capture non-linear and non-trivial user-item relationships.

### 2.2   Knowledge Graphs-Based Explanation

Let $E$ be a set of entities and $R$ a set of edges labeled with relations. A knowledge graph (KG) is defined as a set of triples as follows:

$$KG = \{(h, r, t) | h, t \in E, r \in R\}, \tag{1}$$

where a triple $(h, r, t)$ indicates that the head entity $h$ and the tail entity $t$ have a relation $r$.

Due to the volume of information contained in a knowledge graph, it can help to generate intuitive and more personalized explanations for recommended items. Knowledge graph-based explainable recommender systems have been explored in [3,6,10,14]. For example, Wang et al. [14] proposed an end-to-end explainable recommender system by simulating the ripples propagating on the surface of the water on the knowledge graph. Ma et al. [10] proposed a joint learning model that integrates explainable rule induction in a knowledge graph with a rule-guided neural recommendation model to leveraged machine learning and rule learning.

However, the models in these studies were model-intrinsic. They cannot be applied directly to an efficient and stable existing black-box recommendation model.

## 2.3    Model-Agnostic Based Explanation

Due to its ability to provide explanations for complex black-box recommender systems without affecting underlying recommendation algorithms, the model-agnostic based approach has been the main research direction for explainable recommender systems [11,12,17]. For example, Peake et al. [11] proposed a model-agnostic explainable recommender system based on association rule mining. The recommendation model in the paper is treated as a black-box, because it is based on a matrix factorization method. For each user, the user history and the recommended items constitute a transaction, and the association rules are extracted from the transactions of all users. When the same recommended items can be generated from these association rules, the author uses association rules as explanations of recommended items. Singh et al. [12] proposed another perspective of constructing the model-agnostic recommender system based on learning-to-rank algorithms. A black-box ranker is trained first, and then, an explainable tree-based model is trained with the ranking labels produced by the ranker. Finally, the model-agnostic explanations are generated by the tree-based model.

Despite their efforts to provide high-quality personalized explanations, these studies have largely been unable to provide varying persuasive explanations when facing sparse user-items datasets.

## 3    Proposed System

As shown in Fig. 3, the proposed system consists of two parts, the recommendation module and the explanation module. The recommendation module generates the items to be recommended based on the underlying recommender system. The recommender system is trained by the user-item information and outputs recommended items based on the predicted user preferences. The explanation module takes the recommended item as input and outputs the explanation of why the item was recommended.

The proposed system is model-agnostic; the recommendation module can utilize any mainstream recommender system that generates recommended items based on the user's item interaction history. Therefore, we focus on the explanation mechanisms in our proposed system.

The explanation module generates explanations of recommended items in the following two steps:

1. Item Knowledge Graph Generation.
2. Explanation Generation.

### 3.1    Item Knowledge Graph Generation

A general-purpose open knowledge graph is used in the proposed system. Existing general-purpose open knowledge graphs, such as Wikidata, DBpedia, etc., are

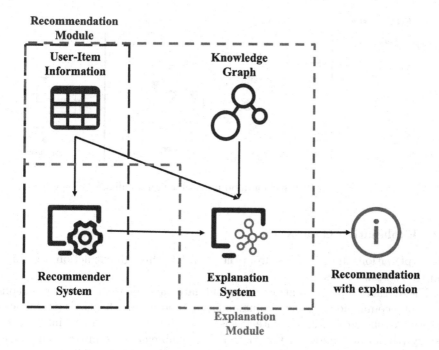

**Fig. 3.** Proposed system

enormous and have more than 100 million connections that contain noise. Therefore, they are not suitable for directly generating explanations of recommended
items. In our method, the relevant parts of a knowledge graph are extracted from
a general-purpose open knowledge graph in order to attain high-quality personalized explanations. The relevant portion of the knowledge graph is referred to
as the *Item Knowledge Graph (Item KG)* in this paper.

We define the Item KG here. The procedure for generating the Item KG is
shown on the left in Fig. 4.

First, a Domain Knowledge Graph (Domain KG), $KG_D$, is extracted from a
general-purpose open knowledge graph, which is represented by

$$KG_D = \{(h, r, t) | h, t \in E, r \in R_D\}, \tag{2}$$

where $R_D$ refers to the relations related to the recommended task. Note that the
method of choosing relations is not unique because explanations can be applied
in a variety of scenarios, even for the same recommendation task.

After extracting a set of entities $E_I$ and a set of relations $R_I$ in user-item
information, we map $E_I$ and add $R_I$ to $KG_D$. All triples unrelated to the items
in the user-item information are removed from $KG_D$

Finally, an Item KG $KG_I$ is constructed with $KD_D$ as follows:

$$KG_I = \{(h, r, t) | h, t \in E \cup E_I, r \in R_D \cup R_I\}. \tag{3}$$

The structure of the Item KG is shown in Fig. 5.

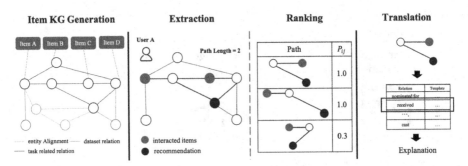

**Fig. 4.** Explanation module (Color figure online)

## 3.2 Explanation Generation

The explanations are generated from the Item KG through extraction, ranking, and translation processes.

The extraction process extracts the candidate paths on the Item KG based on the item recommended by the recommendation module and the user's interaction history. As shown in Fig. 4 (center), the target items in the user's interaction history (the orange nodes in Fig. 4) and the recommended item (the blue node in Fig. 4) are respectively set as start points and an end point. All of the paths with a length of $d$ or less (blue paths in Fig. 4) are extracted from the Item KG as the target user's candidate paths.

Typically, there are multiple paths between the item recommended to a user and his/her history of interacted items. Since it is impractical to use all the paths for generating explanations for only one recommended item, a ranking process is needed in order to choose the most relevant paths for generating effective personalized explanations. The ranking process is based on the user's preference for the entities in candidate paths.

Let $P_{i,j}$ be user $i$'s preference for the entity $E_j$. $P_{i,j}$ is calculated by the following equation:

$$P_{i,j} = \frac{|E_j^*|}{|E_j|}, \tag{4}$$

where $|E_j|$ is the total number of items directly connected to $E_j$ (except for the recommended item), and $|E_j^*|$ is the total number of $E_j$ directly connected to the items with which user $i$ has interacted (except for the recommended item). For example, user A's preferences for the entities shown in Fig. 4 (center) are 1.0 and 0.3. The ranking process selects the top $k$ entities based on $P_{i,j}$. Note that $d$ and $k$ are tuning parameters, which can be changed based on the task and the properties of the Item KG. In general, the persuasiveness of the path decreases as the path length becomes longer. For every selected entity, the proposed system randomly chooses one of the candidate paths as an explanation path.

Finally, the translation process translates the explanation path into a natural language. This is implemented by using templates prepared in advance. A variety

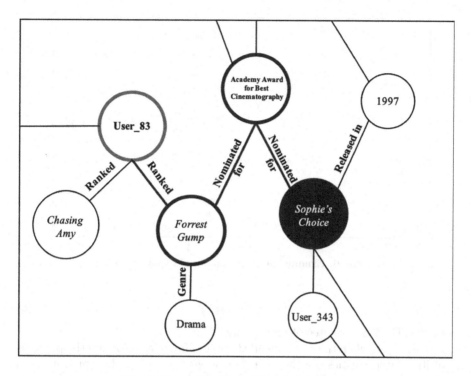

**Fig. 5.** Example of Item Knowledge Graph

of templates are created for each relation in the Item KG, in contrast to the conventional approaches which only use a single template.

An example based on the Item KG is shown in Fig. 5. Suppose that the item recommended to User_83 is movie *Sophie's Choice*, and the path selected is

User_83 -Ranked- *Forrest Gump* -Nominated for - Academy Award for Best Cinematography -Nominated for- *Sophie's Choice*

Using the template "The movie __ was also nominated for __, like the movie __ you viewed before"for the relations Ranked and Nominated for, the explanation of this recommendation can be generated as:

The movie *Sophie's Choice* was also nominated for the Academy Award for Best Cinematography, like the movie *Forrest Gump* which you viewed before.

Moreover, this approach can be extended to the case in which no candidate path exists between the target items in the user's interaction history and recommended items. In such a case, an unpersonalized explanation can be generated based on the popularity of the entities related to a recommended item. Now, we define popularity $Pop_{ij}$ for item $i$'s related entity $E_j \in \{t|(E_i, r, t) \in KG_I\}$. $Pop_{ij}$ is calculated by

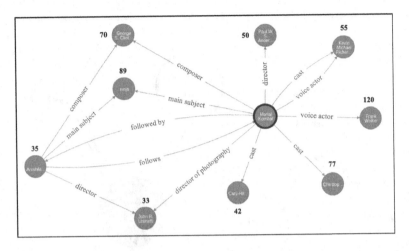

**Fig. 6.** Example of un-personalized explanation

$$Pop_{ij} = deg(E_j), \tag{5}$$

where $deg(E_j)$ is the degree of node $j$ in $KG_I$.

For example, when the recommended movie *Mortal Kombat* and the popularity of its related entities are shown in Fig. 6, entities with high popularity such as *ninja* can be chosen. Then, we use a template to generate an unpersonalized explanation, such as:

How about *Mortal Kombat* whose subject is *ninja*?

## 4    Experiment

### 4.1    Dataset

We used the MovieLens[1] dataset as user interaction history data to train recommendation models and constructed an Item KG. In order to evaluate the proposed model more comprehensively, three different sized MovieLens datasets (MovieLens-100k, MovieLens-1m, MovieLens-20m) are introduced in this experiment. Table 1 shows the detail of MovieLens datasets used in this experiment.

To construct the knowledge graph, we used Wikidata[2] as the basis of the Item KG. We extracted a Movie-related subset KG from the WikiData dataset because the original Wikidata dataset was too large to be processed. The movie-related datasets were extracted from the Wikidata archive dump[3] by using python package WiKiDataSet [5]. The following three steps are performed to extract movie-related entities and relations from the Wikidata archive dump.

---

[1] https://grouplens.org/datasets/MovieLens/.
[2] https://www.wikidata.org/wiki/Wikidata.
[3] https://dumps.wikimedia.org/wikidatawiki/entities/.

Table 1. Details of MovieLens dataset

| | # users | # movies | # ratings | # genres |
|---|---|---|---|---|
| MovieLens-100k | 943 | 1,682 | 100,000 | 19 |
| MovieLens-1m | 6,040 | 3,900 | 1,000,209 | 19 |
| MovieLens-20m | 138,493 | 27,278 | 20,000,263 | 19 |

1. Get the Wikidata entities which are sub-classes of the film topic.
2. Find the lines corresponding to the selected entities in the Wikidata archive dump.
3. Organize the data collected in Step 2 into a triplet format.

## 4.2   Knowledge Graph

Table 2. Knowledge graph statistics

| | # entities | # relations | # triples |
|---|---|---|---|
| Movie-related Dataset | 518,175 | 280 | 3,113,902 |
| Slack Domain Knowledge Graph | 477,462 | 97 | 1,865,064 |
| Strict Domain Knowledge Graph | 451,545 | 52 | 1,580,100 |
| Slack Item Knowledge Graph 100k | 30,163 | 65 | 52,719 |
| Slack Item Knowledge Graph 1m | 43,506 | 68 | 103,344 |
| Slack Item Knowledge Graph 20m | 115,879 | 88 | 438,853 |
| Strict Item Knowledge Graph 100k | 28,189 | 43 | 43,298 |
| Strict Item Knowledge Graph 1m | 40,426 | 45 | 83,490 |
| Strict Item Knowledge Graph 20m | 107,202 | 50 | 351,553 |

Two sets of relations are manually chosen from the movie-related dataset to construct the Domain KG, strict Domain KG, and slack Domain KG. The slack Domain KG contains 97 kinds of relations. We excluded the relations unsuitable for generating explanations of recommendations from the movie-related dataset, such as "box office" and "Australian Classification". The strict Domain KG contains 52 relations. We also excluded the relations that cannot generate a persuasiveness explanation in the experiments on the slack Domain KG. The two relation datasets are shown in Fig. 7 and Fig. 8. The statistics of the Domain KGs can be found in Table 2.

For each slack and strict Domain KG, three Item KGs were constructed based on MovieLens-100k, MovieLens-1m, and MovieLens-20m. We prepared a total of six types of Item KGs for the experiment.

| Relation | Relation | Relation | Relation | Relation | Relation | Relation |
|---|---|---|---|---|---|---|
| director of photo | based on | named after | inspired by | has list | time period | assistant director |
| director | main subject | main category | archives at | list of characters | member of | subclass of |
| screenwriter | nominated for | part of the series | creator | lyrics by | heritage design | copyright status |
| production com | followed by | fabrication method | sport | series spin-off | conflict | influenced by |
| cast member | different from | uses | distribution | cites | scenographer | opposite of |
| producer | movement | place in fictional | publisher | country | participant | item operated |
| country of origin | production design | voice actor | conferred by | editor | location | sponsor |
| narrative location | performer | part of | dedicated to | librettist | recorded at | plot expanded in |
| filming location | costume designer | derivative work | make-up artist | fictional universe | place publication | presented in |
| film editor | follows | characters | narrator | contributors | owned by | manifestation of |
| composer | collection | author | storyboard artist | sound designer | participant of | film script |
| distributor | soundtrack album | depicts | art director | competition class | location | commemorates |
| award received | after a work by | set in period | musical conductor | educated at | operator | |
| has quality | executive producer | theme music | choreographer | catalog | produced by | |

**Fig. 7.** Slack relationship set

| Relation | Relation | Relation | Relation | Relation | Relation | Relation |
|---|---|---|---|---|---|---|
| director of photo | based on | named after | inspired by | | | |
| director | main subject | main category | | | | |
| screenwriter | nominated for | part of the series | creator | | | |
| | followed by | | | series spin-off | conflict | influenced by |
| cast member | | uses | | cites | scenographer | |
| | | place in fictional | | | | |
| country of origin | production design | voice actor | conferred by | editor | | |
| | | part of | dedicated to | librettist | | |
| | | derivative work | | fictional universe | | |
| film editor | follows | characters | narrator | contributors | | |
| composer | | author | | | | film script |
| | | depicts | art director | | | commemorates |
| award received | after a work by | set in period | | educated at | | |
| | | theme music | | catalog | | |

**Fig. 8.** Strict relationship set

To construct the Item KG, we mapped the movies in MovieLens to the entities in the Domain KG using KB4Rec [19], a public domain linked knowledge base dataset. The KB4Rec dataset contains the mapping between the movie ID in MovieLens to the entity ID in freebase [19]. Since the Freebase-Wikidata mapping data[4] is available, MovieLens and Wikidata were integrated by connecting the two datasets. However, not all movies in MovieLens can be mapped to the Wikidata entities due to the different releases of the two datasets. Therefore, we used *movie title* and *release time* in the MovieLens dataset as keywords to complete the remaining mapping. After mapping was completed, the triples that did not contain the movies in MovieLens were removed. Since paths with a length of three or more are not helpful for generating persuasive explanations in the movie domain, we focused on paths with $d = 2$ in this experiment.

Finally, the triples of the Item KG were stored in the Neo4j graph database for further path extraction and translation processing. The paths were extracted with Python through the Neo4j APIs. Table 3 shows part of the templates used in the translation step.

---

[4] https://developers.google.com/freebase.

**Table 3.** Example of template

| Relation | Template |
|----------|----------|
| Award received | How about __? It also received __, like the movie __ which you viewed before |
| Based on | __ and __ are both based on __ |
| Inspired by | __ is inspired by __ |
| Main subject | Remember __? __ has the same topic: __ |

### 4.3 Recommendation Modules

In order to evaluate the model-agnostic properties of the proposed system, we used two conventional recommender systems and two state-of-art recommender systems mentioned in Sect. 2.1 as the recommendation modules in our system. It is very difficult to generate high-quality personalized explanations with only these selected algorithms.

- Item k-nearest neighbor (ItemKNN) [15]
- Simple Algorithm for Recommendation (SAR) [2]
- Neural Collaborative Filtering (NCF) [9]
- Wide and Deep Model (W&D) [7]

Although we have not tested our system in combination with the existing accurate recommender systems, it is reasonable to assume that our system can be applied to any recommendation algorithms.

## 5  Evaluation

### 5.1  Mean Explainability Precision

The explainability of the proposed system is evaluated by mean explainability precision (MEP) [16]. MEP calculates the average proportion of explainable items in the top-$n$ recommended items to the total number of recommended (top-$n$) items for each user to measure the precision of explainability.

$$MEP = \frac{1}{U} \sum_{u=1}^{U} \frac{N_{exp}}{L}, \tag{6}$$

where $N_{exp}$ is the number of explainable items in the top-$n$ recommendations, $L$ is the recommended (top-$n$) items for each user, and $U$ is the number of users.

In our experiment, an explainable item means at least one candidate path exists.

The results of the combined recommender systems and the KGs are shown in Table 4. W&D experiments could not be conducted on Slack Item KG 20m and Strict Item KG 20m due to an out-of-memory error.

As Item KG becomes sparse, MEPs decrease. However, even with Strict Item KG 20m the sparsest Item KG, the proposed system is still able to explain most of the recommended items.

**Table 4.** MEP@20

|                     | ItemKNN | SAR    | NCF    | W&D    |
|---------------------|---------|--------|--------|--------|
| Slack Item KG 100k  | 0.8725  | 0.9941 | 0.9935 | 0.8928 |
| Slack Item KG 1m    | 0.9167  | 0.9753 | 0.9983 | 0.9358 |
| Slack Item KG 20m   | 0.6866  | 0.9516 | 0.9823 | –      |
| Strict Item KG 100k | 0.8145  | 0.9595 | 0.9433 | 0.8621 |
| Slack Item KG 1m    | 0.7878  | 0.9614 | 0.9760 | 0.8958 |
| Strict Item KG 20m  | 0.5970  | 0.9370 | 0.9421 | –      |

## 5.2   Case Study

The results of the case studies verified that the proposed system is able to generate more diverse and high-quality personalized explanations than the conventional model-agnostic method. The following is the result for user no. 53 in MovieLens-100k with strict relations. Table 5 shows the explanations generated by different recommendation modules. Table 6 compares the proposed approach with the existing approach, where SAR is used as the recommendation module. In both tables, the orange and blue titles represent a movie rated by a user and a recommended item, respectively.

**Table 5.** Explanation generated by different recommendation module

|          | Explanation(k=1, d=2)                                                                                   |
|----------|---------------------------------------------------------------------------------------------------------|
| ItemKNN  | Living in Oblivion's cast member Dermot Mulroney is also a cast member in Capycat.                        |
| SAR      | Remember The Fifth Element? Independence Day was also nominated for Satellite Award for Best Visual Effects. |
| NCF      | The Birdcage's cast member Tim Kelleher is also a cast member in Independence Day.                        |
| W&D      | Michael Kahn is the film editor of both Schindler's List and Twister.                                     |

The proposed system is clearly able to generate a higher-quality explanation containing a large volume of information. Compared to the explanations of the recommended items generated by the conventional method, the explanations generated by the proposed system are more persuasive and natural. In addition, the proposed system can create diverse and high-quality explanations even when recommending the same item to the same user.

**Table 6.** Explanation generated by different model-agnostic method

|  | Explanation(k=1, d=2) |
|---|---|
| **Existing Method** | User No.132 and five other users who are similar to you also watched Independence Day.<br><br>Because you watched Men in Black, we recommend Independence Day. |
| **Proposed Method** | Remember The Fifth Element? Independence Day was also nominated for Satellite Award for Best Visual Effects.<br><br>Remember Men in Black's cast member Will Smith? Will Smith is also a cast member in Independence Day.<br><br>Independence Day and Men in Black have the same topic: extraterrestrial life. |

# 6   Conclusion

We proposed a model-agnostic recommendation explanation model that receives a recommendation as input and generates an explanation based on the paths in a knowledge graph. The proposed system showed promise for integrating third-party knowledge bases to enhance the explainability of existing recommender systems. The results of the various evaluations indicated that the proposed system can generate high-quality personalized explanations without affecting the performance of the existing recommender system. Further analysis revealed that the proposed system can provide more diverse explanations compared with the existing model-agnostic method.

A number of factors will be examined in future research. In this paper, the rating scores were treated as equal. However, one star and two stars can be assumed to mean that the user dislikes the movie when the maximum rating is 5 stars. Therefore, the weights of the rating relation in the Item KG should be considered in a future study. Moreover, the proposed system uses templates to translate the KG paths into a natural language. Developments in natural language processing and deep learning may enable translation by text generation. Lastly, although knowledge bases are used in this paper, the reasoning is not leveraged. In the future, a reasoning process can be introduced to enhance the quality of the explanations.

**Acknowledgments.** This work was partly supported by JSPS KAKENHI Grant Numbers 18H03242, 18H03342, and 19H01138.

# References

1. Adomavicius, G., Tuzhilin, A.: Toward the next generation of recommender systems: a survey of the state-of-the-art and possible extensions. IEEE Trans. Knowl. Data Eng. **17**(6), 734–749 (2005). https://doi.org/10.1109/TKDE.2005.99
2. Aggarwal, C.C.: Recommender Systems. Springer, Cham (2016). https://doi.org/10.1007/978-3-319-29659-3
3. Ai, Q., Azizi, V., Chen, X., Zhang, Y.: Learning heterogeneous knowledge base embeddings for explainable recommendation. ArXiv arXiv:1805.03352 (2018)
4. Balog, K., Radlinski, F., Arakelyan, S.: Transparent, scrutable and explainable user models for personalized recommendation. In: SIGIR 2019, pp. 265–274 (2019)
5. Boschin, A., Bonald, T.: WikiDataSets: standardized sub-graphs from WikiData. arXiv:1906.04536 [cs, stat], June 2019. http://arxiv.org/abs/1906.04536
6. Catherine, R., Mazaitis, K., Eskénazi, M., Cohen, W.W.: Explainable entity-based recommendations with knowledge graphs. ArXiv arXiv:1707.05254 (2017)
7. Cheng, H.T., et al.: Wide & deep learning for recommender systems. In: Proceedings of the 1st Workshop on Deep Learning for Recommender Systems DLRS 2016, pp. 7–10. Association for Computing Machinery, New York (2016)
8. Gong, Y., Zhang, Q.: Hashtag recommendation using attention-based convolutional neural network. In: Proceedings of the Twenty-Fifth International Joint Conference on Artificial Intelligence IJCAI 2016, pp. 2782–2788. AAAI Press (2016)
9. He, X., Liao, L., Zhang, H., Nie, L., Hu, X., Chua, T.S.: Neural collaborative filtering. In: Proceedings of the 26th International Conference on World Wide Web WWW 2017, pp. 173–182. International World Wide Web Conferences Steering Committee, Republic and Canton of Geneva, CHE (2017)
10. Ma, W., et al.: Jointly learning explainable rules for recommendation with knowledge graph. In: WWW 2019, pp. 1210–1221 (2019)
11. Peake, G., Wang, J.: Explanation mining: post hoc interpretability of latent factor models for recommendation systems. In: Proceedings of the 24th ACM SIGKDD International Conference on Knowledge Discovery & Data Mining KDD 2018, pp. 2060–2069. Association for Computing Machinery, New York (2018)
12. Singh, J., Anand, A.: Posthoc interpretability of learning to rank models using secondary training data. ArXiv arXiv:1806.11330 (2018)
13. Tintarev, N., Masthoff, J.: Designing and evaluating explanations for recommender systems. In: Ricci, F., Rokach, L., Shapira, B., Kantor, P.B. (eds.) Recommender Systems Handbook, pp. 479–510. Springer, Boston, MA (2011). https://doi.org/10.1007/978-0-387-85820-3_15
14. Wang, H., et al.: RippleNet: propagating user preferences on the knowledge graph for recommender systems. In: Proceedings of the 27th ACM International Conference on Information and Knowledge Management CIKM 2018, pp. 417–426. Association for Computing Machinery, New York (2018)
15. Wang, J., de Vries, A.P., Reinders, M.J.T.: Unifying user-based and item-based collaborative filtering approaches by similarity fusion. In: Proceedings of the 29th Annual International ACM SIGIR Conference on Research and Development in Information Retrieval SIGIR 2006, pp. 501–508. Association for Computing Machinery, New York, NY (2006)
16. Wang, S., Tian, H., Zhu, X., Wu, Z.: Explainable matrix factorization with constraints on neighborhood in the latent space. In: Tan, Y., Shi, Y., Tang, Q. (eds.) DMBD 2018. LNCS, vol. 10943, pp. 102–113. Springer, Cham (2018). https://doi.org/10.1007/978-3-319-93803-5_10

17. Wang, X., Chen, Y., Yang, J., Wu, L., Wu, Z., Xie, X.: A reinforcement learning framework for explainable recommendation. In: 2018 IEEE International Conference on Data Mining (ICDM), pp. 587–596, November 2018. https://doi.org/10.1109/ICDM.2018.00074
18. Zhang, Y., Chen, X.: Explainable recommendation: a survey and new perspectives. ArXiv arXiv:1804.11192 (2018)
19. Zhao, W.X., et al.: Kb4rec: a data set for linking knowledge bases with recommender systems. Data Intell. 1(2), 121–136 (2019)

# Game Theory Based Patent Infringement Detection Method

Weidong Liu[1,2(✉)], Xiaobo Liu[1], Youdong Kong[1], Zhiwei Yang[1],
and Wenbo Qiao[1]

[1] College of Computer Science, Inner Mongolia University, Hohhot, China
`cslwd@imu.edu.cn`, `liuxbimu@gmail.com`, `imukongyd@gmail.com`,
`yangzwimu@gmail.com`, `imu.qiaowb@gmail.com`
[2] Inner Mongolia Key Laboratory of Social Computing and Data Processing,
Hohhot, China

**Abstract.** With fiercely increasing competition of intellectual property, the protection of intellectual property (IP) is paid increasing attention from worldwide. Effective patent infringement detection is the foundation of patent protection. Given the big patent data, manual patent infringement detection is inefficient and error prone. The design of automatic infringement detection method faces some challenges including: (1) how to detect patent infringement by novelty and non-obviousness; (2) which parts of a patent are selected against those of its counterpart patents in infringement detection. To solve the above issues, a game theory based patent infringement detection method is proposed. In the method, both the novelty and the non-obviousness are considered when the patentees take actions. The infringement detection is a game process to find the best action. Our method is compared with the state-of-the-art method on some patent data sets. The results show that the dynamic game method outperforms the baseline method in the evaluation measurement. Such method can be applied to patent examination and patent infringement litigation.

**Keywords:** Game theory · Patent infringement · Infringement detection

## 1 Introduction

As the competition of intellectual property becomes fierce, the patent protection is paid considerable worldwide attention. As the number of patent applications and the amount of patent infringement litigation are increasing, manual patent infringement detection is inefficient and error prone, which is limited by the number of patent examiners and their professionalism. An effective automatic infringement detection method can help applicants make risk averse before the patent application and help examiners shorten the examination time.

Supported by the National Science Foundation of China (grant no. 61801251) and Natural Science Foundation of Inner Mongolia (2018BS06002).

Patent examination often checks three main characteristics including usefulness, novelty and non-obviousness. Among of the three characteristics, the novelty and non-obviousness are used to detect whether one patent infringes other patents as shown in Table 1.

**Table 1.** Principles of infringement testing

| # | $F_A$ | $F_B$ | Supplements | Is Nov | Is Non-Obv. | Dose infringes |
|---|---|---|---|---|---|---|
| 1 | a+b+c | a+b+c | | NO | NO | YES |
| 2 | a+b | a+b+c | c is common sense | YES | NO | YES |
| 3 | a+b | a+b+d | d is no-common sense | YES | YES | NO |
| 4 | a+b+c | a+b | c is common sense | YES | NO | YES |
| 5 | a+b+d | a+b | d is no-common sense | YES | YES | NO |
| 6 | a+b+c | a+b+e | c and e have non-substantial difference | NO | NO | YES |
| 7 | a+b+c | a+b+f | c and f have substantial difference | YES | YES | NO |
| 8 | a+b+c | d+e+f | features are completely different | YES | YES | NO |

Assume the patent application time of the claim A is prior to that of the claim B. Given the features of claim A ($F_A$) and the features of claim B ($F_B$), claim B infringes claim A if $F_B$ has no novelty or non-obviousness compared with $F_A$. Given $F_A = \{a, b\}$ and $F_B = \{a, b, c\}$ in the No. 2 of Table 1, the claim B infringes the claim A since the common scene $c$ broken the non-obviousness of claim B.

The related works on infringement detection methods can be divided into the classification-based methods, the clustering-based methods and the similarity-based methods [1,3–6,9,11]. Compared with the classification-based methods and the clustering-based methods, the similarity-based infringement detection methods are most used in practice, which include top-down representation and bottom-up similarity:

1. Top-down representation [2,9,10,13,16–18]. Some methods represent a patent by a set of keywords, a set of SAO, a set of sentences or a set of claims. Some methods represent a patent by a tree which is consisted of some structured parts (e.g. date, IPC code) and some unstructured parts (e.g. claims).
2. Bottom-up similarity [7–9,13–15]. The bottom-up similarity methods include the similarity of keyword embedding (e.g. word2vec), the similarity of keyword-bag or SAO-bag or claim-bag, the similarity of a dependency tree of claims or sentences. The patent similarity is calculated by summing the weighted similarities bottom up.

However, these methods have the following limitations:

1. Inconsideration of rational decision in patent infringement detection. The previous methods make patent infringement detection without consideration of the rational decision. A suspected infringing patent can be sued by the

infringed patent with the claims as evidence. The suspected infringing patent has the right to select the features in the claims as a suitable evidence against the infringed patent. It is a non-cooperative game process and each parts always do what is best for himself/herself.

2. Incomplete specification about infringement risk and its evidences. The previous methods make patent infringement detection without the specification of infringement evidence about which claims or features infringe the counterpart patent. In fact, a concrete specification about patent infringement should include the infringement risk and some concrete evidences.

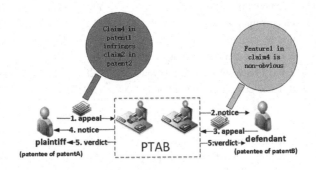

**Fig. 1.** Patent infringement detection process

Figure 1 shows a process of patent infringement detection in inter partes review. In the process, the suit begins with the patentee of the infringed patent (plaintiff) suing the infringing patent (defendant) with the claims as evidence. Then the patentee of infringing patent appeals for denying the infringement by the features in the infringing patent. The judge makes a final decision based on the evidences submitted and issues a verdict to both parties. The patentee of the infringed patent is playing non-cooperative game with the patentee of infringing patent, resulting in infringement risk and some evidences [12].

To predict patent infringement risk, we propose a game theory based infringement detection method.

1. To represent the hierarchical knowledge in patent, we propose a representation of patent, which is a tree consisting of claims and features in the patent.
2. To discover the infringement risk and give a specification, a non-cooperative dynamic game is playing for the patent infringement detection.

The remainder of the paper is organized as follows. In Sect. 2, we introduce the preliminaries including some basic definitions and problem formal definition. In Sect. 3, we propose the dynamic game model for the infringement detection. Experimental results are presented in Sect. 4. We give the conclusion and future work in Sect. 5.

## 2    Preliminaries and Problem Statement

### 2.1    Primary Knowledge

To make infringement detection, we give the patent representation which consists of some claims and features in the patent.

**Definition 1: Patent Representation, PR**

$$PR(p_i) = (p_i(c_{i,j}(f_{i,j,k})|0 \leq k < |c_{i,j}|)|0 \leq j < |p_i|) \tag{1}$$

where $p_i$ denotes the $i^{th}$ patent; $p_i(c_{i,j}|0 \leq j < |p_i|)$ denotes a patent consisted of the claims in the patent; $|p_i|$ denotes the claim number in $p_i$; $c_{i,j}(f_{i,j,k}|0 \leq k < |c_{i,j}|)$ denotes a claim consisted of the features in the claim; $|c_{i,j}|$ denotes the feature number in $c_{i,j}$.

Infringement patent pairs denote some pairs of an infringed patent and an infringing patent to be detected, which are represented by,

**Definition 2: Infringement Patent Pairs, IPP**

$$IPP = \{ipp_i|0 \leq i < |IPP|\} \tag{2}$$

$$ipp_i = \{p_i^m|0 \leq m \leq 1\} \tag{3}$$

where $ipp_i$ denotes the $i^{th}$ infringement patent pair; $|IPP|$ denotes the number of patent pairs; $p$ with $m = 0$ denotes an infringed patent; $p$ with $m = 1$ denotes an infringing patent.

To detect infringement risk of an infringement patent pair by a dynamic game, we construct a patent claim game tree for the infringement patent pair. In the game tree, the strategy set of patent $p_i^0$ consists of claim pairs, and the strategy set of patent $p_i^1$ consists of technical features in claims.

Given two patents $p_i^0$ and $p_i^1$ in $ipp_i$, its patent claim game tree is defined by,

**Definition 3: Patent Claim Game Tree, PCGT**

$$PCGT(p_i^0, p_i^1) = \{(c_{i,j}^0, c_{i,k}^1), f_{i,k,z}^1, \sigma(f_{i,k,z}^1), v^0((c_{i,j}^0, c_{i,k}^1), f_{i,k,z}^1), v^1((c_{i,j}^0, c_{i,k}^1), f_{i,k,z}^1)$$
$$|1 \leq j \leq |p_i^0|, 1 \leq k \leq |p_i^1|, 1 \leq z \leq |c_{i,k}^1|\} \tag{4}$$

where $(c_{i,j}^0, c_{i,k}^1)$ denotes the claim pair as the choice of $p_i^0$; $f_{i,k,z}^1$ denotes the feature of $c_{i,k}^1$ as the choice of $p_i^1$; $\sigma(f_{i,k,z}^1)$ denotes the weight of $f_{i,k,z}^1$ to be selected from $c_{i,k}^1$; $v^0((c_{i,j}^0, c_{i,k}^1), f_{i,k,z}^1)$ denotes the payoff of $p_i^0$ to select $(c_{i,j}^0, c_{i,k}^1)$ when $p_i^1$ selects $f_{i,k,z}^1$; $v^1((c_{i,j}^0, c_{i,k}^1), f_{i,k,z}^1)$ denotes the payoff of $p_i^1$ to select $f_{i,k,z}^1$ when $p_i^0$ selects $(c_{i,j}^0, c_{i,k}^1)$.

Figure 2 shows a patent claim game tree for the two patents $p_i^0$ and $p_i^1$ in $ipp_i$.

We list the notations of the above definitions in Table 2, which is thoroughly used in this paper.

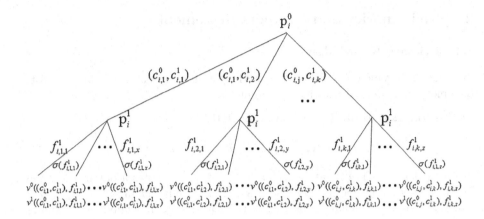

**Fig. 2.** Patent claim game tree

**Table 2.** Symbols and their meanings

| Symbols | Description |
|---------|-------------|
| $p_i$ | The $i^{th}$ patent |
| $|p_i|$ | The claim number in $p_i$ |
| $c_{i,j}$ | The $j^{th}$ claim in the $p_i$ |
| $|c_{i,j}|$ | The feature number in $c_{i,j}$ |
| $f_{i,j,x}$ | The $x^{th}$ feature in the claim $c_{i,j}$ |
| $PR(p_i)$ | Patent representation of $p_i$ |
| $ipp_i = \{p_i^0, p_i^1\}$ | An infringement patent pair including an infringed patent $p_i^0$ and an infringing patent $p_i^1$ |
| IPP | A set of infringement patent pairs |
| $PCGT(p_i^0, p_i^1)$ | A patent claim game tree of $p_i^0, p_i^1$ |
| $\sigma(f_{i,k,z}^1)$ | The weight of $f_{i,k,z}^1$ to be selected from $c_{i,k}^1$ |
| $v^0((c_{i,j}^0, c_{i,k}^1), f_{i,k,z}^1)$ | $v^0((c_{i,j}^0, c_{i,k}^1), f_{i,k,z}^1)$ denotes the payoff of $p_i^0$ to select $(c_{i,j}^0, c_{i,k}^1)$ when $p_i^1$ selects strategy $f_{i,k,z}^1$ |
| $(c_{i,j*}^0, c_{i,k*}^1)$ | The best choice for $p_i^0$ in dynamic game |
| $f_{i,k*,z*}^1$ | The best choice for $p_i^1$ in the dynamic game |

## 2.2   Problem Statement

Given a patent infringement patent pair $ipp_i = \{p_i^0, p_i^1\}$, the task of this paper is to predict which claim is the best choice for $p_i^0/p_i^1$ against his/her counterpart and the infringement risk. To solve the problem, a dynamic game based method is proposed to calculate infringement risk, which is shown in Fig. 3.

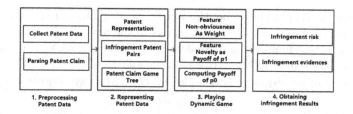

**Fig. 3.** The game process of patent infringement detection

In Fig. 3, the game process of patent infringement detection including preprocessing patent data, representing patent data, playing a dynamic game and obtaining infringement results.

We mainly focus on how to play a dynamic game and obtain the infringement results in the patent claim game tree (PCGT) by the Nash equilibrium in the dynamic game Eq. 5.

$$((c_{i,j*}^0, c_{i,k*}^1), f_{i,k*,z*}^1, v^0((c_{i,j*}^0, c_{i,k*}^1), f_{i,k*,z*}^1), v^1((c_{i,j*}^0, c_{i,k*}^1), f_{i,k*,z*}^1)) \qquad (5)$$

where $(c_{i,j*}^0, c_{i,k*}^1)$ denotes the best choice for infringed patent; $f_{i,k*,z*}^1$ denotes the best choice for infringing patent; $v^0((c_{i,j*}^0, c_{i,k*}^1), f_{i,k*,z*}^1)$ denotes the payoff of $p_i^0$ when $p_i^0$ selects $(c_{i,j*}^0, c_{i,k*}^1)$ and $p_i^1$ selects $f_{i,k*,z*}^1$; $v^1((c_{i,j*}^0, c_{i,k*}^1), f_{i,k*,z*}^1)$ denotes the payoff of $p_i^1$ when $p_i^1$ selects $f_{i,k*,z*}^1$ and $p_i^0$ selects $(c_{i,j*}^0, c_{i,k*}^1)$.
and

$$(c_{i,j*}^0, c_{i,k*}^1) = \underset{(c_{i,j}^0, c_{i,k}^1)}{\arg\max}(v^0((c_{i,j}^0, c_{i,k}^1), \sigma_{i,k}^1)) \qquad (6)$$

and

$$v^0((c_{i,j}^0, c_{i,k}^1), \sigma_{i,k}^1) = \sum_{f_{i,k,z}^1 \in c_{i,k}^1} \sigma(f_{i,k,z}^1) \times v^0((c_{i,j}^0, c_{i,k}^1), f_{i,k,z}^1) \qquad (7)$$

and

$$f_{i,k*,z*}^1 = \underset{f_{i,k*,z}^1}{\arg\max}(v^1((c_{i,j*}^0, c_{i,k*}^1), f_{i,k*,z}^1)) \qquad (8)$$

## 3    Proposed Model

To obtain the infringement risk and the evidence in Eq. 5, we model infringement detection process into the dynamic game in game trees. In the game, the action of infringed patent is to select a claim pair which is consisted of a claim from infringing patent and a claim from infringed patent to sue the infringing patent. The action of infringing patent is to select a feature from the claim to contradict the infringed patent. In the game, the weight of infringing patent action equals to feature non-obviousness, the payoff equals to the novelty.

Since the patent mainly is consisted of claims and features, which can be represented by Claim2vec and Feature2vec. In the following sections, the embedding representation of an object $O$ is denoted by $\overrightarrow{O}$. Given two objects $O_1$ and $O_2$, the distance between them is denoted by $dist(\overrightarrow{O_1}, \overrightarrow{O_2})$.

For a claim $c^1_{i,j}$ of $p^1_i$ in an infringement patent pair $ipp_i = \{p^0_i, p^1_i\}$, the non-obviousness of a feature in the claim is calculated by,

**Definition 1: Feature Non-obviousness as the Weight in Patent Claim Game Tree**

$$\sigma(f^1_{i,k,x}) = \min_{f^1_{i,k,x-} \in c^1_{i,k}} (dist(\overrightarrow{f^1_{i,k,x}}, \overrightarrow{f^1_{i,k,x-}})) \qquad (9)$$

where $f^1_{i,k,x}$ denotes a feature from $c^1_{i,k}$ of patent $p^1_i$ in $ipp_i$; $f^1_{i,k,x-}$ denotes the features of $c^1_{i,k}$ other than $f^1_{i,k,x}$; $dist(\overrightarrow{f^1_{i,k,x}}, \overrightarrow{f^1_{i,k,x-}})$ denotes distance between $f^1_{i,k,x}$ and $f^1_{i,k,x-}$.

**Hypothesis 1:** If the infringing patentee does not consider the claim novelty in infringement detection, he/she will select the most non-obvious feature from the claim in claim pair against that of his/her counterpart claim. The most non-obvious feature $f^1_{i,k,x+}$ is,

$$f^1_{i,k,x+} = \arg\max_{f^1_{i,k,x}}(\sigma(f^1_{i,k,x})) \qquad (10)$$

For a claim pair $(c^0_{i,j}, c^1_{i,k})$ in an infringement patent pair $ipp_i = \{p^0_i, p^1_i\}$, the novelty of a feature in the claim $c^1_{i,k}$ is calculated by Eq. 11 which equals to the payoff of $p^1_i$ in Patent Claim Game Tree as well.

**Definition 2: Feature Novelty as the Payoff of $p^1_i$ in Patent Claim Game Tree**

$$v^1((c^0_{i,j}, c^1_{i,k}), f^1_{i,k,z}) = \min_{f^0_{i,j,x} \in c^0_{i,j}} (dist(\overrightarrow{f^0_{i,j,x}}, \overrightarrow{f^1_{i,k,z}})) \qquad (11)$$

where $f^0_{i,j,x}$ denotes a feature from $c^0_{i,j}$ of patent $p^0_i$ in $ipp_i$; $f^1_{i,k,z}$ denotes a feature from $c^1_{i,k}$ of patent $p^1_i$ in $ipp_i$; $dist(\overrightarrow{f^0_{i,j,x}}, \overrightarrow{f^1_{i,k,z}})$ denotes distance between $f^0_{i,j,x}$ and $f^1_{i,k,z}$.

**Hypothesis 2:** The patentee will make rational decision in selecting a feature $f^1_{i,k^*,z^*}$ from the claim $c^1_{i,k^*}$ against the counterpart claim in infringement detection when he/she has known the action taken by his/her counterpart.

For a claim pair $(c^0_{i,j}, c^1_{i,k})$ of $p^0_i$ in an infringement patent pair $ipp_i = \{p^0_i, p^1_i\}$, the $p^0_i$ payoff of $(c^0_{i,j}, c^1_{i,k})$ is calculated by,

**Definition 3: Payoff of $p_i^0$ in Patent Claim Game Tree**

$$v^0((c_{i,j}^0, c_{i,k}^1), f_{i,k,z}^1) = (1 - dist(\overrightarrow{c_{i,j}^0}, \overrightarrow{c_{i,k}^1})) \times (1 - v^1((c_{i,j}^0, c_{i,k}^1), f_{i,k,z}^1)) \quad (12)$$

where $v^0((c_{i,j}^0, c_{i,k}^1), f_{i,k,z}^1)$ denotes the payoff of $p_i^0$ to select $(c_{i,j}^0, c_{i,k}^1)$; $dist(\overrightarrow{c_{i,j}^0}, \overrightarrow{c_{i,k}^1})$ denotes the distance between $c_{i,j}^0$ and $c_{i,k}^1$; $v^1((c_{i,j}^0, c_{i,k}^1), f_{i,k,z}^1)$ denotes the payoff of $p_i^1$ to select $f_{i,k,z}^1$ in $c_{i,k}^1$ when $p_i^0$ selects $(c_{i,j}^0, c_{i,k}^1)$.

Each feature in claim is represented by Feature2vec and each claim in patent is represented by Claim2vec. The distance of two features/claims equals to 1 minus their cosine value.

**Hypothesis 3:** The patentee will make rational decision in selecting a claim pair $(c_{i,j^*}^0, c_{i,k^*}^1)$ from the patents against the counterpart patent in infringement detection.

The claim pair $(c_{i,j^*}^0, c_{i,k^*}^1)$ in Eq. 6 and the feature $f_{i,k^*,z^*}^1$ in Eq. 8 are selected by the Algorithm 1 with infringement risk $v^0((c_{i,j^*}^0, c_{i,k^*}^1), f_{i,k^*,z^*}^1)$ and $v^1((c_{i,j^*}^0, c_{i,k^*}^1), f_{i,k^*,z^*}^1)$ in Eq. 5.

---

**Algorithm 1:** dynamic game process for patent infringement detection

---

**Input:** $ipp_i = \{p_i^0, p_i^1\}$
**Output:** $v^0((c_{i,j^*}^0, c_{i,k^*}^1), f_{i,k^*,z^*}^1), v^1((c_{i,j^*}^0, c_{i,k^*}^1), f_{i,k^*,z^*}^1)$
1  Construct a patent-claim-game-tree of $p_i^0$ and $p_i^1$ in Eq. 4 ;
2  Initialize weight and payoff by novelty and non-obvious in Eq. 9, Eq. 11 and Eq. 12 ;
3  Obtaining $(c_{i,j^*}^0, c_{i,k^*}^1)$ by Eq. 6;
4  Obtaining $f_{i,k^*,z^*}^1$ by Eq. 8;
5  Return $v^0((c_{i,j^*}^0, c_{i,k^*}^1), f_{i,k^*,z^*}^1)$ and $v^1((c_{i,j^*}^0, c_{i,k^*}^1), f_{i,k^*,z^*}^1)$;

---

## 4  Experiments

To validate the efficiency of our method, we design experiments as follows.

### 4.1  Experimental Datasets

The patents used in our experiments are shown in Table 3. Due to the limited number of patent infringement cases, there are 700 pairs of infringement patents,1000 pairs of similar patents and 1000 pairs of irrelevant patents which are collected from the united states patent and trademark office(USPTO) and Google Patents as the experimental datasets.

Table 3. The description of experimental data

| Source | USPTO | Google patents | USPTO |
|---|---|---|---|
| Data set | Infringement patent pairs | similar patent pairs | Irrelevant patent pairs |
| Number | 700 | 1000 | 1000 |
| Dose infringe | Yes | No | No |

The data sets are collected by the following steps,

1. Infringement patent pairs are extracted from inter parte review (IPR) documents of Patent Trial and Appeal Board (PTAB) from USPTO website[1] by automatic crawler, text parsing and natural language processing. The inter parte review (IPR) documents specify whether a patent infringes other patents according to its novelty and non-obviousness. The pairs of infringed patents and their infringing patents can be extracted from IPR documents. We randomly extract 700 infringement patent pairs where each pair includes an infringed patent and an infringing patent.
2. Similar patent pairs are extracted from Google Patents where each pair is consisted of a patent and one of its similar/backward-citation patents. Although a patent has similar content with its similar/backward-citation patents, the patent does not infringe. We randomly select 1000 patent pairs where each pair is consisted of a patent and one of its similar patents.
3. Irrelevant patent pairs are extracted from USPTO where each pair is consisted of a patent and one of its irrelevant patents which have different CPC codes. We randomly select 1000 such patent pairs where each pair is consisted of a patent and one of its irrelevant patents.

## 4.2  Baseline Models

We compare our method with the state-of-the-art method: semantic patent claim based model (SPC) [9] which includes the following steps.

For each infringement patent pair:

1. construct a claim dependency tree for the infringing patent and one tree for the infringed patent;
2. make one-to-one mapping from each claim in the infringing dependency tree into the most similar claim with the same depth in the infringed dependency tree;
3. sum the weighted similarities of the one-to-one claim pairs bottom-up.

## 4.3  Evaluation Measurements

The infringement risk is the evaluation measurement in this paper. We will compare the experimental results obtained by our proposed method with that of the baseline method on the infringement risk.

---

[1] https://developer.uspto.gov/ptab-web/#/search/decisions.

## 4.4    Experimental Setups

For each infringement patent pair in infringement patent data set, similar patent data set and irrelevant patent data set, we make experiment as follows.

1. our dynamic game method makes infringement detection in Algorithm 1 where Feature2vec and Claim2vec are used to embed the features and claims. The distance of two features/claims equals to 1-cosine of the two features/claims.
2. the baseline method makes infringement detection by the semantic patent claim based method in Sect. 4.2.
3. we compare the results obtained by our method and the baseline method on evaluation measurement in Sect. 4.3.

## 4.5    Experimental Results

Table 4, Table 5 and Table 6 show the results of dynamic game model and baseline method on infringement, similar and irrelevant patent data set respectively.

Compared with baseline methods, our method positions the patent infringement to the claim infringement as the evidences of infringement.

**Table 4.** Results of our method and the baseline method on infringement patent data set

| Infringement pair | Dynamic game method | | SPC |
|---|---|---|---|
| | Payoff | Evidence | Risk |
| 6948754 | 0.5705 | Claim 18 | 0.0630 |
| 8573666 | 0.3304 | Claim 8 | |
| 7229123 | 0.4870 | Claim 5 | 0.2755 |
| 8573666 | 0.4344 | Claim 8 | |
| 0230989 | 0.6491 | Claim 7 | 0.0274 |
| 8573666 | 0.2259 | Claim 8 | |
| ... | ... | ... | ... |
| 5383978 | 0.3696 | Claim 5 | 0.1256 |
| 6537385 | 0.5000 | Claim 1 | |

For the infringement patent data set, the payoff of infringed patent is higher than that of infringing patent in our method as shown in Table 4. It indicates that it is more beneficial for the infringed patent than for the infringing patent, where infringement occurs.

For the similar patent data set, the payoff of infringed patent is lower than that of infringing patent in our method as shown in Table 5. It indicates that it is more beneficial for the infringing patent than for the infringed patent, where infringement does not occur.

**Table 5.** Results of our method and the baseline method on similar patent data set

| Similar pair | Dynamic game method | | SPC |
|---|---|---|---|
| | Payoff | Evidence | Risk |
| 4519003 | 0.2029 | Claim 6 | 0.0425 |
| 5228077 | 0.6929 | Claim 8 | |
| 4566034 | 0.3586 | Claim 7 | 0.1560 |
| 5228077 | 0.5317 | Claim 21 | |
| 5038401 | 0.1904 | Claim 6 | 0.0767 |
| 5228077 | 0.7459 | Claim 13 | |
| ... | ... | ... | ... |
| 6735551 | 0.1606 | Claim 9 | 0.0726 |
| 7921186 | 0.7830 | Claim 2 | |

**Table 6.** Results of our method and the baseline method on irrelevant patent data set

| Irrelevant pair | Dynamic game method | | SPC |
|---|---|---|---|
| | Payoff | Evidence | Risk |
| 5652542 | 0.3429 | Claim 13 | 0.0066 |
| 7542045 | 0.5431 | Claim 12 | |
| 5652542 | 0.1855 | Claim 1 | 0.0249 |
| 7543318 | 0.7384 | Claim 2 | |
| 5652544 | 0.2397 | Claim 4 | 0.0085 |
| 754331 | 0.7171 | Claim 4 | |
| ... | ... | ... | ... |
| 6757717 | 0.2436 | Claim 28 | 0.0216 |
| 8622832 | 0.7023 | Claim 4 | |

For the irrelevant patent data set, the payoff of infringed patent is lower than that of infringing patent in our method as shown in Table 6. It indicates that it is more beneficial for the infringing patent than for the infringed patent, where infringement does not occur.

The payoff of infringed patent is determined by both the claims and the technical features, and the payoff of infringing patent is only determined by the technical features. We use the payoff of infringing patent to calculate the infringement risk since the determination of patent infringement is technical features. The lower the payoff of infringing patent is, the greater infringement risk of the patent pair is. Therefore the infringement risk equals to 1 minus the payoff of infringing patent.

Table 7 and Fig. 4 show the payoff difference value between the payoff of infringed patent and that of infringing patent obtained by our method. For the infringement patents, the average of payoff difference value is positive. For the

**Table 7.** The payoff difference value between payoff of infringed patent and payoff of infringing patent on data set

| Method | Data set | | | | | | | | |
|---|---|---|---|---|---|---|---|---|---|
| | Infringement | | | Similar | | | Irrelevant | | |
| | Max | Min | Avg | Max | Min | Avg | Max | Min | Avg |
| dynamic game | 0.9202 | −0.8998 | 0.0265 | 0.9152 | −0.9936 | −0.0851 | 0.8716 | −0.9678 | −0.3466 |

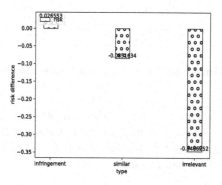

**Fig. 4.** The average payoff difference value between payoff of infringed patent and payoff of infringing patent obtained by the dynamic game method

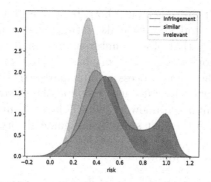

**Fig. 5.** The density plot of infringement risk for dynamic game method

no infringement patents (similar patents and irrelevant patents), the average of payoff difference is negative. The results indicate that infringed patents are more likely to win the game when infringement occurs and infringing patents are more likely to win the game when infringement does not occurs in our method.

Figure 5 and Fig. 6 show the density plot of infringement risk for the dynamic game method and the baseline method. The dynamic game method gives high infringement risk on the infringement patent data set, medium infringement risk on the similar patent data set and low infringement risk on the irrelevant data set. Baseline method gives almost the same infringement risk on the patent

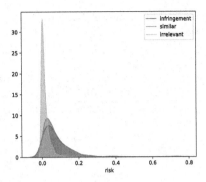

**Fig. 6.** The density plot of infringement risk for baseline method

data set. The results indicate that our method outperforms in distinguishing the infringement patents from the patent data set, but the baseline method does not perform well in distinguishing the infringement patents from the patent data set especially from the similar patent data.

## 5    Conclusions and Future Work

In research on patent infringement, how to model the novelty and non-obviousness of patents and how to make rational decision on which parts of a patent are infringed by that of its counterpart are challenging issues. To solve the above issues, we propose a game theory based patent infringement detection method. Inspired by the game process in inter partes review, the method integrated the measurement of infringement risk and the discovery of the infringement evidences into a unified framework, which plays dynamic game. To the best of my knowledge, this is the first study using game theory on patent infringement detection. Compared with the previous studies, the contributions of our method are summarized as follow.

1. innovation in consideration of rational decision in patent infringement. The method simulates the non-cooperative game process in which the parties of patent infringement always select what is best for himself/herself against its counterpart, resulting in infringement risk and the evidences.
2. innovation in an unified automatic infringement detection framework. The framework includes the collection of infringement/non-infringement patent data, the design of automatic infringement detection method, the automatic experimental validation.
3. innovation in standard data set collection. Before our work, there is no standard infringement/non-infringement patent data set and a few practical reference data is collected by manually. We propose a method to automatically collect infringement/non-infringement patent data from the inter parte review (IPR)/the backward citation patents.

We compare our method with a state-of-the-art method as the baseline method on patent infringement/non-infringement data sets. The results show that the dynamic game method outperforms the baseline method in patent infringement detection.

Some users may consider how to design a unsupervised classification for patent infringement and how to detect whether there is a infringed patent or not for a patent in an existing database and so on. These problems will be addressed in our future works.

# References

1. Abbas, A., Zhang, L., Khan, S.U.: A literature review on the state-of-the-art in patent analysis. World Pat. Inf. **37**, 3–13 (2014)
2. Cascini, G., Zini, M.: Measuring patent similarity by comparing inventions functional trees. In: Cascini, G. (ed.) CAI 2008. TIFIP, vol. 277, pp. 31–42. Springer, Boston, MA (2008). https://doi.org/10.1007/978-0-387-09697-1_3
3. Deng, N., Chen, X., Xiong, C.: Semantic similarity calculation of TCM patents in intelligent retrieval based on deep learning. In: Barolli, L., Hellinckx, P., Natwichai, J. (eds.) 3PGCIC 2019. LNNS, vol. 96, pp. 472–481. Springer, Cham (2020). https://doi.org/10.1007/978-3-030-33509-0_44
4. Gerken, J.M., Moehrle, M.G.: A new instrument for technology monitoring: novelty in patents measured by semantic patent analysis. Scientometrics, **91**(3), 645–670 (2012)
5. Helmers, L., Horn, F., Biegler, F., Oppermann, T., Müller, K.-R.: Automating the search for a patents prior art with a full text similarity search. PLoS ONE **14**(3), e0212103 (2019)
6. Hido, S., et al.: Modeling patent quality: a system for large-scale patentability analysis using text mining. Inf. Media Technol. **7**(3), 1180–1191 (2012)
7. Ji, X., Gu, X., Dai, F., Chen, J., Le, C.: Patent collaborative filtering recommendation approach based on patent similarity. In 2011 Eighth International Conference on Fuzzy Systems and Knowledge Discovery (FSKD), vol. 3, pp. 1699–1703. IEEE (2011)
8. Kasravi, K., Risov, M.: Multivariate patent similarity detection. In: 2009 42nd Hawaii International Conference on System Sciences, pp. 1–8. IEEE (2009)
9. Lee, C., Song, B., Park, Y.: How to assess patent infringement risks: a semantic patent claim analysis using dependency relationships. Technol. Anal. Strateg. Manag. **25**(1), 23–38 (2013)
10. Ma, C., Zhao, T., Li, H.: A method for calculating patent similarity using patent model tree based on neural network. In: Ren, J., et al. (eds.) International Conference on Brain Inspired Cognitive Systems, vol. 10989, pp. 633–643. Springer, Cham (2018). https://doi.org/10.1007/978-3-030-00563-4_62
11. Moehrle, M.G., Gerken, J.M.: Measuring textual patent similarity on the basis of combined concepts: design decisions and their consequences. Scientometrics **91**(3), 805–826 (2012)
12. Myerson, R.B.: Game Theory. Harvard University Press, Cambridge (2013)
13. Park, H., Yoon, J., Kim, K.: Identifying patent infringement using SAO based semantic technological similarities. Scientometrics **90**(2), 515–529 (2012)

14. Rodriguez, A., Kim, B., Turkoz, M., Lee, J.-M., Coh, B.-Y., Jeong, M.K.: New multi-stage similarity measure for calculation of pairwise patent similarity in a patent citation network. Scientometrics **103**(2), 565–581 (2015). https://doi.org/10.1007/s11192-015-1531-8

15. Taduri, S., Lau, G.T., Law, K.H., Kesan, J.P.: Retrieval of patent documents from heterogeneous sources using ontologies and similarity analysis. In: 2011 IEEE Fifth International Conference on Semantic Computing, pp. 538–545. IEEE (2011)

16. Wang, X., Ren, H., Chen, Y., Liu, Y., Qiao, Y., Huang, Y.: Measuring patent similarity with SAO semantic analysis. Scientometrics **121**(1), 1–23 (2019)

17. Xu, H., Zeng, W., Gui, J., Qu, P., Zhu, X., Wang, L.: Exploring similarity between academic paper and patent based on latent semantic analysis and vector space model. In: 2015 12th International Conference on Fuzzy Systems and Knowledge Discovery (FSKD), pp. 801–805. IEEE (2015)

18. Yanagihori, K., Tsuda, K.: Verification of patent document similarity of using dictionary data extracted from notification of reasons for refusal. In: 2015 IEEE 39th Annual Computer Software and Applications Conference, vol. 3, pp. 349–354. IEEE (2015)

# Unveiling Relations in the Industry 4.0 Standards Landscape Based on Knowledge Graph Embeddings

Ariam Rivas[1]([✉]), Irlán Grangel-González[2], Diego Collarana[3], Jens Lehmann[3], and Maria-Esther Vidal[1,4]

[1] L3S, Leibniz University of Hannover, Hanover, Germany
ariam.rivas@tib.eu
[2] Robert Bosch Corporate Research GmbH, Renningen, Germany
irlan.grangelgonzalez@de.bosch.com
[3] University of Bonn and Fraunhofer IAIS, Bonn, Germany
{diego.collarana.vargas,jens.lehmann}@iais.fraunhofer.de
[4] TIB Leibniz Information Centre for Science and Technology, Hanover, Germany
maria.vidal@tib.eu

**Abstract.** Industry 4.0 (I4.0) standards and standardization frameworks have been proposed with the goal of *empowering interoperability* in smart factories. These standards enable the description and interaction of the main components, systems, and processes inside of a smart factory. Due to the growing number of frameworks and standards, there is an increasing need for approaches that automatically analyze the landscape of I4.0 standards. Standardization frameworks classify standards according to their functions into layers and dimensions. However, similar standards can be classified differently across the frameworks, producing, thus, interoperability conflicts among them. Semantic-based approaches that rely on ontologies and knowledge graphs, have been proposed to represent standards, known relations among them, as well as their classification according to existing frameworks. Albeit informative, the structured modeling of the I4.0 landscape only provides the foundations for detecting interoperability issues. Thus, graph-based analytical methods able to exploit knowledge encoded by these approaches, are required to uncover alignments among standards. We study the relatedness among standards and frameworks based on community analysis to discover knowledge that helps to cope with interoperability conflicts between standards. We use knowledge graph embeddings to automatically create these communities exploiting the meaning of the existing relationships. In particular, we focus on the identification of similar standards, i.e., communities of standards, and analyze their properties to detect unknown relations. We empirically evaluate our approach on a knowledge graph of I4.0 standards using the Trans* family of embedding models for knowledge graph entities. Our results are promising and suggest that relations among standards can be detected accurately.

© Springer Nature Switzerland AG 2020
S. Hartmann et al. (Eds.): DEXA 2020, LNCS 12392, pp. 179–194, 2020.
https://doi.org/10.1007/978-3-030-59051-2_12

# 1   Introduction

The international community recognizes Industry 4.0 (I4.0) as the fourth industrial revolution. The main objective of I4.0 is the creation of *Smart Factories* by combining the Internet of Things (IoT), Internet of Services (IoS), and Cyber-Physical Systems (CPS). In smart factories, humans, machines, materials, and CPS need to communicate intelligently in order to produce individualized products. To tackled the problem of interoperability, different industrial communities have created standardization frameworks. Relevant examples are the Reference Architecture for Industry 4.0 (RAMI4.0) [1] or the Industrial Internet Connectivity Framework (IICF) in the US [17]. Standardization frameworks classify, and align industrial standards according to their functions. While being expressive to categorize existing standards, standardization frameworks may present divergent interpretations of the same standard. Mismatches among standard classifications generate semantic interoperability conflicts that negatively impact on the effectiveness of communication in smart factories.

Database and Semantic web communities have extensively studied the problem of data integration [9,15,21], and various approaches have been proposed to support data-driven pipelines to transform industrial data into actionable knowledge in smart factories [13,23]. Ontology-based approaches have also contributed to create a shared understanding of the domain [16], and specifically Kovalenko and Euzenat [15] have equipped data integration with diverse methods for ontology alignment. Furthermore, Lin *et al.* [18] identify interoperability conflicts across domain specific standards (e.g., RAMI4.0 model and the IICF architecture), while works by Grangel-Gonzalez *et al.* [10–12] show the relevant role that Descriptive Logic, Datalog, and Probabilistic Soft Logic play in liaising I4.0 standards. Certainly, the extensive literature in data integration provides the foundations for enabling the semantic description and alignment of "similar" things in a smart factory. Nevertheless, finding alignments across I4.0 requires the encoding of domain specific knowledge represented in standards of diverse nature and standardization frameworks defined with different industrial goals. We rely on state-of-the-art knowledge representation and discovery approaches to embed meaningful associations and features of the I4.0 landscape, to enable interoperability.

We propose a knowledge-driven approach first to represent standards, known relations among them, as well as their classification according to existing frameworks. Then, we utilize the represented relations to build a latent representation of standards, i.e., embeddings. Values of similarity metrics between embeddings are used in conjunction with state-of-the-art community detection algorithms to identify patterns among standards. Our approach determines relatedness among standards by computing communities of standards and analyzing their properties to detect unknown relations. Finally, the *homophily* prediction principle is performed in each community to discover new links between standards and frameworks. We asses the performance of the proposed approach in a data set of 249 I4.0 standards connected by 736 relations extracted from the literature.

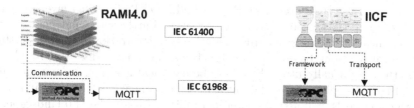

**Fig. 1. Motivating Example.** The RAMI4.0 and IICF standardization frameworks are developed for diverse industrial goals; they classify standards in layers according to their functions, e.g., OPC UA and MQTT under the communication layer in RAMI4.0, and OPC UA and MQTT in the framework and transport layers in IICF, respectively. Further, some standards, e.g., IEC 61400 and IEC 61968, are not classified yet.

The observed results suggest that encoding knowledge enables for the discovery of meaningful associations. Our contributions are as follows:

1. We formalize the problem of finding relations among I4.0 standards and present *I4.0RD*, a knowledge-driven approach to unveil these relations. *I4.0RD* exploits the semantic description encoded in a knowledge graph via the creation of embeddings, to identify then communities of standards that should be related.
2. We evaluate the performance of *I4.0RD* in different embeddings learning models and community detection algorithms. The evaluation material is available[1].

The rest of this paper is organized as follows: Sect. 2 illustrates the interoperability problem presented in this paper. Section 3 presents the proposed approach, while the architecture of the proposed solution is explained in Sect. 4. Results of the empirical evaluation of our methods are reported in Sect. 5 while Sect. 6 summarizes the state of the art. Finally, we close with the conclusion and future work in Sect. 7.

## 2    Motivating Example

Existing efforts to achieve interoperability in I4.0, mainly focus on the definition of standardization frameworks. A standardization framework defines different layers to group related I4.0 standards based on their functions and main characteristics. Typically, classifying existing standards in a certain layer is not a trivial task and it is influenced by the point of view of the community that developed the framework. RAMI4.0 and IICF are exemplar frameworks, the former is developed in Germany while the latter in the US; they meet specific I4.0 requirements of certain locations around the globe. RAMI4.0 classifies the standards OPC UA and MQTT into the Communication layer, stating this, that both standards are

---

[1] https://github.com/i40-Tools/I40KG-Embeddings.

similar. Contrary, IICF presents OPC UA and MQTT at distinct layers, i.e., the framework and the transport layers, respectively. Furthermore, independently of the classification of the standards made by standardization frameworks, standards have relations based on their functions. Therefore, IEC 61400 and IEC 61968 that are usually utilized to describe electrical features, are not classified at all. Figure 1 depicts these relations across the frameworks RAMI4.0 and IICF, and the standards; it illustrates interoperability issues in the I4.0 landscape.

Existing data integration approaches rely on the description of the characteristics of entities to solve interoperability by discovering alignments among them. Specifically, in the context of I4.0, semantic-based approaches have been proposed to represent standards, known relations among them, as well as their classification according to existing frameworks [4, 6, 18, 19]. Despite informative, the structured modeling of the I4.0 landscape only provides the foundations for detecting interoperability issues.

We propose *I4.0RD*, an approach capable of discovering relation over I4.0 knowledge graphs to identify unknown relations among standards. Our proposed methods exploit relations represented in an I4.0 knowledge graph to compute the similarity of the modeled standards. Then, an unsupervised graph partitioning method determines the communities of standards that are similar. Moreover, *I4.0RD* explores communities to identify possible relations of standards, enhancing, thus, interoperability.

## 3    Problem Definition and Proposed Solution

We tackle the problem of unveiling relations between I4.0 standards. We assume that the relations among standards and standardization frameworks like the ones shown in Fig. 2(a), are represented in a knowledge graph named I4.0 KG. Nodes in a I4.0 KG correspond to standards and frameworks; edges represent relations among standards, as well as the standards grouped in a framework layer. An I4.0KG is defined as follows:

Given sets $V_e$ and $V_t$ of entities and types, respectively, a set $E$ of labelled edges representing relations, and a set $L$ of labels. An I.40KG is defined as $\mathcal{G} = (V_e \cup V_t, E, L)$:

- The types Standard, Frameworks, and Framework Layer belong to $V_t$.
- I4.0 standards, frameworks, and layers are represented as instances of $V_e$.
- The types of the entities in $V_e$ are represented as edges in $E$ that belong to $V_e \times V_t$.
- Edges in $E$ that belong to $V_e \times V_e$ represent relations between standards and their classifications into layers according to a framework.
- *RelatedTo, Type, classifiedAs, IsLayerOf* correspond to labels in $L$ that represent the relations between standards, their type, their classification into layers, and the layers of a framework, respectively.

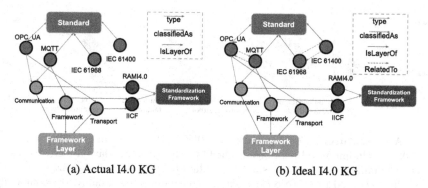

(a) Actual I4.0 KG                    (b) Ideal I4.0 KG

**Fig. 2. Example of I4.0KGs.** Figure 2a shows known relationships among standards to Framework Layer and Standardization Framework. While Fig. 2b depicts all the ideal relationships between the standards expressed with the property relatedTo. Standards OPC UA and MQTT are related, as well as the standards IEC 61968 and IEC 61400. Our aim is discovering relations relatedTo in Fig. 2b.

## 3.1 Problem Statement

Let $\mathcal{G}' = (V_e \cup V_t, E', L)$ and $\mathcal{G} = (V_e \cup V_t, E, L)$ be two I4.0 knowledge graphs. $\mathcal{G}'$ is an *ideal* knowledge graph that contains all the *existing relations* between standard entities and frameworks in $V_e$, i.e., an oracle that knows whether two standard entities are related or not, and to which layer they should belong; Fig. 2 (b) illustrates a portion of an ideal I4.0KG, where the relations between standards are explicitly represented. $\mathcal{G} = (V_e \cup V_t, E, L)$ is an *actual* I4.0KG, which only contains a portion of the relations represented in $\mathcal{G}'$, i.e., $E \subseteq E'$; it represents those relations that are known and is not necessarily complete. Let $\Delta(E', E) = E' - E$ be the set of relations existing in the ideal knowledge graph $\mathcal{G}'$ that are not represented in $\mathcal{G}$. Let $\mathcal{G}_{comp} = (V_e \cup V_t, E_{comp}, L)$ be a *complete* knowledge graph, which includes a relation for each possible combination of elements in $V_e$ and labels in $L$, i.e., $E \subseteq E' \subseteq E_{comp}$. Given a relation $e \in \Delta(E_{comp}, E)$, the problem of discovering relations consists of determining whether $e \in E'$, i.e., if a relation represented by an edge $r = (e_i \ l \ e_j)$ corresponds to an existing relation in the ideal knowledge graph $\mathcal{G}'$. Specifically, we focus on the problem of discovering *relations* between standards in $\mathcal{G} = (V_e \cup V_t, E, L)$. We are interested in finding the maximal set of relationships or edges $E_a$ that belong to the ideal I4.0KG, i.e., find a set $E_a$ that corresponds to a solution of the following optimization problem:

$$\underset{E_a \subseteq E_{comp}}{\operatorname{argmax}} |E_a \cap E'|$$

Considering the knowledge graphs depicted in Figs. 2 (a) and (b), the problem addressed in this work corresponds to the identification of edges in the ideal knowledge graph that correspond to unknown relations between standards.

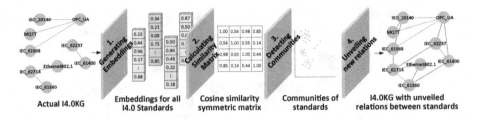

| Actual I4.0KG | Embeddings for all I4.0 Standards | Cosine similarity symmetric matrix | Communities of standards | I4.0KG with unveiled relations between standards |

**Fig. 3. Architecture.** *I4.0RD* receives an I4.0KG and outputs an extended version of the I4.0KG including novel relations. Embeddings for each standard are created using the Trans* family of models, and similarity values between embeddings are computed; these values are used to partition standards into communities. Finally, the homophily prediction principle is applied to each community to discover unknown relations.

## 3.2 Proposed Solution

We propose a relation discovery method over I4.0KGs to identify unknown relations among standards. Our proposed method exploits relations represented in an I4.0KG to compute similarity values between the modeled standards. Further, an unsupervised graph partitioning method determine the parts of the I4.0KG or communities of standards that are similar. Then, the *homophily* prediction principle is applied in a way that similar standards in a community are considered to be related.

## 4    The *I4.0RD* Architecture

Figure 3 presents *I4.0RD*, a pipeline that implements the proposed approach. *I4.0RD* receives an I4.0KG $\mathcal{G}$, and returns an I4.0KG $\mathcal{G}'$ that corresponds to a solution of the problem of discovering relations between standards. First, in order to compute the values of similarity between the entities an I4.0KG, *I4.0RD* learns a latent representation of the standards in a high-dimensional space. Our approach resorts to the Trans* family of models to compute the embeddings of the standards and the cosine similarity measure to compute the values of similarity. Next, community detection algorithms are applied to identify communities of related standards. METIS [14], KMeans [3], and SemEP [24] are methods included in the pipeline to produce different communities of standards. Finally, *I4.0RD* applies the *homophily* principle to each community to predict relations or alignments among standards.

### 4.1    Learning Latent Representations of Standards

*I4.0RD* utilizes the Trans* family of models to compute latent representations, e.g., vectors, of entities and relations in an I4.0 knowledge graph. In particular, *I4.0RD* utilizes TransE, TransD, TransH, and TransR. These models differ on the representation of the embeddings for the entities and relations (Wang et al. [26]). Suppose $e_i$, $e_j$, and $p$, denote the vectorial representation of two entities related by the labeled edge $p$ in an I4.0 knowledge graph. Furthermore, $\|x\|_2$ represents the Euclidean norm.

TransE, TransH, and TranR represent the entity embeddings as $(e_i, e_j \in \mathbb{R}^d)$, while TransD characterizes the entity embeddings as: $(e_i, w_{e_i} \in \mathbb{R}^d - e_i, w_{e_j} \in \mathbb{R}^d)$. As a consequence of different embedding representations, the scoring function also varies. For example, TransE is defined in terms of the score function $\|e_i + p - e_j\|_2^2$, while $\|M_p e_i + p - M_p e_j\|_2^2$ defines TransR[2]. Furthermore, TransH score function corresponds to $\|e_{i\perp} + d_p - e_{j\perp}\|_2^2$, where the variables $e_{i\perp}$ and $e_{j\perp}$ denote a projection to the hyperplane $w_p$ of the labeled relation p, and $d_p$ is the vector of a relation-specific translation in the hyperplane $w_p$. To learn the embeddings, *I4.ORD* resorts to the PyKeen (Python KnowlEdge EmbeddiNgs) framework [2]. As hyperparameters for the models of the Trans* family, we use the ones specified in the original papers of the models. The hyperparameters include embedding dimension (set to 50), number of epochs (set to 500), batch size (set to 64), seed (set to 0), learning rate (set to 0.01), scoring function (set to 1 for TransE, and 2 for the rest), margin loss (set to 1 for TransE and 0.05 for the rest). All the configuration classes and hyperparameters are open in GitHub[3].

## 4.2   Computing Similarity Values Between Standards

Once the algorithm–Trans* family–that computes the embeddings reaches a termination condition, e.g., the maximum number of epochs, the I4.0KG embeddings are learned. As the next step, *I4.ORD* calculates a *similarity symmetric matrix* between the embeddings that represent the I4.0 standards. Any distance metric for vector spaces can be utilized to calculate this value. However, as a proof of concepts, *I4.ORD* applies the Cosine Distance. Let $u$ be an embedding of the Standard-A and $v$ an embedding of the Standard-B, the similarity score, between both standards, is defined as follows:

$$cosine(u, v) = 1 - \frac{u.v}{\|u\|_2 \|v\|_2}$$

After building the *similarity symmetric matrix*, *I4.ORD* applies a threshold to restrict the similarity values. *I4.ORD* relies on percentiles to calculate the value of such a threshold. Further, *I4.ORD* utilizes the function Kernel Density Estimation (KDE) to compute the probability density of the cosine similarity matrix; it sets to zero the similarity values lower than the given threshold.

## 4.3   Detecting Communities of Standards

*I4.ORD* maps the problem of computing groups of potentially related standards to the problem of community detection. Once the embeddings are learned, the

---

[2] $M_p$ corresponds to a projection matrix $M_p \in \mathbb{R}^{d x k}$ that projects entities from the entity space to the relation space; further $p \in \mathbb{R}^k$.

[3] https://github.com/i40-Tools/I4.0KG-Embeddings

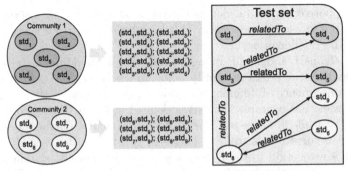

(a) Application of the Homophily Prediction Principle

(b) Known Relations used
to determine discovered relations between standards

**Fig. 4. Discovering Relations Between Standards.** (a) The homophily prediction principle is applied on two communities, as a result, 16 relations between standards are found. (b) Six out of the 16 found relations correspond to meaningfully relations.

standards are represented in a vectorial way according to their functions preserving their semantic characteristics. Using the embeddings, *I4.0RD* computes the similarity between the I4.0 standards as mentioned in the previous section. The values of similarity between standards are utilized to partition the set of standards in a way that standards in a community are highly similar but dissimilar to the standards in other communities. As proof of concept, three state-of-the-art community detection algorithms have been used in *I4.0RD*: SemEP, METIS, and KMeans. They implement diverse strategies for partitioning a set based on the values of similarity, and our goal is to evaluate which of the three is more suitable to identify meaningful connections between standards.

### 4.4   Discovering Relations Between Standards

New relations are discovered in this step; the *homophily* prediction principle is applied over each of the communities and all the standards in a community are assumed to be related. Figure 4 depicts an example where new relations are computed from two communities; unknown relations correspond to connections between standards in a community that did not existing in the input I4.0 KG.

## 5   Empirical Evaluation

We report on the impact that the knowledge encoded in I4.0 knowledge graphs has in the behavior of *I4.0RD*. In particular, we asses the following research questions:

**RQ1)** Can the semantics encoded in I4.0 KG empower the accuracy of the relatedness between entities in a KG?

**RQ2)** Does a semantic community based analysis on I4.0KG allow for improving the quality of predicting new relations on the I4.0 standards landscape?

**Experiment Setup:** We considered four embedding algorithms to build the standards embedding. Each of these algorithms was evaluated independently. Next, a similarity matrix for the standards embedding was computed. The similarity matrix is required for applying the community detection algorithms. In our experiments, three algorithms were used to compute the communities. That means twelve combinations between embedding algorithms and community detection algorithms to be evaluated. To assure statistical robustness, we executed 5-folds cross-validation with one run.

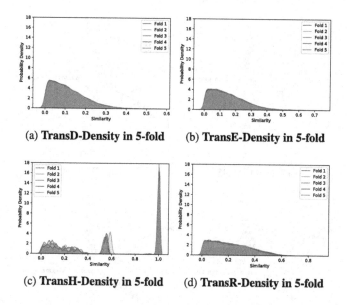

(a) **TransD-Density in 5-fold**    (b) **TransE-Density in 5-fold**

(c) **TransH-Density in 5-fold**    (d) **TransR-Density in 5-fold**

**Fig. 5. Probability density of each fold per Trans\* methods.** Figures 5a, 5b, and 5d show that all folds have values close to zero, i.e., with embeddings created by TransD, TransE, and TransR the standards are very different from each other. However, TransH (cf. Fig. 5c), exploits properties of the standards and generates embeddings with a different distribution of similarity, i.e., values between 0.0 and 0.6, as well as values close to 1.0. According to known characteristics of the I4 standards, the TransH distribution of similarity better represents their relatedness.

**Thresholds for Computing Values of Similarity:** Figure 5 depicts the probability density function of each fold for each embedding algorithm. Figures 5a and 5b show the values of the folds of TransD and TransE where all the similarity values are close to 0.0, i.e., all the standards are different. Figure 5d suggests that all the folds have similar behavior with values between 0.0 and 0.5. Figure 5c shows a group of standards similar with values close to 1.0 and the rest of the standards between 0.0 and 0.6. The percentile of the similarity matrix is computed with a threshold of 0.85. That means all values of the similarity matrix which are less than the percentile computed, are filled with 0.0 and then, these two standards are dissimilar. After analyzing the probability density of each fold (cf. Fig. 5), the thresholds of TransH and TransR are set to 0.50 and 0.75, respectively. The reason is because the two cases with a high threshold find all similar standards. In the case of TransH, there is a high density of values close to 1.0; it indicates that for a threshold of 0.85, the percentile computed is almost 1.0. the values of the similarity matrix less than the threshold are filled with 0.0; values of 0.0 represent that the compared standards are not similar.

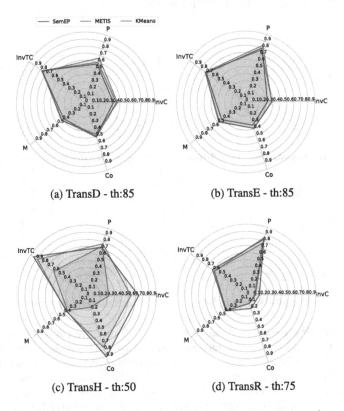

**Fig. 6. Quality of the generated communities.** Communities evaluated in terms of prediction metrics with thresholds (th) of 0.85, 0.50, and 0.75 using the SemEP, METIS, and KMeans algorithms. In this case higher values are better. Our approach exhibits the best performance with TransH embeddings and a threshold of 0.50 for computing the similarity matrix, i.e., Figure (c). SemEP achieves the highest values in four of the five evaluated parameters.

**Metrics:** the following metrics are used to estimate the quality of the communities from the I4.0KG embeddings.

a) **Conductance (InvC):** measures relatedness of entities in a community, and how different they are to entities outside the community [7]. The inverse of Conductance is reported: $1 - Conductance(K)$, where $K = \{k_1, k_2, ...., k_n\}$ the set of standards communities obtained by the cluster algorithm, and $k_i$ are the computed clusters.

b) **Performance (P):** sums up the number of intra-community relationships, plus the number of non-existent relationships between communities [7].

c) **Total Cut (InvTC):** sums up all similarities among entities in different communities [5]. The Total Cut values are normalized by dividing the sum of the similarities between the entities. The inverse of Total Cut is reported as follows: $1 - NormTotalCut(K)$

d) **Modularity (M):** is the value of the intra-community similarities between the entities divided by the sum of all the similarities between the entities, minus the sum of the similarities among the entities in different communities, in case they are randomly distributed in the communities [22]. The value of the Modularity is in the range of $[-0.5, 1]$, which can be scaled to $[0, 1]$ by computing: $\frac{Modularity(K)+0.5}{1.5}$.

e) **Coverage (Co):** compares the fraction of intra-community similarities between entities to the sum of all similarities between entities [7].

**Implementation:** Our proposed approach is implemented in Python 2.7 and integrated with the PyKeen (Python KnowlEdge EmbeddiNgs) framework [2], METIS 5.1[4], SemEP[5], and Kmeans[6]. The experiments were executed on a GPU server with ten chips Intel(R) Xeon(R) CPU E5-2660, two chips GeForce GTX 108, and 100 GB RAM.

***RQ1*** - *Corroborating the accuracy of relatedness between standards in I40KG.* To compute accuracy of *I4.0RD*, we executed a five-folds cross-validation procedure. To that end, the data set is divided into five consecutive folds shuffling the data before splitting into folds. Each fold is used once as validation, i.e., test set while the remaining fourth folds form the training set. Figure 6 depicts the best results are obtained with the combination of the TransH and SemEP algorithms. The values obtained for this combination are as follows: **Inv. Conductance** (0.75), **Performance** (0.77), **Inv. Total Cut** (0.95), **Modularity** (0.36), and **Coverage** (0.91).

***RQ2*** - *Predicting new relations between standards.* In order to assess the second research question, the data set is divided into five consecutive folds. Each fold comprises 20% of the relationships between standards. Next, the precision measurement is applied to evaluate the main objective is to unveil uncovered associations and at the same time to corroborate knowledge patterns that are already known.

As shown in Fig. 7, the best results for the property `relatedTo` are achieved by TransH embeddings in combination with the SemEP and KMeans algorithm.

---

[4] http://glaros.dtc.umn.edu/gkhome/metis/metis/download.

[5] https://github.com/SDM-TIB/semEP.

[6] https://scikit-learn.org/stable/modules/generated/sklearn.cluster.KMeans.html.

The communities of standards discovered using the techniques TransH and SemEP contribute to the resolution of interoperability in I4.0 standards. To provide an example of this, we observed a resulting cluster with the standards *ISO 15531* and *MTConnect*. The former provides an information model for describing manufacturing data. The latter offers a vocabulary for manufacturing equipment. It is important to note that those standards are not related to the training set nor in I40KG. The membership of both standards in the cluster means that those two standards should be classified together in the standardization frameworks. Besides, it also suggests to the creators of the standards that they might look after possible existing synergies between them. This example suggests that the techniques employed in this work are capable of discovering new communities of standards. These communities can be used to improve the classification that the standardization frameworks provide for the standards.

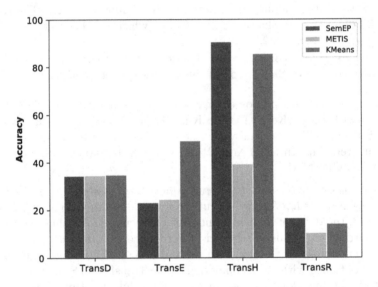

**Fig. 7.** *I4.0RD* accuracy. Percentage of the test set for the property `relatedTo` is achieved in each cluster. Our approach exhibits the best performance using TransH embedding and with the SemEP algorithm reaching an accuracy by up to 90%.

## 5.1   Discussion

The techniques proposed in this paper rely on known relations between I4.0 standards to discover novel patterns and new relations. During the experimental study, we can observe that these techniques could group together not only standards that were known to be related, but also standards whose relatedness was implicitly represented in the I40KG. This feature facilitates the detection of high-quality communities as reported in Fig. 6, as well as for an accurate discovery of relations between standards (cf. Fig. 7). As observed, the accuracy of the approach can be benefited from the application of state-of-the-art algorithms of the Trans* family, e.g., TransH. Additionally, the strategy employed by SemEP that allows for positioning in the same communities highly similar standards, leads our approach into high-quality discoveries. The combination of both techniques TransH and SemEP allows discovering communities with high quality.

To understand why the combination of TransH and SemEP produces the best results, we analyze in detail both techniques. TransH introduces the mechanism of projecting the relation to a specific hyperplane [27], enabling, thus, the representation of relations with cardinality many to many. Since the materialization of transitivity and symmetry of the property relatedTo corresponds to many to many relations, the instances of this materialization are taken into account during the generation of the embeddings, specifically, during the translating operation on a hyperplane. Thus, even thought semantics is not explicitly utilized during the computation of the embeddings, considering different types of relations, empowers the embeddings generated by TransH. Moreover, it allows for a more precise encoding of the standards represented in I4.0KG. Figure 5c illustrates groups of standards in the similarity intervals $[0.9, 1.0], [0.5, 0.6]$, and $[0.0, 0.4]$. The SemEP algorithm can detect these similarities and represent them in high-precision communities. The other three models embeddings TransD, TransE, and TransR do not represent the standards in the best way. Figures 5a, 5b, 5d report that several standards are in the similarity interval $[0.0, 0.3]$. This means that no community detection algorithm could be able to discover communities with high quality. Reported results indicate that the presented approach enables – in average – for discovering communities of standards by up to 90%. Although these results required the validation of experts in the domain, an initial evaluation suggest that the results are accurate.

## 6   Related Work

In the literature, different approaches are proposed for discovering communities of standards as well as to corroborate and extend the knowledge of the standardization frameworks. Zeid et al. [28] study different approach to achieve interoperability of different standardization frameworks. In this work, the current landscape for smart manufacturing is described by highlighting the existing standardization frameworks in different regions of the globe. Lin et al. [18] present similarities and differences between the RAMI4.0 model and the IIRA architecture. Based on the study of these similarities and differences authors proposed a functional alignment among layers in RAMI4.0 with the functional domains and crosscutting functions in IIRA. Monteiro et al. [20] further report on the comparison of the RAMI4.0 and IIRA frameworks. In this work, a cooperation model is presented to align both standardization frameworks. Furthermore, mappings between RAMI4.0 IT Layers and the IIRA functional domain are established. Another related approach is that outlined in [25]. Moreover, the IIRA and RAMI4.0 frameworks are compared based on different features, e.g., country of origin, source organization, basic characteristics, application scope, and structure. It further details where correspondences exist between the IIRA viewpoints and RAMI4.0 layers. Garofalo *et al.* [8] outline KGEs for I4.0 use cases. Existing techniques for generating embeddings on top of knowledge graphs are examined. Further, the analysis of how these techniques can be applied to the I4.0 domain is described; specifically, it identifies the predictive maintenance, quality control, and context-aware robots as the most promising areas to apply the combination of KGs with embeddings. All the approaches mentioned above are limited to describe and characterize existing knowledge in the domain. However, in our view, two directions need to be consider to enhance the knowledge in the domain; 1) the use of a KG based approach to encode the semantics; and 2) the use of machine learning techniques to discover and predict new communities of standards based on their relations.

## 7    Conclusion

In this paper, we presented the *I4.0RD* approach that combines knowledge graphs and embeddings to discover associations between I4.0 standards. Our approach resorts to I4.0KG to discover relations between standards; I4.0KG represents relations between standards extracted from the literature or defined according to the classifications stated by the standardization frameworks. Since the relation between standards is symmetric and transitive, the transitive closure of the relations is materialized in I4.0KG. Different algorithms for generating embeddings are applied on the standards according to the relations represented in I4.0KG. We employed three community detection algorithms, i.e., SemEP, METIS, and KMeans to identify similar standards, i.e., communities of standards, as well as to analyze their properties. Additionally, by applying the homophily prediction principle, novel relations between standards are discovered. We empirically evaluated the quality of the proposed techniques over 249 standards, initially related through 736 instances of the property `relatedTo`; as this relation is symmetric and transitive, its transitive closure is also represented in I4.0KG with 22,969 instances of `relatedTo`. The Trans* family of embedding models were used to identify a low-dimensional representation of the standards according to the materialized instances of `relatedTo`. Results of a 5-fold cross validation process suggest that our approach is able to effectively identify novel relations between standards. Thus, our work broadens the repertoire of knowledge-driven frameworks for understanding I4.0 standards, and we hope that our outcomes facilitate the resolution of the existing interoperability issues in the I4.0 landscape. As for the future work, we envision to have a more fine-grained description of the I4.0 standards, and evaluate hybrid-embeddings and other type of community detection methods.

**Acknowledgments.** Ariam Rivas is supported by the German Academic Exchange Service (DAAD). This work has been partially funded by the EU H2020 Projects IASIS (GA 727658) and LAMBDA (GA 809965).

# References

1. Adolphs, P., et al.: Structure of the Administration Shell. Status report, ZVEI and VDI (2016)
2. Ali, M., Jabeen, H., Hoyt, C.T., Lehmann, J.: The keen universe: an ecosystem for knowledge graph embeddings with a focus on reproducibility and transferability (in press)
3. Arthur, D., Vassilvitskii, S.: K-means++: the advantages of careful seeding. In: Proceedings of the Eighteenth Annual ACM-SIAM Symposium on Discrete Algorithms. SODA 2007, pp. 1027–1035. Society for Industrial and Applied Mathematics, Philadelphia (2007)
4. Bader, S.R., Grangel-González, I., Tasnim, M., Lohmann, S.: Structuring the industry 4.0 landscape. In: 24th IEEE International Conference on Emerging Technologies and Factory Automation, ETFA, Zaragoza, Spain, 10–13 September, pp. 224–231 (2019)
5. Buluç, A., Meyerhenke, H., Safro, I., Sanders, P., Schulz, C.: Recent advances in graph partitioning. In: Kliemann, L., Sanders, P. (eds.) Algorithm Engineering. LNCS, vol. 9220, pp. 117–158. Springer, Cham (2016). https://doi.org/10.1007/978-3-319-49487-6_4
6. Chungoora, N., et al.: Towards the ontology-based consolidation of production-centric standards. Int. J. Prod. Res. **51**(2), 327–345 (2013)
7. Gaertler, M., Erlebach, T.: Clustering, pp. 178–215. Springer, Heidelberg (2005)
8. Garofalo, M., Pellegrino, M.A., Altabba, A., Cochez, M.: Leveraging knowledge graph embedding techniques for industry 4.0 use cases. CoRR abs/1808.00434 (2018)
9. Golshan, B., Halevy, A.Y., Mihaila, G.A., Tan, W.: Data integration: after the teenage years. In: Proceedings of the 36th ACM SIGMOD-SIGACT-SIGAI Symposium on Principles of Database Systems, PODS 2017, Chicago, IL, USA, 14–19 May 2017, pp. 101–106 (2017)
10. Grangel-González, I., et al.: The industry 4.0 standards landscape from a semantic integration perspective. In: 22nd IEEE International Conference on Emerging Technologies and Factory Automation, ETFA, Limassol, Cyprus, 12–15 September, pp. 1–8 (2017)
11. Grangel-González, I., et al.: Alligator: a deductive approach for the integration of industry 4.0 standards. In: Blomqvist, E., Ciancarini, P., Poggi, F., Vitali, F. (eds.) EKAW 2016. LNCS (LNAI), vol. 10024, pp. 272–287. Springer, Cham (2016). https://doi.org/10.1007/978-3-319-49004-5_18
12. Grangel-González, I., et al.: Knowledge graphs for semantically integrating cyber-physical systems. In: Hartmann, S., Ma, H., Hameurlain, A., Pernul, G., Wagner, R.R. (eds.) DEXA 2018. LNCS, vol. 11029, pp. 184–199. Springer, Cham (2018). https://doi.org/10.1007/978-3-319-98809-2_12
13. Hodges, J., García, K., Ray, S.: Semantic development and integration of standards for adoption and interoperability. IEEE Comput. **50**(11), 26–36 (2017)
14. Karypis, G., Kumar, V.: A fast and high quality multilevel scheme for partitioning irregular graphs. SIAM J. Sci. Comput. **20**(1), 359–392 (1998)
15. Kovalenko, O., Euzenat, J.: Semantic matching of engineering data structures. Semantic Web Technologies for Intelligent Engineering Applications, pp. 137–157. Springer, Cham (2016). https://doi.org/10.1007/978-3-319-41490-4_6
16. Lelli, F.: Interoperability of the time of industry 4.0 and the Internet of Things. Future Internet **11**(2), 36 (2019)

17. Lin, S.W., et al.: The Industrial Internet of Things Volume G1: Reference Architecture. White Paper IIC:PUB:G1:V1.80:20170131, Industrial Internet Consortium (2017)
18. Lin, S.W., et al.: Reference Architectural Model Industrie 4.0 (RAMI 4.0). Technical report, Industrial Internet Consortium and Plattform Industrie 4.0 (2017)
19. Lu, Y., Morris, K.C., Frechette, S.: Standards landscape and directions for smart manufacturing systems. In: IEEE International Conference on Automation Science and Engineering, CASE, Gothenburg, Sweden, 24–28 August, pp. 998–1005 (2015)
20. Monteiro, P., Carvalho, M., Morais, F., Melo, M., Machado, R., Pereira, F.: Adoption of architecture reference models for industrial information management systems. In: International Conference on Intelligent Systems (IS), pp. 763–770. IEEE (2018)
21. Mountantonakis, M., Tzitzikas, Y.: Large-scale semantic integration of linked data: A survey. ACM Comput. Surv. **52**(5), 103:1–103:40 (2019)
22. Newman, M.E.J.: Modularity and community structure in networks. Proc. Nat. Acad. Sci. **103**(23), 8577–8582 (2006)
23. O'Donovan, P., Leahy, K., Bruton, K., O'Sullivan, D.T.J.: An industrial big data pipeline for data-driven analytics maintenance applications in large-scale smart manufacturing facilities. Journal of Big Data **2**(1), 1–26 (2015). https://doi.org/10.1186/s40537-015-0034-z
24. Palma, G., Vidal, M.-E., Raschid, L.: Drug-target interaction prediction using semantic similarity and edge partitioning. In: Mika, P., et al. (eds.) ISWC 2014. LNCS, vol. 8796, pp. 131–146. Springer, Cham (2014). https://doi.org/10.1007/978-3-319-11964-9_9
25. Velasquez, N., Estevez, E., Pesado, P.: Cloud computing, big data and the industry 4.0 reference architectures. J. Comput. Sci. Technol. **18**(03), e29 (2018)
26. Wang, Q., Mao, Z., Wang, B., Guo, L.: Knowledge graph embedding: a survey of approaches and applications. IEEE Trans. Knowl. Data Eng. **29**(12), 2724–2743 (2017)
27. Wang, Z., Zhang, J., Feng, J., Chen, Z.: Knowledge graph embedding by translating on hyperplanes. In: AAAI (2014)
28. Zeid, A., Sundaram, S., Moghaddam, M., Kamarthi, S., Marion, T.: Interoperability in smart manufacturing: research challenges. Machines **7**(2), 21 (2019)

# On Tensor Distances for Self Organizing Maps: Clustering Cognitive Tasks

Georgios Drakopoulos[1]([✉])(iD), Ioanna Giannoukou[2](iD), Phivos Mylonas[1](iD), and Spyros Sioutas[3]

[1] Department of Informatics, Ionian University,
Tsirigoti Sq. 7, 49100 Kerkyra, Greece
{c16drak,fmylonas}@ionio.gr
[2] Department of Management Science and Technology, University of Patras,
26504 Patras, Greece
igian@upatras.gr
[3] CEID, University of Patras, 26504 Patras, Greece
sioutas@ceid.upatras.gr

**Abstract.** Self organizing maps (SOMs) are neural networks designed to be in an unsupervised way to create connections, learned through a modified Hebbian rule, between a high- (the input vector space) and a low-dimensional space (the cognitive map) based solely on distances in the input vector space. Moreover, the cognitive map is segmentwise continuous and preserves many of the major topological features of the latter. Therefore, neurons, trained using a Hebbian learning rule, can approximate the shape of any arbitrary manifold provided there are enough neurons to accomplish this. Moreover, the cognitive map can be readily used for clustering and visualization. Because of the above properties, SOMs are often used in big data pipelines. This conference paper focuses on a multilinear distance metric for the input vector space which adds flexibility in two ways. First, clustering can be extended to higher order data such as images, graphs, matrices, and time series. Second, the resulting clusters are unions of arbitrary shapes instead of fixed ones such as rectangles in case of $\ell_1$ norm or circles in case of $\ell_2$ norm. As a concrete example, the proposed distance metric is applied to an anonymized and open under the Creative Commons license cognitive multimodal dataset of fMRI images taken during three distinct cognitive tasks. Keeping the latter as ground truth, a subset of these images is clustered with SOMs of various configurations. The results are evaluated using the corresponding confusion matrices, topological error rates, activation set change rates, and intra-cluster distance variations.

**Keywords:** SOM · Cognitive maps · Dimensionality reduction · Hebbian learning · Multilinear distance · Tensor algebra · Higher order data · Multi-aspect clustering · Cognitive tasks · fMRI imaging

© Springer Nature Switzerland AG 2020
S. Hartmann et al. (Eds.): DEXA 2020, LNCS 12392, pp. 195–210, 2020.
https://doi.org/10.1007/978-3-030-59051-2_13

# 1   Introduction

Self organizing maps (SOMs) are grids of computational neurons which can efficiently learn through a modified Hebbian process to approximate any higher dimensional manifold, called the *input vector space* and represented by a set of data points or data vectors, with a low dimensional space termed the *cognitive map*, which is one- or two-dimensional. This map can be then employed for clustering and visualization purposes. For this reason in a typical data science pipeline SOMs can be placed either at the beginning in order to discover outliers or at the end in order to cluster or visualize high dimensional results.

The approximation and the associated properties are made feasible by rearranging the namely connections between the input vector space and the cognitive map. Distances in the former are reflected to the latter with some loss of minor information, provided the cognitive map contains sufficient neurons to capture the desired input vector features in the geometry of the cognitive map. This ability as well as the segmentwise continuity of the latter can be attributed to the training process and to the distance metric in the input vector space, which can be theoretically arbitrary to a great extent allowing thus the progressive construction of variable shaped clusters. This is a distinct advantage over clustering algorithms which are based on fixed shaped regions. Nonetheless, most common distance metrics do not take advantage of this.

The primary research objective of this conference paper is the development of a multilinear distance metric for the data points of the input space. This lays the groundwork for more flexible cluster shapes in the cognitive map, as they can be the union of arbitrary shapes instead of fixed ones, clustering higher order data points such as graphs and matrices, and for multi-aspect clustering where each object in the input space can be represented by multiple but distinct data points. In order to evaluate the above, they have been tested with an open dataset, under the Creative Commons license from *openneuro.org*, containing fMRI images taken during three cognitive tasks conducted by elderly people. Since ground truth data is available, the confusion matrices have been derived for SOMs of various configurations. Moreover, the corresponding topological error rates, the activation set change rates, and the intra-cluster distance variations have been computed for each epoch during the cognitive map formulation.

The remaining of this work is structured as follows. In Sect. 2 the relevant scientific literature regarding the applications of blockchains, including process mining and IoT, is briefly summarized. Section 3 overviews the various SOM aspects. Section 4 overviews the clustering evaluation metrics, which can be either SOM-specific or generic. The proposed family of tensor-based distance metrics is proposed in Sect. 5, whereas the activation set variation is introduced in Sect. 6 where also the experimental results are given and commented on. Future research directions are described in Sect. 7. Tensors are represented by capital calligraphic letters, matrices with capital boldface, vectors with small boldface, and scalars with small letters. Vectors are assumed to be columns unless stated otherwise. When a function or a set depends on a number of parameters, then the latter are shown in the respective definition after a colon. Finally, Table 1 summarizes the notation of this work.

**Table 1.** Notation of this conference paper.

| Symbol | Meaning |
|---|---|
| $\overset{\triangle}{=}$ | Definition or equality by definition |
| $\{s_1, \ldots, s_n\}$ | Set with elements $s_1, \ldots, s_n$ |
| $(t_1, \ldots, t_n)$ | Tuple with elements $t_1, \ldots, t_n$ |
| $|S|$ | Set or tuple cardinality |
| $\text{loc}(\mathbf{s})$ | Grid location for data point $\mathbf{s}$ |
| $\text{invloc}(u)$ | Set of data points assigned to neuron $u$ |
| $\text{weight}(u)$ | Synaptic weights of neuron $u$ |
| $\text{neighb}(u_1, u_2)$ | Indicator function of neighboring neurons $u_1$ and $u_2$ |
| $\text{vec}(\mathcal{T})$ | Vector operator for tensor $\mathcal{T}$ |

## 2   Previous Work

SOMs have been proposed in [11] for dimensionality reduction, clustering, and visualization and since then their popularity remains unabated. The latter can be mainly attributed to the preservation of the topological properties of the original high dimensional space in the final low dimensional one [8,10]. Among the SOM applications can be found climatology models [7], knowledge discovery in gene expressions [25], and the novel clustering of ECG complexes [14]. A clustering method for strings based on SOMs is presented in [13]. Finding patterns with SOMs in collections of Web documents is explored in [9]. SOMs have been also used as a visualization tool in finance clustering long feature vectors [2]. Additionally, due to SOM popularity, a number of implementations exists. In [24] a Simulink library aiming at FPGAs is proposed, while [23] discusses a parallel SOM library based on the MapReduce paradigm.

Tensors have found multiple applications in fields such as signal processing [1], computer vision [22], and deep learning [29]. Unsupervised methods for face recognition based on the separation of higher order subspaces are described in [28] and extended in [20]. The term-document matrix of information retrieval is extended to a term-keyword-document tensor in [4] in order to exploit the added semantic value of keywords. Face recognition in the context of emotion discovery is discussed in [27]. In [6] a genetic algorithm is proposed for discovering communities based on linguistic and geographical criteria. Multilinear discriminant analysis is a higher order clustering approach based on tensor algebra [16]. A distinction between its primary approaches with an application to gait recognition is given in [19]. Finally, an LMS-based update scheme for tensor entries has been used in adaptive system identification as shown in [5].

# 3 Cognitive Maps

## 3.1 Input and Coordinate Spaces

In an SOM each neuron is placed on a one- or two-dimensional grid, with the latter being more common. In the following analysis it will be assumed that a two dimensional $p \times q$ grid is used. Without loss of generality, the following concepts can be applied also to one dimensional girds [12,15]. Once the training process is complete, the SOM is a cognitive map, namely a low dimensional representation of the data point space with the former preserving topological similarities and neighborhoods of the latter. This technique is believed to be performed by the hippocampus of the human brain in order to construct mental representations of the physical or perceived world [3,21,26].

Let $n$ denote the number of available data vectors or data points and $d$ their fixed dimensionality. The latter holds true even when the data points are of variable length such as strings. In this case, $d$ is taken to be the maximum string length.

**Definition 1 (Input vector space).** *The input space $\mathcal{V}$ is the d-dimensional space spanned by the set of the n data points $S \triangleq \{s_k\}_{k=1}^{n}$.*

In this work each data vector is assumed to consist of $d$ numerical features extracted from the dataset of Sect. 6:

$$s_k \triangleq \begin{bmatrix} s_{k,1} & s_{k,2} & \cdots & s_{k,d} \end{bmatrix}^T \tag{1}$$

**Definition 2 (Coordinate space).** *The coordinate space $\mathcal{C}$ is a $p \times q$ two dimensional discrete space composed of the $m = pq$ locations of the neurons.*

Each neuron $u_k$ represents a location vector in $\mathcal{C}$ as follows:

$$u_k \leftrightarrow \begin{bmatrix} x_k & y_k \end{bmatrix}^T \in \mathcal{C}, \quad 0 \leq k \leq m - 1 \tag{2}$$

Since multiple data vectors can be mapped to the same neuron, $\text{loc}\,(\cdot)$ is not a function as its inverse returns the set of data vectors $S_k \subseteq S$ defined as:

$$\text{invloc}\,(u_k) \triangleq \left\{ s_{k_1}, \ldots, s_{k_{f_k}} \right\} \triangleq S_k \tag{3}$$

**Definition 3 (Neighboring neurons).** *Any neuron adjacent in $\mathcal{C}$ to a given neuron $u_i$ is termed to be a neighbor of $u_i$.*

The designation of neighborhood depends on the grid number of dimensions, grid shape, and neighborhood strategy. Thus, in a square grid the neighboring neurons can form a cross, a square, or a hexagon around $u_i$. In this work the first choice is used. The following indicator function codifies the above:

$$\text{neighb}\,(u_i, u_j) \triangleq \begin{cases} 1, u_i \text{ and } u_j \text{ are neighboring} \\ 0, \text{otherwise} \end{cases} \tag{4}$$

Therefore, for each neuron $u_i$ the neighborhood set $\Gamma(u_i)$ is defined as follows:

$$\Gamma(u_i) \triangleq \{u_j \in \mathcal{C} \mid \text{neighb}(u_i, u_j) = 1\} \tag{5}$$

Each neuron $u_i$ in the grid has its own vector of synaptic weights $\mathbf{w_i}$:

$$\text{weight}(u_i) \triangleq \mathbf{w_i} \in \mathcal{V} \tag{6}$$

## 3.2 Distance Metrics and Weight Function

SOMs require three functions in order to construct the final cognitive map, namely the distance metrics of $\mathcal{V}$ and $\mathcal{C}$ and the weight function. As its name suggests, the distance metrics in $d(\cdot, \cdot)$ used between data points and synaptic weights. Common choices, depending on the nature of data points, are: $\ell_2$, $\ell_1$, or $\ell_0$ norms for numerical vectors, Hamming distance for binary vectors, Tanimoto coefficient for sets, and Levenshtein distance for strings.

Common distance metrics $g(\cdot, \cdot)$ in $\mathcal{C}$ include the triangular, square, circular, and Gaussian distances. Notice that it is not necessary to use the same distance metric in both $\mathcal{V}$ and $\mathcal{C}$. In fact, because these spaces are of different nature, a given metric may not be even applicable in one of these spaces.

Finally, the weight function $h(\cdot, \cdot)$, which is usually normalized such that its maximum value equals one, depends on the location of two neurons in the grid. In fact, in most cases the weight depends on some form of the grid distance, not to be confused with the function $g(\cdot, \cdot)$ defined earlier, between two neurons and not on their absolute locations. In certain scenaria $h(\cdot, \cdot)$ and $g(\cdot, \cdot)$ are related, by they do not need to be so in the general case. In this work these functions are independent of each other. Common options include:

- **Constant:** Each neuron in the proximity set receives the same weight.
- **Gaussian:** The Gaussian kernel has a smooth decay rate which drops fast enough to ensure the clusters remain relatively compact. It is defined as:

$$h(u_i, u_j; \sigma_0) \triangleq \frac{1}{\sigma_0 \sqrt{2\pi}} \exp\left(-\frac{\|u_i - u_j\|_1^2}{2\sigma_0^2}\right) \tag{7}$$

- **Triangular:** This function has a linear decay rate since:

$$h(u_i, u_j; \gamma_0) \triangleq 1 - \frac{\|u_i - u_j\|_1}{\gamma_0} \tag{8}$$

- **Circular:** The weight function forms a semicircle around neuron $u_j$:

$$h(u_i, u_j; \rho_0) \triangleq \frac{\sqrt{\rho_0^2 - \|u_i - u_j\|_2^2}}{\rho_0} \tag{9}$$

The weight function also plays a central role in the formation for each neuron of its proximity set, which is defined as follows:

**Definition 4 (Proximity set).** *The proximity set* $\Delta(u_i; \xi_0)$ *for a neuron* $u_i$ *given a threshold* $\xi_0$ *is the set of neurons for which* $h(\cdot, \cdot)$ *remains above* $\xi_0$:

$$\Delta(u_i; \xi_0) \triangleq \{u_j \in C \mid h(u_i, u_j) \geq \xi_0\} \tag{10}$$

The proximity set $\Delta(u_i; \xi_0)$ of a given neuron $u_i$ differs from its the neighborhood $\Gamma(u_i)$ introduced in Definition 3. The former corresponds to the effective area of a cluster around $u_i$ whereas the latter to the central area of that cluster. Therefore, $\Delta(u_i; \xi_0)$ always includes $\Gamma(u_i)$ but extends beyond that. The limiting case where the two sets coincide is very rarely used in practice, if at all.

### 3.3 Learning Rate

The learning rate $\eta$ of an SOM controls as in other neural network architectures the convergence rate of the training process.

- **Constant:** Under this policy the learning rate is a fixed constant $\eta_0$.
- **Exponential decay:** In order to ensure that early epochs receive a larger weight in comparison to the later ones, the following scheme can be used:

$$\eta[r; \alpha_0, \gamma_0] = \alpha_0 \exp(-\gamma_0 r) \tag{11}$$

- **Cosine decay:** This option is based on the following function:

$$\eta[r; \alpha_0, r_0] = \alpha_0 \cos\left(\frac{\pi r}{2r_0}\right), \quad 0 \leq r \leq r_0 - 1 \tag{12}$$

- **Inverse polynomial:** This rate attempts to achieve a smoother decay as:

$$\eta[r; \alpha_0, \beta_0, \gamma_0, \delta_0] = \frac{\delta_0}{(\alpha_0 r + \beta_0)^{\gamma_0}} \tag{13}$$

- **Inverse logarithm:** Finally, this choice yields the formula:

$$\eta[r; \alpha_0, \beta_0, \delta_0] \triangleq \frac{\delta_0}{a_0 \log(r) + \beta_0} \tag{14}$$

The learning rate can be allowed to vary within the same epoch or remain fixed and change only with each epoch. In the former case, the learning rate can very well be a product of two factors, one taking into account the epoch and the other depending on the projection within the epoch. As a general note, when the learning parameter is allowed to change within the same epoch, then it should be combined with a policy which does not preserve the order which the input vectors are projected to the SOM. Here $\eta$ remains constant during each epoch.

### 3.4 Grid Dimensions

In the general case there is no straightforward criterion to determine the grid size. This work follows the empirical rule proposed in [10] which takes into consideration only the total number of data points:

$$p = q = \lceil 5\sqrt{n} \rceil \tag{15}$$

## 3.5  Training Process

The synaptic weights of the winning neuron $u^*$ are updated in each projection of the epoch $r$ based on a modified Hebbian update rule as follows:

$$\Delta \text{weight}(u^*)[r] = \eta[r] \cdot d(\mathbf{s_i}, \text{weight}(u^*)) \tag{16}$$

The winning neuron $u^*$ for the specific input vector $\mathbf{s_i}$ is the neuron whose weight vector is closest in $\mathcal{V}$ to $\mathbf{s_i}$:

$$u^* \stackrel{\triangle}{=} \operatorname{argmin}\{d(\mathbf{s_i}, \text{weight}(u))\} \tag{17}$$

Notice that the distance metric in (17) is computed in $\mathcal{V}$ and not in $\mathcal{C}$ since data points belong in the former. Additionally, $u^*$ is added to the set $D[r]$ of activated neurons for that given epoch is formed, assuming that $D[r]$ is empty at the beginning of the epoch:

$$D[r] \leftarrow D[r] \cup u^* \tag{18}$$

Additionally, besides the winning neuron $u^*$ the synaptic weights of each neighboring neuron $u_i \in \Delta(u^*; \xi_0)$ are also updated as follows:

$$\Delta \text{weight}(u_j)[r] = \eta[r] \cdot h(u^*, u_j) \cdot d(\mathbf{s_k}, \text{weight}(u_j)) \tag{19}$$

Both updates (16) and (19) drive the weights of the corresponding neurons closer to the data point $\mathbf{s_k}$, gradually creating a cluster. Neurons in $\Delta(u^*; \xi_0)$ receive a reduced update because of the added factor of the weight function.

# 4  Performance Metrics

Perhaps the most common and intuitive SOM error metric is the topological error. As local coherence preservation is of primary importance in the final cognitive map, it is only natural to measure the average probability that a data point is assigned to the periphery of a cluster.

**Definition 5 (Topological error rate).** *The topological error $T[r]$ for epoch $r$ is the ratio of the number of data points which are mapped to $\Delta(u^*) \setminus \Gamma(u^*)$.*

$$T[r] \stackrel{\triangle}{=} \frac{1}{n} \sum_{\mathbf{s_k} \in S} |\{\operatorname{loc}(\mathbf{s_k}) \notin \{u \cup \Gamma(u) \mid u \in D[r]\}\}| \tag{20}$$

Another SOM performance metric is the activation set evolution, essentially the number of clusters formed at each epoch and the centroids thereof. The latter is not static but rather dynamic and it is controlled by the data points.

**Definition 6 (Activation set change rate).** *The change of activation set is the percentage of the activation set cardinalty during the current epoch.*

$$K[r] \stackrel{\triangle}{=} \frac{|D[r]| - |D[r-1]|}{|D[r]|}, \quad r \geq 1 \tag{21}$$

Clustering performance is usually difficult to evaluate. One common metric for assessing clustering is the average intra-clustering distance, namely the arithmetic mean of the average distsance of each data point from the corresponding centroid. In the case of the SOMs this metric takes the following form.

**Definition 7 (Intra-cluster distance).** *The intra-cluster distance is the average of the average distance of each data point from its respective centroid.*

$$Q\,[r] \triangleq \frac{1}{|D\,[r]|} \sum_{u_i \in D[r]} \frac{1}{|C_i|} \sum_{u_j \in C_i} g\,(u_j, u_i) \qquad (22)$$

## 5   Tensor Distance

Here the multilinear distance metric, based on work previously done [17], is introduced. First, an auxiliary definition is in order.

**Definition 8 (Tensor vector operator).** *The tensor vector operator maps a P-th order tensor $T \in \mathbb{R}^{I_1 \times I_2 \times \dots \times I_P}$ to a vector $\mathbf{v} \in \mathbb{R}^N$ by stacking from top to bottom the columns of dimension $I_k$ from left to right. Also $N = I_1 \cdot \dots \cdot I_P$.*

The proposed distance is defined as follows:

**Definition 9 (Multilinear distance).** *The multilinear distance metric is defined as the squared root weighted quadratic form of the vectorized difference of the data points where the latter are represented as tensors.*

$$J\,(\mathcal{X}, \mathcal{Y}; \mathbf{G}) \triangleq \left( \mathrm{vec}\,(\mathcal{X} - \mathcal{Y})^T\, \mathbf{G}\, \underbrace{\mathrm{vec}\,(\mathcal{X} - \mathcal{Y})}_{\mathbf{s}} \right)^{\frac{1}{2}} = \sqrt{\mathbf{s}^T \mathbf{G} \mathbf{s}} \qquad (23)$$

Metric $J$ includes bilinear and quadratic functions as special cases. This allows more flexibility as different tensor components can receive different weights.

The weight matrix $\mathbf{G}$ is defined in [17] elementwise as the Gaussian kernel whose values depend only on the locations of the elements of $\mathbf{s}$ involved. Although this resulted in a matrix with non-negative entries, a better way is to define $\mathbf{G}$ as positive definite, meaning that the value of the quadratic function under the square root in (23) is guaranteed to be positive.

Moreover, it makes sense to select a $\mathbf{G}$ tailored to the underlying nature of the data points. Since in the experiments the data points are stacked images, the following two properties of the discrete cosine transform (DCT) will be exploited: First, since images consist of real values, the transformation spectrum will also contain only real values. Second this spectrum will be very sparse, with non-zero values concentrated along the low spatial frequencies. Therefore, subtracting and weighing through a linear filter only these values instead of every pixel saves tremendous computational time.

The original data points are three stacked fMRI images for each subject:

$$\mathbf{X}_k \triangleq \left[\ \mathbf{X}_{k,1}\ |\ \mathbf{X}_{k,2}\ |\ \mathbf{X}_{k,3}\ \right] \qquad (24)$$

Therefore, $\mathbf{G}$ is built as a matrix product involving the discrete DCT matrix $\mathbf{D}$, the linear filtering matrix $\mathbf{F}$, the identity matrix $\mathbf{I}$, and the $3 \times 1$ auxiliary vector $\mathbf{1}_{3,1}$ whose elements equal 1 as follows:

$$\mathbf{G} \triangleq (\mathbf{1}_{3,1} \otimes \mathbf{D})^T \cdot (\mathbf{I} \otimes \mathbf{F}) \cdot (\mathbf{1}_{3,1} \otimes \mathbf{D}) \tag{25}$$

In Eq. (25) $\otimes$ denotes the Kronecker tensor product which is used to create the following structured matrices:

$$\mathbf{1}_{3,1} \otimes \mathbf{D} = \begin{bmatrix} \mathbf{D} \\ \mathbf{D} \\ \mathbf{D} \end{bmatrix} \quad \mathbf{I} \otimes \mathbf{F} = \begin{bmatrix} \mathbf{F} & & \\ & \mathbf{F} & \\ & & \mathbf{F} \end{bmatrix} \tag{26}$$

The linear filter matrix $\mathbf{F}$ has been selected to be the first order lowpass or smoothing Butterworth filter.

The advantages of using a higher order distance metric such as that of Eq. (23) are the following:

– It enables multi-aspect clustering, meaning that the various objects of the input vector space can be clustered in more than one ways. Thus, different aspects of the same object may contribute to clustering. Alternatively, this can be interpreted as that each object is represented by multiple attribute vectors, which is the basis for many advanced clustering algorithms.
– There is more flexibility in defining weights for the various attribute vectors or aspects, which can help understand the underlying clustering dynamics.
– Multilinear metrics are suitable for evaluating the distance between higher order data such as graphs, tensors, time series, and images in a natural way.

## 6  Results

The anonymized dataset for the experiments was first published in [18] as part of a project for studying the ageing human brain and uploaded to *openneuro.org* under the Creative Commons license[1]. Following the warnings of the respository, the files pertaining to one subject were ignored. According to the dataset authors, the latter consists of 34 participants (33 in our case), 14 male and 20 (19 in our case) female, all right handed with no health problems. Each participant was presented for three seconds an image of neutral or negative affective polarity selected from the International Affective Picture System and then was issued an audio instruction to *suppress, maintain,* or *enhance* their emotions. These are the three cognitive tasks, respectively denoted as **T1**, **T2**, and **T3**.

Again according to [18], the fMRI images used here were collected with the following parameters. There were 30 sagittal images with each such image having a 3 mm thickness with repetition time (TR) and echo time (TE) being respectively equal to 2000 msec and 30 msec, a slice gap of 33%, a field of view (FoV)

---

[1] Dataset doi:10.18112/openneuro.ds002366.v1.0.0.

of $192 \times 192$ mm$^2$, and a spatial resolution of 3 mm. From these images were kept three, one each at 10%, 50%, and 90% of the block length. Finally, after normalization, registration, and correcting for magnetic field inhomogenetity and head motion, spatial preprocessing took place with a Gaussian kernel of 6 mm FWHM. Table 2 has the total and marginal frequency distribution for the three cognitive tasks and the image polarity. It follows that not only the three classes but also their subclasses are well balanced.

**Table 2.** Distribution of tasks.

|          | T1 | T2 | T3 | Subtotal |
|----------|----|----|----|----------|
| Neutral  | 6  | 5  | 5  | 16       |
| Negative | 6  | 6  | 5  | 17       |
| Subtotal | 12 | 11 | 10 | 33       |

The configuration of an SOM is represented by a tuple $c_k$:

$$c_k \stackrel{\triangle}{=} (p, q, d\,(\cdot, \cdot), g\,(\cdot, \cdot), h\,(\cdot, \cdot), \eta) \tag{27}$$

The possible options for each field of $c_k$ are given in Table 3. Out of the 27 possible combinations of metrics for $\mathcal{V}$, $\mathcal{C}$, and learning parameter rate, in total 9 were chosen with emphasis given to the learning parameter rate.

**Table 3.** SOM configuration options.

| $d\,(\cdot, \cdot)$ | $g\,(\cdot, \cdot)$ | $h\,(\cdot, \cdot)$ | $\eta$ |
|---------------------|---------------------|---------------------|--------|
| T: tensor           | G: Gaussian         | G: Gaussian         | S: cosine |
| L2: $\ell_2$ norm   | C: circular         | C: circular         | E: exponential |
| L1: $\ell_1$ norm   | R: triangle         | R: triangle         | P: inverse polynomial |

The values used in each function as needed by each SOM configuration are shown in Table 4. Recall that $p_0$ and $q_0$ are the SOM tableau dimensions, as

**Table 4.** Values for each configuration parameter.

| Param. | Value           | Param.       | Value |
|--------|-----------------|--------------|-------|
| $r_0$  | 25              | S            | $\alpha_0$: 1.0 |
| G      | $\sigma_0$: 4   | E            | $(\alpha_0, \gamma_0)$: $(1.0, 0.5)$ |
| C      | $\rho_0$: 5     | P            | $(\alpha_0, \beta_0, \gamma_0, \delta_0)$: $(1.0, 1.0, 2, 1.0)$ |
| R      | $\gamma_0$: 0.2 | $p_0, q_0$   | 30, 30 |

determined by (15), and $r_0$ is the fixed number of epochs used. Notice that $p_0$ and $q_0$ were selected to be identical resulting in square tableau, but this is by no means mandatory.

The actual configurations used in the experiments along with their numbering are shown in Table 5. As it can be seen, the exact same number of combinations was used with the $\ell_1$ norm, the $\ell_2$ norm, and the proposed multilinear metric. Configurations achieving the best confusion matrix in terms of maximizing its trace, one for each distance metric in $\mathcal{V}$, are marked in boldface.

**Table 5.** SOM configurations.

| # | Configuration | # | Configuration | # | Configuration |
|---|---|---|---|---|---|
| 1 | $(p_0, q_0, L1, C, C, E)$ | 10 | $(p_0, q_0, L2, C, C, E)$ | 19 | $(p_0, q_0, T, C, C, E)$ |
| 2 | $(p_0, q_0, L1, C, C, P)$ | 11 | $(p_0, q_0, L2, C, C, P)$ | 20 | $(p_0, q_0, T, C, C, P)$ |
| 3 | $(p_0, q_0, L1, C, C, S)$ | 12 | $(p_0, q_0, L2, C, C, S)$ | 21 | $(p_0, q_0, T, C, C, S)$ |
| 4 | $(p_0, q_0, L1, R, R, E)$ | 13 | $(p_0, q_0, L2, R, R, E)$ | 22 | $(p_0, q_0, T, R, R, E)$ |
| 5 | $(p_0, q_0, L1, R, R, P)$ | 14 | $(p_0, q_0, L2, R, R, P)$ | 23 | $(p_0, q_0, T, R, R, P)$ |
| 6 | $(p_0, q_0, L1, R, R, S)$ | 15 | $(p_0, q_0, L2, R, R, S)$ | 24 | $(p_0, q_0, T, R, R, S)$ |
| 7 | $(p_0, q_0, L1, G, G, E)$ | 16 | $(p_0, q_0, L2, G, G, E)$ | 25 | $(p_0, q_0, T, G, G, E)$ |
| 8 | $\mathbf{(p_0, q_0, L1, G, G, P)}$ | 17 | $(p_0, q_0, L2, G, G, P)$ | 26 | $(p_0, q_0, T, G, G, P)$ |
| 9 | $(p_0, q_0, L1, G, G, S)$ | 18 | $\mathbf{(p_0, q_0, L2, G, G, S)}$ | 27 | $\mathbf{(p_0, q_0, T, G, G, S)}$ |

**Table 6.** Best (left, configuration 8) and average confusion matrix for $\ell_1$.

| Task(%) | T1 | T2 | T3 | T1 | T2 | T3 |
|---|---|---|---|---|---|---|
| T1 | 78.66 | 8.33 | 13.00 | 77.11 | 11.89 | 11.00 |
| T2 | 10.66 | 75.00 | 14.33 | 11.00 | 74.50 | 14.50 |
| T3 | 4.00 | 7.00 | 79.00 | 5.20 | 8.20 | 76.60 |

Tables 6, 7, and 8 are the confusion matrices when the SOM distance metric is the $\ell_1$ norm, the $\ell_2$ norm, and that of Eq. (23) respectively for the SOM configuration which achieved the least misclassification rate, in other words it maximized the trace of the corresponding confusion matrix. In all three cases the best case entries are distinct from but not very far from their average case countrparts. This points to the best case scenario being actually achievable and not being merely a fortuitous but rare result.

Moreover, from the entries of Tables 6, 7, and 8 follows that the proposed multilinear distance metric consistently results in lower misclassification rates. This can be attributed to the additional flexibility offered by it, in terms of partitioning the data input space to arbitrary regions and of discovering latent

**Table 7.** Best (left, configuration 18) and average confusion matrix for $\ell_2$.

| Task(%) | T1 | T2 | T3 | T1 | T2 | T3 |
|---|---|---|---|---|---|---|
| T1 | 79.66 | 8.00 | 12.33 | 78.00 | 7.50 | 14.50 |
| T2 | 9.00 | 81.00 | 10.00 | 7.30 | 79.40 | 13.30 |
| T3 | 7.66 | 8.33 | 83.00 | 5.00 | 13.50 | 81.50 |

**Table 8.** Best (left, config. 27) and average confusion matrix for multilinear metric.

| Task(%) | T1 | T2 | T3 | T1 | T2 | T3 |
|---|---|---|---|---|---|---|
| T1 | 91.01 | 6.99 | 2.00 | 88.50 | 11.16 | 1.33 |
| T2 | 2.33 | 85.33 | 12.33 | 2.00 | 84.00 | 14.00 |
| T3 | 5.00 | 8.66 | 87.33 | 3.00 | 11.50 | 86.50 |

**Table 9.** Performance scores vs configuration.

| # | Scores | # | Scores | # | Scores |
|---|---|---|---|---|---|
| 1 | 7.58, 0.66, 15.11, 1.15 | 10 | 6.22, 0.74, 12.54, 1.13 | 19 | 6.02, 0.82, 10.34, 7.22 |
| 2 | 7.61, 0.64, 15.43, 1.13 | 11 | 6.71, 0.73, 13.53, 1.11 | 20 | 6.17, 0.81, 10.84, 7.21 |
| 3 | 7.65, 0.61, 15.04, 1.19 | 12 | 6.43, 0.69, 13.42, 1.14 | 21 | 6.39, 0.79, 10.72, 7.28 |
| 4 | 8.56, 0.68, 14.48, 1.17 | 13 | 6.25, 0.76, 12.52, 1.12 | 22 | 5.82, 0.83, 9.11, 7.25 |
| 5 | 8.71, 0.67, 14.94, 1.15 | 14 | 6.36, 0.73, 12.55, 1.12 | 23 | 5.97, 0.81, 9.34, 7.23 |
| 6 | 8.62, 0.64, 14.65, 1.13 | 15 | 6.35, 0.74, 12.67, 1, 11 | 24 | 6.08, 0.81, 9.56, 7.22 |
| 7 | 6.93, 0.71, 13.43, 1.13 | 16 | 6.23, 0.77, 12.13, 1.13 | 25 | 5.35, 0.84, 8.56, 7.24 |
| 8 | 7.03, 0.67, 13.46, 1.11 | 17 | 6.55, 0.74, 12.62, 1.11 | 26 | 5.55, 0.82, 8.99, 7.25 |
| 9 | 7.11, 0.69, 13.77, 1.12 | 18 | 6.42, 0.75, 12.37, 1.13 | 27 | 5.47, 0.81, 8.84, 7.22 |

features in the image triplets. Also, since there were divisions by 33, the total number of subjects, fractions like 0.33 and 0.66 were very frequent.

In Table 6 there is a trend for **T2** to be confused with **T3** and vice-versa. Specifically, the entries denoting a confusion between **T2** and **T3** and vice versa were significantly higher than those recording confusion between **T1** and **T3** and vice versa as well as **T1** and **T3** and vice versa.

From Table 7 is clear that $\ell_2$ norm outperforms $\ell_1$. This is an indication that spheres may be a better way to partition the data input space, perhaps due to their isotropic curvature. The strong confusion between **T2** and **T3** continues.

Table 8 contains the best values both in the optimal and in the mean case. The confusion trend between **T2** and **T3** is also present.

Table 9 shows the average topological error rate, the average absolute activation set change rate, the average inter-cluster distance variation, and the mean wallcklock time (in seconds) in this order for each configuration. Each SOM configuration was executed eleven times, with the first time being considered a test and thus not contributing at all to the time and performance measurements presented here. The numbering is exactly the same as that of Table 5.

Since the SOM training is computationally intensive, it is a reasonable indicator of the true computational time. Although the proposed metric is more expensive, its cost is not prohibitive. There is minimum variation among the time measurements for each distance metric with the $\ell_1$ and $\ell_2$ norms achieving very similar scores.

Figure 1 depicts the mean topological error rate scores of Table 9 in order to obtain a intuitive evaluation of the indicative performance of each configuration. The remaining three scores were omitted to avoid cluttering.

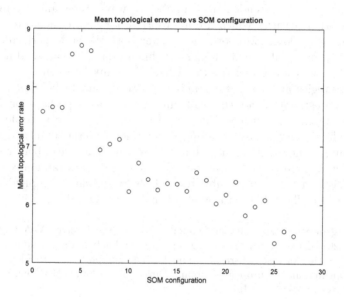

**Fig. 1.** Mean topological error vs SOM configuration.

Certain conclusions can be drawn regarding the various SOM configurations. Partitioning the cognitive map to Gaussian regions leads to better performance scores, followed by spherical and rectangular regions. The same can be said about the cosine decay rate. Also, the tensor distance metric yield systematically better scores at all three performance metrics.

The conclusion which can be drawn about the three cognitive tasks is that there is a systematic trend for confusion between **T2** and **T3** as in all cases the confusion between these two was consistently higher regardless of metric distance. This may indicate a tendency for negative emotions to persist in older adults, a view also shared by [18].

## 7 Conclusions and Future Work

This conference paper describes the application of SOMs to the clustering of fMRI image triplets in order to distinguish between three cognitive tasks. The SOMs configurations had different combinations of distance metrics for input and neuron space, learning rate decay rate, and neighbourhood shape. The three distance metrics for the data input space were the $\ell_1$ norm, the $\ell_2$ norm, and a multilinear metric. The distinction between the cognitive tasks is in compliance with established results. Moreover, the proposed metric achieves lower topological error, intra-cluster distance, and activation set change rate, whereas it resulted in a confusion matrix with the highest trace.

This work can be extended in a number of ways. First and foremost, the selection of other features which are appropriate for high dimensional cultural datasets can be explored. Moreover, evaluating both the SOM performance and the activation set variation with larger benchmark datasets can shed more light in their behavior. Concerning the SOM itself, the link between the data point selection strategies and the total number of epochs should be explored. Also, an approriate estimation for the grid dimensions, based perhaps on information theoretic measures such as the AIC, BIC, or MDL, is a possible research direction. The development of training termination criteria connected with the overall quality of the cognitive map should be sought. The role of bias in ensuring that no inactive neurons exist, especially for large grids with randomly initalized synaptic weights is another possible topic. How SOMs can be applied to online clustering, especially in big data pipelines, is worth researching.

**Acknowledgment.** This conference paper is part of the Interreg V-A Greece-Italy Programme 2014-2020 project "Fostering capacities and networking of industrial liaison offices, exploitation of research results, and business support" (ILONET), co-funded by the European Union, European Regional Development Funds (ERDF), and by the national funds of Greece and Italy.

## References

1. Cichocki, A., et al.: Tensor decompositions for signal processing applications: from two-way to multiway component analysis. IEEE Signal Process. Mag. **32**(2), 145–163 (2015)
2. Deboeck, G., Kohonen, T.: Visual Explorations in Finance with Self-Organizing Maps. Springer, Heidelberg (2013). https://doi.org/10.1007/978-1-4471-3913-3
3. Downs, R.M., Stea, D.: Cognitive maps and spatial behavior: process and products. Adaline (1973)
4. Drakopoulos, G., Kanavos, A., Karydis, I., Sioutas, S., Vrahatis, A.G.: Tensor-based semantically-aware topic clustering of biomedical documents. Computation **5**(3), 34 (2017). https://doi.org/10.3390/computation5030034
5. Drakopoulos, G., Mylonas, P., Sioutas, S.: A case of adaptive nonlinear system identification with third order tensors in TensorFlow. In: INISTA, pp. 1–6 (2019). https://doi.org/10.1109/INISTA.2019.8778406

6. Drakopoulos, G., et al.: A genetic algorithm for spatiosocial tensor clustering. EVOS **1**(11) (2019). https://doi.org/10.1007/s12530-019-09274-9
7. Hewitson, B., Crane, R.G.: Self-organizing maps: applications to synoptic climatology. Clim. Res. **22**(1), 13–26 (2002)
8. Kangas, J.A., Kohonen, T.K., Laaksonen, J.T.: Variants of self-organizing maps. IEEE Trans. Neural Netw. **1**(1), 93–99 (1990)
9. Kaski, S., Honkela, T., Lagus, K., Kohonen, T.: WEBSOM-self-organizing maps of document collections. Neurocomputing **21**(1–3), 101–117 (1998)
10. Kiviluoto, K.: Topology preservation in self-organizing maps. In: ICNN, vol. 1, pp. 294–299. IEEE (1996)
11. Kohonen, T.: The self-organizing map. Proc. IEEE **78**(9), 1464–1480 (1990)
12. Kohonen, T.: Exploration of very large databases by self-organizing maps. In: ICNN. vol. 1, pp. PL1-PL6. IEEE (1997)
13. Kohonen, T., Somervuo, P.: Self-organizing maps of symbol strings. Neurocomputing **21**(1–3), 19–30 (1998)
14. Lagerholm, M., Peterson, C., Braccini, G., Edenbrandt, L., Sornmo, L.: Clustering ECG complexes using Hermite functions and self-organizing maps. IEEE Trans. Biomed. Eng. **47**(7), 838–848 (2000)
15. Lampinen, J., Oja, E.: Clustering properties of hierarchical self-organizing maps. J. Math. Imaging Vis. **2**(2–3), 261–272 (1992)
16. Li, Q., Schonfeld, D.: Multilinear discriminant analysis for higher-order tensor data classification. TPAMI **36**(12), 2524–2537 (2014)
17. Liu, Y., Liu, Y., Zhong, S., Chan, K.C.: Tensor distance based multilinear globality preserving embedding: a unified tensor based dimensionality reduction framework for image and video classification. Expert Syst. Appl. **39**(12), 10500–10511 (2012)
18. Lloyd, W., Morriss, J., Macdonald, B., Joanknecht, K., Sigurd, J., van Reekum, C.: Longitudinal change in executive function is associated with impaired top-down frontolimbic regulation during reappraisal in older adults. bioRxiv (2019)
19. Lu, H., Plataniotis, K., Venetsanopoulos, A.: Uncorrelated multilinear discriminant analysis with regularization for gait recognition. In: Biometrics Symposium, pp. 1–6. IEEE (2007)
20. Lu, H., Plataniotis, K.N., Venetsanopoulos, A.N.: Uncorrelated multilinear discriminant analysis with regularization and aggregation for tensor object recognition. IEEE Trans. Neural Netw. **20**(1), 103–123 (2008)
21. O'Keefe, J., Nadel, L.: The Hippocampus as a Cognitive Map. Clarendon Press, Oxford (1978)
22. Shashua, A., Hazan, T.: Non-negative tensor factorization with applications to statistics and computer vision. In: ICML, pp. 792–799. ACM (2005)
23. Sul, S.J., Tovchigrechko, A.: Parallelizing BLAST and SOM algorithms with MapReduce-MPI library. In: International Symposium on Parallel and Distributed Processing, pp. 481–489. IEEE (2011)
24. Tisan, A., Cirstea, M.: SOM neural network design- a new Simulink library based approach targeting FPGA implementation. Math. Comput. Simul. **91**, 134–149 (2013)
25. Törönen, P., Kolehmainen, M., Wong, G., Castrén, E.: Analysis of gene expression data using self-organizing maps. FEBS Lett. **451**(2), 142–146 (1999)
26. Tversky, B.: Cognitive maps, cognitive collages, and spatial mental models. In: Frank, A.U., Campari, I. (eds.) COSIT 1993. LNCS, vol. 716, pp. 14–24. Springer, Heidelberg (1993). https://doi.org/10.1007/3-540-57207-4_2

27. Vasilescu, M.A.O., Terzopoulos, D.: Multilinear analysis of image ensembles: TensorFaces. In: Heyden, A., Sparr, G., Nielsen, M., Johansen, P. (eds.) ECCV 2002. LNCS, vol. 2350, pp. 447–460. Springer, Heidelberg (2002). https://doi.org/10.1007/3-540-47969-4_30
28. Yan, S., Xu, D., Yang, Q., Zhang, L., Tang, X., Zhang, H.J.: Multilinear discriminant analysis for face recognition. IEEE Trans. Image Process. **16**(1), 212–220 (2006)
29. Yu, D., Deng, L., Seide, F.: The deep tensor neural network with applications to large vocabulary speech recognition. IEEE Trans. Audio Speech Lang. Process. **21**(2), 388–396 (2012)

# Machine Learning

# MUEnsemble: Multi-ratio Undersampling-Based Ensemble Framework for Imbalanced Data

Takahiro Komamizu[✉], Risa Uehara, Yasuhiro Ogawa, and Katsuhiko Toyama

Nagoya University, Nagoya, Japan
taka-coma@acm.org

**Abstract.** Class imbalance is commonly observed in real-world data, and it is still problematic in that it hurts classification performance due to biased supervision. Undersampling is one of the effective approaches to the class imbalance. The conventional undersampling-based approaches involve a single fixed sampling ratio. However, different sampling ratios have different preferences toward classes. In this paper, an undersampling-based ensemble framework, MUEnsemble, is proposed. This framework involves weak classifiers of different sampling ratios, and it allows for a flexible design for weighting weak classifiers in different sampling ratios. To demonstrate the principle of the design, in this paper, three quadratic weighting functions and a Gaussian weighting function are presented. To reduce the effort required by users in setting parameters, a grid search-based parameter estimation automates the parameter tuning. An experimental evaluation shows that MUEnsemble outperforms undersampling-based methods and oversampling-based state-of-the-art methods. Also, the evaluation showcases that the Gaussian weighting function is superior to the fundamental weighting functions. In addition, the parameter estimation predicted near-optimal parameters, and MUEnsemble with the estimated parameters outperforms the state-of-the-art methods.

**Keywords:** Imbalanced classification · Resampling · Undersampling · Ensemble · Gaussian function

## 1 Introduction

Class imbalance is still a crucial problem in real-world applications that degrades classification performance, especially with minority classes. Class imbalance refers to a situation with datasets in which the number of examples in a class is much larger than that in the other classes. The large difference in terms of the numbers of examples causes classifiers to be biased toward the majority class. Class imbalance has been observed and dealt with in various domains, such as the clinical domain [5], economic domain [25] and agricultural domain [28], and in software engineering [26] and computer networks [11].

© Springer Nature Switzerland AG 2020
S. Hartmann et al. (Eds.): DEXA 2020, LNCS 12392, pp. 213–228, 2020.
https://doi.org/10.1007/978-3-030-59051-2_14

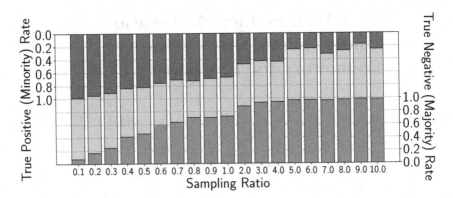

**Fig. 1.** The small sampling ratios prefer the minority, vice versa.

Resampling is one of the effective solutions for class imbalance, and it has been widely studied [3,7,19,29]. Resampling techniques can be roughly classified into two categories: oversampling (e.g., SMOTE [7] and SWIM [29]) and undersampling (e.g., EasyEnsemble [19] and RUSBoost [27]). Undersampling is a simple and powerful resampling technique for dealing with class imbalance [10]. Not only single-shot undersampling but also combining multiple undersampled datasets in an ensemble manner have been applied for the problem [15,19,27].

A preliminary survey on various datasets on the effects of different undersampling ratios, shown in Fig. 1, indicated that different undersampling ratios have different preferences toward classes. Sampling ratio refers to the ratio of the sampled majority size over the minority size. In this paper, the minority class and the majority class are regarded as the positive class and the negative class, respectively. A sampling ratio 1.0 means that the majority examples are randomly selected so that the number of sampled examples equals that of the minority examples. A ratio below 1.0 means that the number of sampled majority examples is below that of the minority examples. In this paper, this is called *excessive undersampling* and its antonym is *moderate undersampling*. The figure depicts the true positive and negative ratios for different sampling ratios in the Abalone dataset. It indicates that classifiers learned with excessively undersampled datasets favor the minority class and those by moderately undersampled datasets favor the majority class and 1.0 may not be best balanced ratio.

This paper is an attempt to combine classifiers in different sampling ratios by using an ensemble framework, **MUEnsemble**. First, the majority are undersampled with different sampling ratios. To utilize the large portion of majority examples, multiple undersampled datasets are generated for each sampling ratio, and a base ensemble classifier is learned by using the multiple datasets. Then, the overall classifier is an ensemble of the base classifiers.

To weight the weak classifiers at different sampling ratios, MUEnsemble includes a programmable balancing function that calculates the number of weak classifiers depending on the sampling ratio. In this paper, three fundamental functions (*constant, convex, concave*) and an advanced function (*Gaussian*) are designed as the balancing functions. Also, to reduce effort spent on setting parameters, a grid search is applied to optimize the parameters of the functions.

MUEnsemble was evaluated by using 19 publicly available real-world datasets. It was compared with the state-of-the-art methods, and it outperformed these methods. In addition, three lessons were learned. (1) The excessive undersampling has positive effect, (2) the Gaussian function is the best balancing function due to its flexibility, and (3) the grid search finds near-optimal parameters.

The contributions of this paper are abridged as follows.

- **Multi-ratio Undersampling-based Ensemble Framework:** In this paper, an ensemble framework, MUEnsemble, for imbalanced data is proposed. MUEnsemble employs multiple undersampling ratios including excessive undersampling. It learns an ensemble classifier for each sampling ratio and combines the ensemble classifiers among sampling ratios in an ensemble manner. It has a programmable function, called the balancing function, for weighting classifiers in different sampling ratios.
- **Investigation on Balancing Function:** MUEnsemble controls the number of votes from different sampling ratios by changing the number of weak classifiers for each sampling ratio. In this paper, three fundamental functions and one function based on the Gaussian function are introduced. Due to its flexibility, the Gaussian function performs the best.
- **Experimentation on Various Public Datasets:** MUEnsemble is compared with the state-of-the-art resampling methods by using various datasets that are publicly available. The datasets include various imbalance ratios in a wide range of domains. The experiment demonstrates that excessive undersampling has a positive effect, and the Gaussian balancing function with grid search optimization outperforms the state-of-the-art methods.

The rest of this paper is organized as follows. Section 2 introduces related resampling-based approaches for imbalanced data. Section 3 explains the proposed framework, MUEnsemble, and four balancing functions. Section 4 demonstrates the effectiveness of MUEnsemble over the state-of-the-art methods, and the effects of the excessive undersampling, the balancing functions, and the parameter optimization are discussed. Finally, Sect. 5 concludes this paper.

## 2    Related Work: Resampling Approaches

To deal with class imbalance, there are basically three groups of approach, namely, resampling, cost-adjustment [9,16], and algorithm modification [1,32]. Resampling is the most commonly used, because it has been shown to have a robust performance and is applicable to any classifiers. Resampling approaches are roughly classified into two categories, namely, oversampling and undersampling.

## 2.1  Oversampling-Based Approaches

A simple oversampling approach is to randomly copy minority examples so that the numbers of minority and majority examples become the same. This approach easily causes overfitting. To cope with the overfitting problem, oversampling approaches generate synthetic minority examples that are close to the minority. SMOTE [7] is the most popular synthetic oversampling method. It generates synthetic minority examples based on the nearest neighbor technique. Since SMOTE does not take majority examples into consideration, the generated examples can easily overlap with majority examples. This degrades the classification performance. To overcome the weakness of SMOTE, more recent approaches have incorporated majority examples into the resampling process. SMOTE-Tomek [2] and SMOTE-ENN [3] employ data cleansing techniques including the removal of Tomek links [31] and Edit Nearest Neighbours [33]. Along this line, there are more advanced approaches (e.g., ADASYN [13], borderline-SMOTE [12], and SVM-SMOTE [23]). One of the state-of-the-art synthetic oversampling approaches is SWIM [4,29]. SWIM utilizes the density of each minority example with respect to the distribution of majority examples in order to generate synthetic minority examples.

## 2.2  Undersampling-Based Approaches

Undersampling-based approaches can be classified into three categories: example selection, iterative learning, and ensemble. Example selection is an approach to choosing majority examples that are expected to contribute to better classification. Major approaches choose majority examples hard to distinguish from minority examples. NearMiss [22] samples majority examples close to the minority examples. Instance hardness [30] is a hardness property that indicates the likelihood that an example is misclassified.

Iterative learning is a learning method that gradually changes majority examples. For each iteration, iterative learning approaches remove a part of the majority examples. BalanceCascade [19] is an iterative method that removes correctly classified majority examples. RUSBoost [27] is a weighted random undersampling approach for removing majority examples that are likely to be classified correctly. EUSBoost [20] introduces a cost-sensitive weight modification and an adaptive boundary decision strategy to improve the model performance. Trainable Undersampling [24] is the state-of-the-art in this category. It trains a classifier by reinforcement learning.

The ensemble approach is to combine multiple weak classifiers each of which is learned on individual undersampled training data in an ensemble manner. Ensemble of Undersampling [15] is one of the earlier ensemble methods using undersampled training data. EasyEnsemble [19] is an ensemble-of-ensemble approach that ensembles AdaBoost classifiers for each undersampled training data. In [10], a comprehensive experiment is reported on ensemble and iterative approaches. It shows that RUSBoost and EasyEnsemble are the best performing ensemble methods, and they outperform oversampling-based approaches.

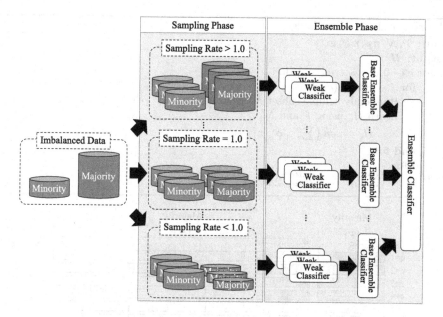

**Fig. 2.** An Overview of MUEnsemble

MUEnsemble is classified in the ensemble category. The distinctions of MUEnsemble over existing undersampling approaches are as follows. First, MUEnsemble is the first ensemble approach combining multiple sampling ratios. Second, it includes excessive undersampling, which has not been considered among these approaches. Third, it arranges the numbers of weak classifiers, while other ensemble methods fix the numbers to be constant.

## 3    MUEnsemble

MUEnsemble is an ensemble framework that involves multiple sampling ratios. Figure 2 is an overview of MUEnsemble. MUEnsemble consists of two phases: the *sampling phase* and *ensemble phase*. In the sampling phase, first, the number $n$ of sampling ratios is determined. Second, for each sampling ratio, the input data are undersampled several times to make a batch of undersampled datasets. In the ensemble phase, each batch of undersampled datasets is used for learning a *base* ensemble classifier. Last, the base ensemble classifiers for all sampling ratios are combined into a classifier in an ensemble manner. In MUEnsemble, there are two major parameters: the set of sampling ratios and the weights on the base classifiers. In this paper, the weights are considered to be dominant factors in classification performance. Therefore, to decide a set of sampling ratios, MUEnsemble employs a simple splitting strategy. For the weights of the base classifiers, in this paper, the number of votes (the number of weak classifiers, in other words) in a sampling ratio is used as the weight. For extensibility,

**Algorithm 1.** MUEnsemble

**Input:** $\mathcal{P},\mathcal{N},n_\mathcal{P},n_\mathcal{N},w$
**Output:** $H$
1: **for** $c \leftarrow 0$ to $n_\mathcal{P} + n_\mathcal{N} + 1$ **do**
2:     **for** $i \leftarrow 0$ to $B(c)$ **do**
3:         Randomly sample subset $\mathcal{N}'$ from $\mathcal{N}$ s.t. $|\mathcal{N}'| : |\mathcal{P}| = R(c, n_\mathcal{P}, n_\mathcal{N}) : 1$
4:         Learn $H_i^c$ using $\mathcal{P}$ and $\mathcal{N}'$
$$H_i^c = sgn\left(\sum_{j=1}^{w} \alpha_{i,j}^c h_{i,j}^c\right)$$
5:     **end for**
6: **end for**
7: $H = sgn\left(\sum_{c=1}^{n_\mathcal{P}+n_\mathcal{N}+1} \sum_{i=1}^{B(c)} \sum_{j=1}^{w} \alpha_{i,j}^c h_{i,j}^c\right)$

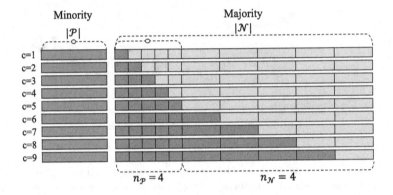

**Fig. 3.** Splitting Strategy

MUEnsemble adjusts the weights by using a programmable *balancing function* $B$. Since the balancing function can be implemented by many functions, in this paper, four functions are implemented.

Algorithm 1 displays the pseudo-code of MUEnsemble. Let $\mathcal{P}$ and $\mathcal{N}$ denote sets of minority examples and majority examples, respectively. At sampling step $c$, $\mathcal{N}$ is undersampled to $\mathcal{N}'$ so that $\frac{|\mathcal{N}'|}{|\mathcal{P}|} = R(c, n_\mathcal{P}, n_\mathcal{N})$, where $R(\cdot)$, $n_\mathcal{P}$, and $n_\mathcal{N}$ are a sampling ratio and the numbers of the sampling ratios below and above 1, respectively, in the splitting strategy. For $\mathcal{N}'$, MUEnsemble trains the $i$-th base ensemble classifier $H_i^c$, which consists of $w$ weak classifiers, by AdaBoost learning as $H_i^c = sgn\left(\sum_{j=1}^{w} \alpha_{i,j}^c h_{i,j}^c\right)$, where $sgn$ is the sign function, $h_{i,j}^c$ is the $j$-th weak classifier, and $\alpha_{i,j}^c$ is its weight. The number of base classifiers at the $c$-th step is determined by balancing function $B(c)$, and the classifiers are made into an ensemble over all sampling steps, $n_\mathcal{P} + n_\mathcal{N} + 1$, as follows.

$$H = sgn\left(\sum_{c=1}^{n_\mathcal{P}+n_\mathcal{N}+1} \sum_{i=1}^{B(c)} \sum_{j=1}^{w} \alpha_{i,j}^c h_{i,j}^c\right)$$

## 3.1    Sampling Ratio Determination

MUEnsemble employs the splitting strategy for the sampling ratio determination. It takes the numbers $n_\mathcal{P}$ and $n_\mathcal{N}$ to determine the number $|\mathcal{N}'|$ of sampled majority examples. $n_\mathcal{P}$ (resp. $n_\mathcal{N}$) is the number of blocks that equally split the minority (resp. majority) example size. Figure 3 illustrates an example of the splitting strategy. The majority example size is divided into $n_\mathcal{P} + n_\mathcal{N} + 1$ blocks ($n_\mathcal{P} = n_\mathcal{N} = 4$ in the example).

Given $n_\mathcal{P}$ and $n_\mathcal{N}$, the sampling ratio determination function $R$ calculates the sampling ratio as a cumulative ratio of blocks as follows.

$$R(c, n_\mathcal{P}, n_\mathcal{N}) = \begin{cases} \frac{c}{n_\mathcal{P}+1} & \text{if } c \le n_\mathcal{P} \\ 1 & \text{if } c = n_\mathcal{P} + 1 \, , \\ 1 + \frac{IR}{n_\mathcal{N}}(c - n_\mathcal{P}) & \text{if } c > n_\mathcal{P} + 1 \end{cases} \tag{1}$$

where $IR = \frac{|\mathcal{N}|}{|\mathcal{P}|}$ is the imbalance ratio, which is the number of majority examples over that of minority examples. In Fig. 3, the sampling ratios of the corresponding iteration $c$ are the blue-shaded areas. The blue-shaded and orange-shaded amounts for the minority and majority examples are used in the training.

## 3.2    Balancing Function

The balancing function is a programmable function for deciding the weights on base classifiers of sampling ratios. To investigate the base choices for the balancing function, in this paper, a constant function, two quadratic functions (convex and concave), and a Gaussian function are introduced.

The constant function is a function that equally weights base ensemble classifiers. In particular, it provides the same number $b_s$ of weak classifiers for all $c$, which is defined as follows.

$$B_{cons}(c) = b_s \tag{2}$$

The idea of the convex function is that the combining of classifiers leaned at extremely high and low sampling ratios is expected to be harmonically performed. On the basis of the expectation, the function provides larger numbers of weak classifiers when the sampling ratios are further from 1. The convex function is defined as follows.

$$B_{conv}(c) = \lfloor a_v(c - (n_\mathcal{P} + 1))^2 + b_v \rfloor , \tag{3}$$

where $a_v$ is a quadratic coefficient and $b_v$ is a constant of the smallest number of weak classifiers. For the split in Fig. 3, the number of classifiers at $c = 5$ is $b_v$, and the convex function gives the largest numbers at $c = 1$ and $c = 9$.

In contrast, the idea of the concave function is that weak classifiers with sampling ratios around 1 are in severe classification conditions. Therefore, these classifiers are more trustable than those with sampling ratios further from 1. This function is defined as follows.

$$B_{conc}(c) = \lfloor -a_c(c - (n_\mathcal{P} + 1))^2 + b_c \rfloor , \tag{4}$$

where $a_c$ is a quadratic coefficient, and $b_c$ is a constant of the largest number of classifiers. For the split in Fig. 3, the numbers of classifiers at $c = 1$ and $c = 9$ are $b_c$, and the concave function gives the largest number at $c = 5$.

The Gaussian function is a more flexible function than these fundamental functions. The coordinates of the peak of these fundamental functions are not easy to change due to their few flexibility. Therefore, the peak of these functions should be carefully changed along with the quadratic coefficients. However, different datasets can have different preferences toward sampling ratios. To capture the differences, a Gaussian function is introduced as follow.

$$B_{gauss}(c) = \left\lfloor a_g \cdot exp\left(\frac{(c-\mu)^2}{2\sigma^2}\right) \right\rfloor , \tag{5}$$

where $\mu$ and $\sigma^2$ are the tunable parameters, coefficient $a_g$ controls the number of weak classifiers, and $\mu$ is responsible for where the peak is.

Since the parameters $\mu$ and $\sigma^2$ in the Gaussian function are not easy to determine intuitively, in this paper, a grid search-based parameter optimization is employed. The optimization is performed in a $k$-fold cross-validation manner. First, a training dataset is split into $k$ folds. Second, one fold is used as a validation dataset and the rest of the folds are used for training MUEnsemble with a particular combination of $\mu$ and $\sigma^2$. Third, possible combinations of $\mu$ and $\sigma^2$ are examined by changing the validation fold. Last, the combination of $\mu$ and $\sigma^2$ that achieves the best classification performance on average in the $k$ trials is chosen as the optimal parameters.

## 4     Experimental Evaluation

In this paper, MUEnsemble is evaluated to answer the following questions:

**Q1:** Does excessive undersampling have a positive effect?
**Q2:** What is a good strategy for the balancing function?
**Q3:** Does the parameter estimation find optimal parameters?
**Q4:** Does MUEnsemble outperform baseline methods?

### 4.1     Settings

**Metric.** In this paper, *gmean*, which is an evaluation metric for imbalanced classification was employed. It is the geometric mean, $gmean = \sqrt{TPR \cdot TNR}$, of the true positive rate, $TPR$, and the true negative rate, $TNR$. $TPR$ (resp. $TNR$) corresponds to how accurately minority (resp. majority) examples are classified. Compared with other conventional metrics like *precision*, *recall* and *accuracy*, *gmean* is regarded as a robust evaluation metric for classification on imbalanced data [17]. Experimental processes were randomly repeated 10 times to ensure the accuracy of the estimation of *gmean* scores.

**Datasets.** This evaluation was performed on the publicly available datasets that have a wide range of dataset features. Table 1 summarizes the datasets. D1–D10

Table 1. Classification datasets

| ID | Dataset (binary classes if multi-class) | #dim. | #major | #minor | IR |
|---|---|---|---|---|---|
| D1 | Abalone (9 v. 18) | 8 | 689 | 42 | 16.4 |
| D2 | Anuran Calls (Lept. v. Bufo.) | 22 | 4,420 | 68 | 65.0 |
| D3 | Covertype (2 v. 5) | 54 | 283,301 | 9,493 | 29.8 |
| D4 | default of credit card clients | 23 | 23,364 | 6,636 | 3.5 |
| D5 | HTRU2 | 8 | 16,259 | 1,639 | 9.9 |
| D6 | Online Shoppers Purchasing Intention | 18 | 10,422 | 1,908 | 5.5 |
| D7 | Polish companies bankruptcy | 64 | 41,314 | 2,091 | 19.8 |
| D8 | Spambase | 56 | 2,788 | 1,813 | 1.5 |
| D9 | Wine Quality – Red ((3, 4) v. others) | 11 | 1,536 | 63 | 24.4 |
| D10 | Wine Quality – White (7 v. 3) | 11 | 880 | 20 | 44.0 |
| D11 | Churn Modelling | 9 | 7,963 | 2,037 | 3.9 |
| D12 | Credit Card Fraud Detection | 30 | 284,315 | 492 | 577.9 |
| D13 | ECG Heartbeat – Arrhythmia (N v. F) | 187 | 90,589 | 803 | 112.8 |
| D14 | Financial Distress | 85 | 3,536 | 136 | 26.0 |
| D15 | LoanDefault LTFS AV | 39 | 182,543 | 50,611 | 3.6 |
| D16 | Mafalda Opel – Driving Style | 14 | 9,530 | 2,190 | 4.4 |
| D17 | Mafalda Peugeot – Driving Style | 14 | 12,559 | 678 | 18.5 |
| D18 | Rain in Australia | 20 | 110,316 | 31,877 | 3.5 |
| D19 | Surgical | 24 | 10,945 | 3,690 | 3.0 |

were obtained from the UCI Machine Learning Repository [8] and D11–D19 were obtained from the Kaggle Dataset[1]. These datasets include few categorical attributes, and when categorical attributes exist, they are dictionary-encoded to numeric attributes[2]. Some of the datasets were for the multi-class classification task, and two of the classes in the datasets were selected to be for the bi-class classification task, which are represented in the dataset column in brackets. The datasets had various characteristics in terms of dimensionality (#dim.), the numbers of majority and minority examples (#major, #minor), and the imbalance ratio (IR).

**Baselines.** MUEnsemble was compared with simple baselines, popular synthetic oversampling approaches and undersampling-based approaches, listed as follows.

- ORG: classification without resampling
- RUS: random undersampling with sampling ratio of 1
- SMT: SMOTE [7]
- ADA: ADASYN [13]

---

[1] https://www.kaggle.com/datasets/.

[2] Coping with categorical attributes is out of the scope of this paper.

– SWIM [4, 29]
– EE: EasyEnsemble [19]
– RBST: RUSBoost [27]

ORG and RUS are simple baselines, where ORG classifies without resampling and RUS classifies with undersampling, for which the sampling ratio is set to 1. SMOTE, ADASYN, and SWIM are popular or recent synthetic oversampling methods, where SWIM is the latest method and can cope with highly imbalanced data. The classification algorithm for these baselines is the CART decision tree [6]. EasyEnsemble and RUSBoost are undersampling-based ensemble methods, where EasyEnsemble is a closer algorithm to MUEnsemble except for the inclusion of various sampling ratios, and RUSBoost is an iterative learning-based ensemble method. Note that trainable undersampling [24] is one of the state-of-the-art undersampling-based methods; however, under the evaluation metric *gmean*, its performance was inferior to the other baseline methods in a preliminary experiment. This is because of the different metrics used in [24], that is, precision and recall-based metrics. As mentioned above, such metrics are not suitable for the class imbalance problem. Therefore, trainable undersampling was not included in this experimental evaluation. Implementations of RUS, SMOTE, ADASYN, EasyEnsemble, and RUSBoost were in the imbalanced-learn library [18], and that of SWIM was obtained from the authors[3]. Classifiers were learned using scikit-learn (v. 0.20.3)[4].

**Parameters.** The parameters of MUEnsemble were set as follows. $n_\mathcal{P} = n_\mathcal{N} = 10$, $a_v = a_c = \frac{1}{5}$, $a_g = 20$, $b_s = b_v = 10$, $b_c = 50$, and $w = 50$. For the optimization of the Gaussian function, the number of folds for the cross-validation $k$ was set to 3, and the parameter spaces of $\mu$ and $\sigma^2$ were set as $\mu \in \{2, 4, 6, 8, 10, 12, 14, 16, 18\}$ and $\sigma^2 \in \{\frac{1}{8}, \frac{1}{2}, 1, 2, 5, 10, 20, 30, 50\}$. The parameters of the baselines were the default values.

### 4.2    Results

**Q1: Does the excessive undersampling have a positive effect?** Conventionally, the excessive undersampling has not been considered because of two reasons: (1) combining multiple sampling ratios has not been studied, and (2) sampling under the number of minority examples seems to not be effective. Therefore, this paper is the first attempt with excessive undersampling. The effect of the excessive undersampling was observable when limiting the number of weak classifiers for sampling ratios below 1 in MUEnsemble. To this end, the constant function (Eq. 2) is modified as a step function:

$$B_{step}(c) = \mathbb{I}(c)B_{cons}(c), \tag{6}$$

where $\mathbb{I}(c)$ is an indicator function that returns 0 if $c < n_\mathcal{P} + 1$ and 1 otherwise. MUEnsemble implementations based on the constant function (Eq. 2, denoted

---

[3] https://github.com/cbellinger27/SWIM.
[4] https://scikit-learn.org/.

**Table 2.** Effect of excessive undersampling. The best scores are boldfaced.

| Func. | D1 | D2 | D3 | D4 | D5 | D6 | D7 | D8 | D9 | D10 | D11 | D12 | D13 | D14 | D15 | D16 | D17 | D18 | D19 |
|---|---|---|---|---|---|---|---|---|---|---|---|---|---|---|---|---|---|---|---|
| Step | .616 | .908 | .425 | .624 | .910 | .784 | .844 | .916 | .505 | .419 | .711 | .855 | .688 | .703 | .207 | .461 | .523 | .721 | .762 |
| Cns | **.732** | **.956** | **.778** | **.694** | **.931** | **.844** | **.907** | **.917** | **.643** | **.664** | **.761** | **.939** | **.896** | **.859** | **.592** | **.771** | **.712** | **.765** | **.767** |

**Table 3.** Comparison of balancing functions. The best scores are boldfaced.

| Func. | D1 | D2 | D3 | D4 | D5 | D6 | D7 | D8 | D9 | D10 | D11 | D12 | D13 | D14 | D15 | D16 | D17 | D18 | D19 | Avg. | Rank |
|---|---|---|---|---|---|---|---|---|---|---|---|---|---|---|---|---|---|---|---|---|---|
| Cns | .732 | .956 | .778 | .694 | .931 | .844 | .907 | .917 | .643 | .664 | .761 | **.939** | .896 | .859 | .592 | .771 | .712 | .765 | .767 | .796 | 2.9 |
| Cnv | .674 | .955 | .779 | **.700** | .934 | .846 | .904 | .914 | .549 | .689 | .758 | .938 | .897 | .847 | .577 | .786 | .713 | .766 | .760 | .789 | 3.0 |
| Cnc | .751 | .956 | .777 | .689 | .930 | .843 | .906 | **.919** | .685 | .665 | **.762** | .938 | .897 | .858 | .591 | .746 | .708 | .764 | .770 | .798 | 3.0 |
| Gauss | **.753** | **.971** | **.808** | **.700** | **.936** | **.849** | **.908** | **.919** | **.705** | **.700** | **.762** | **.939** | **.900** | **.862** | **.593** | **.789** | **.791** | **.767** | **.803** | **.813** | **1.0** |

Cns) and the step function (Eq. 6, denoted Stp) were compared to observe the effect of the excessive undersampling.

Table 2 shows that the excessive undersampling had clear positive effects on classifying the imbalanced data. Cns outperformed Step for the all datasets with clear gaps. On individual datasets, the excessive undersampling helped to improve the classification performance, the only exceptions being D8 and D19, which showed small gaps.

**Q2: What is a good strategy for the balancing function?** Table 3 showcases the *gmean* scores of MUEnsemble using the four balancing functions on the datasets, where the Gaussian function was used together with the grid search-based optimization. In the table, the two right-most columns show the average *gmean* scores and average ranks among the functions over the datasets, respectively. The Gaussian function performed the best for all datasets. The constant, convex, and concave functions were comparable. The concave function performed relatively better than the constant and concave functions in terms of *gmean* scores. In contrast, the constant function was better than the convex and concave functions in terms of ranks. These facts indicate that the concave function whose center is fixed at the middle sampling ratio is not always suitable, and the Gaussian function with optimization can flexibly adjust the central coordinate for the function.

**Q3: Does the parameter estimation find optimal parameters?** In the grid search-based optimization, the parameters are examined in the training data; thus, the best parameters in the training data are not always optimal in the test data as well. To evaluate the optimization quality, in this paper, optimal parameters were explored by using all training data with every parameter combination of $\mu$ and $\sigma^2$. The optimal parameters in the test data were recorded and compared with those estimated.

Table 4 and Table 5 report the *gmean* scores and the parameters of MUEnsemble in terms of the estimated parameters and the optimal parameters, respectively. As Table 4 illustrates, MUEnsemble with the estimated parameters performed: close to that with the optimal parameters except for D1. The esti-

**Table 4.** Effect of optimization. The differences larger than 0 are boldfaced.

| Method | D1 | D2 | D3 | D4 | D5 | D6 | D7 | D8 | D9 | D10 | D11 | D12 | D13 | D14 | D15 | D16 | D17 | D18 | D19 |
|---|---|---|---|---|---|---|---|---|---|---|---|---|---|---|---|---|---|---|---|
| Estimated | .753 | .971 | .808 | .701 | .936 | .849 | .908 | .919 | .705 | .735 | .762 | .938 | .900 | .862 | .593 | .789 | .791 | .767 | .803 |
| Optimal | .772 | .971 | .809 | .701 | .936 | .849 | .910 | .919 | .705 | .735 | .762 | .939 | .900 | .865 | .593 | .789 | .791 | .767 | .803 |
| Diff | **.019** | .000 | **.001** | .000 | .000 | .000 | **.002** | .000 | .000 | .000 | .000 | **.001** | .000 | **.003** | .000 | .000 | .000 | .000 | .000 |

**Table 5.** Estimated and optimal Parameters ($\mu$ and $\sigma^2$). The estimated parameters equal to the optimal parameters are boldfaced.

| Method | D1 | D2 | D3 | D4 | D5 | D6 | D7 | D8 | D9 | D10 |
|---|---|---|---|---|---|---|---|---|---|---|
| Estimated | (8, 2) | (10, 50) | **(6, 50)** | **(6, $\frac{1}{8}$)** | (8, 30) | (4, 50) | (8, 10) | (10, $\frac{1}{8}$) | **(12, 30)** | (8, $\frac{1}{8}$) |
| Optimal | (10, 2) | (4, 50) | **(6, 50)** | **(6, $\frac{1}{8}$)** | (6, 30) | (6, 30) | (8, $\frac{1}{2}$) | (14, $\frac{1}{8}$) | **(12, 30)** | (8, $\frac{1}{8}$) |

| Method | D11 | D12 | D13 | D14 | D15 | D16 | D17 | D18 | D19 |
|---|---|---|---|---|---|---|---|---|---|
| Estimated | **(10, 5)** | (8, 20) | **(8, 30)** | (10, 1) | **(10, 50)** | **(10, $\frac{1}{8}$)** | **(8, 50)** | **(8, 50)** | **(14, 50)** |
| Optimal | **(10, 5)** | (8, 30) | **(8, 30)** | (10, $\frac{1}{2}$) | **(10, 50)** | **(10, $\frac{1}{8}$)** | **(8, 50)** | **(8, 50)** | **(14, 50)** |

mated parameters shown in Table 5 indicate that the optimal parameters were estimated for 11 out of the 19 datasets. Note that the *gmean* scores reported in Table 4 are rounded to the third decimal; therefore, there may be negligible differences between Estimated and Optimal when the parameters are different. It is noteworthy that even though the optimization failed to estimate the optimal parameters, the *gmean* scores were almost optimal.

Table 6 showcases the standard deviations of *gmean* scores for the optimization for the estimated and optimal parameters. In most of the datasets, the deviations were small, except D1. The deviations for D1 for both the estimated and optimal parameters were both above 1 point. This is why the grid search-based optimization failed to estimate the optimal parameters or other better parameters. That is, the standard deviations indicate that the best parameters varied in the different subsets of the training data.

**Table 6.** Standard deviation of *gmea* scores in the optimzation.

| Method | D1 | D2 | D3 | D4 | D5 | D6 | D7 | D8 | D9 | D10 | D11 | D12 | D13 | D14 | D15 | D16 | D17 | D18 | D19 |
|---|---|---|---|---|---|---|---|---|---|---|---|---|---|---|---|---|---|---|---|
| Estimated | .104 | .033 | .005 | .007 | .009 | .009 | .016 | .009 | .069 | .094 | .011 | .016 | .014 | .032 | .003 | .010 | .015 | .003 | .014 |
| Optimal | .159 | .023 | .005 | .007 | .009 | .009 | .016 | .010 | .091 | .094 | .012 | .016 | .014 | .032 | .003 | .010 | .015 | .003 | .017 |

**Q4: Does MUEnsemble outperform baseline methods?** Table 7 demonstrates that :MUEnsemble outperforms baseline methods. In 15 out of 19 datasets, MUEnsemble performed the best and ranked 1.4 on average for the datasets among the approaches. EasyEnsemble (EE) was the closest approach to MUEnsemble in the sense that EasyEnsemble ensembles multiple weak classifiers with a fixed undersampling ratio. The differences in *gmean* scores between MUEnsemble and EasyEnsemble show the effect of the involvement of multiple sampling ratios. The oversampling approaches were inferior to the undersampling approaches including MUEnsemble.

Table 7. Comparison over baselines. The best scores are boldfaced.

| Dataset | ORG | Oversampling | | | Undersampling | | | MUEnsemble |
|---------|-----|------|------|------|------|------|------|-----------------|
| | | SMT | ADA | SWIM | RUS | RBST | EE | Gauss (optimal) |
| D1 | .580 | .675 | .671 | .642 | .670 | .577 | .741 | **.753** (.772) |
| D2 | .915 | .931 | .897 | .909 | .925 | .954 | .963 | **.971** (.971) |
| D3 | .891 | .924 | .916 | .747 | **.928** | .852 | .798 | .808 (.809) |
| D4 | .581 | .585 | .584 | .580 | .616 | .528 | .689 | **.701** (.701) |
| D5 | .896 | .910 | .908 | .906 | .907 | .897 | .930 | **.936** (.936) |
| D6 | .713 | .733 | .739 | .709 | .790 | .731 | .845 | **.849** (.849) |
| D7 | .810 | .829 | .834 | .760 | .854 | .786 | **.908** | .908 (.910) |
| D8 | .900 | .900 | .898 | .896 | .896 | **.931** | .916 | .919 (.919) |
| D9 | .420 | .467 | .473 | .519 | .624 | .436 | .680 | **.705** (.705) |
| D10 | .475 | .444 | .574 | .666 | .616 | .412 | .662 | **.735** (.735) |
| D11 | .642 | .652 | .647 | .642 | .678 | .619 | .761 | **.762** (.762) |
| D12 | .876 | .877 | .865 | .917 | .905 | .895 | .937 | **.938** (.939) |
| D13 | .822 | .859 | .853 | .829 | .883 | .831 | .895 | **.900** (.900) |
| D14 | .546 | .548 | .576 | .562 | .775 | .606 | **.863** | .862 (.865) |
| D15 | .466 | .474 | .476 | .442 | .538 | .463 | .592 | **.593** (.593) |
| D16 | .708 | .755 | .737 | .724 | **.794** | .702 | .779 | .789 (.789) |
| D17 | .760 | .780 | .771 | .757 | .770 | .747 | .710 | **.791** (.791) |
| D18 | .677 | .690 | .689 | .678 | .714 | .641 | .762 | **.767** (.767) |
| D19 | **.803** | .787 | .760 | **.803** | .785 | .761 | .760 | **.803** (.803) |
| Avg. | .710 | .727 | .730 | .720 | .772 | .704 | .800 | **.815** (.817) |
| Ranks | 6.1 | 4.3 | 5.0 | 5.8 | 3.5 | 6.2 | 2.8 | **1.4** (– ) |

## 4.3  Lessons Learned

– **Excessive undersampling and combining multiple sampling ratios improve classification performances.** A novelty of MUEnsemble is that it enables excessive undersampling to be included, and in the evaluation, it was shown that the inclusion had positive effects. The comparison between MUEnsemble and EasyEnsemble showed that the involvement of multiple sampling ratios of undersampling improved the classification performance.
– **The Gaussian balancing function is flexible, so MUEnsemble with the estimated parameters performed the best.** The Gaussian function is flexible in terms of arranging the central coordinates and deviations of the weights of the base ensemble classifiers. Therefore, it performed the best among the proposed balancing functions. Grid search-based optimization successfully estimated the optimal parameters and the parameters that were close to the best performances.
– **MUEnsemble outperformed the baseline methods.** The experimental evaluation indicated that the undersampling approaches were more prefer-

able than the oversampling ones. Among the undersampling approaches, MUEnsemble was superior to the other approaches.

## 5    Conclusion

In this paper, an ensemble classification framework, MUEnsemble, was proposed for the class imbalance problem, by which multiple undersampling with different sampling ratios are applied. In an experiment, the Gaussian balancing function proved its flexibility, and the grid search-based optimization helped in finding the optimal parameters. Since there were still datasets (e.g., D15) for which none of the approaches could perform well, investigating reasons such as *small disjuncts* [14] and the *overlap* of example distributions of the majority and the minority [21] and combining solutions for the degradation reasons into MUEnsemble are promising future directions.

**Acknowledgements.** This work was partly supported by JSPS KAKENHI Grant Number JP18K18056 and the Kayamori Foundation of Informational Science Advancement.

## References

1. Bao, H., Sugiyama, M.: Calibrated surrogate maximization of linear-fractional utility in binary classification. CoRR abs/1905.12511 (2019). http://arxiv.org/abs/1905.12511
2. Batista, G.E.A.P.A., Bazzan, A.L.C., Monard, M.C.: Balancing training data for automated annotation of keywords: a case study. In: II Brazilian Workshop on Bioinformatics, pp. 10–18 (2003)
3. Batista, G.E.A.P.A., Prati, R.C., Monard, M.C.: A study of the behavior of several methods for balancing machine learning training data. SIGKDD Explor. **6**(1), 20–29 (2004)
4. Bellinger, C., Sharma, S., Japkowicz, N., Zaïane, O.R.: Framework for extreme imbalance classification: SWIM—sampling with the majority class. Knowl. Inf. Syst. **62**(3), 841–866 (2019). https://doi.org/10.1007/s10115-019-01380-z
5. Bhattacharya, S., Rajan, V., Shrivastava, H.: ICU mortality prediction: a classification algorithm for imbalanced datasets. In: AAAI 2017, pp. 1288–1294 (2017)
6. Breiman, L., Friedman, J.H., Olshen, R.A., Stone, C.J.: Classification and Regression Trees. Wadsworth (1984)
7. Chawla, N.V., Bowyer, K.W., Hall, L.O., Kegelmeyer, W.P.: SMOTE: synthetic minority over-sampling technique. J. Artif. Intell. Res. **16**, 321–357 (2002)
8. Dua, D., Graff, C.: UCI Machine Learning Repository (2019). http://archive.ics.uci.edu/ml
9. Elkan, C.: The foundations of cost-sensitive learning. IJCAI **2001**, 973–978 (2001)
10. Galar, M., Fernández, A., Tartas, E.B., Sola, H.B., Herrera, F.: A review on ensembles for the class imbalance problem: bagging-, boosting-, and hybrid-based approaches. IEEE Trans. Syst. Man Cybern. Part C **42**(4), 463–484 (2012)
11. Gómez, S.E., Hernández-Callejo, L., Martínez, B.C., Sánchez-Esguevillas, A.J.: Exploratory study on class imbalance and solutions for network traffic classification. Neurocomputing **343**, 100–119 (2019)

12. Han, H., Wang, W.-Y., Mao, B.-H.: Borderline-SMOTE: a new over-sampling method in imbalanced data sets learning. In: Huang, D.-S., Zhang, X.-P., Huang, G.-B. (eds.) ICIC 2005. LNCS, vol. 3644, pp. 878–887. Springer, Heidelberg (2005). https://doi.org/10.1007/11538059_91
13. He, H., Bai, Y., Garcia, E.A., Li, S.: ADASYN: adaptive synthetic sampling approach for imbalanced learning. IJCNN **2008**, 1322–1328 (2008)
14. Jo, T., Japkowicz, N.: Class imbalances versus small disjuncts. SIGKDD Explor. **6**(1), 40–49 (2004)
15. Kang, P., Cho, S.: EUS SVMs: ensemble of under-sampled SVMs for data imbalance problems. In: King, I., Wang, J., Chan, L.-W., Wang, D.L. (eds.) ICONIP 2006. LNCS, vol. 4232, pp. 837–846. Springer, Heidelberg (2006). https://doi.org/10.1007/11893028_93
16. Krawczyk, B., Wozniak, M., Schaefer, G.: Cost-sensitive decision tree ensembles for effective imbalanced classification. Appl. Soft Comput. **14**, 554–562 (2014)
17. Kubat, M., Matwin, S.: Addressing the curse of imbalanced training sets: one-sided selection. In: ICML 1997, pp. 179–186 (1997)
18. Lemaître, G., Nogueira, F., Aridas, C.K.: Imbalanced-learn: a python toolbox to tackle the curse of imbalanced datasets in machine learning. J. Mach. Learn. Res. **18**(17), 1–5 (2017)
19. Liu, X., Wu, J., Zhou, Z.: Exploratory undersampling for class-imbalance learning. IEEE Trans. Syst. Man Cybern. Part B **39**(2), 539–550 (2009)
20. Lu, W., Li, Z., Chu, J.: Adaptive ensemble undersampling-boost: a novel learning framework for imbalanced data. J. Syst. Softw. **132**, 272–282 (2017)
21. Luengo, J., Fernández, A., García, S., Herrera, F.: Addressing data complexity for imbalanced data sets: analysis of SMOTE-based oversampling and evolutionary undersampling. Soft. Comput. **15**(10), 1909–1936 (2011)
22. Mani, I., Zhang, I.: kNN approach to unbalanced data distributions: a case study involving information extraction. In: ICML 2003 Workshop on Learning from Imbalanced Datasets, vol. 126 (2003)
23. Nguyen, H.M., Cooper, E.W., Kamei, K.: Borderline over-sampling for imbalanced data classification. IJKESDP **3**(1), 4–21 (2011)
24. Peng, M., et al.: Trainable undersampling for class-imbalance learning. In: AAAI 2019, pp. 4707–4714 (2019)
25. Pozzolo, A.D., Caelen, O., Johnson, R.A., Bontempi, G.: Calibrating probability with undersampling for unbalanced classification. In: SSCI 2015, pp. 159–166 (2015)
26. Rodríguez, D., Herraiz, I., Harrison, R., Dolado, J.J., Riquelme, J.C.: Preliminary comparison of techniques for dealing with imbalance in software defect prediction. In: EASE 2014, pp. 43:1–43:10 (2014)
27. Seiffert, C., Khoshgoftaar, T.M., Hulse, J.V., Napolitano, A.: RUSBoost: a hybrid approach to alleviating class imbalance. IEEE Trans. Syst. Man Cybern. Part A **40**(1), 185–197 (2010)
28. Sharififar, A., Sarmadian, F., Minasny, B.: Mapping imbalanced soil classes using Markov chain random fields models treated with data resampling technique. Comput. Electron. Agric. **159**, 110–118 (2019)
29. Sharma, S., Bellinger, C., Krawczyk, B., Zaïane, O.R., Japkowicz, N.: Synthetic oversampling with the majority class: a new perspective on handling extreme imbalance. In: ICDM 2018, pp. 447–456 (2018)
30. Smith, M.R., Martinez, T., Giraud-Carrier, C.: An instance level analysis of data complexity. Mach. Learn. **95**(2), 225–256 (2013). https://doi.org/10.1007/s10994-013-5422-z

31. Tomek, I.: Two modifications of CNN. IEEE Trans. Syst. Man Cybern. **SMC-6**(11), 769–772 (1976)
32. Wang, H., Gao, Y., Shi, Y., Wang, H.: A fast distributed classification algorithm for large-scale imbalanced data. In: ICDM 2016, pp. 1251–1256 (2016)
33. Wilson, D.L.: Asymptotic properties of nearest neighbor rules using edited data. IEEE Trans. Syst. Man Cybern. **2**(3), 408–421 (1972)

# The Linear Geometry Structure of Label Matrix for Multi-label Learning

Tianzhu Chen[1,2], Fenghua Li[1,2], Fuzhen Zhuang[3,4], Yunchuan Guo[1,2],
and Liang Fang[1(✉)]

[1] Institute of Information Engineering, Chinese Academy of Sciences,
Beijing, China
fangliang@iie.ac.cn
[2] School of Cyber Security, University of Chinese Academy of Sciences,
Beijing, China
[3] Key Lab of Intelligent Information Processing of Chinese Academy of Sciences
(CAS), Institute of Computing Technology, CAS, Beijing 100190, China
[4] Xiamen Data Intelligence Academy of ICT, CAS, Beijing, China

**Abstract.** Multi-label learning annotates a data point with the relevant labels. Under the low-rank assumption, many approaches embed the label space into the low-dimension space to capture the label correlation. However these approaches usually have weak prediction performance because the low-rank assumption is usually violated in real-world applications. In this paper, we observe the fact that the linear representation of row and column vectors of label matrix does not depend on the rank structure and it can capture the linear geometry structure of label matrix (LGSLM). Inspired by the fact, we propose the LGSLM classifier to improve the prediction performance. More specifically, after rearranging the columns of a label matrix in decreasing order according to the number of positive labels, we capture the linear representation of the row vectors of the compact region in the label matrix. Moreover, we also capture the linear and sparse representation of column vectors using the $L_1$-norm. The experimental results for five real-world datasets show the superior performance of our approach compared with state-of-the-art methods.

**Keywords:** Multi-label learning · Linear representation · Sparse representation

## 1 Introduction

Multi-label learning [12,18,20] usually annotates a data object with multiple relevant labels. For example, a web page describing the Divine Comedy may be annotated with the tags "14th-century Christian texts" and "epic poems in Italian"; an article that discusses physical education can be annotated with education and fitness tags; an image that shows a beach and a city can be annotated with beach and city tags. Presently, multi-label learning is applied to

© Springer Nature Switzerland AG 2020
S. Hartmann et al. (Eds.): DEXA 2020, LNCS 12392, pp. 229–244, 2020.
https://doi.org/10.1007/978-3-030-59051-2_15

various applications ranging from web page retrieval to document classification and image annotation.

Considerable literature has showed that a better prediction performance can be achieved in multi-label learning through capturing label correlation [12,18,20]. Based on the rank structure of label matrix, researchers usually exploit the embedding method [2,5,11,18,21,22] to capture the label correlation. Generally, the embedding method can be divided into two categories: the linear embedding method and the nonlinear embedding method. The key idea of the linear approach [11,18,21,22] is to exploit a low-rank transform to project the high-dimensional label space onto a low-dimensional space. Subsequently, the prediction is performed on the low-dimensional label space and then the original labels are recovered. However the label correlation learned by the linear embedding method does not capture the intrinsic structure among the labels because the low-rank assumption is often violated in practice.

To address this problem, after grouping all instances into the clusters, Bhatia et al. [2] exploited nonlinear embedding to compress the label matrix of every cluster into a small-dimension label space. By preserving the pairwise distances between a label vector and the nearest label vectors, [2] compressed the label space into a small label space and then learned the regressor in the embedding space. By leveraging the fact that nearest neighbors have been preserved during training, [2] used a k-nearest neighbor (kNN) classifier in the embedding space to predict test instances. Based on the framework of [2,5] exploited the Skip-Gram Negative Sampling approach to embed the label space of every cluster into low-dimension space. However, a small dimension [2,5] was used to embed the label space that ignored the overall structure of the label matrix, thereby leading to information loss.

By observing the fact that the linear representation of row and column vectors of label matrix does not depend on the rank structure and it can capture the linear geometry structure of label matrix (LGSLM), in this paper, we propose the LGSLM classifier to capture the more intrinsic geometry structure of the label matrix. More specifically, after rearranging the columns in decreasing order according to the number of 1-valued entries, we segment the label matrix into multiple regions and select the compact region to capture the row representation. In the column representation, we exploit the $L_1$ norm to constrain the linear representation, which captures the sparse representation. Further, we adjust the label value to capture the linear structure of larger region. The experimental results for five real-world datasets show the superior performance of our approach. The details of our contributions are mentioned below.

(1) By representing the row and column vectors, we propose a novel classifier, LGSLM, which captures the linear geometry structure of a label matrix. Further, we adjust the value of positive labels and linearly represent the relatively compact regions of the label matrix. To the best of our knowledge, we are the first to apply linear representation to multi-label learning.

(2) To reduce the computational complexity of row representation, after transforming the row representation optimization as a linear system, we decom-

pose the system matrix into a block diagonal matrix and obtain the multiple subsystems, which reduces the computational complexity from $O(l^6)$ to $O(l^4)$ (refer to Theorem 1).

(3) Experiments on five real-world datasets show that our approach has the best prediction performance, which means that the linear representation of a label matrix is more intrinsic than the low-rank constraint.

## 2   Related Work

As mentioned in the previous section, under the low-rank assumption, many existing approaches capture the label correlation using linear embedding methods. For example, based on principal component analysis, Tai et al. [15] proposed principal label space transformation by compressing the label vector. Further, Zhang et al. [21] simultaneously compressed the label space and feature space using canonical correlation analysis that extract the most predictable directions in the label space to characterize the label dependency. Subsequently, to make output codes predictable and discriminative, Zhang et al. [22] developed maximum margin output coding (MMOC) to learn output coding that is not only close to the correct codeword but is also far from any correct codeword. MMOC requires an expensive decoding procedure to recover the multiple labels of each testing instance, which becomes unacceptable when there are many labels. To avoid decoding completely, Liu et al. [11] proposed large margin metric learning with k-nearest neighbor constraints. Yu et al. [18] decomposed the regression matrix into two low-rank matrices, compressed the high-dimension feature space into low-dimension space and learned the classifier on the low-dimension feature space. However, these approaches have one common problem: because the label matrix becomes sparse, the low-rank assumption can be easily violated, thereby leading to degradation in the testing performance.

To address this problem, after grouping all instances into clusters, Bhatia et al. [2] embedded the label space of every cluster into a small space by preserving the pairwise distances between a label vector and the nearest label vectors. This breaks the traditional low-rank assumption and boosts classification accuracy. Based on the framework of [2,5] exploited the Skip-Gram Negative Sampling (SGNS) approach to embed the label space of every cluster into low-dimension space. Although the method of [2,5] captured the distance structure of every cluster, the small embedding neglected the overall label correlation. To address this problem, by simultaneously embedding the overall label space into a low-dimensional latent space, Shen et al. [12] exploited the manifold regularization to preserve the semantic similarity structure between the embeddings in the hamming space and the original label space. Additionally, after extracting the additional sparse component from the label matrix as outliers, Xu et al. [17] captured the low-rank structure of the residual matrix, thus validating the classical low-rank principle in multi-label learning. For the long-tail recommendation, [10] decomposed the overall items into a low-rank part and a sparse part. The low-rank part and the sparse part are respectively exploited to recommend the short-head items and the long-tail items. However, the small embedding dimension used in [2,12] not only ignored the overall structure of the label matrix but also led to information loss.

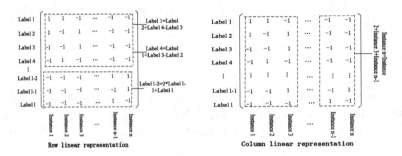

**Fig. 1.** In left figure, the row vector of label $l-2$ can be linearly represented by the row vectors of label $l-1$ and label $l$. In right figure, the column vector of instance $n$ can be represented by the column vector of instance 2, instance 3 and instance $n-1$.

## 3    Proposed Model

Let $\mathscr{D} = \{(\boldsymbol{x}_1, \boldsymbol{y}_1), (\boldsymbol{x}_2, \boldsymbol{y}_2), \cdots, (\boldsymbol{x}_n, \boldsymbol{y}_n)\}$ be the training data where instance $\boldsymbol{x}_i = (x_{i1}, \cdots, x_{id}) \in \mathbb{R}^d$ is a feature vector of $d$ dimensions and $\boldsymbol{y}_i = (y_{i1}, \cdots, y_{il}) \in \{-1, +1\}^l$ is a label vector of $l$ dimensions. $y_{ij} = 1$ means that instance $\boldsymbol{x}_i$ is associated with the $j$-th label (i.e. the positive label), $y_{ij} = -1$ means the $j$-th label does not exist in $\boldsymbol{x}_i$ (i.e. the negative label). In detail, given $n$ training instances (represented by matrix $\boldsymbol{X} = [\boldsymbol{x}_1, \boldsymbol{x}_2, \cdots, \boldsymbol{x}_n] \in \mathbb{R}^{d \times n}$) and label matrix (represented by matrix $\boldsymbol{Y} = [\boldsymbol{y}_1, \boldsymbol{y}_2, \cdots, \boldsymbol{y}_n] \in \{-1, +1\}^{l \times n}$), our goal is to find an regression matrix $\boldsymbol{M} \in \mathbb{R}^{l \times d}$ to accurately predict the label vectors for unseen instances via computing the function $f(\boldsymbol{x}) = \boldsymbol{M}\boldsymbol{x}$. More specially, when $\boldsymbol{M}_i \boldsymbol{x}$ is greater than or equal to 0, the predicted $i$-th label of instance $\boldsymbol{x}$ is positive label; otherwise, the predicted $i$-th label of instance $\boldsymbol{x}$ is negative label.

### 3.1    Linear Representation of Row Vectors

The motivation for our framework comes from the following three observations. **Observation 1:** Every row/column vector of a label matrix can usually be linearly represented (or approximately represented) by the other row/column vectors. The linear representation, which does not depend on the rank structure of the label matrix, captures the linear correlation of several labels. **Observation 2:** The linear representation, which is usually obtained by minimizing certain distance of row/column vectors, has more influence to the negative labels because there exists plenty of negative labels and a small number of positive labels in dataset. **Observation 3:** After rearranging the columns of label matrix in decreasing order of the number of positive labels into a new matrix, the compact region of the new matrix has a more remarkable linear structure. In this section, we capture three observations by linearly representing the row/column vectors of the compact region of label matrix.

In linear representation, four critical factors (i.e. the number of row dimensions, the number of column dimensions, the number of positive labels and the number of negative labels) have a strongly influence on the representative ability. For example, if the row dimension is not much less than the column dimension, one row vectors can usually be accurately represented by the other row vectors; when the number of positive labels is much less than the number of the negative labels, the linear representation have an more influence on the negative labels. Thus, by selecting the compact region which has the biggest fraction of positive labels to negative labels, we measure the relatively accurate linear representation discussed as follows.

After rearranging the columns of a label matrix in decreasing order according to the number of positive labels, without loss of generality, we assume that the new label matrix $Y$ is segmented into the formulation $Y = \{Y_1^r, Y_2^r, \cdots, Y_S^r\}$ where the row is written as r, $Y_s^r = \{y_{s_1}, y_{s_2}, \cdots, y_{s_{k_s}}\}$, $k_s$ is the number of instances from the $s$-th region, $Y_i^r \cap Y_j^r = \varnothing$ and $\bigcup_i Y_i^r = Y$.

For any local region $Y_s^r$, the row linear representation $R_s^r$ is as follows,

$$R_s^r Y_s^r = Y_s^r, R_{s_{(ii)}}^r = 0, \|R_s^r\|_2^2 < K_0. \tag{1}$$

where $\|R_s^r\|_2^2$ is the regularization term and $R_{s_{(ii)}}^r = 0$ denotes the diagonal elements to be 0, which eliminates the trivial solution of writing a label function as a linear combination of itself. In formulation $R_s^r Y_s^r = Y_s^r$, $R_s^r = [r_1, r_2, \cdots, r_{l_s}] \in \mathbb{R}^{l_s \times l_s}$ is the linear representation of $Y_s^r$, where $l_s$ is the number of labels.

To guarantee the consistency between the row representation of real label matrix and the row representation of predicted label matrix, we constrain $R_s^r M X_s^r = M X_s^r$ where $X_s^r$ is the corresponding feature matrix to the local label matrix $Y_s^r$. Here, we usually select the $Y_1^r$ to capture the row linear representation.

If the row dimension is much less than the column dimension, row vectors may be not well represented by the other row vectors. Here, we capture the column linear representation to complementary the shortcoming. As the column dimension becomes larger, one column vector may be represented by small fraction of column vectors so that the column linear representation is sparse. Without loss of generality, we assume that the label matrix $Y$ is segmented into the set $Y = \{Y_1^c, Y_2^c, \cdots, Y_P^c\}$ where the column is written as c, $Y_p^c = \{y_{p_1}, y_{p_2}, \cdots, y_{p_{k_p}}\}$, $k_p$ is the number of the $p$-th set, $Y_i^c \cap Y_j^c = \varnothing$ and $Y_i^c \cup Y_j^c = Y$. For any local region $Y_p^c$, we have the following column linear representation $R_p^c$,

$$Y_p^c R_p^c = Y_p^c, R_{p_{(ii)}}^c = 0, \|R_k^c\|_0 < K_1. \tag{2}$$

In formulation $Y_p^c R_p^c = Y_p^c$, $Y_p^c = [r_1, r_2, \cdots, r_{l_p}] \in \mathbb{R}^{l \times l_p}$ is the linear correlation of $Y_p^c$ where constraint $R_{k_{ii}}^c = 0$ eliminates the trivial solution of writing a label function as a linear combination of itself. $||R_p^c||_0$ is the number of non-zero elements of $R_p^c$.

Since $||R_p^c||_0$ tends to increase the optimization complexity, we employ one popular L1-norm minimization (to encourage sparsity) [4] to relax the constraints in problem (2), such that

$$Y_p^c R_p^c = Y_p^c, R_{p_{ii}}^c = 0, ||R_p^c||_1 < K_1. \tag{3}$$

To guarantee the consistency between the linear representation of real label matrix and the linear representation of predicted label, we constrain the following equality,

$$Re = \{MX_p^c R_p^c = MX_p^c | p \in P\}.$$

where $X_p^c$ is the corresponding feature matrix to the local label matrix $Y_p^c$.

### 3.2   The Strengthen of Linear Representation and Unified Objective Function

In the represented region, there exists plenty of negative labels and a small number of positive labels. In such situation, the linear representation, which is measured by minimizing a distance of row/column vectors (referred to problem (1)), has more influence on the negative labels than the positive labels. This usually makes the linear representation be only effective in relatively small region. To address the problem, we adjust the value of positive labels to effectively capture the linear geometry structure of larger region. Motivated by empirical observation, we model the weight of the positive label value as a sigmoidal function of $\log N_l$ [8],

$$\bar{Y}_{ij} = 1 + Ce^{-Alog(N_l+B)}, if \ Y_{ij} = 1.$$
$$C = (logN - 1)(B + 1)^A$$

where $A$ and $B$ are two fixed parameters. In our experiment, we fix the value of $A$ and $B$ as 0.1.

We use the square loss to evaluate the discrepancy between $\bar{Y}$ and the predicted label matrix $MX$. Combining the row linear representation and the column linear representation, converting the problem into an unconstrained optimization problem by using a penalty method, we obtain the following optimization problem,

$$\min_M L(\bar{Y}, MX) = ||\bar{Y} - MX||_2^2 + \lambda_1 ||M||_2^2$$
$$+\lambda_2 \sum_{s \in S} ||R_s^r MX_s^r - MX_s^r||_2^2 + \lambda_3 \sum_{p \in P} ||MX_p^c R_p^c - MX_p^c||_2^2 \tag{4}$$

where $\lambda_1$ is an parameter to control regularization, penalty parameters $\lambda_2 > 0$, $\lambda_3 > 0$ are used to control the weight of row/column linear representation, respectively. Parameters $\lambda_2$ and $\lambda_3$ should be large to ensure that both $R_s^r MX_s^r = MX_s^r$ and $MX_p^c R_p^c = MX_p^c$ are satisfied.

## 4   Optimization Algorithm

In this section, we first present the optimization method for problem (1) and problem (3). Subsequently, we divide the objective function (4) into two cases and present the corresponding optimization methods.

### 4.1   Solving $R^r$

Problem (1) can be transformed into the following form,

$$\min_{R_s^r} \|R_s^r Y_s^r - Y_s^r\|_2^2 + c_1 \|R_s^r\|_2^2 + c_2 \sum_{i=1}^{l} \|R_{s_{ii}}^r\|_2^2. \tag{5}$$

where $l$ is the number of labels and $(c_2 \rightarrow \infty)$ makes $R_{s_{ii}}^r$ approach to 0, which avoids writing a linear representation as a linear combination of itself. Problem (5) can usually be transformed as solving one $l^2 \times l^2$ linear system, which requires the complexity of $O(l^6)$. To reduce the computation complexity, we decompose the linear system into $l$ number of $l \times l$ linear sub-systems, which requires the complexity of $O(l^4)$.

**Theorem 1.** The problem (5) is equivalent to the set of sub-linear systems:

$$(2Y_s^r Y_s^{rT} + 2c_1 I + 2c_2 I_{ii})(R_s^r)_{i,:}^T = (2Y_s^r Y_s^{rT})_{i,:}^T,$$
$$i = 1, 2, \cdots, l.$$

where $(-)_{i,:}$ is the $i$-th row of the corresponding matrix or vector, $(-)^T$ is the transposition of the corresponding matrix and vector, $I_{ii}$ is an $l \times l$ matrix whose element are 0 except the $ii$ entry is 1.

Thus, the solution of the linear system is $\mathbf{vec}(R_s^r) = 2diag\{A_1^{-1}, A_2^{-1}, \cdots, A_l^{-1}\}\mathbf{vec}(Y_s^r Y_s^{rT})$ where $\mathbf{vec}(R_s^r)$ and $\mathbf{vec}(Y_s^r Y_s^{rT})$ are two vectors formed by the row vectors of $R_s^r$ and $Y_s^r Y_s^{rT}$, $A_i = 2Y_s^r Y_s^{rT} + 2c_1 I + 2c_2 I_{ii}$.

**Proof:** Taking the derivation of (5) w.r.t. $R_s$ and setting it zero, we have the following form,

$$R_s(2Y_s Y_s^T + 2c_1 I) + 2c_2 diag(R_{s_{i,i}}) = 2Y_s Y_s^T \tag{6}$$

Problem (6) is an $l^2 \times l^2$ linear system. Let $vec(R_s)$ be an vector formed by the row vectors of $R_s$. In the problem, the $i$-th row of $R_s$ forms the following equality $(R_s)_{i,:}(2Y_s Y_s^T + 2c_1 I) + 2c_2 R_{ii} = (2Y_s Y_s^T)_{i,:}$. The equality is equivalent to the following sub-linear system,

$$(2Y_s Y_s^T + 2c_1 I + 2c_2 I_{ii})(R_s)_{i,:}^T = (2Y_s Y_s^T)_{i,:}^T$$

The solution of the sub-linear system is $A_i^{-1}(2Y_s Y_s^T)_{i,:}^T$ where $A_i = (2Y_s Y_s^T + 2c_1 I + 2c_2 I_{ii})$. By traversing all rows of linear representation $R_s$, we obtain the block matrix $A = diag\{A_1^{-1}, A_2^{-1}, \cdots, A_l^{-1}\}$ and the solution $vec(R_s) = 2diag\{A_1^{-1}, A_2^{-1}, \cdots, A_l^{-1}\}vec(Y_s Y_s^T)$.

## 4.2  Solving $R^c$

Problem (3) can be transformed into the following form,

$$\min_{R_p^c} ||Y_p^c R_p^c - Y_p^c||_2^2 + c_1||R_p^c||_2^2 + c_2 \sum_{i=1}^{n_p} ||R_{p_{ii}}^c||_2^2 + c_3||R_p^c||_1. \qquad (7)$$

where $n_p$ is the number of represented column vectors and $(c_2 \to \infty)$ makes $R_{p_{ii}}$ approach to 0. In problem (7), the non-smoothness of $||R_c^p||_1$ makes the standard proxy function-based optimization methods unsuitable. Thus, we exploit the alternating direction method of multipliers and soft thresholding operation [6] to optimize the problem.

By introducing an auxiliary variable $R_c^p = Z_1$, we transform problem (7) into the following formulation,

$$\min_{R_p^c, Z_1, \Upsilon_1} ||Y_p^c R_p^c - Y_p^c||_2^2 + c_1||R_p^c||_2^2 + c_2 \sum_{i=1}^{n_k} ||R_{p_{ii}}^c||_2^2$$

$$+\beta||R_p^c - Z_1 + \Upsilon_1||_2^2 + c_3||Z_1||_2 \qquad (8)$$

where $\Upsilon_1 \in \mathbb{R}^{l \times n_k}$ is the positive Lagrange multiplier, $\beta$ is a positive number. Our method includes three key steps (i.e., updating $R_p^c, Z_1, \Upsilon_1$, respectively).

**Update $R_p^c$.** If $Z_1$ and $\Upsilon_1$ is fixed, $R_p^c$ is updated by the following problem,

$$\min_{R_p^c} ||Y_p^c R_p^c - Y_p^c||_2^2 + c_1||R_p^c||_2^2 + c_2 \sum_{i=1}^{n_k} ||R_{p_{ii}}^c||_2^2 + \beta||R_p^c - Z_1 + \Upsilon_1||_2 \qquad (9)$$

Taking the derivation of (9) w.r.t. $R_p^c$ and setting it zero, we obtain the form $(2Y_p^{cT}Y_p^c + 2c_1 I)R_p^c + 2c_2 diag(R_p^c) = 2Y_p^{cT}Y_p^c + 2\beta(Z_1 - \Upsilon_1)$ which can be solved by the Theorem 1 after transposing it.

**Update $Z_1$.** If $R_p^c$ and $\Upsilon_1$ is fixed, $Z_1$ is updated by solving the following problem,

$$\min_{Z_1} \beta||R_p^c - Z_1 + \Upsilon||_1 + c_3||Z_1||_1$$

which can be re-written into a convenient form:

$$\min_{Z_1} ||R_p^c - Z_1 + \Upsilon||_1 + \frac{c_3}{\beta}||Z_1||_1$$

The optimal $Z_1$ can be obtained via soft thresholding operation $Z = soft(R_p^c + \Upsilon, \frac{c_3}{\beta})$ [6] where $soft(a, b) = sign(a)max(|a| - b, 0)$.

**Updating $\Upsilon_1$.** If $R_p^c$ and $Z_1$ is fixed, $\Upsilon_1$ is updated by the following form,

$$\Upsilon_1 = \Upsilon_1 + (R_p^c - Z_1)$$

### 4.3  The Optimization of Objective Function

In this section, we divide the objective into two cases. In case 1, we only consider the row representation of overall datasets. In case 2, we consider the mixture of row and column representation. In the following, we illustrate the optimization of two cases, respectively.

**Row Representation of Overall Dataset.** When the linear representation only contains the row representation of overall datasets, the optimization problem is the following formulation,

$$\min_{M} L(\bar{Y}, MX) = ||\bar{Y} - MX||_2^2 + \lambda_1 ||M||_2^2 + \lambda_3 ||(R^r - I)MX||_2^2 \quad (10)$$

Taking the derivation of (10) w.r.t. $M$ and setting it zero, we obtain the form $(2I + 2\lambda_3(R^r - I)^T(R^r - I))MXX^T + 2\lambda_1 M$. Due to symmetry of $Q_1 = 2I + 2\lambda_2 I + 2\lambda_3(R^r - I)^T(R^r - I)$ and $XX^T$, we decompose $Q_1$ and $XX^T$ into $Q_1 = UAU^T$, $XX^T = VBV^T$ where $U$, $V$ are the eigen vectors, and $A$ and $B$ are diagonal matrices. By substituting $Q_1$ and $2XX^T$, and multiplying $U^T$ and $V$ from left to right on both sides, we obtain the following form,

$$AU^T MVB + 2\lambda_1 U^T MV = 2U^T \bar{Y} X^T V \quad (11)$$

Let $U^T MV = \bar{M}$, then (11) becomes $A\bar{M}B + 2\lambda_1\bar{M} = U^T QV$. Thus, we can obtain $M$ as $M = U\bar{M}V^T$ where $\bar{M}_{ij} = \frac{(U^T QV)_{ij}}{A_{ii}B_{jj} + 2\lambda_1}$ denotes the $(i, j)$-th entry of $\bar{M}$.

**The Mixture of Row and Column Representation.** By introducing an auxiliary variable $M = Z_3$, we transform the objective function into the following formulation,

$$\min_{M} L(\bar{Y}, MX) = ||\bar{Y} - MX||_2^2 + \lambda_1 ||M||_2^2 + \lambda_3 ||(R_s^c - I)Z_3 X_s||_2^2$$

$$+\lambda_3 \sum_{p \in P} ||MX_p R_p^c - MX_p||_2^2 + \frac{\beta}{2}||M - Z_3||_2^2 + tr(\Upsilon_3(M - Z_3)) \quad (12)$$

**Solving M:** By fixing the variables $Z_3$ and $\Upsilon_3$, taking the derivation of (12) w.r.t. $M$, and setting the derivates zero, we have the form $M((2 + 2\lambda_2)XX^T + 2\lambda_3 I + 2\lambda_3 \sum_{p \in P} X(R_p^c - I)(R_p^c - I)^T X^T) = 2\bar{Y} X^T$. And the solution is $M = (2\bar{Y} X^T)((2 + 2\lambda_2)XX^T + 2\lambda_3 I + 2\lambda_3 \sum_{p \in P} X_p (R_p^c - I)(R_p^c - I)^T X^T)^{-1}$.

**Solving $Z_3$:** By fixing the variables $M$ and $\Upsilon_3$, we have the following formulation,

$$\min_{Z_3} L(\bar{Y}, MX) = \lambda_3 ||(R_s^r - I)Z_3 X_i||_2^2 + \frac{\beta}{2}||M - Z_3||_2^2 + tr(\Upsilon_3(M - Z_3))$$

Take the derivation of (4.3) w.r.t. $Z_3$. By setting the derivates zero, we have the similar formulation with Case 1.

**Update $\Upsilon_3$:** By fixing the variables $Z_3$ and $M$, we can update $\Upsilon_3$ by $\Upsilon_3 = \Upsilon_3 + \beta(M - Z_3)$.

## 5    Generalization Error Bounds

In this section we analyze excess risk bounds for our learning model with the row representation of overall dataset. Our model is characterized by a distribution $\mathscr{D}$ on the feature space and the label space $X \times Y$. Let $n$ samples $\{(\boldsymbol{x}_1, \boldsymbol{y}_1), (\boldsymbol{x}_2, \boldsymbol{y}_2), \cdots, (\boldsymbol{x}_n, \boldsymbol{y}_n)\}$ be drawn i.i.d. from the distribution $\mathscr{D}$, which is denoted by $D \sim D^n$.

Given this training data, we perform ERM to learn $\widehat{M} \in M$ by the following formulation,

$$\inf_{\boldsymbol{RM}\,\boldsymbol{X}=\boldsymbol{M}\,\boldsymbol{X}} \widehat{L}(M) = \frac{1}{n} \sum_{k=1}^{n} \mathscr{L}(y_i, f(x_i, M))$$

where $\widehat{L}(M)$ is the empirical risk of the predictor $M$. Our goal would be to show that $\widehat{M}$ has good generalization properties, i.e.

$$\inf L(\widehat{M}) \leq \inf_{\boldsymbol{RM}\,\boldsymbol{X}=\boldsymbol{M}\,\boldsymbol{X}} \widehat{L}(M) + \epsilon.$$

where $L(M) = \mathbb{E}_{x,y,l}[\![L(y_i^l, f_i^l(x_i; M))]\!]$ is the population risk of a predictor.

The Rademacher complexity, which is an effective way to measure the complexity of a function class, is exploited to find the generalization error bound of the learning algorithm by standard approaches [1].

**Definition 1.** Let $X = \{x_1, \cdots, x_n\} \in X^n$ be independent samples and $F$ be a class of functions mapping from $X$ to $R$. The empirical Rademacher compxlexity of $F$ is defined as,

$$\widehat{R}_n(F) = \mathbb{E}_\sigma[\sup_{f \in F} |\frac{2}{n} \sum_{j=1}^{n} \sigma_i f(x_i)| x_1, \cdots, x_n]$$

where $\sigma = (\sigma_1, \cdots, \sigma_n)$ are independent uniform $\pm 1$-valued random variables. The Rademacher compxlexity of $F$ is,

$$\widehat{R}_n(F) = \mathbb{E}_{x,\sigma}[\sup_{f \in F} |\frac{2}{n} \sum_{j=1}^{n} \sigma_i f(x_i)|]$$

Thus, the Rademacher compxlexity of our algorithm is as follows,

$$\widehat{R}_n(F) = \frac{2}{n} \mathbb{E}_{x,\sigma}[\sup_{M \in F} \sum_{i=1}^{n} \sigma_i \sum_{j=1}^{l} M_{j,:} x_i]$$

where $M_{j,:}$ is the j-th row of $M$. Let $\widehat{x}$ denote the weighted summarization $\sum_{i=1}^{L} \sigma_i x_i$ and $\widehat{X}$ be the stacker of $L$ copies of $\widehat{x}$. Thus, we simply the the multi-label Rademacher complexity as,

$$\widehat{R}_n(F) = \frac{2}{n} \mathbb{E}_{x,\sigma}[\sup_{M \in F} M\widehat{X}]$$

**Table 1.** Statistics of five datasets.

| Datasets | #Dataset | #Training | #Testing | #Features | #Labels | #Card-label | #Domain |
|---|---|---|---|---|---|---|---|
| CAL500 | 502 | 452 | 50 | 68 | 174 | 26.044 | Music |
| COREL5K | 5,000 | 4,500 | 500 | 499 | 374 | 3.522 | Images |
| DELICIOUS | 16,105 | 14,495 | 1,610 | 500 | 983 | 9.020 | Text |
| EUR-Lex | 19,348 | 17,413 | 1,935 | 5000 | 3,993 | 5.310 | Text |
| NUS-WIDE-V | 269,648 | 161,789 | 117,859 | 128 | 81 | 1.869 | Images |

**Theorem 2.** The proposed algorithm learns $M$ over $n$ training points with $l$ labels i.i.d. sampled from distribution $\mathscr{D} = X \times Y$. For any data point, we assume that $\|x\| < \Lambda$. The learning algorithm encourages $RMX = MX$ which usually satisfies the $\|RMX - MX\|_2^2 < \epsilon_0$ ($\epsilon_0$ is a constant). We also assume $(R - I)^{-1}$ to be upper bounded by $\delta$. Then, the Rademacher complexity of F is,

$$\widehat{R}_n(F) \leq \frac{2\delta\epsilon}{n}.$$

**Proof.** The learning algorithm constrains $RMX = MX$ to capture the linear representation, which is usually not satisfied. Instead of equation constraint, we have $\|RMX - MX\|_2^2 < \epsilon_0$ been satisfied. By multiplying $(R - I)^{-1}$ left on $\|RMX - MX\|_2^2 < \epsilon_0$, we have the following formulation $\|MX\|_2^2 < (R - I)^{-1}\epsilon_0$. According to equation $\widehat{R}_n(F) = \frac{2}{n}\mathbb{E}_{x,\sigma}[\sup_{M \in F} M\widehat{X}]$, we prove $\widehat{R}_n(F) \leq \frac{2\|(R-I)^{-1}\|\epsilon}{n}$.

## 6   Experiment

In this section, we perform experiments to evaluate the effectiveness of our method. After introducing datasets and methodology, we compare our method with two representative class of methods.

### 6.1   Datasets

The experiments are conducted on five real-world multi-label datasets[1], namely CAL500, COREL5K, DELICIOUS, EUR-Lex, NUS-WIDE-V.

1) Cal500: In this dataset, 4500 songs by different artists are annotated with 174 labels to represent their genres, instruments, emotions, and other related concepts [16].
2) COREL5K: It contains about 5000 images from Stock Photo CDs which are annotated with 374 labels [3].
3) DELICIOUS: It contains textual data of Web pages along with 983 tags extracted from the del.icio.us social book marking site [7].

---

[1] http://mulan.sourceforge.net/datasets-mlc.html.

4) EUR-Lex: It collects documents on European Union law [23]. There are several EuroVoc descriptors, directory codes, and types of subject matter to describe the labels.
5) NUS-WIDE-V: In this version, it represents the images as 128-D cVLAD+ features described in [14] (Table 2).

**Table 2.** Predictive performance comparison on five real-world datasets with the embedding method of label matrix. The best results are in bold. N/A denotes not available.

| Datasets | Example-F1 ↑ | | | | | | | |
| --- | --- | --- | --- | --- | --- | --- | --- | --- |
| | CCA | kNN | ML-kNN | LM-kNN | SLEEC | CoH | RROD (ours) | MRCR (ours) |
| CAL500 | 0.3520 | 0.3561 | 0.3216 | 0.3511 | 0.3099 | 0.3602 | 0.4501 | **0.4606** |
| COREL5K | N/A | 0.0223 | 0.0178 | 0.1295 | 0.0839 | 0.1996 | 0.2721 | **0.2943** |
| DELICIOUS | N/A | 0.1878 | 0.1518 | 0.2553 | 0.2197 | 0.2745 | **0.4024** | 0.3645 |
| EUR-Lex | N/A | 0.3409 | 0.3005 | 0.3818 | 0.3705 | 0.3915 | **0.4257** | 0.4252 |
| NUS-WIDE-V | N/A | 0.1382 | 0.0331 | 0.0982 | 0.1703 | 0.2838 | 0.3428 | **0.3429** |
| Datasets | MIcro-F1↑ | | | | | | | |
| | CCA | kNN | ML-kNN | LM-kNN | SLEEC | CoH | RROD (ours) | MRCR (ours) |
| CAL500 | 0.3537 | 0.3593 | 0.3184 | 0.3542 | 0.3077 | 0.3664 | 0.4500 | **0.4607** |
| COREL5K | N/A | 0.0321 | 0.0278 | 0.1670 | 0.1215 | 0.2127 | 0.2675 | **0.2929** |
| DELICIOUS | N/A | 0.2154 | 0.1738 | 0.3104 | 0.2532 | 0.3011 | **0.4044** | 0.3626 |
| EUR-Lex | N/A | 0.4011 | 0.3489 | 0.4344 | 0.4235 | **0.4689** | 0.0933 | 0.0348 |
| NUS-WIDE-V | N/A | 0.2772 | 0.2826 | 0.2093 | 0.3249 | 0.3282 | 0.3634 | **0.3638** |
| Datasets | Macro-F1 ↑ | | | | | | | |
| | CCA | kNN | ML-kNN | LM-kNN | SLEEC | CoH | RROD (ours) | MRCR (ours) |
| CAL500 | 0.0917 | 0.0971 | 0.0534 | 0.1001 | 0.0462 | 0.1173 | **0.1826** | 0.1698 |
| COREL5K | N/A | 0.052 | 0.0086 | 0.0266 | 0.0240 | 0.0423 | **0.0483** | 0.0433 |
| DELICIOUS | N/A | 0.0550 | 0.0481 | 0.1383 | 0.0702 | 0.0993 | **0.1851** | 0.1128 |
| EUR-Lex | N/A | 0.0877 | 0.0635 | 0.0919 | 0.0805 | **0.1053** | 0.0376 | 0.0212 |
| NUS-WIDE-V | N/A | 0.0682 | 0.0159 | 0.0171 | 0.0437 | 0.0498 | **0.0809** | 0.0803 |

## 6.2 Methodology

**Baselines:** To demonstrate the performance of our method, we compare it with two representative approaches: the label-matrix embedding method and the regression-matrix constraining method. The label-matrix embedding method contains CCA (Zhang and Schneider 2011) [21], kNN, ML-kNN (Zhang and Zhou 2007) [19], LM-kNN (Liu and Tsang 2015) [11], SLEEC (Bhatia et al. 2015) [2], COH (Xiaobo Shen et al. 2018) [12]. The regression-matrix constraining method contains LEML(Yu et al. 2014) [18], SLRM (Jing et al. 2015) [9], REML(Xu et al. 2016) [17]. In LM-kNN, we set $\eta = 0.4$ and $C = 10$ by following the experimental settings [11]. In SLEEC, following the original settings [2], we set the number of clusters as $\frac{n}{6000}$ and the number of learners as 15. In COH [12], by following the original settings, the regularization parameter is empirically set to 100 for the two NUS-WIDE datasets, and 1 for the other datasets. In LEML [18], we set the rank as 50, 30, 170,1100, 70, 70 on datasets CAL, CORE5K, DELICIOUS, EUR-Lex, NUS-WIDE-V. In SLRM (Jing et al. 2015), we consider the $N$ nearest neighbor graph of every instance to model the distance structure among data. In REML [2], we set the rank of regression matrix to be similar to LEML [18].

**Evaluation Metric:** Three metrics (i.e., Example-F1, Micro-F1, Macro-F1) are often used to evaluate the performance of multi-label learning and their detailed definitions can be found in Ref [13].

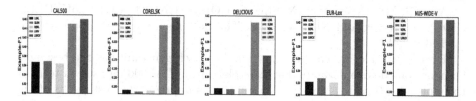

**Fig. 2.** Predictive performance of comparison in term of Example-F1 on five real-datasets with the method of low-rank constraint on regression matrix.

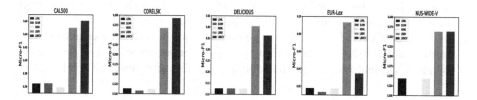

**Fig. 3.** Predictive performance of comparison in term of Micro-F1 on five real-datasets with the method of low-rank constraint on regression matrix.

## 6.3 Experimental Results

**Experimental Settings.** In this experiment, we consider two schemes: the row representation of overall datasets (RROD) and the mixture of row and column representation (MRCR). In RROD, we set the parameters $\lambda_1 = 1$ and $\lambda_2 = 1.3$ on every dataset. By adding the column representation into RROD, we obtain the MRCR and set the parameters $\lambda_1 = 1$, $\lambda_2 = 1.3$ and $\lambda_3 = 5$ on every datasets. More specially, on CAL dataset, the column representation of overall dataset is computed. In dataset CORE, DELICIOUS and EUR-Lex, the one-third of instances are selected to compute the column representation. In NUS-WIDE-V, we only select 1500 number of instances to compute the column representation.

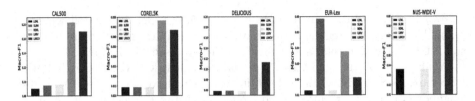

**Fig. 4.** Predictive performance of comparison in term of Macro-F1 on five real-datasets with the method of low-rank constraint on regression matrix.

**Experimental Analysis on Embedding Methods.** In Table 1, in term of Example-F1, Micro-F1, Macro-F1, we present the prediction performances of CCA, kNN, ML-kNN, LM-kNN, SLEEC, COH on CAL500, COREL5K, DELICIOUS, EUR-Lex and NUS-WIDE-V. The best results of each approach are marked in bold.

From this table, we can see that, our approach generally has the best performance on the five datasets. For example, on the CAL500 dataset, in term of Example-F1, Micro-F1, Macro-F1, our approach improves the best results of the baselines by 10.04%, 9.43% and 6.53%. This means that the linear representation may capture the intrinsic geometry structure of label matrix. The detailed descriptions are as follows.

1) Our schemes outperform than the SLEEC and COH which both break the low-rank structure of label matrix by preserving the distance structure of label matrix. This shows that linear representation may keep the more information than the small embedding in SLEEC and COH.
2) In EUR-Lex and NUS-WIDE-V, the performance of MRCR does not outperform than RROD. This shows that the column representation may be covered by row representation, thus the column representation does not work.

**Experimental Analysis on Low-Rank Constraint.** In this section, we compare the proposed RROD and MRCR methods with LEML [18], SLRM [9], REML [17], which directly capture the rank structure of regression matrix. In REML, we do not evaluate the performance of NUS-WIDE-V because a $N \times N$ matrix is too large to be stored. Figure 2, Fig. 3 and Fig. 4 respectively show the prediction performance of Example-F1, Micro-F1, and Macro-F1 on five datasets. From the three Figures, we can see that, our approach usually has the best performance on five datasets. This again shows that linear representation may keep more information than the embedding of feature space. This shows that our scheme outperforms than the methods of the rank-constraint of regression matrix.

## 7 Conclusion

We propose the LGSLM classifier to capture the intrinsic geometry structure of label matrix. More specially, we measure the row linear representation of overall datasets. In the column representation, by segmenting the label matrix into sub-regions and constraining the $L_1$ norm of the linear representation, we obtain the sparse structure of linear representation. Experiment results on five real datasets show the superior performances of our approach.

**Acknowledgement.** This work is supported by the National Key Research and Development Program of China (No. 2016QY06X1203), the Strategic Priority Research Program of the Chinese Academy of Sciences (No. XDC02040400), the National Natural Science Foundation of China (No. U1836203) and Shandong Provincial Key Research and Development Program (2019JZZY20127).

# References

1. Bartlett, P.L., Mendelson, S.: Rademacher and Gaussian complexities: risk bounds and structural results. J. Mach. Learn. Res. **3**(Nov), 463–482 (2002)
2. Bhatia, K., Jain, H., Kar, P., Varma, M., Jain, P.: Sparse local embeddings for extreme multi-label classification. In: Proceedings of NeurIPS, pp. 730–738 (2015)
3. Duygulu, P., Barnard, K., de Freitas, J.F.G., Forsyth, D.A.: Object recognition as machine translation: learning a lexicon for a fixed image vocabulary. In: Heyden, A., Sparr, G., Nielsen, M., Johansen, P. (eds.) ECCV 2002. LNCS, vol. 2353, pp. 97–112. Springer, Heidelberg (2002). https://doi.org/10.1007/3-540-47979-1_7
4. Elhamifar, E., Vidal, R.: Sparse subspace clustering. In: Proceedings of ICCV, pp. 2790–2797 (2002)
5. Gupta, V., Wadbude, R., Natarajan, N., Karnick, H., Jain, P, Rai, P.: Distributional semantics meets multi-label learning. In: Proceedings of AAAI, pp. 3747–3754 (2019)
6. Donoho, D.L.: De-noising by soft-thresholding. IEEE Trans. Inf. Theory **41**(3), 613–627 (1995)
7. Tsoumakas, G., Katakis, I.: Multi-label classification: an overview. Int. J. Data Warehouse. Min. **3**(3), 1–13 (2007)
8. Jain, H., Prabhu, Y., Varma, M.: Extreme multi-label loss functions for recommendation, tagging, ranking and other missing label applications. In: Proceedings of SIGKDD, pp. 935–944 (2016)
9. Jing, L., Yang, L.: Semi-supervised low-rank mapping learning for multi-label classification. In: Proceedings of CVPR, pp. 1483–1491 (2015)
10. Li, J., Lu, K., Huang, Z., Shen, H.Y.: Two birds one stone: on both cold-start and long-tail recommendation. In: Proceedings of ACM MM, pp. 898–906 (2017)
11. Liu, W., Tsang, I.W.: Large margin metric learning for multi-label prediction. In: Proceedings of AAAI (2015)
12. Shen, X., Liu, W., Tsang, I.W., Sun, Q.S., Ong, Y.S.: Compact multi-label learning. In: Proceedings of AAAI (2018)
13. Sorower, M.S.: A literature survey on algorithms for multi-label learning. Oregon State University, vol. 18, pp. 1–25 (2010)
14. Spyromitros, X.E., Papadopoulos, S., Kompatsiaris, I.Y., Tsoumakas, G., Vlahavas, I.: A comprehensive study over VLAD and product quantization in large-scale image retrieval. IEEE Trans. Multimedia **16**(6), 1713–1728 (2014)
15. Tai, F., Lin, H.T.: Multi-label classification with principal label space transformation. Neural Comput. **24**(9), 2508–2542 (2012)

16. Turnbull, D., Barrington, L., Torres, D., Lanckriet, G.: Semantic annotation and retrieval of music and sound effects. IEEE Trans. Audio Speech Lang. Process. **16**(2), 467–476 (2008)
17. Xu, C., Tao, D., Xu, C.: Robust extreme multi-label learning. In: Proceedings of SIGKDD, pp. 1275–1284 (2016)
18. Yu, H.F., Jain, P., Kar, P., S. Dhillon, I.: Large-scale multi-label learning with missing labels. In: Proceedings of ICML, pp. 593–601 (2014)
19. Zhang, M.L., Zhou, Z.H.: ML-KNN: a lazy learning approach to multi-label learning. Pattern Recogn. **40**(7), 2038–2048 (2007)
20. Zhang, M.L., Zhou, Z.H.: A review on multi-label learning algorithms. IEEE Trans. Knowl. Data Eng. **26**(8), 1819–1837 (2013)
21. Zhang, Y., Schneider, J.: Multi-label output codes using canonical correlation analysis. In: Proceedings of AISTATS, pp. 873–882 (2011)
22. Zhang, Y., Schneider, J.: Maximum margin output coding. In: Proceedings of ICML (2012)
23. Mencia, E.L., Fürnkranz, J.: Efficient pairwise multilabel classification for large-scale problems in the legal domain. In: Proceedings of Joint European Conference on Machine Learning and Knowledge Discovery in Databases, pp. 50–65 (2002)

# Expanding Training Set for Graph-Based Semi-supervised Classification

Li Tan[1,2] , Wenbin Yao[1,2(✉)] , and Xiaoyong Li[3]

[1] School of Computer Science,
Beijing University of Posts and Telecommunications, Beijing 100876, China
yaowenbin@bupt.edu.cn
[2] Beijing Key Laboratory of Intelligent Telecommunications
Software and Multimedia,
Beijing University of Posts and Telecommunications, Beijing 100876, China
[3] School of Cyberspace Security,
Beijing University of Posts and Telecommunications, Beijing 100876, China

**Abstract.** Graph Convolutional Networks (GCNs) have made significant improvements in semi-supervised learning for graph structured data and have been successfully used in node classification tasks in network data mining. So far, there have been many methods that can improve GCNs, but only a few works improved it by expanding the training set. Some existing methods try to expand the label sets by using a random walk that only considers the structural relationships or selecting the most confident predictions for each class by comparing the softmax scores. However, the spatial relationships in low-dimensional feature space between nodes is ignored. In this paper, we propose a method to expand the training set by considering the spatial relationships in low-dimensional feature space between nodes. Firstly, we use existing classification methods to predict the pseudo-label information of nodes, and use such information to compute the category center of nodes which has the same pseudo label. Then, we select the k nearest nodes of the category center to expand the training set. At last, we use the expanded training set to reclassify the nodes. In order to further verify our proposed method, we randomly select the same number of nodes to expand the training set, and use the expanded training set to reclassify nodes. Comprehensive experiments conducted on several public data sets demonstrate effectiveness of the proposed method over the state-of-art methods.

**Keywords:** Graph structured data · Semi-supervised learning · Graph Convolutional Networks · Node classification

Supported by NSFC-Guangdong Joint Found (U1501254) and the Joint Fund of NSFC-General Technology Fundamental Research (U1836215).

© Springer Nature Switzerland AG 2020
S. Hartmann et al. (Eds.): DEXA 2020, LNCS 12392, pp. 245–258, 2020.
https://doi.org/10.1007/978-3-030-59051-2_16

# 1 Introduction

Various complex data can conveniently be represented as graphs and are widely used in real life, such as social networks, citation networks, protein molecules networks, chemical molecules networks. As the explosive growth of data generated by various ways such as Internet has caused great trouble to data analysis, we need to uncover potentially useful information in data. The classification, a basic problem, has become a hot topic in data mining. Methods of node classification using multiple types of relationships between nodes, such as the labels of nodes, the structural and the attributes relationships between nodes. The methods which using relationships between nodes are verified effectively improved node classification performance, but the network data is sparse which limiting the use of the relationships between nodes. In addition, since there is a large amount of unlabeled data, and collecting labeled data is expensive and inefficient. Therefore, the semi-supervised classification of graph data has attracted the attention of many researchers. For example, in the citation networks, the nodes represent papers, and the edges represents the citation relationships between the papers, the features of each node is the bag-of-words representation of the corresponding paper, the semi-supervised node classification for citation networks refers to classification of a large number of papers without labels by using a small number of labeled papers with the relationship of citations and attributes among them.

The graph convolutional networks [1], which passes the information of graph node attributes to his neighbors, breaks through the previous convolutional neural network's limitation in Euclidean space and extends the convolutional neural network to non-Euclidean space. The performance has greatly improved in contrast with previous methods using directly connection or weighted connection to combine the structural and attribute relationships between nodes. However, graph convolutional networks [1] also have some shortcomings. For example, the first-order similarity and second order similarity between nodes are considered but the high-order similarity between nodes is ignored. And overfitting is easy to occur. Nevertheless, most of the problems is caused by the small number of training set. Therefore, in this paper, we propose a novel method to expand the training set.

We propose a method to expand the training set which consider the spatial relationships in low-dimensional feature space between nodes. At first, we use GCN or other classification methods to obtain the low dimensional dense feature vector representations and pseudo-label of nodes. Then, we look for the category center according to the density of nodes which have the same pseudo-label, and select the k nearest nodes to the category center to expand the training set. At last, using the expanded training set to reclassify nodes through the same classification method as above. In order to further verify the effectiveness of our proposed method, we use the same method to classification and obtain the pseudo-labelling of nodes and then randomly select the same number of nodes to expand the training set. finally, using the expanded training set to reclassification. Our model gains a considerable improvement on several public datasets over the state-of-art methods.

In summary, our main contributions can be summarized as follows:

(1) An effective method for expanding the training set is proposed. The category center is calculated based on the density of nodes that have the same pseudo-label obtained by pre-classification, selecting the k nearest nodes of the category center to expand training set.
(2) The method to expand the training set can be applied not only in GCN [1], but also in other methods and fields. This method can be seen as a framework applied to other methods.
(3) Extensive experiments on a variety of public datasets and show that the proposed method constantly outperforms other state-of-art approaches.

The rest of this paper is organized as follows. The second part introduces the related works. The third part discusses the proposed method in detail. The fourth part introduces experiment establishment and the experimental results. Finally, we will briefly summarize this paper.

## 2   Related Work

In this section, at first, we review the graph convolutional networks and its application in semi-supervised node classification tasks, and then review clustering algorithms based on density of nodes, and finally introduce graph-based semi-supervised node classification.

### 2.1   Graph Convolutional Networks

In the past few years, the frequency of using graph convolutional networks has increased rapidly. The essence of these methods is attributes information transmission, which is mainly divided into two branches, spectral-based approaches and spatial-based approaches. The spectral-based approaches are mainly based on the spectral graph theory to define parameterized filters. [2] first defines the convolution operation in the Fourier domain, but are limited to expand to large graphs as the approaches require great amounts of computation. ChebNet [3] improves [2] by approximate the K-polynomial filters by means of a Chebyshev. GCN [1] further simplifies the convolution operation by using first-order similarity of graph nodes instead of k-order polynomials. DGCN [4] computes the positive pointwise mutual matrix through random walks, and uses this matrix and Laplacian regularized adjacency matrix by sharing weights to capture local and global information of nodes [5]. Obtained the pseudo-labels of the nodes through GCN pre-classification, and then expand the validation set with the pure structural information by randomly walk or by selecting the most confident predictions for each class by comparing the softmax scores. Finally, it uses the extended validation set to retrain the model. Spatial-based approaches are not covered in this paper and will not be introduced here.

As far as we know, we are the first to improve GCN [1] by expanding the training set. Although [5] also proposed a method to expand the validation set, but it use a random walk that only considers the structural relationship or by selecting the most confident predictions for each class by comparing the softmax scores to expand the dataset, which have not considered the spatial information between nodes.

## 2.2 Clustering Algorithm Based on Density

The density-based clustering algorithm considers the density of nodes during the clustering process. DBSCAN [6] is a well-known density-based clustering algorithm. It traverses all nodes according to the specified distance $e$ to find all core nodes to produce a set which contains some core nodes, and then randomly select a node from this set and to find all reachable nodes according to a density threshold $\rho$ to composite a category. If this category has nodes in core nodes set, removing them from the set. Repeating the above process until all core nodes are traversed. Density-ratio [7] algorithm solve the problem that may find multiple category centers while the nodes are same category due to density distribution. [8] proposed two central ideas. The one is the cluster centers belonging to same category has higher density. The other one is the cluster centers of different categories should be as far as possible. The cluster center calculated by these two ideas. Mean shift [9] randomly initializes some cluster centers and calculate the density of nodes, and then the cluster centers shift along the direction of increasing density until the distance between new category center and previous category center less than the specified threshold $min_{dist}$, and this category center is the final cluster center.

## 2.3 Graph-Based Semi-supervised Classification

In this paper, we consider graph-based semi-supervised node classification tasks. Suppose $Y_L$ represents the nodes set with labels and $Y_u$ represents the nodes set without labels, while the number of $Y_u$ is much larger than $Y_L$, and graph-based semi-supervised classification is predicting the labels of $Y_u$ by using the $Y_L$.

Graph-based semi-supervised classification has been very popular in the past few decades. Due to the relationships between edges, using fewer nodes with label can achieve good performance. Some algorithms such as spectral graph transducer algorithm [10], label propagation algorithm [11], and iterative classification algorithm [12] are all based on the assumption that structurally similar nodes have a high probability of having the same label.

However, the above semi-supervised classification algorithm only relies on the structural relationships in the graph nodes and ignores the attribute relationships that are as important as the structural relationships. There are also have some algorithms combine the structural relationships and the attributes relationships such as manifold regularization (LapSVM) [13] regularize support vector machine with Laplacian regularization. Planetoid [14] regularized neural network by using category labels and the semantic information of nodes.

# 3   The Model

In this section, we first simply introduce the graph convolutional networks, and then we introduce the overview of our proposed method, finally, we will describe the process of expanding the training set in detail.

## 3.1   Preliminaries

The Graph Convolutional Networks [1] has received extensive attention since it was proposed, many methods have tried to improve it and our method also based on it. Given a graph $G = (V, E, X)$, where $V$ and $E$ is the set of $n$ nodes and $e$ edges respectively, $X \in R^{n \times k}$ represents the attributes contained in each node, $X_i \in R^{1 \times k}$ represents the attributes of node $i$. On the $i$-th layer of the GCN, $h^{i-1}$ represents the attributes be fed into neural networks, the output $h^i$ is defined as:

$$h^i = \sigma(\widehat{A} h^{i-1} W^{i-1}) \tag{1}$$

$A$ represents the adjacency matrix, $\widehat{A}$ represents Laplacian regularization of matrix $A$ and shows the structural relationships between nodes, $\widehat{A} = D^{-\frac{1}{2}}(A + I)D^{-\frac{1}{2}}$, $I \in R^{n \times n}$ is identity matrix, $D$ is diagonal matrix and represent the degree of nodes and $h^0 = X$, $\sigma$ is the activation function and in this paper we use RELU, $W^{i-1}$ represents the trainable weights.

The loss function of the GCN [1] is the cross-entropy, the output of the GCN is $h^i = Z_{lf} \in R^{n \times F}$, $Y_{lf}$ is the label set of the training set, and the loss function is:

$$loss = - \sum_{l \in Y_L} \sum_{f=1}^{F} Y_{lf} ln Z_{lf} \tag{2}$$

## 3.2   Overview of the Models

In view of the good classification performance of GCN, to make better use of the advantages of GCN and make up for the shortcomings of GCN, we naturally thought of co-training [15] structure, combining the structural information and attribute information as well as the spatial relationships in low-dimensional feature space between nodes to find the most likely k nodes of each category to expand training set.

We will mainly introduce the overview of our proposed method. Firstly, we use GCN [1] to obtain the pseudo-labels of nodes and calculate the density between nodes which have same pseudo-label, and then find the category center by using Mean Shift [9], using the low-dimensional representation of the category center to compute the k nearest nodes which has the same pseudo-labelling to expand the training set, and repeat the above process until the accuracy of node classification decreases. Algorithm 1 is the overview of the algorithm.

**Algorithm 1.** The overview of the model

---

**Require:** $X \in R^{n \times f}$ represents attributes of nodes,$n$ is the number of nodes

   $A \in R^{n \times n}$ is the adjacency matrix of graph, $n$ is the number of nodes

   $d$ is the dimensions of embedding

   $k$ is the number of nodes in each category to expand the training set onetime

**Ensure:** $h^l \in R^{n \times d}$ is the final embedding and represent the classification result

1: Randomly initialize the weights of each hidden layer $W^0, W^1, \ldots, W^k$

2: Computing $\widehat{A} = D^{-\frac{1}{2}}(A+I)D^{-\frac{1}{2}}$, $D_i = \sum_{0 \leq j \leq n} A_{(i,j)}$

3: $pre_{acc} = 0$

4: **while true do**

5:    Classification algorithms such as GCN to calculate the classification accuracy
      $accuracy$ and the pseudo-label of each node

6:    **if** $accuracy > pre_{acc}$ **then**

7:       $pre_{acc} = accuracy$

8:       save model

9:       $add\_node(k, h^{l+1}, h^l)$

10:   **else**

11:      break

12:   **end if**

13: **end while**

---

As described in the Algorithm 1, the third line's $pre_{acc}$ represents the accuracy of the previous node classification, and the initial accuracy rate is zero. The fifth line represents the execution of the GCN [1], where $h^l$ represents the embedding obtained from the penultimate layer, which is low-dimensional feature representation combining the attributes and structural relationships of nodes, $h^{l+1}$ represents the final classification result. Lines 6–12 is the process of expanding the training set until the accuracy of node classification will not increase. The final model is the model we need.

### 3.3 Expanding the Training Set

In this section, we mainly introduce this model how to expand the training set. The intuitive idea is to use the pseudo-labels obtained by GCN pre-classification, and then take the means of the embedding of node which have the same pseudo-label as the center of this category. Then, calculating the Euclidean distance between center and other nodes have same pseudo-label, selecting the k closest nodes to extend the training set. But this method has some shortcomings, which is hardly used to the U-shaped node distribution, circular-shaped node distribution and so on, the accuracy of nodes expand to training set is lower and the accuracy of node classification will be lower.

To address the above challenges and select the category center more accurately. The embedding of nodes obtained by GCN pre-classification effectively combines the attribute relationships and structural relationships of nodes. Besides, embedding is the representations of nodes in the dense feature space which the nodes of the same category have smaller distances, and the distances of different categories of nodes will be farther. In order to capture the spatial relationships between the nodes, the spatial distance between the nodes is used to calculate the category center. At first, We randomly select a node as the category center $c$ for each category, and then calculate the Euclidean distance between the embedding of category center $c$ and node $i$ has the same pseudo-label.

$$distance = ||embedding_c - embedding_i|| \tag{3}$$

Based on the distance between the nodes, calculating density of nodes have same pseudo-label. We use high-dimensional Gaussian kernel function to calculate the density of the nodes $\rho(x_i)$. The *bandwidth* is a parameter to calculate the density between nodes.

$$\rho(x_i) = \frac{1}{bandwidth \times \sqrt{2\pi}} \times e^{-\frac{1}{2} \times (\frac{distance}{bandwidth})^2} \tag{4}$$

Similar to the gradient descent, the category center is continuously move towards density increasing direction and update the category center until the distance between new category center and previous category center less than the distance threshold $min_{dist}$.

$$\partial\rho(x_i) = \frac{1}{bandwidth \times \sqrt{2\pi}} \partial(e^{-\frac{1}{2} \times (\frac{distance}{bandwidth})^2}) \tag{5}$$

$$\rho(x_i) = \rho(x_i) + \alpha \times \partial\rho(x_i) \tag{6}$$

As described in Algorithm 2, lines 6–15 represents the process of updating the category center until the distance between new category center and previous category center less than the distance threshold $min_{dist}$, and lines 16–21 represents the process of finding the k closest nodes to the category center. The loss function is cross-entropy, which is the same as used in GCN [1] and described in 3.1.

## 4    Experiments

In this section, we present the results of several experiments to verify the performance of our method in graph-based semi-supervised node classification tasks. Firstly, we introduce the data sets we used in the experiments, and then simply list the comparison methods, finally, we present the experimental results and low-dimensional visualization of nodes and discuss the advantages and disadvantages of this method.

### 4.1    Experimental Settings

**Datasets:** For comparison purposes, we used the same dataset [1,14] used in the previous works, which contains three citation network datasets, Cora, Citeseer, and PubMed. The Table 1 gives the overview of this three datasets.

---

**Algorithm 2.** add_node(k,prediction,embedding)

---

**Require:** $k$ The number of nodes in each category expanding to training set onetime
   $prediction$ Pseudo-labelling of each node
   $embedding$ Low-dimensional feature representation of each node
   $bandwidth$ Parameter for calculating the density between nodes
   $min_{dist}$ The threshold of calculating category center
**Ensure:** The nodes selected to expand the training set
 1: Counting the number of nodes in each class $class_{num}$
 2: **for** each $label_i \in label_{num}$ **do**
 3:    Selecting low-dimensional embedding of nodes have same pseudo-label
 4:    Randomly initialize a category center $c$
 5:    Calculating the representations' Euclidean distance between the category center
      and the nodes have same pseudo-label by equations (3)
 6:    **while true do**
 7:       Calculating the density of nodes according to the Gaussian kernel function by
         equations (4)
 8:       According to the equations (5) to calculating the density derivative
 9:       Update the category center $c_{new}$ by equations (6)
10:       Calculating the Euclidean distance $dist$ between new category center $c_{new}$
         and previous category center $c$
11:       **if** $dist < min_{dist}$ **then**
12:          break
13:       **end if**
14:       Calculating the representations' Euclidean distance between the category cen-
         ter and the nodes have same pseudo-label by equations (3)
15:    **end while**
16:    **for** each $class_j \in class_{num}$ **do**
17:       Selecting the k smallest distance between category center and other nodes
         have same pseudo-label
18:       **if** $node \notin trainingset$ **then**
19:          add it to training set
20:       **end if**
21:    **end for**
22: **end for**

---

The nodes in Table 1 indicate paper, the edges indicate the citation relationship between papers, the classes indicates the categories of the nodes, the
features of each node are the bag-of-words representation of the corresponding
paper, label rate indicates the ratio of labeled nodes to total nodes, where each
category contains 20 nodes with label in training set.

**Baselines:** Since different baselines use different strategies to embed graphics
knowledge, we guarantee that our baseline set is highly diverse. Therefore, the
following method was chosen.

Deepwalk [16]: By randomly walking along the graph, multiple different
paths can be generated. By treating paths as "sentences" and each node as

**Table 1.** Statistics of the datasets

| Datasets | Nodes | Edges | Features | Classes | Label rate |
|----------|-------|-------|----------|---------|------------|
| Cora     | 2708  | 5429  | 1433     | 7       | 0.052      |
| Citeseer | 3327  | 4732  | 3707     | 6       | 0.036      |
| PubMed   | 19717 | 44338 | 500      | 3       | 0.003      |

"word", deepWalk [16] summarizes the language modeling techniques from word sequences to paths in the graph.

Iterative classification algorithm (ICA) [12]: In conjunction with two logistic regression classifiers, one for local node features alone and one for relational classification using local features and an aggregation operator as described in [17], we first train the local classifier using all labeled training set nodes and use it to bootstrap class labels of unlabeled nodes for relational classifier training. We run iterative classification (relational classifier) with a random node ordering for 10 iterations on all unlabeled nodes.

Planetoid [14]: Inspired by the Skip-gram model [12] from NLP, Planetoid [14] through positive and negative sampling to obtain the information representation of the network. During the sampling process, the structural and the attribute relationships were all considered.

GCN [1]: The Graph Convolutional Networks (GCN) [1] embedding of nodes combine the structural information and attribute information by attribute information transmission.

DIGCN [6]: DIGCN uses GCN [1] to pre-classify the nodes, use the random walk method or selecting the most confident predictions for each class by comparing the softmax scores or combine the two methods to expand the nodes. And use the expanded nodes to re-use GCN for classification. We select the best performance from the above three methods for following experimental comparison.

GCN-rand: To further verify the effectiveness of the proposed method, we use the same process as proposed method, but the nodes to expand the training set were randomly selected from the nodes have the same pseudo-label.

**Parameter Settings:** We use GCN [1] as a classifier to obtain pseudo-label of nodes. For comparison, the parameter setting involved in GCN [1] are the same as the parameters in the GCN open source code [18]. The bandwidth in Eq. (4)(5)(6) is 3, and for different value of $k$ on the different dataset is shown in the following Table 2.

As described by Table 2, the value k is the number of nodes added to the training set for each category in each dataset onetime. Since there are fewer data nodes and have rich attribute information in Cora and Citeseer, the embeddings obtained by GCN can better reflect the spatial relationships between nodes. According to the spatial relationships can more accurately add more nodes at onetime when expanding the training set. There are more nodes in the PubMed

**Table 2.** Parameter settings

| Datasets | k |
|----------|---|
| Cora | 10 |
| Citeseer | 10 |
| PubMed | 3 |

while the attribute information is few. can not reflect the spatial relationship between nodes well. Therefore, if too many nodes are added at a time, the classification accuracy rate may decrease, so fewer nodes are added at a time.

### 4.2 Experimental Results and Analysis

**Node Classification.** In order to verify the effectiveness of our method, we first compare the nodes classification accuracy with existing state-of-art methods. To further exclude accidents, we use the same process but randomly select the same number of nodes with same pseudo-label to expand the training set nodes, and compare the classification accuracy again.

(1) We compare the node classification tasks with existing state-of-art methods. We selected 20 nodes with label from each category for semi-supervised node classification, and used the accuracy of classification as a measure. The detailed results are shown in Table 3, the bold numbers indicate the highest accuracy.

**Table 3.** Accuracy of node classification (in percent)

| Method | Cora | Citeseer | PubMed |
|--------|------|----------|--------|
| DeepWalk | 67.2 | 43.2 | 65.3 |
| ICA | 75.1 | 69.1 | 73.9 |
| Planetoid | 75.7 | 64.7 | 77.2 |
| GCN | 81.5 | 70.3 | 79.0 |
| DIGCN | 81.7 | 71.2 | 79.2 |
| Our Method | **82.8** | **74.1** | **79.6** |

The results shown in Table 3 are encouraging, that is to say, our proposed method outperforms all baselines on all 3 datasets. In particular, the random walk-based method performs poorly on all three datasets. For example, the classification accuracy of DeepWalk [16] is lowest on the Citeseer dataset. From the superiority of Planetoid [14] and GCN over ICA [12] and DeepWalK [16] on all three data sets, it can be seen that the attributes and structural information of nodes play an equally important role in the performance of node classification. The performance of GCN [1] on all three datasets is better than Planetoid [14]. Although they all take the influence of structural information and attribute information between nodes into account on classification performance, the way of GCN used to attribute information transmission can better combine the attributes and the structural information between nodes, and can prevent the negative impact caused by the random walk boundary. There is indeed a certain improvement in the accuracy of node classification but the performance is not obvious in [6] by expanding the dataset, because it either uses a simply random walk that only considers the structural relationship to expand the data set, or select the most confident predictions for each class by comparing the softmax scores to expand the dataset, which did not consider the spatial information of nodes. It can be clearly found that our proposed method has significant improvements over existing methods.

(2) In order to exclude accidents of our proposed method and to further verify the correctness of the nodes we added and the importance of considering spatial relationships in low-dimensional vector of nodes, we randomly select the same number of nodes with the same pseudo-label to add training set for comparison.

**Table 4.** Accuracy of node classification (in percent)

| Method | Cora | Citeseer | PubMed |
|--------|------|----------|--------|
| GCN-rand | 80.7 | 69.5 | 78.5 |
| GCN | 81.5 | 70.3 | 79.0 |
| Our Method | **82.8** | **74.1** | **79.6** |

As shown in Table 4, the node classification results of GCN-rand performance are worse than the original GCN on all three datasets. Our method is superior to GCN-rand and GCN on all three datasets, which reflects the correctness of added nodes and the importance of the spatial information of the embedding to node classification.

**Visualization.** In order to further show the relationship in low-dimensional feature space of nodes, we use t-SNE [19] to visualize the embedding into two-dimensional space. Each node as a point in a two-dimensional space, and use different colors mark different category of nodes. Therefore, good visualization should be as close as possible to nodes with the same label (same color), and as far away as possible to nodes with different labels (different color).

As shown in above visualizations, The embedding is mapped into the two-dimensional space obtained by GCN, whose boundaries between different categories are not obvious, and there are some obvious gaps between the nodes

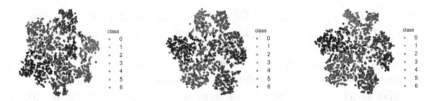

**Fig. 1.** t-SNE [19] visualization the Cora by using GCN, GCN-rand, and our proposed method by using 20 nodes with labels for each category. The left is the visualization using GCN, and the right is the visualization using the method we proposed, and the median is the visualization using GCN-rand.

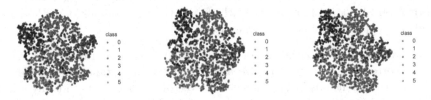

**Fig. 2.** t-SNE [19] visualization the Citeseer by using GCN, GCN-rand, and our proposed method by using 20 nodes with labels for each category. The left is the visualization using GCN, and the right is the visualization using the method we proposed, and the median is the visualization using GCN-rand.

have same category in multiple classes, which not well capture the spatial information between nodes. The visualization obtained by GCN-rand since partially incorrect nodes with the wrong label and expand to training set result some misclassification of boundary nodes. While our method effectively solved problems.

The proposed method has achieved more obvious improvements on the first two datasets, Cora and Citeseer. While PubMed has fewer attributes, fewer categories, and a large number of nodes so this data set does not improve significantly. This is a disadvantage of the method we proposed, when the embedding learned by pre-classification can not better represent the spatial information of the categories, this method will can not achieve obvious improvements (Figs. 1 and 2).

**Impact of the Number of Node Used to Expand Training Set.** Due to space issues, we only discuss the impact of accuracy of the final node classification in the citeseer.

As shown in Fig. 3, As the number of nodes added to the training set increases, the accuracy of node classification increases accordingly. When it reaches a certain peak, there is a slight ups and downs. If the number of nodes added to training set at once continues to increase, the rate gradually decreases. When too many nodes are added to the training set onetime, there will be a greater probability added wrong nodes.

There also have a peak in other two datasets. Therefore, too many of nodes add to training set at onetime can not improve the performance of node

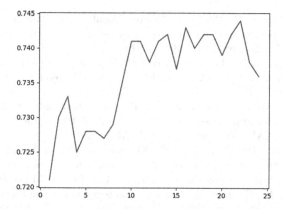

**Fig. 3.** This is the citeseer increasing the number of nodes at one time and the classification correct rate. The abscissa indicates the number of nodes added to the training set at one time, and the ordinate indicates the classification correct rate.

classification, PubMed has fewer added nodes because it contains less attribute information, Spatial relationships are difficult to capture. Therefore, when more nodes are added at one time, the probability of node errors is greater. Therefore, when 1–5 nodes are added at one time, the accuracy of node classification is better.

## 5   Conclusions

In this paper, a graph-based semi-supervised classification algorithm by expanding training set has proposed. Using pseudo-labels obtained by pre-classifying and find the central node of each category based on density of nodes which have same pseudo-label, and calculating the distance between the category center, and then selects the k closest nodes to the category center of each category to expand the training set. The proposed method improves the GCN from another angle. There is little work improved GCN by expanding training set not only combine the structural information and attribute information of nodes but also take the spatial information in low-dimensional feature space between nodes into consideration. Compared to the previous works, we propose improvements to GCN and obtain a better performance. Last but not least, our method can be used not only in the improvement of GCN, but also in other methods or other fields, to expand the training set to achieve better classification results.

## References

1. Kipf, T., Max W.: Semi-supervised classification with graph convolutional networks. In: 5th International Conference on Learning Representations (ICLR 2017), pp. 1–14 (2017)
2. Joan, B., Wojciech, Z., Arthur S., Yann, L.: Spectral networks and locally connected networks on graphs. In: 2nd International Conference on Learning Representations (ICLR 2014) (2014)

3. Michaël, D., Xavier, B., Pierre, V.: Convolutional neural networks on graphs with fast localized spectral filtering. In: 29th Advances in Neural Information Processing Systems(NIPS 2016), pp. 3844–3852 (2016)
4. Chenyi, Z., Qiang, M.: Dual graph convolutional networks for graph-based semi supervised classification. In 27th World Wide Web Conference (WWW 2018), pp. 499–508 (2018)
5. Qimai, L., Zhichao, H., Xiao-ming, W.: Deeper insights into graph convolutional networks for semi-supervised learning. In: 32nd AAAI Conference on Artificial Intelligence (AAAI 2018), pp. 3538–3545. ACM (2018)
6. Martin, E., Hans-Peter, K., Jörg, S., Xiaowei, X.: Density-based spatial clustering of applications with noise. In: 12nd International Conference on Knowledge Discovery and Data Mining (KDD 1996), pp. 226–231. ACM (1996)
7. Zhu, Y., Ting, K., Carman, M.: Density-ratio based clustering for discovering clusters with varying densities. Pattern Recogn. **2016**, 3844–3852 (2016)
8. Alex, R., Alessandro, L.: Clustering by fast search and find of density peaks. Science **344**, 1492 (2014)
9. Fukunaga, K., Hostetler, L.: The estimation of the gradient of a density function, with applications in pattern recognition. IEEE Trans. Inf. Theory **21**(1), 32–40 (1975)
10. Macqueen, J.: Some methods for classification and analysis of multi variate observations. In: Berkeley Symposium on Mathematical Statistics and Probability, pp. 281–297 (1967)
11. Joachims, T.: Transductive learning via spectral graph partitioning. In: ICML, pp. 290–297. ACM (2003)
12. Zhu, X., Ghahramani, Z., Lafferty, J.: Semi-supervised learning using Gaussian fields and harmonic functions. In: ICML 2003, pp. 912–919. ACM (2003)
13. Qing, L., Lise, G.: Link-based classification. In: International Conference on Machine Learning (ICML), vol. 3, pp. 496–503 (2003)
14. Belkin, M., Niyogi, P., Sindhwani, V.: Manifold regularization: a geometric framework for learning from labeled and unlabeled examples. J. Mach. Learn. Res. **7**, 2434 (2006)
15. Blum, A., Mitchell, T.: Combining labeled and unlabeled data with co-training. In: 11th Annual Conference on Computational Learning Theory, pp. 92–100. ACM (1998)
16. Perozzi, B., Al-Rfou, R., Skiena, S.: Deepwalk: online learning of social representations. In: 20th International Conference on Knowledge Discovery and Data Mining (KDD 2014), pp. 701–710. ACM (2014)
17. Prithviraj, S., Galileo, N., Mustafa, B., Lise, G., Brian, G., Tina, E.: Collective classification in network data. AI Mag. **29**(3), 93 (2008)
18. Kipf, T., Max, W.: Semi-supervised classification with graph convolutional networks. GCN open source code. https://github.com/tkipf/gcn
19. Van, L., Maaten, D., Hinton, G.: Visualizing data using t-SNE. J. Mach. Learn. Res. **9**(2579–2605), 85 (2008)

# KPML: A Novel Probabilistic Perspective Kernel Mahalanobis Distance Metric Learning Model for Semi-supervised Clustering

Chao Wang, Yongyi Hu, Xiaofeng Gao$^{(\boxtimes)}$, and Guihai Chen

Shanghai Key Laboratory of Data Science, Department of Computer Science
and Engineering, Shanghai Jiao Tong University, Shanghai, China
{wangchao.2014,huyongyi0903}@sjtu.edu.cn
{gao-xf,gchen}@cs.sjtu.edu.cn

**Abstract.** Metric learning aims to transform features of data into another based on some given distance relationships, which may improve the performances of distance-based machine learning models. Most existing methods use the difference between the distance of similar pairs and that of dissimilar pairs as loss functions for training. This kind of loss function may lack interpretability since people can only observe the distance or the difference of the distance, a number with no bounds, but have no idea about how large or small it is. To provide more explanation of these metric learning models, in this paper, we propose the probabilistic theoretical analysis of metric learning, design a special loss function, and propose the Kernelized Probabilistic Metric Learning (KPML) approach. With all the distance values transformed into probabilities, we can, therefore, compare and explain the results of the model. Besides, to effectively make use of both the labeled and unlabeled data to enhance the performance of semi-supervised clustering, we propose a KPML-based approach that leverages metric learning and semi-supervised learning effectively in a novel way. Finally, we use our model to do experiments about kNN-based semi-supervised clustering and the results show that our model significantly outperforms baselines across various datasets.

**Keywords:** Metric learning · Semi-supervised clustering · Kernel trick

## 1 Introduction

Recently, the question about how to evaluate the distance or similarity between two objects has become a hot topic since it is widely used in the machine learning

This work was supported by the National Key R&D Program of China [2018YFB1004700]; the National Natural Science Foundation of China [61872238, 61972254]; the Tencent Joint Research Program, and the Open Project Program of Shanghai Key Laboratory of Data Science (No. 2020090600001). The authors also would like to thank Mingding Liao for his contribution on the early version of this paper.

S. Hartmann et al. (Eds.): DEXA 2020, LNCS 12392, pp. 259–274, 2020.
https://doi.org/10.1007/978-3-030-59051-2_17

and data mining field, such as retrieval [33,40], classification [2,20] and clustering [3,12]. For each problem, we can manually select some features and their combinations to get a suitable distance function. But it costs lots of time and is very hard to generalize. So we pay attention to *metric learning* [13] which tries to find a general way to get suitable mapping functions of features for a specific problem.

An informal definition of metric learning is: given two objects $\mathbf{x}$ and $\mathbf{y}$ and some information about the distance between $\mathbf{x}$ and $\mathbf{y}$, we need to determine the best distance function $d(\mathbf{x}, \mathbf{y})$. In the absence of prior knowledge, most distance-based classifiers use simple Euclidean distances to measure the dissimilarities between examples represented as vector inputs. Euclidean distance metrics, however, do not capitalize on any statistical regularities that might be estimated from a large training set of labeled examples. Thus, one way of metric learning is to learn a mapping function $f$, then the new distance function $d(\mathbf{x}, \mathbf{y})$ equals to the Euclidean distance between $f(\mathbf{x})$ and $f(\mathbf{y})$. In the linear case, the mapping function $f(\mathbf{x})$ can be represented by $L\mathbf{x}$ and to achieve convexity, there are different constraints on $L$ in different models, such as positive semi-definiteness. Also, we have nonlinear methods, which means $f$ is non-linear and we can get $f$ in different ways such as the kernel trick or neural networks. In fact, as shown by many researchers [6,7,23], kNN classification can be significantly improved by learning a distance metric from labeled examples. Even a simple linear transformation of input features has been shown to yield much better kNN classifiers [7,23].

The largest margin nearest neighbor (LMNN) algorithm [31] is one of the most popular metric learning methods, which always yields competitive results for many classification tasks because it captures the local information of the data set. Similarly, many existing methods use the distance of similar points and the difference between the distance of similar pairs and distance of dissimilar pairs as loss functions. However, this kind of loss function may lack interpretability since all the terms are just numbers with no bounds. It is hard to understand, explain, or compare the value of each term considering that the dimensions and variances of the features are different among different datasets. For instance, if the distance between two users is 2 in social networks, it is hard to say intuitively whether the distance is large or small.

To give more interpretability for the metric learning model, in this paper, we propose the *Kernelized Probabilistic Metric Learning (KPML)* model, a probability-based model in which each term has its probabilistic meaning. First, we initialize every element of the transferring matrix $L$ with zero-mean Gaussian priors. It can be shown that this prior can be considered as an $L_2$ regularizer, which is widely known to be able to reduce the complexity of $L$ and prevent the model from overfitting. Then, for any two data points $\mathbf{x_i}$ and $\mathbf{x_j}$, we assume the distance between them also obeys a Gaussian distribution whose mean is $\|L\mathbf{x_i} - L\mathbf{x_j}\|^2$. The intuition behind this assumption is that the ideal distance between two data points is mainly determined by $L$, $x_i$, and $x_j$ but have some noise. This additional noise is similar to a soft margin which allows the model to ignore some outliers of training data and improves the generalization ability.

After that, we apply the idea of LMNN into our model through: for triples $(\mathbf{x_i},$ $\mathbf{x_j}, \mathbf{x_k})$, we can calculate the probability $p(d_{i,k} - d_{i,j} > k)$ which means the probability that the distance between dissimilar data points $(\mathbf{x_i}, \mathbf{x_k})$ is larger than the distance between similar data points $(\mathbf{x_i}, \mathbf{x_j})$ by $k$, where $k$ is a parameter. Finally, with all we have from above, we can get the objective function of $L$ and use the *Gradient Descent Optimization* to find the best $L$. To solve the gradient problem we met during training the model, we approximate the gradient function under some conditions with numerical analysis and obtain better results with the approximated gradient function. In addition, the kernel trick can also be applied to our model so that it can catch the non-linear relationship between data points. We use our model to do experiments about kNN classification prediction and the results show that our model significantly outperforms baselines across various data sets.

## 2  Related Work

### 2.1  Semi-supervised Clustering

Cluster analysis, as one of the most important technologies of data mining, has been developing various algorithms continuously [1,36], which is widely applied in a variety of application scenarios, such as social network analysis [18], community detection [30], and knowledge discovery [39]. Clustering and classification are two of the most important categories of machine learning, and their major difference is whether the dataset is labeled or not. The objective of clustering is to put similar data points (according to a specific metric in extracted feature space) into the same clusters and to separate those with highly distinct features.

Initially, clustering, as a branch of unsupervised learning, only processes the unlabeled data. The unsupervised clustering technique has drawn a tremendous amount of research attention, and many clustering methods have been proposed in the past. These clustering methods can be generally categorized into three types: (1) Feature learning based methods. This kind of method tries to find more discriminative features by using data dimension reduction techniques or subspace learning techniques. (2) Metric learning based methods [11]. These methods aim to learn an appropriate distance metric for the training data. Under the learned distance metric, similar samples can be grouped together, and meanwhile, dissimilar data points are separated apart. (3) Graph-based clustering. This kind of method partitions the data into different classes according to their pairwise similarities. Recently, deep learning technique has achieved great success in many fields due to its superiority of learning capacity, and some deep learning based methods [4] have also been used to solve clustering problems.

Generally speaking, how to extract useful features and learn an appropriate metric for high-dimensional data without any supervised information is a challenging task. Consequently, some supervised clustering algorithms [8] have been proposed to improve the clustering result. However, most of these methods have great limitations in real practical applications, because it is almost impossible for all data having labels. At the same time, tagging enough samples manually requires a large number of human resources, and it is also unrealistic. In fact, in most of the real-world applications, we can only obtain limited labeled data while most of the data are unlabeled. Based on the above problems, more and more semi-supervised based clustering methods have emerged recently. These methods adjust the learning framework through limited label data so that the clustering process can be executed in the supervised framework, which greatly improves the clustering performance and have a wide range of application scenarios.

## 2.2  Metric Learning

One of the earliest metric learning methods is MMC [34] whose main idea is minimizing the sum of similar objects' distances and maximizing the sum of dissimilar objects' distances. Schultz and Joachims present their method [22] which includes a squared Frobenius norm regularizer and uses relative distance constraints with the assumption that the target matrix is diagonal. Kwok and Tsang propose their method that utilizes Frobenius regularization and distance constraints as well without the diagonal target matrix restriction[14]. Shalev-Shwartz et al. show how metric learning applied to the online setting with their Frobenius regularizer method [24].

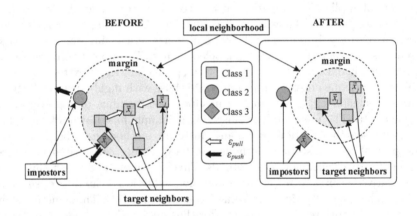

**Fig. 1.** Illustration of an input's ($x_i$) neighborhood before (left) versus after (right). Arrows indicate the gradients on distances arising from the optimization of the cost function [31]. After training, three neighbors which have the same label with $x_i$ (target neighbors) lie within a smaller radius and two data points with different labels from $x_i$ (imposters) are outside this smaller radius.

Recently, in the metric learning field, there have been several directions. Weinberger et al. propose Large-Margin Nearest Neighbors (LMNN) model [31, 32] which is a combination of the trace regularizer [34] and relative distance constraints. And it has become the origin of many other models. There are models about feature transformation: Zhai et al. deal with Hamming distance with a hash function [37]. Bohne et al. use GMM method [5]. There are models about new loss function: Nguyen et al. apply Jeffrey divergence in their model [5]. Ye et al. utilize KL divergence [35]. There are models about optimization algorithms: Li et al. accomplish the optimization via eigenvector [15]. Zhang et al. propose an efficient stochastic optimization algorithm [38].

## 2.3   LMNN Review

In this section we will give a review about LMNN since we have applied its idea into our model. Weinberger et al. propose LMNN model which has become one of the most popular methods in the metric learning field. The intuition behind it is: a data point should have the same label as its neighbors and those points which have different labels should be far from the given point. Figure 1 shows the main idea of LMNN; The term *target neighbor* means: for the input $x_i$, we have $k$ target neighbors-that is, k other data points with the same label $y_i$ which we hope they have minimal distance to $x_i$. The term *imposters* refers to the nearest data points which have different labels from $x_i$.

LMNN's cost function has two terms. The first one penalizes large distances between each data point and its target neighbors. The second one penalizes small distance between each data point and its imposters. The goal of LMNN is as (1).

$$\min_{L \geq 0} \quad \sum_{i,j} \|Lx_i - Lx_j\|^2$$
$$+ \lambda \sum_{i,j,k} [1 + \|Lx_i - Lx_j\|^2 - \|Lx_i - Lx_k\|^2]_+ \tag{1}$$

where $\lambda$ is a positive constant to weigh two cost terms, $x_j$ is *target neighbors* of $x_i$ and $x_k$ is *imposters* of $x_i$. $[ \cdot ]_+ = \max(\cdot, 0)$ denotes the standard hinge loss. The advantages of LMNN include:

1. The target matrix $L$ of LMNN contains a lot of statistical information of data which be widely applied into many other machine learning areas, sucha as classification, feature extraction and dimension reduction [17,29].
2. The optimization problem of LMNN is very similar to the classic soft-margin Support Vector Machine (SVM) [21] problem. So it is also relatively easy to utilize kernel technique to get the non-linear version of LMNN. As a result, LMNN can be applied to more generalized problems [9,28]
3. With LMNN, we can get very competitive results for many other problems like clustering since it utilizes the local information of the data [10,41]

However, there are also some drawbacks to the LMNN model. LMNN tries to pull *target neighbors* closer to $x_i$ and to push *imposters* far from $x_i$ at the same time. And it implements these two competitive actions by using a hyper-parameter to weigh the cost terms of pulling and pushing. So people need to finely tune the hyper-parameter to achieve high performance on different tasks which is time-consuming. We can see that targets of pulling and pushing in LMNN are both to make the neighborhood of the given point $\mathbf{x_i}$ more clear and pure. These two similar terms would increase the complexity and make it difficult to find the best solution. As a result, we think one term is enough to accomplish the goal since their effects are the same and for this reason, we only apply the second term of LMNN's cost function into our model.

## 3    Kernelized Probabilistic Metric Learning

### 3.1    Problem Definition and Notation

Let $\{D = (\mathbf{x_i}, y_i)\}^n$ denote a set of labeled data inputs $\mathbf{x_i} \in \mathbb{R}^n$ with discrete class labels $y_i$. Our goal is to find the best transferring matrix $L \in \mathbb{R}^{m \times n}$ which can be used to get the distance between two inputs as $d_{i,j} = ||L\mathbf{x_i} - L\mathbf{x_j}||^2$.

### 3.2    Initialization

We place zero-mean spherical Gaussian priors with variance $\sigma_1^2$ on transferring matrix $L$, in other words $p(L \mid \sigma_1^2) = \prod_{i=1}^{m} \prod_{j=1}^{n} \mathcal{N}(L_{i,j}|0, \sigma_1^2)$. Now we can get the priori probability of the transferring matrix $L$:

$$P(L)_{\text{prior}} = p(L \mid \sigma_1^2) = \prod_{i=1}^{m} \prod_{j=1}^{n} \frac{1}{\sqrt{2\pi}\sigma_1} e^{-\frac{L_{i,j}^2}{2\sigma_1^2}} \tag{2}$$

### 3.3    Probabilistic Theoretical Analysis

Since we have the transferring matrix $L$ to map data inputs from an $n$ dimension into an $m$ dimension space, the expected distance between two data inputs $\mathbf{x_i}, \mathbf{x_j}$ is $||L\mathbf{x_i} - L\mathbf{x_j}||^2$.

Then we define the conditional distribution over the observed distance $d_{i,j}$:

$$P(d_{i,j} \mid L, D) = \mathcal{N}(d_{i,j} \mid ||Lx_i - Lx_j||^2, \sigma_2^2). \tag{3}$$

The intuition behind this assumption is that the ideal distance between two data points are mainly determined by $L$, $x_i$, and $x_j$ but have some noise. This additional noise is similar to a soft margin which allows model to ignore some outliers of training data and improves the generalization ability. The higher the hyper-parameter $\sigma_2$, the more robustness and the less learning ability the model has. It is a classical trade-off between variance and accuracy.

After that, we can get the conditional distribution over the difference between $d_{i,j}$ and $d_{i,k}$:

$$P(d_{i,k} - d_{i,j} \mid L, D)$$
$$= \mathcal{N}(d_{i,k} - d_{i,j} \mid ||Lx_i - Lx_k||^2 - ||Lx_i - Lx_j||^2, 2\sigma_2^2) \tag{4}$$

Equation (4) can be obtained directly from Eq. (3) with the assumption that $d_{i,k}$ and $d_{i,j}$ are independent. Because the distribution of the difference between two independent normal distribution variables is still a normal distribution with the mean as the difference of their means and the variance as the sum of their variances.

Now let the input data be triples of $(x_i, x_j, x_k)$, where the label $y_i$ is same as the label $y_j$ and the label $y_i$ is different from the label $y_k$. Inspired by the constraint terms of LMNN, we define that $d_{i,k} - d_{i,j}$ should be larger than 1. According to the distribution in Eq. (4), for a data input triple $(x_i, x_j, x_k)$, we can get the likelihood probability that $d_{i,k} - d_{i,j}$ is larger than 1:

$$P(d_{i,k} - d_{i,j} > 1|L) = \int_1^\infty \left( \frac{1}{2\sqrt{\pi}\sigma_2} e^{\frac{(t-(d_1-d_2))^2}{-4\sigma_2^2}} \right) dt \tag{5}$$

where $d_1 = ||Lx_i - Lx_k||^2$ and $d_2 = ||Lx_i - Lx_j||^2$.

Based on Eq. (5), we can calculate the likelihood probability of dataset D including all the triples $(x_i, x_j, x_k)$:

$$P_{\text{likelihood}}(D|L) = \prod_{i,j,k} \left( \int_1^\infty \left( \frac{1}{2\sqrt{\pi}\sigma_2} e^{\frac{(t-(d_1-d_2))^2}{-4\sigma_2^2}} \right) dt \right) \tag{6}$$

According to the prior Eq. (2) and likelihood Eq. (6), we can get the posterior probability of $L$ from Bayes formula:

$$P(L|D)_{\text{posterior}} = P(L)_{\text{prior}} \times P(D|L)_{\text{likelihood}} =$$
$$\prod_{i=1}^m \prod_{j=1}^n \frac{1}{\sqrt{2\pi}\sigma_1} e^{-\frac{L_{i,j}^2}{2\sigma_1^2}} \times \prod_{i,j,k} \left( \int_1^\infty \left( \frac{1}{2\sqrt{\pi}\sigma_2} e^{\frac{(t-(d_1-d_2))^2}{-4\sigma_2^2}} \right) dt \right) \tag{7}$$

Since the evidence term in Bayes formula is only a constant and will not influence the results, we ignore it and only let prior time likelihood to get the posterior probability of L. For simplicity, we let $f(L) = P(L|D)_{\text{posterior}}$.

### 3.4   Cost Function

At that time, considering that we should maximize the posterior probability and the terms in it are multiplied, we minimize $-\ln f(L)$:

$$-\ln f(L) = -\sum_{i,j,k} \left( \ln \int_1^\infty \frac{1}{2\sqrt{\pi}\sigma_2} e^{\frac{(t-(d_1-d_2))^2}{-4\sigma_2^2}} dt \right) + C \tag{8}$$

This equation is a probabilistic explanation of the intuition that the prior of $L$ should be zero-mean spherical Gaussian distribution.

Now, we choose $-\ln f(L)$ as the cost function. And we can get the following optimization problem:

$$\min_{L} -\sum_{i,j,k}(\ln\int_{1}^{\infty}\frac{1}{2\sqrt{\pi}\sigma_2}e^{-\frac{(t-(d_1-d_2))^2}{4\sigma_2^2}}\,dt)$$

$$s.t.\quad i\neq j\neq k$$
$$y_i = y_j$$
$$y_i \neq y_k$$

(9)

where $d_1 = ||Lx_i - Lx_k||^2$ and $d_2 = ||Lx_i - Lx_j||^2$.

The constraints indicate that all the triples $(\mathbf{x_i}, \mathbf{x_j}, \mathbf{x_k})$ where the label $y_i$ is same with the label $y_j$ and the label $y_i$ is different from the label $y_k$ are considered.

## 3.5  Optimization

In the last section, we obtain the cost function and the optimization problem. Now, we will use the *Gradient Descent Optimization* method to get the best transferring matrix $L$. The following is the calculation procedure:

For simplicity, let $A = \int_{1}^{\infty}\frac{1}{2\sqrt{\pi}\sigma}e^{-\frac{(t-(d_1-d_2))^2}{4\sigma_2^2}}\,dt$ and $p = \frac{t-d_1+d_2}{2\sigma_2}$.

$$\frac{\partial -\ln f(L)}{\partial L}$$

$$= -\sum_{i,j,k}\frac{1}{A}\int_{1}^{\infty}e^{-\frac{(t-(d_1-d_2))^2}{4\sigma_2^2}}(-\frac{1}{2\sigma_2^2})(t-d_1+d_2)dt \times \frac{\partial(d_2-d_1)}{\partial L}$$

$$= \sum_{i,j,k}\frac{1}{A}\int_{1}^{\infty}e^{-p^2}\frac{p}{\sigma_2}dt \times \frac{\partial(d_2-d_1)}{\partial L}$$

$$= \sum_{i,j,k}\frac{1}{A}\int_{\frac{1-d_1+d_2}{2\sigma_2}}^{\infty}e^{-p^2}2pdp \times \frac{\partial(d_2-d_1)}{\partial L}$$

$$= \sum_{i,j,k}\frac{1}{A}\int_{\frac{1-d_1+d_2}{2\sigma_2}}^{\infty}e^{-p^2}d(p^2) \times \frac{\partial(d_2-d_1)}{\partial L}$$

$$= \sum_{i,j,k}\frac{1}{A}e^{-(\frac{1-d_1+d_2}{2\sigma_2})^2} \times \frac{\partial(d_2-d_1)}{\partial L}$$

$$= \sum_{i,j,k}\frac{1}{A}e^{-(\frac{1-d_1+d_2}{2\sigma_2})^2} \times 2L((x_j-x_i)(x_j-x_i)^T - (x_k-x_i)(x_k-x_i)^T)$$

Though we get the gradient, there is no analytical solution for integration of normal distribution. We have to calculate the value of $A$ by numerical integration. It causes another numerical problem. When $d_2 - d_1$ is too large, the problem happened on the term $\frac{1}{A}e^{-(\frac{1-d_1+d_2}{2\sigma_2})^2}$, where we can see the mean of the distribution over $t$, i.e. the distribution $e^{-(\frac{t-(d_1-d_2)}{2\sigma_2})^2}$, would be far from 1 since $d_1 - d_2$ would be a large negative number. Then, the fraction term would be very large which means the gradient is large and the training is unstable.

To solve this numerical problem, when $d_2 - d_1 > \rho$, we modify the gradient as following:

$$\frac{\partial - \ln f(L)}{\partial L} = \sum_{i,j,k} 2L((x_j - x_i)(x_j - x_i)^T - (x_k - x_i)(x_k - x_i)^T) \qquad (10)$$

The modification of the gradient function is that when $d_2 - d_1 > \rho$, we approximately make $A \approx e^{-(\frac{1-d_1+d_2}{2\sigma_2})^2}$. It can be seen as a kind of gradient clip: we just approximately let the fraction be 1 and focus on the effects of other terms. Empirically, we set $\rho$ as $4\sigma_2 - 1$, i.e. when $e^{-(\frac{1-d_1+d_2}{2\sigma_2})^2} < e^{-2}$. With this modification, the model can be finally stably trained and we obtain desired results.

## 3.6 Kernel Trick

In the previous section, we propose a general model for learning global linear metrics. According to the terms in the gradient function, we find that it is convenient to use the *Kernel Trick* [25,27] to obtain the non-linear method in order to generalize our model.

We denote $\phi(\mathbf{x_i})$ as a non-linear mapping function and denote kernel function as $\mathcal{K}(\mathbf{a}, \mathbf{b}) = <\phi(\mathbf{a}), \phi(\mathbf{b})> = \phi(\mathbf{a})\phi(\mathbf{b})^T$, where $\phi(\cdot) : \mathbb{R}^l \to \mathbb{R}^{l'}$ transforms the input features into higher-dimension features. Then, we can get $d_1 = ||L\phi(x_i) - L\phi(x_k)||^2$ and $d_2 = ||L\phi(x_i) - L\phi(x_j)||^2$. The gradient function is as following:

$$\frac{\partial - \ln f(L)}{\partial L} = \begin{cases} \sum_{i,j,k} \frac{1}{A} e^{-(\frac{1-d_1+d_2}{2\sigma_2})^2} 2LB & \text{, when } d_2 - d_1 \leq 4\sigma_2 - 1 \\ \\ \sum_{i,j,k} 2LB & \text{, when } d_2 - d_1 \leq 4\sigma_2 - 1 \end{cases} \qquad (11)$$

where $B = (\mathcal{K}(x_j, x_j) + \mathcal{K}(x_k, x_i) + \mathcal{K}(x_i, x_k) - \mathcal{K}(x_j, x_i) - \mathcal{K}(x_i, x_j) - \mathcal{K}(x_k, x_k))$. With the *Kernel Trick*, kernelization of linear metric learning can make it possible to apply the metric learning to fields where feature spaces are high but where some efficient kernels can be developed. Our experiments results show the power of the *Kernel Trick*.

## 3.7 Semi-supervised Clustering Based on KPML

In this subsection, we introduce a novel labeling strategy based on kNN updating strategy [16] to transform the unlabeled data into labeled data to solve the semi-supervised clustering problem.

As shown in Fig. 2, all the labeled data are classified into $C$ clusters with the KPML method. To make full use of the features of unlabeled data, we try to add

$k * C$ new unlabeled data to the labeled dataset each time. The main process of our label updating strategy is as follows.

**Step 1:** Compute the center of each cluster according to the labeled data, like the k-means algorithm.

**Step 2:** Search the $k$ nearest unlabeled data from the center of labeled data in each cluster, and then update their attributes from unlabeled data to **weak** labeled data.

**Step 3:** For every weak labeled data point, if more than $\alpha\%$ of its nearest k-neighbors labels are the same as its, we can convert it to labeled data, otherwise, we convert it back to unlabeled data.

**Step 4:** If the new labeled data reaches the threshold number $\beta$, retrain the KPML model, otherwise, repeat the above process. If the maximum number of iterations $R$ is reached, all weakly labeled data will be labeled directly in the next cycle.

In our model, $\alpha, \beta, R$ are hyper-parameters that we set it in advance according to different tasks. In order to simplify the model, we can directly use the kNN algorithm instead of parameter $\alpha$ in step 3.

## 4    Experiments

### 4.1    Dataset Description

There are benchmark datasets in our experiments. Two of them are the faces image datasets: ORL[1] and YALE[2]. One of them are the handwriting dataset called USPS[3]. Three of them is the datasets from UCI Machine learning Repository[4]: Wine, Iris and Isolet are taken from the UCI Machine learning Repository. All the six data sets are often adopted as benchmark data sets for distance metric learning in recent works. In these experiments, the principal component analysis is employed to reduce the feature dimension with the reservation of 90% of singular value. Each experiments are repeated by 100 times.

### 4.2    Experiment Setup

In the experiments, four methods are employed to be compared with KPML, which are margin nearest neighbor algorithm (LMNN), Information Theory Metric Learning (ITML), Distance metRIc learning Facilitated by disTurbances (DRIFT) [35], supervised Distance Metric Learning with Jeffrey divergence [19] (DMLMJ). The program codes of DRIFT and DMLMJ are provided by corresponding authors. The prior parameters $\theta_1$ and $\theta_2$ are set as 0.01 to reserve the generality. The kernel is set as classical RBF kernel: $\mathcal{K}(u,v) = e^{\frac{-\|u-v\|^2}{\theta_3}}$ where $\theta_3$

---

[1] http://www.cl.cam.ac.uk/research/dtg/attarchive/facedatabase.html.
[2] http://vision.ucsd.edu/content/yale-face-database.
[3] https://www.csie.ntu.edu.tw/cjlin/libsvmtools/datasets/multiclass.html.
[4] http://www.ics.ucl.edu/mlean/MLRepository.html.

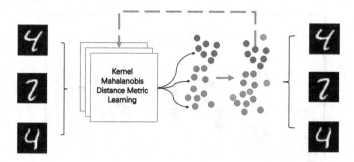

**Fig. 2.** The framework of the semi-supervised clustering with kernel Mahalanobis distance metric learning. The dataset on the left (handwritten digits) is labeled data and the dataset on the right is unlabeled data. We train the KPML model with the labeled data at first. Then we use our label updating strategy to add unlabeled data into labeled data. We alternately train KPML model and label data, and finally, we can learn labels of all data.

is set based on the variance of each dataset. The triple used in the metric learning model is randomly selected from the datasets. For semi-supervised learning, 30% of data in each datasets are random selected as the labeled data and others are treated as unlabeled data.

### 4.3   Experiment Results

In the experiments, the classic model LMNN and ITML, and the state-of-the-art model DRIFT and DMLMJ are compared with our model KPML. The result is showed in Table 1. In Table 1, PML is the simplified version of KPML, which do not use the kernel trick. The results in Table 1 show that our model outperforms the baselines in most of cases, which is highlighted in blod font. Moreover, PML performs worse than the start-of-the-art baselines, which indicates the importance of the kernel method in KPML.

**Table 1.** Accuracy comparison of KPML and the several baselines

| Datasets | LMNN | ITML | DRIFT | DMLMJ | PML | KPML |
|----------|------|------|-------|-------|-----|------|
| ORL | 75.1% | 75.0% | 80.1% | 83.2% | 76.1% | **83.4%** |
| YALE | 78.9% | 76.5% | **82.1%** | 81.0% | 72.4% | 81.8% |
| USPS | 80.5% | 81.5% | 84.9% | 86.1% | 84.4% | **87.9%** |
| WINE | 93.1% | 92.2% | 92.9% | 94.2% | 93.7% | **95.6%** |
| IRIS | 92.3% | 95.1% | 93.5% | 95.3% | 95.1% | **95.9%** |
| ISOLET | 79.5% | 81.9% | 91.7% | **92.3%** | 83.1% | 92.0% |

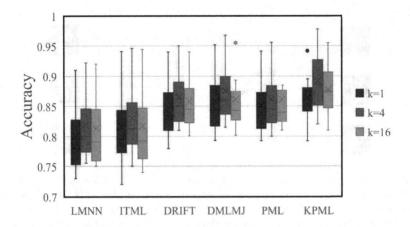

**Fig. 3.** Accuracy comparison of KPML and the several baselines with different $k$. Gray and blue dots are outliers (Color figure online)

Figure 3 shows the results of KPML and several baselines with different $k$ in kNN. The results indicate that it has better performance when $k = 4$.

To test the performance of our method when dealing with noisy information, we compare our KPML method with LMNN, ITML, DMLMJ, and DRIFT. As shown in Fig. 4, the results on noise-free datasets are filled with dark color, and the increases in error rates when training with noisy counterparts are denoted using a light color. LMNN and ITML show conspicuous poorer performance than the last four models in the performance. Both DRIFT and DMLMJ get worse results than the KPML one. In summary, the performance comparison validates the robustness of KPML, which strengthens the advantage of KPML in an unknown scenario.

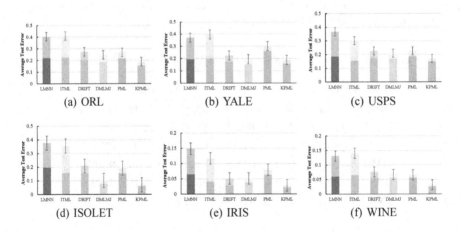

**Fig. 4.** The results with noise datasets

### 4.4  Experiments on Different Applications

In this subsection, we introduce the experiments on different datasets in detail. It is proved that our model has a wide range of applications.

**Small Data Sets with Few Classes:** The wine and iris datasets are small data sets, with less than 500 training examples and just three classes, taken from the UCI Machine Learning Repository. On data sets of this size, a distance metric can be learned in a matter of seconds. The results were averaged over 100 experiments with different random 50/50 splits of each data set.

**Face Recognition:** The ORL face recognition data set contains 400 grayscale images of 40 individuals in 10 different poses. We downsampled the images from to 38 × 31 pixels and used PCA to obtain 30-dimensional eigenfaces [26]. Training and test sets were created by randomly sampling 7 images of each person for training and 3 images for testing. The task involved 40-way classification essentially, recognizing a face from an unseen pose.

**Spoken Letter Recognition:** The Isolet data set from UCI Machine Learning Repository has 6238 examples and 26 classes corresponding to letters of the alphabet. We reduced the input dimensionality by projecting the data onto its leading 172 principal components-enough to account for 95% of its total variance.

**Handwritten Digit Recognition:** The USPS dataset of handwritten digits contains 20000 images with the size of 16 × 16. We used PCA to reduce their dimensionality by retaining only the first 96 principal components.

## 5  Conclusion

In this paper, we propose KPML, a novel kernel probabilistic perspective large margin metric learning model. Different from LMNN, we present the probabilistic analysis to the metric learning and produce the new cost function from the Probabilistic Kernel Perspective, which provides a new direction for metric learning. We set the zero-mean Gaussian priors to transfer matrix in metric learning and formulate the process of metric learning from the viewpoint of probability. Experiments on different kinds and sizes of datasets prove the effectiveness of KPML.

## References

1. Afzalan, M., Jazizadeh, F.: An automated spectral clustering for multi-scale data. Neurocomputing **347**, 94–108 (2019). https://doi.org/10.1016/j.neucom.2019.03.008
2. Baccour. L., Alimi, A.M, John, R.I: Intuitionistic fuzzy similarity measures and their role in classification. JIIS, **25**(2), 221–237 (2016). http://www.degruyter.com/view/j/jisys.2016.25.issue-2/jisys-2015-0086/jisys-2015-0086.xml

3. Belesiotis, A., Skoutas, D., Efstathiades, C., Kaffes, V., Pfoser, D.: Spatio-textual user matching and clustering based on set similarity joins. Very Large Data Bases J. (VLDBJ) **27**(3), 297–320 (2018). https://doi.org/10.1007/s00778-018-0498-5

4. Bhatnagar, B.L., Singh, S., Arora, C., Jawahar, C.V.: Unsupervised learning of deep feature representation for clustering egocentric actions. In: International Joint Conference on Artificial Intelligence (IJCAI), pp. 1447–1453 (2017). https://doi.org/10.24963/ijcai.2017/200

5. Bohne, J., Ying, Y., Gentric, S., Pontil, M.: Learning local metrics from pairwise similarity data. Pattern Recognit. **75**, 315–326 (2018). https://doi.org/10.1016/j.patcog.2017.04.002

6. Domeniconi, C., Gunopulos, D., Peng, J.: Large margin nearest neighbor classifiers. IEEE Trans. Neural Netw. Learn. Syst. **16**(4), 899–909 (2005). https://doi.org/10.1109/TNN.2005.849821

7. Goldberger, J., Roweis, S.T., Hinton, G.E., Salakhutdinov, R.: Neighbourhood components analysis. In: Annual Conference on Neural Information Processing Systems (NeurIPS), pp. 513–520 (2004). http://papers.nips.cc/paper/2566-neighbourhood-components-analysis

8. Haponchyk, I., Uva, A., Yu, S., Uryupina, O., Moschitti, A.: Supervised clustering of questions into intents for dialog system applications. In: Conference on Empirical Methods in Natural Language Processing (EMNLP), pp. 2310–2321 (2018). https://doi.org/10.18653/v1/d18-1254

9. He, Y., Chen, W., Chen, Y., Mao, Y.: Kernel density metric learning. In: IEEE International Conference on Data Mining (ICDM), pp. 271–280 (2013). https://doi.org/10.1109/ICDM.2013.153

10. Hsieh, C.K., Yang, L., Cui, Y., Lin, T.Y., Belongie, S., Estrin, D.: Collaborative metric learning. In: International Conference on World Wide Web (WWW), pp. 193–201 (2017). https://doi.org/10.1145/3038912.3052639

11. Kalintha, W., Ono, S., Numao, M., Fukui, K.: Kernelized evolutionary distance metric learning for semi-supervised clustering. In: AAAI Conference on Artificial Intelligence (AAAI), pp. 4945–4946 (2017). http://aaai.org/ocs/index.php/AAAI/AAAI17/paper/view/14714

12. Kang, Z., Peng, C., Cheng, Q.: Twin learning for similarity and clustering: a unified kernel approach. In: AAAI Conference on Artificial Intelligence (AAAI), pp. 2080–2086 (2017). http://aaai.org/ocs/index.php/AAAI/AAAI17/paper/view/14569

13. Kulis, B., et al.: Metric learning: A survey. Found. Trends® Mach. Learn. **5**(4), 287–364 (2013). https://doi.org/10.1561/2200000019

14. Kwok, J.T., Tsang, I.W.: Learning with idealized kernels. In: ACM International Conference on Machine Learning (ICML), pp. 400–407 (2003). http://www.aaai.org/Library/ICML/2003/icml03-054.php

15. Li, D., Tian, Y.: Global and local metric learning via eigenvectors. Knowl. Based Syst. **116**, 152–162 (2017). https://doi.org/10.1016/j.knosys.2016.11.004

16. Li, X., Yin, H., Zhou, K., Zhou, X.: Semi-supervised clustering with deep metric learning and graph embedding. World Wide Web (WWW) **23**(2), 781–798 (2020). https://doi.org/10.1007/s11280-019-00723-8

17. Li, Y., Tian, X., Tao, D.: Regularized large margin distance metric learning. In: IEEE International Conference on Data Mining (ICDM), pp. 1015–1022 (2016). https://doi.org/10.1109/ICDM.2016.0129

18. Liu, G., Zheng, K., Wang, Y., Orgun, M.A., Liu, A., Zhao, L., Zhou, X.: Multi-constrained graph pattern matching in large-scale contextual social graphs. In: IEEE International Conference on Data Engineering (ICDE), pp. 351–362 (2015). https://doi.org/10.1109/ICDE.2015.7113297

19. Nguyen, B., Morell, C., Baets, B.D.: Supervised distance metric learning through maximization of the Jeffrey divergence. Pattern Recognit. **64**, 215–225 (2017). https://doi.org/10.1016/j.patcog.2016.11.010

20. Polat, K.: Similarity-based attribute weighting methods via clustering algorithms in the classification of imbalanced medical datasets. Neural Comput. Appl. **30**(3), 987–1013 (2018). https://doi.org/10.1007/s00521-018-3471-8

21. Schölkopf, B., Smola, A.J.: Learning with Kernels: support vector machines, regularization, optimization, and beyond. In: Adaptive Computation and Machine Learning Series. MIT Press (2002). http://www.worldcat.org/oclc/48970254

22. Schultz, M., Joachims, T.: Learning a distance metric from relative comparisons. In: Advances in Neural Information Processing Systems (NeurIPS), pp. 41–48 (2003). http://papers.nips.cc/paper/2366-learning-a-distance-metric-from-relative-comparisons

23. Shalev-Shwartz, S., Singer, Y., Ng, A.Y.: Online and batch learning of pseudometrics. In: International Conference on Machine Learning (ICML) (2004). https://doi.org/10.1145/1015330.1015376

24. Shalev-Shwartz, S., Singer, Y., Ng, A.Y.: Online and batch learning of pseudometrics. In: ACM International Conference on Machine Learning (ICML), p. 94 (2004). https://doi.org/10.1145/1015330.1015376

25. Shawe-Taylor, J., Cristianini, N.: Kernel Methods for Pattern Analysis. Cambridge University Press (2004). https://doi.org/10.1017/CBO9780511809682. https://kernelmethods.blogs.bristol.ac.uk/

26. Simard, P.Y., LeCun, Y., Denker, J.S.: Efficient pattern recognition using a new transformation distance. In: Annual Conference on Neural Information Processing Systems (NeurIPS), pp. 50–58 (1992)

27. Smola, A.J.: Learning with Kernels. Citeseer (1998). http://d-nb.info/955631580

28. Song, K., Nie, F., Han, J., Li, X.: Parameter free large margin nearest neighbor for distance metric learning. In: AAAI Conference on Artificial Intelligence (AAAI), pp. 2555–2561 (2017). http://aaai.org/ocs/index.php/AAAI/AAAI17/paper/view/14616

29. St Amand, J., Huan, J.: Sparse compositional local metric learning. In: ACM SIGKDD International Conference on Knowledge Discovery and Data Mining (KDD). pp. 1097–1104 (2017). https://doi.org/10.1145/3097983.3098153

30. Wang, Q., Yin, H., Hu, Z., Lian, D., Wang, H., Huang, Z.: Neural memory streaming recommender networks with adversarial training. In: ACM SIGKDD International Conference on Knowledge Discovery & Data Mining(KDD), pp. 2467–2475 (2018). https://doi.org/10.1145/3219819.3220004

31. Weinberger, K.Q., Blitzer, J., Saul, L.K.: Distance metric learning for large margin nearest neighbor classification. In: Advances in neural information processing systems (NeurIPS), pp. 1473–1480 (2005). http://papers.nips.cc/paper/2795-distance-metric-learning-for-large-margin-nearest

32. Weinberger, K.Q., Saul, L.K.: Distance metric learning for large margin nearest neighbor classification. J. Mach. Learn. Res. **10**, 207–244 (2009). https://dl.acm.org/citation.cfm?id=1577078

33. Wu, Y., Wang, S., Huang, Q.: Online asymmetric similarity learning for cross-modal retrieval. In: IEEE Conference on Computer Vision and Pattern Recognition, CVPR, pp. 3984–3993 (2017). https://doi.org/10.1109/CVPR.2017.424

34. Xing, E.P., Jordan, M.I., Russell, S.J., Ng, A.Y.: Distance metric learning with application to clustering with side-information. In: Advances in Neural Information Processing Systems (NeurIPS), pp. 505–512 (2002). http://papers.nips.cc/paper/2164-distance-metric-learning-with-application-to-clustering

35. Ye, H.J., Zhan, D.C., Si, X.M., Jiang, Y.: Learning mahalanobis distance metric: Considering instance disturbance helps. In: International Joint Conference on Artificial Intelligence (IJCAI), pp. 3315–3321 (2017). https://doi.org/10.24963/ijcai.2017/463

36. Yu, S., Chu, S.W., Wang, C., Chan, Y., Chang, T.: Two improved k-means algorithms. Appl. Soft Comput. **68**, 747–755 (2018). https://doi.org/10.1016/j.asoc.2017.08.032

37. Zhai, D., Liu, X., Chang, H., Zhen, Y., Chen, X., Guo, M., Gao, W.: Parametric local multiview hamming distance metric learning. Pattern Recognit. **75**, 250–262 (2018). https://doi.org/10.1016/j.patcog.2017.06.018

38. Zhang, J., Zhang, L.: Efficient stochastic optimization for low-rank distance metric learning. In: AAAI Conference on Artificial Intelligence (AAAI), pp. 933–940 (2017). http://aaai.org/ocs/index.php/AAAI/AAAI17/paper/view/14373

39. Zheng, K., Zheng, Y., Yuan, N.J., Shang, S., Zhou, X.: Online discovery of gathering patterns over trajectories. IEEE Trans. Knowl. Data Eng. **26**(8), 1974–1988 (2014). https://doi.org/10.1109/TKDE.2013.160

40. Zhu, H., Long, M., Wang, J., Cao, Y.: Deep hashing network for efficient similarity retrieval. In: AAAI Conference on Artificial Intelligence (AAAI), pp. 2415–2421 (2016). http://www.aaai.org/ocs/index.php/AAAI/AAAI16/paper/view/12039

41. Zuo, W., Wang, F., Zhang, D., Lin, L., Huang, Y., Meng, D., Zhang, L.: Distance metric learning via iterated support vector machines. IEEE Trans. Image Process. **26**(10), 4937–4950 (2017). https://doi.org/10.1109/TIP.2017.2725578

# Semantic Web and Ontologies

# Indexing Data on the Web: A Comparison of Schema-Level Indices for Data Search

Till Blume[1](✉)(iD) and Ansgar Scherp[2](iD)

[1] Kiel University, Kiel, Germany
tbl@informatik.uni-kiel.de
[2] Ulm University, Ulm, Germany
ansgar.scherp@uni-ulm.de

**Abstract.** Indexing the Web of Data offers many opportunities, in particular, to find and explore data sources. One major design decision when indexing the Web of Data is to find a suitable index model, i.e., how to index and summarize data. Various efforts have been conducted to develop specific index models for a given task. With each index model designed, implemented, and evaluated independently, it remains difficult to judge whether an approach generalizes well to another task, set of queries, or dataset. In this work, we empirically evaluate six representative index models with unique feature combinations. Among them is a new index model incorporating inferencing over RDFS and owl:sameAs. We implement all index models for the first time into a single, stream-based framework. We evaluate variations of the index models considering sub-graphs of size 0, 1, and 2 hops on two large, real-world datasets. We evaluate the quality of the indices regarding the compression ratio, summarization ratio, and F1-score denoting the approximation quality of the stream-based index computation. The experiments reveal huge variations in compression ratio, summarization ratio, and approximation quality for different index models, queries, and datasets. However, we observe meaningful correlations in the results that help to determine the right index model for a given task, type of query, and dataset.

## 1   Introduction

Graph indices are well-established to efficiently manage large heterogeneous graphs like the Web of Data. In general, one can distinguish instance-level indices and schema-level indices for the Web of Data. Instance-level indices focus on finding specific data instances [8,11,17], e.g., searching for a specific book by its title such as "Towards a clean air policy". In contrast, schema-level indices (short: *SLI*) support structural queries, e.g., searching for data instances with the property *dct:creator* and RDF type *bibo:book* [7]. An *SLI model* defines how and which combinations of types and properties are indexed, i.e., how data instances are summarized and which queries are supported by the index. In the past, various SLI models have been developed for different tasks such as data exploration [1,12,14,16], query size estimation [13], vocabulary terms recommendation [15], related entity retrieval [5], data search [7], and others. The task

© Springer Nature Switzerland AG 2020
S. Hartmann et al. (Eds.): DEXA 2020, LNCS 12392, pp. 277–286, 2020.
https://doi.org/10.1007/978-3-030-59051-2_18

of data search is to find (sub-)graphs on the Web that match a given schema structure. Search systems like LODatio [7], LODeX [1], Loupe [12], and LODatlas [14] rely on SLI to offer a search for relevant data sources or exploration of data sources. The problem is that all SLI models were designed, implemented, and evaluated for their individual task only, using different queries, datasets, and metrics. Our hypothesis is that there is no SLI model that fits all tasks and that the performance of the specific SLI depends on the specific types of queries and characteristics of the datasets. However, so far only very limited work has been done on understanding the behavior of SLI models in different contexts. With each SLI model evaluated independently in a specific context, it remains difficult to judge whether an approach generalizes well to another task or not. In other words, it is not known which SLI model can be used for which contexts, tasks, and datasets.

To fill this gap, we conduct an extensive empirical evaluation of representative SLI models. To this end, we have for the first time defined and implemented the features of existing SLI models in a common framework available on GitHub. Based on the discussion of related works, we chose six SLI models with unique feature combinations, which were developed for different tasks, to understand and compare their behavior for the data search task. We empirically investigate the behavior of each selected SLI model in three variants, where we index subgraphs of 0, 1, and 2 hop lengths.

The empirical evaluation consists of two sets of experiments. In a first set of experiments, we analyze the relative size of the computed SLI compared to the original dataset (compression ratio) and the number of schema elements in the SLI compared to the number of data instances in the dataset (summarization ratio). The second set of experiments quantifies the quality of a stream-based computation of the SLI for large datasets obtained from the Web of Data. The stream-based approach is designed to scale to graphs of arbitrary sizes by observing the graph over a stream of edges with fixed window size. Inherently, this approach introduces inaccuracies in the SLI computation by potentially extracting incomplete schema structures due to limited window size [10]. Our experiments show huge variations in compression ratio, summarization ratio, and approximation quality for the SLI models. However, we also observe strong positive and negative correlations between the three metrics. These insights shed light on the behaviors of SLI models for different datasets and queries that help to determine the right SLI model for a given task and dataset.

## 2    Discussion of Schema-Level Index Models and Features

Various SLI models were defined, which capture different schema structures and are defined using different theoretical graph models [4]. For example Konrath et al. [10] use the RDF graph model but Ciglan et al. [5] use the labelled property graph (LPG) model [5]. As demonstrated by Ciglan et al. [5], RDF graphs can be easily transformed into LPGs. Thus, we can compare different schema structures of approaches defined using RDF and LPG. Table 1 is an overview of SLI models

**Table 1.** Nine index models (left column) and their features (top row). Features marked with X are fully supported, (X) are partially supported, and - are not supported.

| Index Model / Feature | Property sets | Type sets | Neighbor information | Path information | k-bisimulation | Incoming property set | OR combination | Related properties | RDF Schema | SameAs |
|---|---|---|---|---|---|---|---|---|---|---|
| Characteristic Sets [13] | X | - | - | X | - | X | - | - | - | - |
| SemSets [5] | X | - | X | X | - | - | - | - | - | - |
| Weak Property Clique [6] | X | - | - | X | X | X | X | X | X | - |
| ABSTAT [16] | X | X | X | X | - | - | - | - | (X) | - |
| LODex [1] | X | X | X | X | - | - | - | - | - | - |
| Loupe [12] | X | X | X | X | - | - | - | - | - | - |
| SchemEX [10] | X | X | X | X | - | - | - | - | - | - |
| TermPicker [15] | X | X | X | - | - | - | - | - | - | - |
| SchemEX+U+I | X | X | X | X | - | - | - | - | X | X |

reported in the literature and the specific features they support to define schema structures. These features are the use of property sets [1,5,6,10,12,13,15,16], use of type sets [1,6,10,12,15,16], use of neighbor information [1,5,6,10,12,15, 16], use of path information [1,5,6,10,12,13,16], use of $k$-bisimulation [6], use of incoming property sets [6,13], use of OR combination of feature [6], use of transitively co-occurring (related) properties [6], and use of inferred information from RDF Schema properties [6,16]. In addition, we propose inferencing over owl:sameAs as a new feature. A detailed discussion of the SLI models is presented in our extended report [3]. In summary, we can state that there exists a variety of SLI models that capture different schema structures and are suitable for different tasks. SLI models are designed, implemented, and evaluated independently for their specific tasks. There has never been a systematic comparison of SLI models nor has it been investigated how different SLI models, queries, and datasets influence the results of querying on the Web of Data.

## 3  Experimental Study

We implement all features discussed in Sect. 2 into a single framework. Our framework allows flexibly combining these features to define SLI models and is available on GitHub along with descriptive dataset statistics, queries, and raw results (https://github.com/t-blume/fluid-framework).

We select six representative index models based on our analysis in Sect. 2. These SLI models are Characteristic Sets, Weak Property Clique, SemSets,

TermPicker, SchemEX, and SchemEX+U+I. We select Characteristic Sets, Weak Property Clique, SemSets, TermPicker since they provide unique feature combinations. Furthermore, we select SchemEX since the SLI model shares the same base schema structure with LODex, Loupe, and ABSTAT (see Table 1). Finally, we select SchemEX+U+I since it uses `owl:sameAs` and full RDFS reasoning. This covers also the RDFS type hierarchy inferencing of ABSTAT and the RDFS reasoning provided by the compact graph summaries. For each of the six selected SLI models, we apply the $k$-height parameterization feature proposed by Tran et al. [18]. Reasonable values for the height parameterization $k$ are 0, 1, and 2 [18]. Thus, in total, we compare 18 unique index models.

We use two datasets crawled from the Web of Data. The first dataset is called *TimBL-11M*. It contains about 11 million triples crawls in a breadth-first search starting from a single seed URI, the FOAF profile of Tim Berners-Lee [10]. The second dataset *DyLDO-127M* is the first crawl provided by the Dynamic Linked Data Observatory (DyLDO), which contains $127M$ triples [9]. The DyLDO-127M dataset is a breadth-first crawl from about $95k$ seed URIs. We provide detailed dataset descriptions in our extended report [3]. Both datasets are reasonably large with $11M$ triples and $127M$ triples, while still allowing to compute a gold standard for the stream-based index computation (Sect. 5). A gold standard is created by loading the entire dataset into the main memory and computing the indices with no window size limit.

## 4    Experiment 1: Compression and Summarization Ratio

We evaluate the index size for the selected indices over the two datasets mentioned above. The size of an index refers to the number of triples when stored as an RDF graph. We compare the number of triples in the index to the number of triples in the dataset (**compression ratio**). Furthermore, we compare the number of schema elements in the index to the number of data instances in the dataset (**summarization ratio**). This ratio gives an idea of how well the defined schema structure can summarize data instances on the Web of Data. For the compression and summarization ratios, we use exact indices. This means we loaded the complete data graph into the main memory before we started the index computation process.

The results of the experiments regarding the compression ratio and summarization ratio are documented in Table 2. As one can see, there is a huge variety in terms of how well indices compress and summarize the data. For the TimBL-11M dataset, SemSets' compression ratio (with $k = 1$) is about 10 times larger than all other indices except for Weak Property Cliques (only about 5 times larger). For the DyLDO-127M dataset, SemSets' compression ratio (with $k = 1$) is up to 75 times larger. Additionally, there is no increase in index size from $k = 1$ to $k = 2$, but a more than ten-times increase from $k = 0$ to $k = 1$. A similar increase appears for the summarization ratio. SemSets is the only index that uses neighbor information but not neighbor type sets, i.e., they compare the object URIs $o$ of each $(s, p, o)$ triple. In contrast, the other indices either ignore

**Table 2.** Results from the analysis of the **compression ratio** and **summarization ratio** of the six selected SLI models (with height parameter $k \in \{0, 1, 2\}$). $\#t$ is the number of triples in millions (M) in the SLI and in brackets below the ratio compared to the number of triples in the dataset (**compression ratio**). $\#e$ is the number of schema elements in thousands (T) in the SLI and in brackets below the ratio compared to the number of instances in the dataset (**summarization ratio**). As datasets, we use the TimBL-11M (top) and DyLDO-127M datasets (bottom).

| | Index model | $k = 0$ | | $k = 1$ | | $k = 2$ | |
|---|---|---|---|---|---|---|---|
| | | $\#t$ | $\#e$ | $\#t$ | $\#e$ | $\#t$ | $\#e$ |
| TimBL-11M | Characteristic sets | na (na) | na (na) | 0.7M (6.5%) | 9.6T (1.4%) | 1.6M (14.6%) | 37.2T (5.5%) |
| | Weak property clique | na (na) | na (na) | 1.9M (17.9%) | 74 (<0.1%) | 1.1M (9.9%) | 50 (<0.1%) |
| | SemSets | 0.3M (2.9%) | 2.8T (0.4%) | 7.6M (69.2%) | 139.0T (20.6%) | 7.6M (69.2%) | 139.0T (69.2%) |
| | SchemEX | 0.3M (2.9%) | 2.8T (0.4%) | 0.8M (6.9%) | 12.0T (1.8%) | 1.4M (12.5%) | 27.7T (4.1%) |
| | TermPicker | 0.3M (2.9%) | 2.8T (0.4%) | 0.7M (6.5%) | 10.8T (1.6%) | 1.8M (16.0%) | 37.3T (5.5%) |
| | SchemEX +U+I | 0.4M (3.8%) | 3.1T (0.5%) | 0.8M (7.1%) | 11.3T (1.7%) | 1.8M (15.9%) | 31.0T (4.6%) |
| DyLDO-127M | Characteristic Sets | na (na) | na (na) | 0.6M (0.5%) | 23.0T (0.3%) | 2.1M (1.7%) | 112.8T (1.6%) |
| | Weak property clique | na (na) | na (na) | 14.8M (9.0%) | 394 (<0.1%) | 25.1M (19.7%) | 102 (<0.1%) |
| | SemSets | 4.1M (3.2%) | 46.6T (0.7%) | 45.3M (35.6%) | 1733.5T (25.0%) | 45.3M (35.6%) | 1733.5T (25.0%) |
| | SchemEX | 4.1M (3.2%) | 46.6T (0.7%) | 15.7M (12.3%) | 254.5T (3.6%) | 19.8M (15.6%) | 431.1T (6.1%) |
| | TermPicker | 4.1M (3.2%) | 46.6T (0.7%) | 11.1M (8.7%) | 238.4T (3.4%) | 25.4M (19.9%) | 559.1T (7.9%) |
| | SchemEX+U+I | 8.5M (6.7%) | 53.0T (0.8%) | 19.9M (15.7%) | 249.5T (3.5%) | 22.9M (18.0%) | 466.9T (6.6%) |

objects or consider their type sets only. SemSets has a summarization ratio of 20%–25%, i.e., on average 4–5 data instances share the same schema structure. The smallest index, Characteristic Sets, has a summarization ratio of 0.3%, i.e., about 330 data instances share the same schema structure. A notable exception is the Weak Property Clique, which shows the most condensed summarization (summarization ratio of less than 0.1%). However, the combination of either incoming or outgoing related properties in Weak Property Cliques leads to a considerably large compression ratio. Weak Property Clique indices are more than twice the size of Characteristic Sets indices.

When considering the semantics of RDFS and *owl:sameAs* in SchemEX+ U+I, the index size increases compared to SchemEX by about 3% more triples. Despite being a larger index in terms of the number of triples, for $k = 1$ fewer schema elements are computed when including the semantics of RDFS and *owl:sameAs*. For $k = 0$ and $k = 2$, SchemEX+U+I requires more schema elements than SchemEX to summarize the data instances.

In summary, including the semantics of *owl:sameAs* and RDF Schema increases the size of the index. However, it can reduce the number of schema elements. Furthermore, using weak equivalences leads to a handful of schema elements with a considerably large size summarizing all data instances.

# 5    Experiment 2: Stream-Based Index Computation

In this experiment, we are interested in how well queries of varying complexity can be supported by the indices if the SLI is computed over a stream of graph edges. Motivated from stream-databases, the idea is to consider the triples in the datasets as a stream that is observed in windows of sizes $1k$, $100k$, and $200k$. This allows us to scale the computation to in principle arbitrary sized input graphs [10]. However, the approach produces approximation errors since only a fraction of the data graph is kept simultaneously in the main memory, while the remainder is not yet known or inaccessible. Thus, we potentially extract incomplete schema structures.

Regarding the index model of SchemEX+U+I, we evaluate two variants in this experiment: The RDFS inferencing requires an additional data structure during the computation process, the so-called schema graph [2]. This schema graph is constructed from the triples using RDFS `range`, `domain`, `subClassOf`, or `subPropertyOf`. With the domain, range, and hierarchical types/properties information, we infer additional types and properties for the remaining data instances. In one version called SchemEX+U+oI, the RDFS information is extracted and inferred *on-the-fly*. Here, we construct the schema graph simultaneously to the index computation. The advantage is that only one pass over the dataset is needed. However, since the schema graph is built while the index is computed information may be missing for the inferencing. In the other version called SchemEX+U+pI, we first extract all RDFS information in a *pre-processing* step to construct the schema graph. The advantage is that the inferencing of triples is conducted on the complete schema graph only. The drawback is that two passes over the dataset are needed.

A central challenge for this experiment is the choice of queries to be executed over the indices. Here, we follow the work by Konrath et al. [10] who conducted a data-driven query generation for the evaluation of approximate graph indices. This means the queries are generated from the actual data instances in the datasets, i.e., their combination of types and properties. We distinguish two types of queries, simple queries (SQ) and complex queries (CQ). Simple queries search for data instances that have a common type set (or in the case of SemSets a common set of objects). In contrast, complex queries search for data instances that match the complete schema structure defined by the specific index model, e.g., include property paths over 2 hops for Characteristic Sets with $k = 2$. We execute the simple and complex queries on the SLI computed with fixed window size and on the gold standard SLI. For our data search task, the results of the queries are the two sets $D_{gold}$ and $D_{window}$, which contain the corresponding data source URIs. Following Konrath et al. [10], the approximation quality is measured by comparing $D_{gold}$ and $D_{window}$ using the F1-score.

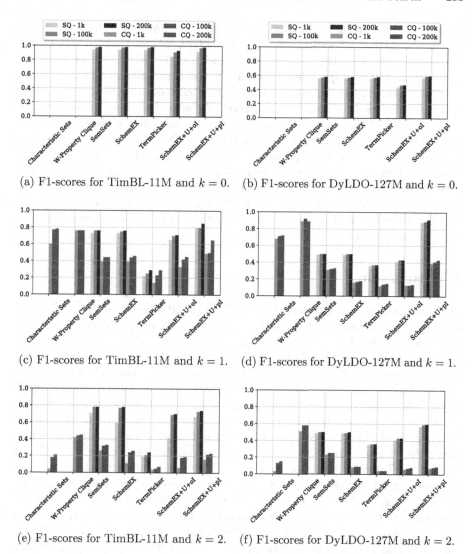

(a) F1-scores for TimBL-11M and $k = 0$.    (b) F1-scores for DyLDO-127M and $k = 0$.

(c) F1-scores for TimBL-11M and $k = 1$.    (d) F1-scores for DyLDO-127M and $k = 1$.

(e) F1-scores for TimBL-11M and $k = 2$.    (f) F1-scores for DyLDO-127M and $k = 2$.

**Fig. 1.** F1-score for simple queries (SQ) and complex queries (CQ) and for window sizes (1k, 100k, 200k). The left column shows the values for the TimBL-11M dataset and the right column the DyLDO-127M dataset, respectively. The influence of the height parameter $k \in \{0, 1, 2\}$ can be seen in the rows from top to bottom.

Figure 1 shows the approximation quality in terms of F1-score for the selected index models. For indices with a height parameter $k = 0$, the simple queries and the complex queries are alike. Moreover, Characteristic Sets and Weak Property Cliques do not use type information (or object information). Thus, simple queries are not available for these index models. From the results of our experiment, we can state that simple queries consistently show higher F1-scores than complex queries. TermPicker and Weak Property Cliques are the only indices that have a higher F1-score on the DyLDO-127M dataset than on the TimBl-11M dataset. As described in Sect. 2, TermPicker is the only index not using the path information feature. This restriction is the only difference in the schema structure compared to SchemEX. Still, TermPicker has a 50% lower F1-score than SchemEX on the TimBL-11M dataset. We also observe an influence of the characteristics of the crawled dataset on the approximation quality. All indices have on average a .15 lower F1-score on the DyLDO-127M dataset compared to the TimBL-11M dataset. In particular, simple queries achieve much lower F1-scores. On average, simple queries have .25 lower F1-scores and complex queries have .04 lower F1-scores on the DyLDO-127M dataset compared to the TimBL-11M dataset. Furthermore, larger window sizes consistently improve F1-scores. In contrast, on-the-fly inferencing lowered the F1-scores in our experiment compared to no inferencing.

## 6     Discussion

Key insights from our experiments are: (1) SLI models perform very differently in terms of compression ratio, summarization ratio, and approximation quality depending on the queries as well as the characteristic of the dataset. (2) The approximation quality of an index computed in a stream-based approach depends on three factors: First, we observe an influence of the characteristics of the crawled dataset. Second, simple queries consistently outperform complex queries. Third, a larger window size typically improves the quality only marginally.

Regarding the first insight, we conducted a detailed analysis to understand the relationship between compression ratio and summarization ratio. We computed the Pearson and Spearman correlation coefficient for the $n = 32$ SLI reported in Table 2. Results of the Pearson correlation indicated that there was a significant relationship between compression ratio and summarization ratio, $r(30) = .84, p < .0001$, and as well as for Spearman, $r_s(30) = .64, p < .0001$. Furthermore, there is a significant negative correlation between summarization ratio and approximation quality of a stream-based computation approach. We computed the Pearson and Spearman correlation coefficient for the three cache sizes $1k$, $100k$, and $200k$. We compared the reported F1-scores for the complex queries (for $k = 0$, we used the simple queries) for each cache size (Fig. 1) to the summarization ratio of the corresponding gold standard index (Table 2).

From the statistical analysis, we can see that a lower summarization ratio leads to a higher F1-score. This means index structures that summarize well, i.e., summarize many data instances to the same schema element, can be computed with high accuracy in a stream-based approach. When we compute correct schema elements in the stream-based approach, for index models with a low summarization ratio, we assign more data instances to the correct schema element than for index models with a high summarization ratio.

We also observe an influence of the characteristics of how the data has been crawled. First, all indices have on average a .15 lower F1-score on the DyLDO-

127M dataset compared to the TimBL-11M dataset. We explain this observation by the different crawling strategies. First, the DyLDO-127M dataset contains nearly 4-times more unique properties and about 11-times more unique types as the TimBL-11M dataset (compare extended dataset description in our extended report [3]). Moreover, the TimBL-11M contains fewer data sources, and data instances are defined in fewer data sources than in the DyLDO-127M dataset. This could be one possible explanation for the overall better performance on the TimBL-11M dataset. The dataset characteristic also influences the size of the index. On average, the compression ratio of indices computed for the TimBL-11M dataset is 14.9% and for the DyLDO-127M dataset, it is 10.5%. Additionally, data instances in the DyLDO-127M dataset have more variety in the number of outgoing properties, but less variety in the number of types. However, the indices using types (SchemEX, TermPicker, SchemEX+U+I) consistently achieve better compression and summarization ratios on the TimBL-11M dataset. The evaluated indices not using types (Characteristic Sets, W-Property Cliques, SemSemts) achieve better compression and summarization ratios on the DyLDO-127M dataset. Thus, the complexity of the combination of type sets and properties seems to be predominately impacted by the number of properties rather than the number of types.

Finally, we observe that inferencing RDF Schema information on-the-fly (SchemEX+U+oI) leads to lower F1-scores than inferencing in a pre-processing step (SchemEX+U+pI). For SchemEX+U+oI, the schema graph information used for inferencing is incomplete until the last triple using an RDFS property is processed. Thus, for SchemEX+U+oI inferencing is another source for approximation errors. However, while including the semantics of `owl:sameAs` and RDFS increases the size of the index, it reduced the number of schema elements in some experiments, i.e., it achieves a better summarization ratio.

# 7   Conclusion

Our empirical evaluations reveal huge variations in compression ratio, summarization ratio, and approximation quality for different index models, queries, and datasets. This confirms our hypothesis that there is no single schema-level index model that equally fits all tasks and that the performance of the SLI model depends on the specific types of queries and characteristics of the datasets. However, we observed meaningful correlations in the results that help to determine the right index model for a given task, type of query, and dataset.

**Acknowledgment.** This research was co-financed by the EU H2020 project MOVING (http://www.moving-project.eu/) under contract no 693092.

# References

1. Benedetti, F., Bergamaschi, S., Po, L.: Exposing the underlying schema of LOD sources. In: Joint IEEE/WIC/ACM WI and IAT, pp. 301–304. IEEE (2015)

2. Blume, T., Scherp, A.: FLuID: a meta model to flexibly define schema-level indices for the web of data. CoRR abs/1908.01528 (2019)
3. Blume, T., Scherp, A.: Indexing data on the web: a comparison of schema-level indices for data search - extended Technical report. CoRR abs/2006.07064 (2020)
4. Čebirić, Š., et al.: Summarizing semantic graphs: a survey. VLDB J. **28**(3), 295–327 (2018). https://doi.org/10.1007/s00778-018-0528-3
5. Ciglan, M., Nørvåg, K., Hluchý, L.: The SemSets model for ad-hoc semantic list search. In: WWW, pp. 131–140. ACM (2012)
6. Goasdoué, F., Guzewicz, P., Manolescu, I.: Incremental structural summarization of RDF graphs. In: EDBT, pp. 566–569. OpenProceedings.org (2019)
7. Gottron, T., Scherp, A., Krayer, B., Peters, A.: LODatio: using a schema-level index to support users infinding relevant sources of linked data. In: K-CAP, pp. 105–108. ACM (2013)
8. Hose, K., Schenkel, R., Theobald, M., Weikum, G.: Database foundations for scalable RDF processing. In: Polleres, A., et al. (eds.) Reasoning Web 2011. LNCS, vol. 6848, pp. 202–249. Springer, Heidelberg (2011). https://doi.org/10.1007/978-3-642-23032-5_4
9. Käfer, T., Abdelrahman, A., Umbrich, J., O'Byrne, P., Hogan, A.: Observing linked data dynamics. In: Cimiano, P., Corcho, O., Presutti, V., Hollink, L., Rudolph, S. (eds.) ESWC 2013. LNCS, vol. 7882, pp. 213–227. Springer, Heidelberg (2013). https://doi.org/10.1007/978-3-642-38288-8_15
10. Konrath, M., Gottron, T., Staab, S., Scherp, A.: SchemEX - efficient construction of a data catalogue by stream-based indexing of linked data. J. Web Sem. **16**, 52–58 (2012)
11. Lei, Y., Uren, V., Motta, E.: SemSearch: a search engine for the semantic web. In: Staab, S., Svátek, V. (eds.) EKAW 2006. LNCS (LNAI), vol. 4248, pp. 238–245. Springer, Heidelberg (2006). https://doi.org/10.1007/11891451_22
12. Mihindukulasooriya, N., Poveda-Villalón, M., García-Castro, R., Gómez-Pérez, A.: Loupe - an online tool for inspecting datasets in the linked data cloud. In: ISWC Posters & Demos, vol. 1486. CEUR-WS.org (2015)
13. Neumann, T., Moerkotte, G.: Characteristic sets: accurate cardinality estimation for RDF queries with multiple joins. In: ICDE, pp. 984–994. IEEE (2011)
14. Pietriga, E., et al.: Browsing linked data catalogs with LODAtlas. In: Vrandečić, D., et al. (eds.) ISWC 2018. LNCS, vol. 11137, pp. 137–153. Springer, Cham (2018). https://doi.org/10.1007/978-3-030-00668-6_9
15. Schaible, J., Gottron, T., Scherp, A.: *TermPicker*: enabling the reuse of vocabulary terms by exploiting data from the linked open data cloud. In: Sack, H., Blomqvist, E., d'Aquin, M., Ghidini, C., Ponzetto, S.P., Lange, C. (eds.) ESWC 2016. LNCS, vol. 9678, pp. 101–117. Springer, Cham (2016). https://doi.org/10.1007/978-3-319-34129-3_7
16. Spahiu, B., Porrini, R., Palmonari, M., Rula, A., Maurino, A.: ABSTAT: ontology-driven linked data summaries with pattern minimalization. In: Sack, H., Rizzo, G., Steinmetz, N., Mladenić, D., Auer, S., Lange, C. (eds.) ESWC 2016. LNCS, vol. 9989, pp. 381–395. Springer, Cham (2016). https://doi.org/10.1007/978-3-319-47602-5_51
17. Tran, T., Haase, P., Studer, R.: Semantic search – using graph-structured semantic models for supporting the search process. In: Rudolph, S., Dau, F., Kuznetsov, S.O. (eds.) ICCS-ConceptStruct 2009. LNCS (LNAI), vol. 5662, pp. 48–65. Springer, Heidelberg (2009). https://doi.org/10.1007/978-3-642-03079-6_5
18. Tran, T., Ladwig, G., Rudolph, S.: Managing structured and semi-structured RDF data using structure indexes. TKDE **25**(9), 2076–2089 (2013)

# CTransE: An Effective Information Credibility Evaluation Method Based on Classified Translating Embedding in Knowledge Graphs

Yunfeng Li[✉], Xiaoyong Li, and Mingjian Lei

Key Laboratory of Trustworthy Distributed Computing and Service (BUPT),
Ministry of Education, Beijing University of Posts and Telecommunications,
Beijing, China
liyfbupt@163.com, lxyxjtu@163.com,
chnleimingjian@163.com

**Abstract.** With the advent of the Big Data era, new network modes have evolved, such as 5G communication and online social networks, resulting in a dramatically increasing amount of information. However, difficulties occur in traditional information credibility evaluation methods, such as manual analysis. It would cost tremendous manpower and time to distinguish the trusted information from the fictitious ones faced with massive data. Therefore, it is urgent and necessary to come up with a more intelligent method to evaluate the credibility of the information. Aiming at the problem of low information quality and the need of efficient assessment in the big data environment, we present an information credibility evaluation method based on knowledge graphs. Firstly, we propose a CTransE model, a translating embedding model based on the classification optimization, which maps entities and relationships into continuous vector space according to scheduled rules. The method reduces the randomness of the algorithm to enhance the stability and accuracy of vector representation. Secondly, we use parameter adaptive adjustment method to optimize the process of stochastic gradient descent. With this approach, we not only obtain a quick convergence to reduce the time cost, but also acquire a better convergence result of knowledge representation compared with previous methods. Finally, we take both ranking and vector distance into account to calculate the information credibility and feedback the most likely information at the same time. Performance on real datasets shows that average ranking has improved about 4% and accuracy in top ten percent has improved more than 13%. Besides, the method also performs well in the field of knowledge completion, database cleaning and so on. It is a breakthrough for applying knowledge graph to quantitative calculation of information credibility evaluation and the method proves to be effective since extensive experiments show that the performance of CTransE is remarkable superior to previous ones on several large-scale knowledge bases.

**Keywords:** Knowledge graph · Information credibility · Knowledge representation

© Springer Nature Switzerland AG 2020
S. Hartmann et al. (Eds.): DEXA 2020, LNCS 12392, pp. 287–300, 2020.
https://doi.org/10.1007/978-3-030-59051-2_19

# 1    Introduction

As of January 2020, more than 4.5 billion people, nearly 60 percent of the world's population, have used the Internet [1]. Meanwhile, social media users have passed the 3.8 billion mark and the latest trends suggest that more than half of the world's total population will use social media by the middle of this year [1]. According to statistics, Internet users spend an average of 6 h and 43 min online every day, which means that more than 40 percent of our waking time is spent on the Internet [1]. It can be observed that the number of network information is also increasing rapidly.

Network information resources demonstrate two significant characteristics, dramatic expansion of scale and disordered structure. The rapid development of social platform and the increase of the transmission channel have even contributed to a higher speed of information dissemination. However, the existing evaluation method are likely to require human effort in terms of calibration, which can hardly adapt to the increasing information. Confronted with vast quantities of information, whose authenticity cannot be guaranteed, to evaluate their credibility more reasonably and efficiently becomes particularly important. Thus, we present an effective information credibility evaluation method based on classified translating embedding in knowledge graphs.

In view that information with large volume and heterogeneous structures is difficult to be effectively used, we adopt knowledge graph as the main storage of massive data and a triple, (subject, predicate, object), is used as the minimum unit of information expression. Though the way of knowledge representation machine learning, we seek the unique vector expression of entities and relationships, so as to make information evaluation become a quantifiable process, which can replace the tedious process of manual retrieval and save time cost and labor cost.

Previous knowledge representation methods [2, 3] have following challenges, weak ability in dealing with many-to-many relationships and the diversity of learning rate for different training sets becomes obvious resulting in high time cost or poor convergence result. We improve a negative sampling method based on entity classification to reduce the randomness and non-negative of corrupted triples, so as to improve the accuracy of the algorithm and the uniqueness of the generated vector representation. Furthermore, we put forward the adaptive change of hyper-parameter. By monitoring the loss function, we use a large learning rate in the early stage to achieve fast convergence and a smaller learning rate in the later stage to achieve better training results.

Traditional information credibility evaluation is mostly a bipartition problem. We transform it into a quantifiable problem to make the result more intuitive and proper through rank-based module. The system not only realizes users' evaluation needs, but also provides the most authentic relevant information.

# 2    Related Work

## 2.1    Knowledge Graph

Knowledge graph [4] is a new knowledge storage format to reproduce entities and their relationships in the objective world in the form of graphs, where nodes correspond to entities and directed edges correspond to their relationships. A statement can be normally performed in a triple as (subject, predicate, object), where the subject and the object are entities in the graph and distinct relationships between them are indicated by the predicates [5]. Knowledge graph is different from traditional graphs with only one single type of entity and a unique from of relationship. Generally, it consists of finite entities with various types and relationships carrying heterogeneous information.

## 2.2 Knowledge Representation Learning

Representation learning [6] is to represent the semantic information of the research object as a dense low dimensional real value vector. Vector representation is a kind of distributes representation. Each dimension in the vector has no clear corresponding meaning, but the semantic information can be represented by synthesizing each dimension to form a vector [7]. By projecting entities or relationships into the vector space, we can express the semantic information of them and calculate the complex semantic relation between entities and relationships efficiently [8]. Structured Embedding projects the head entity vector and the tail entity vector into the corresponding space through two matrices of the relationship and calculates the distance between two vectors [9]. The distance reflects the relevance of two entities and the smaller the distance is, indicating that the two entities have more possibility to have this relationship. Semantic Matching Energy [10, 11] projects entities and relationships to the embedding vector in the input layer, and combines the relationship with the head and tail entity respectively to the hidden layer. Borders et al. propose the TransE model [12], which regards the relationship in the knowledge base as a kind of translation vector between entities. Holographic Embedding [13] proposes to use "circular correlation" of the head and tail entity vectors to represent entities.

## 2.3 Link Prediction

Knowledge credibility evaluation in knowledge graphs can be considered as the process of link prediction, assuming that most existing edges tend to be the true statement and statements to be evaluated is regarded as the missing edges. Early link prediction problem is seen as a binary classification problem with a probability threshold. The potential edges with probability above the threshold tends to be true and those below the threshold are considered as false [14]. Liben-Nowell and Kleinberg proposed the basic definition that inferring the existence of edges between nodes in a graph would be referred to as link prediction. They focused on the temporal prediction problem in a dynamic graph where the relationship at time $t$ was given and the changing at time $t + 1$ was what they wanted [15]. Yu et al. predicted missed protein-protein interactions(PPI) using the topology of the protein interaction network [16]. Clauset et al. proposed a "hierarchical random graph" method, which fits a hierarchical model to all possible dendrograms of a given network and can be used to calculate the possibility of an existing edge [17]. Aouay et al. combined particle swarm algorithm with supervised machine learning strategy to predict connections in social networks [18].

# 3  Method

A knowledge graph is an emerging knowledge memory structure based on triples, in which nodes represent entities and edges represent relationships between connected entities. Each of them carrying heterogeneous information is identified clearly, where an entity is defined by its ontology label and a relationship is defined by the predicate label. With the above assumptions, we formally define a credibility evaluation method as Fig. 1.

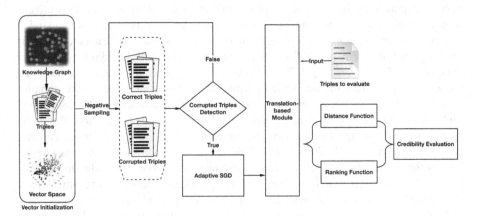

**Fig. 1.**  Flow diagram for credibility evaluation method

## 3.1  Knowledge Graph Definition

A knowledge graph is a directed graph $G = (V, E, O, R, \psi, \varphi)$, where V represents the sets of entities, E presents the sets of directed edges, O represents the set of ontologies, and R represents the set of labeled predicates. The ontology mapping function $\psi(v) = o$, for $v \in V$ and $o \in O$, links an entity vertex to its ontology in the label set. The predicate mapping function $\varphi(e) = r$, for $e \in E$ and $r \in R$, maps an edge to its predicate type.

## 3.2  Translation-Based Module

Given a training set S of triplets $(h, r, t)$ composed of two entities $h, t \in V$ labeled ontology information $l_h, l_t \in O$, and an edge $e \in E$ carrying relationship information $r \in R$, this module utilizes the translation invariance property of word vector and learns vector embeddings of entities and relationships. The pseudocode of the process is as follows. It is assumed that an assertion will be inclined to the credible when the vector of the triplet $(h, r, t)$ satisfies the condition $h + r \approx t$, otherwise the result of $h + r$ should be distinctly different from $t$. The implementation mainly contains three parts: vector initialization, negative sampling, and parameter optimization.

---

**Algorithm 1** Classified Translating Embedding

---

**Input:** Training set $S = \{(h, r, t)\}$, Entities sets $E$, Labels sets $L$, Relations sets $R$, Learning rate $\lambda$, Margin $\gamma$, Embeddings dimension $k$

1: **initialize** for each $r \in R$, $r_j \leftarrow$ uniform $\left(-\frac{6}{\sqrt{k}}, \frac{6}{\sqrt{k}}\right)$ for $j$ in range($k$)

2: $\quad\quad r \leftarrow r/\|r\|$

3: $\quad\quad$ for each $e \in E$, $e_j \leftarrow$ uniform $\left(-\frac{6}{\sqrt{k}}, \frac{6}{\sqrt{k}}\right)$ for $j$ in range($k$)

4: **loop**

5: $\quad$ $e \leftarrow e/\|e\|$ for each relation $e \in E$

6: $\quad$ $S_{batch} \leftarrow$ sample $(S, b)//$ sample a $b$-size minibatch from training set

7: $\quad$ $T_{batch} \leftarrow \emptyset //$ initialize the set of triples and corrupted triples

8: $\quad$ **for** $(h, r, t) \in S_{\text{batch}}$ **do**

9: $\quad\quad$ $(h', r, t) \leftarrow$ uniform $h' \neq h \in E$ or $(h, r, t') \leftarrow$ uniform $t' \neq t \in E$ $\quad\quad //$generate a corrupted triplet

10: $\quad\quad$ **if** $(h', r, t') \notin S_{batch}$ **then** $//$ judge the validity of corrupted triples

11: $\quad\quad\quad$ **if** $L_{\text{h}} = L_{\text{h}'}$ or $L_{\text{t}} = L_{\text{t}'}$ $//$ sort according to the entity label

12: $\quad\quad\quad$ $T_{batch} \leftarrow T_{batch} \cup \{(h, r, t), (h', r, t')\}$

13: **end for**

14: $\quad$ Update embeddings $Loss = \sum\limits_{((h,r,t),(h',r,t')) \in T_{batch}} Hinge\big[d(\boldsymbol{h}+r+t), d(\boldsymbol{h}'+r,t')|\gamma\big]$

15: $\quad$ Update margin $\gamma^{(t)} \leftarrow \gamma^{(t+1)} - \lambda \frac{1}{B} \sum\limits_{Bt+1}^{B(t+1)} \frac{\partial L(\gamma)}{\partial \gamma}$

16: **end loop**

---

**Vector Initialization.** Input contains training sets S, entities sets E, labels sets L, relations sets R, margin $\gamma$, learning rate $\lambda$, embeddings dimension k. S contains credible assertions in the form of triplets $(h, r, t)$, in which $h$ represents the head node, $t$ represents the tail node and $r$ represents the relationship between them. $\gamma$ is the margin-based hyper-parameter to be trained, and $k$ is the dimension of the training vectors. To initialize the entity vector and relationship vector, take a random value in $\left(-\frac{6}{\sqrt{k}}, \frac{6}{\sqrt{k}}\right)$ for each dimension of each vector. All vectors should be normalized after initialization.

**Negative Sampling.** Since representation learning for translating embedding does not have obvious supervisory signals, we introduce the concept of negative sampling to obtain the fast convergence performance. Through this method, negative examples are constructed automatically by replacing one part of the training triplet randomly for each time. The purpose is to maximize the distance between the closest positive and negative samples, namely, to maximize the hyper-parameter $\gamma$.

For a training triplet $(h, r, t)$ from the training set $S_{batch}$, we can get its corrupted triplet $(h', r, t')$ by negative sampling and combine the corrupted triplet with the training triplet to obtain $T_{batch}$.

$$T_{batch} = \{(h, r, t) | (h, r, t) \in S_{batch}\} \cup \{(h', r, t) | h' \in E\} \cup \{(h, r, t') | t' \in E\} \quad (1)$$

To acquire more accurate translating embeddings, we are supposed to filter inappropriate data out of the corrupted triplets. On the one hand, the triplets obtained by substitution are not necessarily negative samples. These corrupted triplets should be removed if they appear in the training set at the same time. On the other hand, substituted entities or relationships should be limited; in other words, substituted entity should have the same ontology as the original entity according to their label information.

**Parameter Optimization.** Our model defines a dissimilarity function $d(h + r, t)$ to measure the distance between $h + r$ and $t$ in either L1 or L2-norm, thereby evaluating the accuracy of translating embeddings. To learn such embeddings, the loss function is defined as,

$$L = \sum_{\{(h,r,t),(h',r,t')\} \in T_{batch}} Hinge(d(h + r, t), d(h' + r, t') | \gamma) \quad (2)$$

where $Hinge(x, x' | margin)$ is designed as a hinge loss function,

$$Hinge(x, x' | margin) = \begin{cases} 0, & x' - x > margin \\ margin + x - x', & x' - x \leq margin \end{cases} \quad (3)$$

We purpose to maximize the margin $\gamma$ and minimize a margin-based ranking criterion. The optimization is carried out by stochastic gradient descent and the $t^{th}$ parameter optimization function is defined as follows,

$$\gamma^{(t)} \leftarrow \gamma^{(t-1)} - \lambda \frac{1}{B} \sum_{t'=Bt+1}^{B(t+1)} \frac{\partial L(\gamma)}{\partial \gamma} \quad (4)$$

where $\gamma^{(t)}$ denotes parameter to be updated at the $t^{th}$ time, $\lambda$ denotes the learning rate, $B$ is a value between 1 and the size of the training set, and $L(\gamma)$ represents the loss function.

We monitor the relative change rate of the loss function. On the one hand, too small learning rate will waste more time cost. On the other hand, too large learning rate may make the model unable to converge. So, we supervise the loss function as signal function. In the early stage of stochastic gradient descent optimization, a large learning rate is used to make the loss function converge rapidly. When there is fluctuation, the learning rate reduces gradually to obtain a suitable learning rate and a better convergence.

To make the convergence more efficient, our model adopts minibatch mode. For each iteration process, use *batch_size* samples for parameter updating instead of traversing all training sets. Using one batch at a time can dramatically reduce the number of iterations required for convergence and make the result closer to the effect of gradient descent. At each main iteration of the algorithm, a small set of triplets sampled from the training set will serve as the training triplets of the minibatch. The iteration stops based on the performance on a validation set.

## 3.3  Rank-Based Module

The vector representation of each entity and relationship is obtained through the translation-based module. For given entities $h$ and $t$, traverse all the relationships in the library, calculate Manhattan Distance or Euclidean Distance and receive a ranking according to the ascending order of distance. The output is in $N * 3$ matrix format, where $N$ represents the total number of relationships, and three columns correspond to Relationship, Distance and Ranking respectively, which is showed in Table 1.

**Table 1.** Output format of rank-based module

| Relationship | Distance | Ranking |
|---|---|---|
| $r_1$ | $D_{min}$ | 1 |
| ... | ... | ... |
| $r$ | $D_r$ | $Rank_r$ |
| ... | ... | ... |
| $r_N$ | $D_{max}$ | $Rank_{max}$ |

## 3.4  Credibility Evaluation

Given a knowledge graph $G$ and a statement of fact $S = (h, r, t)$, which is uncertain for the truth, where subject $h \in V$, object $t \in V$. Credibility evaluation is the process of using a learned translation module of triplets to determine the possibility that the edge $h \xrightarrow{r} t$ appears in the graph. Combined with the rank-based module, we propose a normalized computational formula for credibility evaluation, which is designed as,

$$C_r = \left(1 - \frac{Rank_r}{Rank_{max}}\right) * sin\left(\frac{\pi D_{min}}{2D_r}\right) \tag{5}$$

where the result is normalized in the range from zero to one, and the closer the value gets to one indicates that the statement has a higher credibility.

# 4   Experiment

## 4.1   Data Sets

To research large-scale knowledge graphs with multiple relationships, we adopt typical knowledge graphs WordNet [19] and Freebase [20] to verify our thoughts.

**Freebase.** Freebase, using structured data form, divide data into three layers, Domain, Type and Topic. In Freebase, each item is called a Topic, and the fixed field in each topic is called Property. The same kind of topics are integrated into a Type and all the relevant Types compose a Domain.

**WordNet.** WordNet, based on cognitive linguistics, puts words into a wide range of English vocabulary semantic web according to their meanings. Nouns, verbs, adjectives and adverbs are grouped into sets of cognitive synonyms, each expressing a distinct concept.

We use FB15K, a dense subset of the Freebase and WN18, a subset of WordNet for evaluation. The statistics of the data sets are listed in Table 2.

**Table 2.** Statistics of the dataset

| Dataset | #Relation | #Entity | #Train | #Valid | #Test |
|---------|-----------|---------|--------|--------|-------|
| FB15K | 1,345 | 14,951 | 483,142 | 50,000 | 59,071 |
| WN18 | 18 | 40,493 | 141,442 | 5,000 | 5,000 |

## 4.2   Experimental Setting

We consider the credibility evaluation task as a type of link prediction problem for a statement $S = (h, r, t)$ can be regarded as an edge $h \xrightarrow{r} t$ in the given knowledge graph $G$. The probability that an uncertain statement $S$ is true is equivalent to the probability that the edge $h \xrightarrow{r} t$ is missing in $G$. Undoubtedly, credibility evaluation of a statement is not only limited to the link prediction problem, but also includes the entity prediction problem. They have a similar vector training process, but have the different substituted part. Considering this circumstance, this paper takes the link prediction problem as an instance to validate the feasibility of proposed models and demonstrates the superiority of the algorithm by comparing with the existing algorithm. Furthermore, the paper shows test samples to verify the practicality for statement credibility evaluation.

In this paper, two measurement indexes, MeanRank and Hit@10%, are used to evaluate the feasibility of the algorithm, which is convenient to compare with the traditional algorithm.

MeanRank measures the average rankings of overall predicted ranks. $N$ represents the total number of data.

$$MeanRank = \frac{\sum_{k=1}^{N} Rank_k}{N} \tag{6}$$

Hit@10% measures the proportion of predicted results ranked in the top 10 percent.

$$Hit@10\% = \frac{\sum_{i=1}^{0.1N} Rank_i}{\sum_{j=1}^{N} Rank_j} \tag{7}$$

### 4.3 Model Performance

We trained two models using the code provided by the author. For experiments with CTransE, optimal configurations for each parameter, including vector dimension $k$, learning rate $\lambda$, the margin $\gamma$, dissimilarity measure $d$, and *batch_size* for the mini-batch stochastic gradient descent, are displayed as follows,

**Table 3.** Parameter setting

| Parameter | $k$ | $\lambda$ | $\gamma$ | $d$ | *batch_size* |
|-----------|-----|-----------|----------|-----|--------------|
| WN18 | 20 | 0.01 | 2 | $L_1$ | 800 |
| FB15K | 50 | 0.01 | 1 | $L_1$ | 1000 |

For contrast experiment, we survey other existing prediction models, such as SE, SME(linear), SME(bilinear) and TransE, with dimension $k$ among $\{20, 50\}$, learning rate $\lambda$ among $\{0.001, 0.01, 0.1\}$, margin $\gamma$ among $\{1, 2\}$, dissimilarity measure $d$ among $\{ L_1, L_2 \}$.

FB15K, as the data set of classical link prediction problem, contains varieties of entities and relationships. It can be shown in Fig. 2 that CTransE has a better performance among all test models.

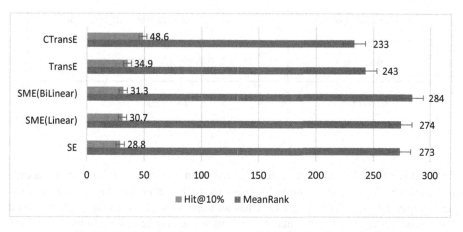

**Fig. 2.** Performance comparison on FB15K for distinct methods

WN18 contains complex relationship cases, which conduces to study multiple relationships (Table 4).

**Table 4.** Model performance

| Dataset | WN18 | |
|---|---|---|
| Index | MeanRank | Hit@10% |
| SE [21] | 1011 | 68.5 |
| SME(LINEAR) [10] | 545 | 65.1 |
| SME(BILINEAR) [10] | 526 | 54.7 |
| TransE [22] | 263 | 75.4 |
| CTransE | 249 | 81.6 |

## 4.4 Credibility Evaluation

**Overall Results.** Table 5 displays an example of what the system shows. It contains the triple entered, the relationship predicted, the ranking of the given relationship and the credibility evaluated. Output is shown as follows,

**Table 5.** Sample output

| Input (Head and Tail) | James Arness (Head) and Minneapolis (Tail) |
|---|---|
| Relationship to Be Evaluated | /people/person/places_lived./people/place_lived/location |
| Predicted Relationship | /people/marriage/location_of_ceremony |
| Ranking | 25 (Among 1345 relationships) |
| Credibility | 0.91 |

**Detailed Results.** Figure 3 compares CTransE with TransE on ranking-based module performance. We collect counting numbers for each certain ranking. It can be obviously observed that CTransE not only has an improvement on average ranking about ten rankings (4%), but also has about 13.5% more data concentrating in top ten percent rankings.

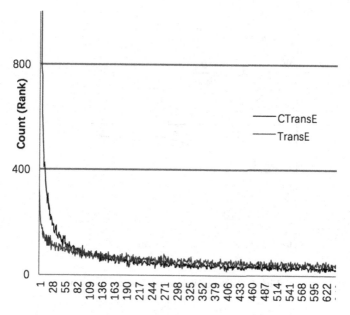

**Fig. 3.** Performance comparison on ranking-based module

In the process of credibility evaluation, two factors, ranking and distance, are both taken into account. $C_r$ represents the credibility of the information, $C_1$ represents the ranking factor and $C_2$ represents the distance factor.

$$C_r = C_1 * C_2 \tag{8}$$

Ranking-based function can be regarded as,

$$C_1 = 1 - \frac{Rank_r}{Rank_{max}} = 1 - \frac{Rank_{min} + r_1}{Rank_{min} + r_2} = \frac{1 - \frac{r_1}{r_2}}{Rank_{min} + 1} \propto \frac{1}{|r|} \tag{9}$$

For each evaluation, $Rank_{min}$ can be considered as a constant. $|r| = \frac{r_1}{r_2}$ measures the proportion of current ranking and maximum ranking. So, credibility is inversely proportional to the ratio, namely better ranking leads to higher credibility.

Distance-based function can be summarized as follows,

$$C_2 = sin\left(\frac{D_{min} * \pi}{2D_r}\right) = sin\left(\frac{D_{min} * \pi}{2(D_{min} + |d|)}\right) \propto \frac{1}{|d|} \tag{10}$$

For each evaluation, $D_{min}$ can be considered as a constant. $|d|$ measures the relative distance between current distance and minimum distance. So, credibility is inversely proportional to the distance, namely shorter distance results in higher credibility.

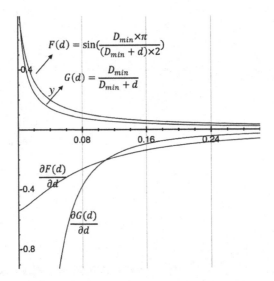

**Fig. 4.** Distance-based function comparison

It can be observed from real cases that only distance factor or ranking factor can hardly measure the credibility accurately. Top 10 percent always have little difference on the distance factor, which means it is not supposed to be the main factor when it approaches zero. From Fig. 4 above, it can be viewed that convert the inverse proportional function to the sine function has a lower gradient and decreases smoothly, resulting in a more reasonable result.

## 5   Conclusions and Future Work

In this paper, we propose an effective information credibility evaluation method based on classified translating embedding in knowledge graphs. The feasibility can be proved by a large amount of experiments. The method solves a key problem, how to evaluate the credibility of information in the big data network environment. CTransE model settles the problem in the process of negative sampling and has a better capacity for understanding complex relationships. The application of parameter adaptive technique increases the accuracy of the measurement and decreases time cost. In the future, we will try to experiment more with different knowledge representation algorithms to obtain a better performance.

**Acknowledgment.** This work was supported by State Grid Corporation's science and technology project "Reliable Analysis and Defense Key Technology Research on Business Security of Distribution Automation System" (No. PDB17201800158) and the NSFC-General Technology Fundamental Research Joint Fund (No. U1836215).

**Availability.** Codes are available at: https://github.com/8218xXXx/Knowledge-Graph.

# References

1. DIGITAL 2020 GLOBAL DIGITAL OVERVIEW. https://wearesocial.cn/wordpress/wp-content/uploads/common/digital2020/digital-2020-global.pdf
2. De Araujo, D.A., Müller, C., Chishman, R., et al.: Information extraction for legal knowledge representation –a review of approaches and trends. Revista Brasilra De Computao Aplicada 6(2) (2014)
3. Riano, D., Peleg, M., Ten, T.A.: Ten years of knowledge representation for health care (2009–2018): topics, trends, and challenges. Artif. Intell. Med. **100**, 101713 (2019)
4. Li, T., Wang, Z.C., Li, H.K.: Development and construction of knowledge graph. J. Nanjing Univ. Technol. **41**(01), 22–34 (2017)
5. Qi, G., Gao, H., Wu, T.X.: Research progress of knowledge map. Inf. Eng. **3**(01), 4–25 (2017)
6. Bengio, Y., Courville, A., Vincent, P.: Representation learning: a review and new perspectives. IEEE Trans. Pattern Anal. Mach. Intell. **35**(8), 1798–1828 (2013)
7. Turian, J., Ratinov, L., Bengio, Y.: Word representations: a simple and general method for semi-supervised learning. In: Proceedings of ACL, pp. 384–394. ACL, Stroudsburg (2010)
8. Lin, Z.Y., Sun, M.S., Lin, Y.K., Xie, R.B.: Research progress of knowledge representation learning. Comput. Res. Dev. **53**(02), 247–261 (2016)
9. Brodes, A., Weston, J., Collobert, R., et al.: Learning structured embeddings of knowledge base. In: Proceedings of AAAI, pp. 301–306. AAAI, Menlo Park (2011)
10. Bordes, A., Glorot, X., Weston, J., et al.: A semantic matching energy function for learning with multi-relational data. Mach. Learn. **94**(2), 233–259 (2014)
11. Bordes, A., Glorot, X., Weston, J., et, al.: Joint learning of words and meaning representations for open-text semantic parsing. In: Proceedings of AISTATS, pp. 127–135. JMLR, Cadiz, Spain (2012)
12. Bordes, A., Usunier, N., Garcia-Duran, A., et al.: Translating embeddings for modeling multi-relational data. In: Proceedings of NIPS, pp. 2787–2795, MIT Press, Cambridge (2013)
13. Nickel, M., Rosasco, L., Poggio, T.: Holographic embeddings of knowledge graphs. arXiv preprint arXiv: 1510. 04935.2015
14. Galileo, N., Lise, G.: Link Prediction. Springer Science. https://doi-org.libezproxy.umac.mo/10.1007/978-1-4899-7687-1_486
15. David, L., Jon, K.: The link prediction problem for social networks. In: Proceedings of the Twelfth International Conference on Information and Knowledge Management, pp. 556–559 November 2003
16. Yu, H., Paccanaro, A., Trifonov, V., Gerstein, M.: Predicting interactions in protein networks by completing defective cliques. Bioinformatics **22**(7), 823–829 (2006)
17. Clauset, A., Moore, C., Newman, M.E.J.: Hierarchical structure and the prediction of missing links in networks. Nature **453**, 98–101 (2008)
18. O'Madadhain, J., Hutchins, J., Smyth, P.: Prediction and ranking algorithms for event-based network data. ACM SIGKDD Explor. Newslett. **7**(2), 23–30 (2005)
19. Miller, G.A.: WordNet: a lexical database for English. Commun. ACM **38**(11), 39–41 (1995)
20. Bollacker, K., Evans, C., Paritosh, P., et al.: Freebase: a collaboratively created graph database for structuring human knowledge. In: Proceedings of KDD, pp. 1247–1250. ACM, New York (2008)

21. Bordes, A., Weston, J., Collobert, R., Bengio, Y.: Learning structured embeddings of knowledge bases. In: Proceedings of the 25th Annual Conference on Artificial Intelligence (AAAI) (2011)
22. Bordes, A., Nicolas, U., Alberto, G.: Translating embeddings for modeling multi-relational data. In: Advances in Neural Information Processing System, NIPS, vol. 26 (2013)

# Updating Ontology Alignment on the Instance Level Based on Ontology Evolution

Adrianna Kozierkiewicz[1], Marcin Pietranik[1]([✉]),
and Loan T. T. Nguyen[2,3]

[1] Faculty of Computer Science and Management,
Wroclaw University of Science and Technology,
Wybrzeze Wyspianskiego 27, 50-370 Wroclaw, Poland
{adrianna.kozierkiewicz,marcin.pietranik}@pwr.edu.pl
[2] School of Computer Science and Engineering,
International University, Ho Chi Minh City, Vietnam
nttloan@hcmiu.edu.vn
[3] Vietnam National University, Ho Chi Minh City, Vietnam

**Abstract.** This paper is devoted to investigating if it is possible to update stale alignments, based solely on the analysis of changes introduced to ontologies participating in the alignment. A change-significance function will be presented, which can be used to indicate whether or not the alignment revalidation is potentially required. Next, an algorithm for revalidation of existing mappings will be proposed. Due to the limited space of this paper, only the instance level will be considered.

**Keywords:** Ontology alignment · Ontology evolution · Knowledge management

## 1 Introduction

Ontologies, in the most common understanding, can be seen as a flexible model of some selected universe of discourse. In other words, they are a convenient way for knowledge representation. Maintaining ontologies is not a daunting task. However, it becomes difficult when ontologies are involved in system integration. It requires communication between two or more ontologies by establishing a set of mappings between them. Such mappings can then be used to "translate" content of one ontology to the content of some other ontology.

This is a seemingly simple problem, referred to as ontology alignment, which was addressed in a variety of publications. If ontologies are fixed this task has many solutions. However, in modern applications, a fixed knowledge model is not sufficient, because different business requirements constantly appear or become stale. Therefore, ontologies should follow those changes, which obviously may result in the established ontology alignment becoming outdated.

© Springer Nature Switzerland AG 2020
S. Hartmann et al. (Eds.): DEXA 2020, LNCS 12392, pp. 301–311, 2020.
https://doi.org/10.1007/978-3-030-59051-2_20

The described situation may be easily addressed by using once again the chosen alignment procedure for evolved ontologies. Such an approach would surely yield good results, but is followed by the cost of the mapping procedure. In the article, we propose a different approach. We claim that it is possible to update the initial alignment between two ontologies based only on the analysis of how participating ontologies changed. This task can be further decomposed into two subproblems.

The first is deciding whether or not the maintained alignment should be re-evaluated. Not all modifications of some ontology during its evolution are equally significant. Therefore, not all of them entail elements of the alignment becoming outdated. For example, some minor changes in some concept's labels are not as major as modifying multiple concepts with their attributes and connecting relations. This paper is devoted solely to the level of instances, therefore, the task can be formally defined as follows: *For a given ontology $O$ in its two consecutive states in time, denoted as $O^{(m)}$ and $O^{(n)}$, one should determine a function $\Psi_I$ representing the degree of significance to which relations have been changed in time.*

The second element of the main task addresses the actual revalidation and it can be formally defined as: *For a given source ontology $O^{(m)}$ in a moment in time denoted as $m$, a target ontology $O'^{(n)}$ in a moment in time denoted as $n$, and an alignment of their instances denoted as $Align(O^{(m)}, O'^{(n)})$, one should provide an algorithm which can update this alignment if the source ontology significantly evolves from the state $O^{(m)}$ to the state $O^{(m+1)}$ according to applied changes.*

The article is structured as follows. Section 2 describes related works found in the literature. Section 3 provides basic notions and definitions further used in the paper. Section 4 contains proposed solutions for the problems defined earlier in the following section. These procedures have been experimentally verified. All of the collected results can be found in Sect. 5. The research plans and a summary are given in Sect. 6.

## 2   Related Works

Ontology alignment is a well-known and widely discussed topic that is still trending [2]. Most of the publications concerning this issue can be divided into two groups: the ones focusing on matching concepts, and the ones focusing on matching instances. A broad overview of the former can be found in [3], which contains a wide array of different solutions to the addressed issue.

In this paper, we focus on aligning ontologies on the latter group. A good introduction to the topic is provided in [6] and [4]. The topic is very important due to emerging unstructured data, that can be modeled as ontology instances [9]. A description of data that can be used as benchmarks for this kind of solution can be found in [7].

One of the basic approaches to ontology alignment evolution is proposed in [13]. Authors base their idea on a high-level language of changes, which is further interpreted as sound global-as-view mappings to produce equivalent rewritings among ontology versions. A solution allows query answering in data integration systems that utilize evolving ontologies. The idea, unlike the approach presented in the following paper, does not analyze the overall change of ontology but focuses on separated elements that have been modified. A more sophisticated approach to change analysis is presented in [12] where authors provided a definition of ontology change history, which logs every alteration applied to the maintained ontology. The content of these logs is then used during the reconciliation of the mapping process.

Authors of [15] address the issue of observing the impact and the adaptation to the evolution of an imported ontology and provide a broad categorization of different evolution scenarios. This approach is different from the one presented in the following paper because it deals with ontology importing, and not updating the previously established bridge between two ontologies.

In [8] ontology alignment is refined using the analysis of change operations. Authors focus, however, on revalidation of semantic relations in concept mappings, and not on a holistic approach to ontology mapping re-validation (as proposed in this paper). Authors of [17] attempt to simplify the issue, by proposing to process only new concepts, while discarding other changes in ontologies. However, what is worth emphasizing it the fact that described approaches do not require any external knowledge bases. On the other hand, in [5] the maintenance of ontology mappings is built on generating annotations from external resources (e.g. text corpora).

However, the problem of updating the instances' alignment caused by modifications of the instances is still open and very briefly researched. To the best of our knowledge, there are only a handful of publications addressing this issue. One of them is [16] where authors propose a set of heuristic update scenarios based on the classification of change operation that may appear during ontology evolution. However, the topic still needs a broader investigation and one of the approaches is proposed in this paper.

## 3   Basic Notions

We assume the existence of a real-world defined as a pair $(A, V)$, where $A$ is a set of attributes and $V$ is a set of attributes domains. An $(A, V)$-based ontology is defined as a quintuple:

$$O = (C, H, R^C, I, R^I) \tag{1}$$

where $C$ denotes a set of concepts; $H$ denotes a concepts' hierarchy; $R^C$ is a set of relations between concepts; $I$ denotes a set of instances' identifiers; $R^I$ is a set of relations between concepts' instances.

Given a concept $c$, we define its instances from the set $I^c$ as a tuple $i = (id^i, v_c^i)$ where: $id^i$ is an instance identifier, $v_c^i$ is a function with a signature: $v_c^i : A^c \rightarrow V^c$. For simplicity, we write $i \in c$ which can be read that the instance $i$ belongs to the concept $c$.

We also define two auxiliary functions, that are used further to simplify manipulating sets. The first, for a given concept $c$ returns a set containing identifiers of instances assigned to it: $Ins(c) = \{id^i | (id^i, v_c^i) \in I^c\}$. The second function $Ins^{-1}$ returns a set of concepts to which an instance with some identifier belongs: $Ins^{-1}(i) = \{c | c \in C \land i \in c\}$.

Ontology evolution introduces a notion of time into ontologies. A superscript $O^{(m)} = (C^{(m)}, H^{(m)}, R^{C(m)}, I^{(m)}, R^{I(m)})$ is used to denote the ontology $O$ in a selected moment in time $m$. A symbol $\prec$ is denotes a fact that $O^{(m-1)}$ is an earlier version of $O$ than $O^{(m)}$ ($O^{(m-1)} \prec O^{(m)}$). This approach can is further adapted to particular elements of the ontology, e.g. $c^{(m-1)} \prec c^{(m)}$ denotes that a concept $c$ has at least two versions, and $c^{(m-1)}$ is an earlier one.

To follow changes applied to the ontology $O$ on the instance level we have defined two additional functions $diff_C$ and $diff_I$. The first function, is fed with two successive states of the ontology $O^{(m-1)}$ and $O^{(m)}$ ($O^{(m-1)} \prec O^{(m)}$) and returns three sets containing concepts added, deleted and modified. It is defined below:

$$diff_C(O^{(m-1)}, O^{(m)}) = \Big\langle new_C(C^{(m-1)}, C^{(m)}),$$
$$del_C(C^{(m-1)}, C^{(m)}), \qquad (2)$$
$$alt_C(C^{(m-1)}, C^{(m)}) \Big\rangle$$

where: $new_C(C^{(m-1)}, C^{(m)}) = \Big\{ c \Big| c \in C^{(m)} \land c \notin C^{(m-1)} \Big\}$; $del_C(C^{(m-1)},$ $C^{(m)}) = \Big\{ c \Big| c \in C^{(m-1)} \land c \notin C^{(m)} \Big\}$; $alt_C(C^{(m-1)}, C^{(m)}) = \Big\{ (c^{(m-1)}, c^{(m)}) |$ $c^{(m-1)} \in C^{(m-1)} \land c^{(m)} \in C^{(m)} \land c^{(m-1)} \prec c^{(m)} \land (A^{c^{(m-1)}} \neq A^{c^{(m)}} \lor V^{c^{(m-1)}} \neq V^{c^{(m)}} \lor I^{c^{(m-1)}} \neq I^{c^{(m)}}) \lor ctx(c^{(m-1)}) \neq ctx(c^{(m)}) \Big\}$.

A function $diff_I$ takes as input two consecutive states of the ontology $O$ and returns a triple of functions that return sets describing instances that have been added to the ontology, instances removed from the ontology, and instances that have been altered. Formally, for two ontology states $O^{(m-1)}$ and $O^{(m)}$ such that $O^{(m-1)} \prec O^{(m)}$:

$$diff_I(I^{(m-1)}, I^{(m)}) = \Big\langle new_I(I^{(m-1)}, I^{(m)}), del_I(I^{(m-1)}, I^{(m)}) \Big\rangle \qquad (3)$$

where: $new_I(I^{(m-1)}, I^{(m)}) = \Big\{ i \Big| i \in I^{(m)} \land i \notin I^{(m-1)} \Big\}$; $del_I(I^{(m-1)}, I^{(m)}) = \Big\{ i \Big| i \in I^{(m-1)} \land i \notin I^{(m)} \Big\}$.

We do not consider alterations, because elements of the set $I$ contains atomic instances' identifiers. Therefore, their modification only appears within particular concepts. For more details of the formal foundations of our work, please refer to one of our previous articles, for example, [10] and [14].

## 4    Updating Ontology Alignment on an Instance Level

Having two ontologies $O_1 = (C_1, H_1, R^{C_1}, I_1, R^{I_1})$ and $O_2 = (C_2, H_2, R^{C_2}, I_2, R^{I_2})$ a concept alignment between them is a set: $AL_C(O_1, O_2) = \{(c_1, c_2, \lambda_C(c_1, c_2), r) | c_1 \in C_1 \wedge c_2 \in C_2 \wedge \lambda_C(c_1, c_2) \geq T_C\}$ where: $\lambda_C$ is a degree to which concept $c_1$ can be aligned to concept $c_2$, and $T_C$ is some assumed threshold.

For short we write $(c_1, c_2) \in AL_C(O_1, O_2)$ to indicate that a concept $c$ can be aligned to concept $c'$ to some degree $\lambda_C(c_1, c_2)$ higher that the assumed threshold $T_C$.

The alignment on the instance level between two ontologies is defined as a set of sets of alignments of instances belonging to two aligned concepts: $AL_I(O_1, O_2) = \{AL_{O_1, O_2}(c_1, c_2) | (c_1, c_2) \in AL_C(O_1, O_2)\}$. A single element $AL_{O_1, O_2}(c_1, c_2)$ that can be understood as an alignment of instances of concepts $c_1$ and $c_2$ is defined as follows:
$AL_{O_1, O_2}(c_1, c_2) = \{(i_1, i_2, \lambda_I(v_{c_1}^{i_1}, v_{c_2}^{i_2})) | i_1 \in I^{c_1} \wedge i_2 \in I^{c_2} \wedge \lambda_I(v_{c_1}^{i_1}, v_{c_2}^{i_2})) \geq T_I\}$,
where $\lambda_I$ is a degree to which an instance $i_1$ can be aligned an instance $i_2$, and $T_I$ is some assumed threshold.

Having an ontology $O = (C, H, R^C, I, R^I)$ in its two subsequent states $O^{(m-1)}$ and $O^{(m)}$, such that $O^{(m-1)} \prec O^{(m)}$, and a concept difference function $diff_C$ a function calculating **a degree of change significance on the level of instances** must meet the following postulates:

- **P1.** $\Psi_I(C^{(m-1)}, C^{(m)}) = 0 \iff diff_C(C^{(m-1)}, C^{(m)}) = \langle \phi, \phi, \phi \rangle$
- **P2.** $\Psi_I(C^{(m-1)}, C^{(m)}) = 1 \iff \bigcup_{c \in C^{(m-1)}} Ins(c) \cap \bigcup_{c \in C^{(m)}} Ins(c) = \phi$

**P1** considers a situation in which no modifications at all have been applied to the maintained ontologies. If nothing changed, then none of the elements of the alignment carries the risk of being outdated. Therefore, the significance of a change is minimal.

**P2** addresses the opposite situation. If the change significance is maximal, then the ontology has been completely modified on the instance level. This situation occurs if all instances from the *(m-1)*-th state have been removed and any instance in the *m*-th state is new.

Having the above postulates in mind we define the $\Psi_I$ with a signature $\Psi_I : C^{(m-1)} \times C^{(m)} \rightarrow [0, 1]$ as follows:

$$\Psi_I(C^{(m-1)}, C^{(m)}) = \frac{\displaystyle\sum_{c \in new_C(C^{(m-1)}, C^{(m)})} |Ins(c)| + \sum_{c \in del_C(C^{(m-1)}, C^{(m)})} |Ins(c)|}{\displaystyle\sum_{c \in C^{(m)}} |Ins(c)| + \sum_{c \in del_C(C^{(m-1)}, C^{(m)})} |Ins(c)|} +$$

$$+ \frac{\displaystyle\sum_{c \in alt_C(C^{(m-1)}, C^{(m)})} |Ins(c)|}{\displaystyle\sum_{c \in C^{(m)}} |Ins(c)| + \sum_{c \in del_C(C^{(m-1)}, C^{(m)})} |Ins(c)|}$$

$$\tag{4}$$

Having the above function defined it is simple to confront it with an assumed threshold, which allows deciding if the alignment designated between the tracked ontology and some other ontology needs revalidation because one of them evolved. If such a requirement appears, then the algorithm described in the next section should be used.

Lets assume that there exist two ontologies $O_1^{(m-1)} = (C_1^{(m-1)}, H_1^{(m-1)}, R_{C_1}^{(m-1)}, I_1^{(m-1)}, R_{I_1}^{(m-1)})$ and $O_2^{(n)} = (C_2^{(n)}, H_2^{(n)}, R_{C_2}^{(n)}, I_2^{(n)}, R_{I_2}^{(n)})$, from which the first evolved from the state $(m-1)$ to the state $m$. The procedure (presented in details in Algorithm 1) that updates the alignment $AL_I(O_1^{(m-1)}, O_2^{(n)})$ to its new state $AL_I(O_1^{(m)}, O_2^{(n)})$ needs to be performed after updating ontology alignment on the concept level. It takes three elements as an input:

1. The alignment level between two ontologies on the instance level in earlier state $(m-1)$ $AL_I(O_1^{(m-1)}, O_2^{(n)}) = \{AL_{O_1^{(m-1)}, O_2^{(n)}}(c_1, c_2) | (c_1, c_2) \in AL_C(O_1^{(m-1)}, O_2^{(n)})\}$
2. The alignment on the concept level in the current state $AL_C(O_1^{(m)}, O_2^{(n)})$
3. A difference function which describes what changed in ontology $O_1$ during its evolution $diff_C(O_1^{(m-1)}, O_1^{(m)})$

## 5   Experimental Verification

Our experiments have been conducted on benchmark datasets provided by the Ontology Alignment Evaluation Initiative (OAEI) [1]. From available datasets, the SABINE (which consists of two ontologies: target and source) has been chosen to detect the degree of similarity between pairs of instances. We also decided to base our experiment on LogMap [11], which is an ontology alignment and alignment repair system. More importantly, LogMap earned high positions in subsequent OAEI campaigns. To simulate the ontology evolution process, we apply some modifications in source ontology from SABINE datasets.

The first aim of our experiments was to verify a function $\Psi_I$ as a tool for detecting the significance of changes in evolving ontology and assessment of the quality of procedures for updating ontology alignment on the instance level. It

---

**Algorithm 1:** Updating ontology alignment on the instance level

**Input** : $AL_I(O_1^{(m-1)}, O_2^{(n)}), AL_C(O_1^{(m)}, O_2^{(n)}), diff_C(O_1^{(m-1)}, O_1^{(m)})$
**Output**: $AL_I(O_1^{(m)}, O_2^{(n)})$

1  **begin**
2    $\widetilde{del} := \{AL_{O_1,O_2}(c_1,c_2)|AL_{O_1,O_2}(c_1,c_2) \in AL_I(O_1^{(m-1)}, O_2^{(n)}) \wedge c_1 \in del_C(O_1^{(m-1)}, O_1^{(m)})\}$
3    $AL_I(O_1^{(m)}, O_2^{(n)}) := AL_I(O_1^{(m-1)}, O_2^{(n)}) \setminus \widetilde{del}$
4    $\widetilde{new} := \{(c_1,c_2)|c_1 \in new_C(O_1^{(m-1)}, O_1^{(m)}) \wedge (c_1,c_2) \in AL_C(O_1^{(m)}, O_2^{(n)})\}$
5    **for** $(c_1,c_2) \in \widetilde{new}$ **do**
6       $AL_{O_1,O_2}(c_1,c_2) := \{(i_1,i_2,\lambda_I(v_{c_1}^{i1}, v_{c_2}^{i2}))|i_1 \in I^{c_1} \wedge i_2 \in I^{c_2} \wedge \lambda_I(v_{c_1}^{i1}, v_{c_2}^{i2})) \geq T_I\}$
7       $AL_I(O_1^{(m)}, O_2^{(n)}) := AL_I(O_1^{(m)}, O_2^{(n)}) \cup \{AL_{O_1,O_2}(c_1,c_2)\}$
8    **end**
9    **for** $c \in alt_C(O_1^{(m-1)}, O_1^{(m)})$ **do**
10       **for** $AL_{O_1,O_2}(c_1,c_2) \in \{AL_{O_1,O_2}(c_1,c_2)|AL_{O_1,O_2}(c_1,c_2) \in AL_I(O_1^{(m)}, O_2^{(n)}) \wedge c = c_1\}$ **do**
11          **if** $A^{c(m-1)} \setminus A^{c(m)} \neq \phi$ **then**
12             **for** $(i,i') \in AL_{O_1,O_2}(c_1,c_2)$ **do**
13                **if** $\lambda_I(i,i') < T_I$ **then**
14                   $AL_{O_1,O_2}(c_1,c_2) := AL_{O_1,O_2}(c_1,c_2) \setminus \{(i,i')\}$
15                **end**
16             **end**
17          **end**
18          **if** $A^{c(m)} \setminus A^{c(m-1)} \neq \phi$ **then**
19             **for** $(i,i') \in \{Ins(c_1) \times Ins(c_2) \setminus AL_{O_1,O_2}(c_1,c_2)\}$ **do**
20                **if** $\lambda_I(v_{c_1}^{i1}, v_{c_2}^{i2})) \geq T_I$ **then**
21                   $AL_{O_1,O_2}(c_1,c_2) := AL_{O_1,O_2}(c_1,c_2) \cup \{(i,i')\}$
22                **end**
23             **end**
24          **end**
25       **end**
26    **end**
27    **for** $i \in del_I(I^{(m-1)}, I^{(m)})$ **do**
28       **for** $c \in Ins^{-1}(i)$ **do**
29          **for**
            $AL_{O_1,O_2}(c_1,c_2) \in \{AL_{O_1,O_2}(c_1,c_2)|AL_{O_1,O_2}(c_1,c_2) \in AL_I(O_1^{(m)}, O_2^{(n)}) \wedge c = c_1\}$
            **do**
30             $\widetilde{del} := \{(i,i')|(i,i') \in AL_{O_1,O_2}(c_1,c_2) \wedge i_1 = i\}$
31             $AL_{O_1,O_2}(c_1,c_2) := AL_{O_1,O_2}(c_1,c_2) \setminus \widetilde{del}$
32          **end**
33       **end**
34    **end**
35    **for** $c \in \{c|c \in C_1^{(m)} \wedge c \in C_1^{(m-1)} \wedge I^{c(m-1)} \neq I^{c(m)} \wedge \exists c' \in C_2^{(n)} : (c,c',\lambda_C(c,c'),r) \in AL_C(O_1^{(m)}, O_2^{(n)})\}$ **do**
36       **for** $AL_{O_1,O_2}(c_1,c_2) \in \{AL_{O_1,O_2}(c_1,c_2)|AL_{O_1,O_2}(c_1,c_2) \in AL_I(O_1^{(m)}, O_2^{(n)}) \wedge c = c_1\}$
         **do**
37          **for** $(i,i') \in AL_{O_1,O_2}(c_1,c_2)$ **do**
38             **if** $\lambda_I(v_{c_1}^{i1}, v_{c_2}^{i2})) < T_I$ **then**
39                $AL_{O_1,O_2}(c_1,c_2) := AL_{O_1,O_2}(c_1,c_2) \setminus \{(i,i')\}$
40             **end**
41          **end**
42          $\widetilde{new} := \{i|i \in I^{(m)} \wedge \neg\exists(i,i',\lambda_I(v_{c_1}^{i1}, v_{c_2}^{i2})) \in AL_{O_1,O_2}(c_1,c_2)\}$
43          **for** $i \in (new_I(I^{(m-1)}, I^{(m)}) \cap Ins(c^{(m)}) \cup \widetilde{new}$ **do**
44             **for** $i' \in Ins(c_2)$ **do**
45                **if** $\lambda_I(v_{c_1}^{i1}, v_{c_2}^{i2})) \geq T_I$ **then**
46                   $AL_{O_1,O_2}(c_1,c_2) := AL_{O_1,O_2}(c_1,c_2) \cup \{(i,i')\}$
47                **end**
48             **end**
49          **end**
50       **end**
51    **end**
52    **return** $AL_I(O_1^{(m)}, O_2^{(n)})$;

has been achieved by showing how different modifications of an ontology that may appear during its evolution can affect its alignments. As a reference, a base alignment between source and target ontology has been set. The source ontology has been modified according to every alteration scenario from Table 1. For the two versions of the source ontology (before and after changes) the $\Psi_I$ has been calculated. Using LogMap a new alignment between modified source and target ontology has been designated. A Dice coefficient measure between the base and new alignment has been calculated. The Dice coefficient is a very popular and intuitive measure allowing to compare dissimilarity between two sets. This value can clearly show changes within ontologies affect mappings between ontologies.

**Table 1.** Different scenarios for ontology pairs, 706 instances in the base ontology

| No. | Description | Dice measure | $\Psi_I$ |
|---|---|---|---|
| 1 | No changes | 0 | 0 |
| 2 | Removing 70 instances | 0.082 | 0.099 |
| 3 | Adding 70 related instances | 0.178 | 0.090 |
| 4 | Modifying 70 instances | 0.001 | 0.099 |
| 5 | Adding and removing 70 instances | 0.203 | 0.180 |
| 6 | Adding and modifying 70 instances | 0.136 | 0.180 |
| 7 | Modifying and removing 70 instances | 0.049 | 0.198 |
| 8 | Adding, removing and modifying 70 instances | 0.152 | 0.271 |
| 9 | Removing 350 instances | 0.393 | 0.496 |
| 10 | Adding 341 instances | 0.468 | 0.326 |
| 11 | Adding 785 instances | 0.837 | 0.526 |
| 12 | Removing 706 instances | 0.861 | 1.000 |
| 13 | Adding 70 not related instances | 0.073 | 0.090 |

In our research, we would like to verify that between the Dice coefficient measure and $\Psi_I$ function exists some correlation. For this purpose, the Shapiro-Wilk test was used to analyze the collected data. For both samples $p - value$ were smaller than $\alpha = 0.05$ thus we rejected the null hypothesis and claim that samples do not come from a normal distribution. Next, the Spearman's rank correlation has been calculated. The obtained result: $p-value$ equal 0.001346 and statistical value 0.788972 allows us to claim that there exists a strong, monotonic relation between the examined samples. Therefore, developed function $\Psi_I$ can be used as a trigger of alignment revalidation in case of a change that may appear during the ontology evolution and can serve as a tool for detecting valid and significant alteration

The second goal of our experiments is to evaluate the proposed procedure for updating ontology alignment on the instance level. To achieve that we have modified the source ontology by adding new related instances according to different scenarios. In the first one, only a few new instances have been added. In consecutive scenarios, the number of alteration has been increased accordingly to Table 2. For modified source ontology and target ontology, new mappings have been determined by our approach and LogMap. The methods have been compared against the number of correctly generated connections between instances.

**Table 2.** Different number of modifications in the source ontology

| No. of changes | Number of mappings found by our approach | Number of maps found by LogMap |
|---|---|---|
| 10% | 464 | 438 |
| 20% | 539 | 495 |
| 30% | 608 | 538 |
| 40% | 678 | 573 |
| 50% | 759 | 628 |
| 60% | 837 | 675 |
| 70% | 905 | 713 |
| 80% | 972 | 753 |
| 90% | 1047 | 781 |
| 100% | 1114 | 834 |

The obtained result allows us to conclude that our approach and LogMap do not give the same alignments of two ontologies. Our method found more correct connections between instances. More importantly, the difference in the number of mappings determined by examined procedures increased with the number of changes made in source ontology. We claim that our approach is more efficient than the application of LogMap. Updating existing alignments (by our procedure) is less expensive than building new mappings from the beginning (by LogMap) in terms of computational complexity. Moreover, a bigger alignment simplifies and reduces the costs of the further integration of two ontologies, thus more correct connections are desirable.

## 6   Future Works and Summary

The paper presents algorithms that can be used to update an ontology alignment on the instance level. This update is required due to the fact, that the mapped ontologies changed over time, therefore, rendering the existing alignment stale. However, not all alterations that appear within ontologies are equally important, thus not all should be followed by updating the alignment.

The method has been experimentally evaluated using well-known datasets provided by the Ontology Alignment Evaluation Initiative. Its usefulness has been proved based on a statistical analysis of the collected results.

Therefore, the paper also presents a function that can be used to evaluate whether or not such an update needs to be performed. This function is used to estimate the significance of change within ontologies and is build on top of the ontology timeline, which is a formal tool for tracking ontology evolution. Its outcome can then be simply confronted with some assumed threshold and used to decide about performing ontology alignment revalidation.

In the future, we plan to focus on developing methods for estimating potential knowledge increase after updating ontology mapping. In other words, estimating how much information has been gained due to the modification of the alignment caused by ontology evolution. We will also perform more extensive experiments that will involve larger ontologies.

**Acknowledgement.** This research project was supported by grant No. 2017/26/ D/ST6/00251 from the National Science Centre, Poland.

# References

1. Achichi, M., et al.: Results of the ontology alignment evaluation initiative 2017. In: OM 2017–12th ISWC Workshop on Ontology Matching, pp. 61–113 (2017). No commercial editor
2. Algergawy, A., et al.: Results of the ontology alignment evaluation initiative 2019. In: CEUR Workshop Proceedings, vol. 2536, pp. 46–85 (2019)
3. Alor-Hernández, G., Sánchez-Cervantes, J.L., Rodríguez-González, A., Valencia-García, R. (eds.): Current Trends in Semantic Web Technologies: Theory and Practice. SCI, vol. 815. Springer, Cham (2019). https://doi.org/10.1007/978-3-030-06149-4
4. Abubakar, M., Hamdan, H., Mustapha, N., Aris, T.N.M.: Instance-based ontology matching: a literature review. In: Ghazali, R., Deris, M.M., Nawi, N.M., Abawajy, J.H. (eds.) SCDM 2018. AISC, vol. 700, pp. 455–469. Springer, Cham (2018). https://doi.org/10.1007/978-3-319-72550-5_44
5. Cardoso, S.D., et al.: Leveraging the impact of ontology evolution on semantic annotations. In: Blomqvist, E., Ciancarini, P., Poggi, F., Vitali, F. (eds.) EKAW 2016. LNCS (LNAI), vol. 10024, pp. 68–82. Springer, Cham (2016). https://doi.org/10.1007/978-3-319-49004-5_5
6. Castano, S., Ferrara, A., Lorusso, D., Montanelli, S.: On the ontology instance matching problem. In: 2008 19th International Workshop on Database and Expert Systems Applications, pp. 180–184. IEEE (2008)
7. Daskalaki E., Flouris G., Fundulaki I., Saveta T: Instance matching benchmarks in the era of linked data. J. Web Semant. **39**, 1–14 (2016)
8. Destro, J.M., dos Reis, J.C., Torres, R.D.S., Ricarte, I.: Evolution-based refinement of cross-language ontology alignments. In: Anais do XXXIV Simpósio Brasileiro de Banco de Dados, pp. 61–72. SBC (2019)
9. Heflin J., Song D.: Ontology instance linking: towards interlinked knowledge graphs. In: Thirtieth AAAI Conference on Artificial Intelligence (2016)
10. Hnatkowska, B., Kozierkiewicz, A., Pietranik, M.: OWL RL to framework for ontological knowledge integration preliminary transformation. In: Asian Conference on Intelligent Information and Database Systems 2020 (2020)

11. Jimenez-Ruiz, E., Grau, B.C., Cross, V.: LogMap family participation in the OAEI 2018. In: Proceedings of the Twelfth International Workshop on Ontology Matching, OM-2018, pp. 187–191 (2018)
12. Khattak, A.M., et al.: Mapping evolution of dynamic web ontologies. Inform. Sci. **303**, 101–119 (2015). https://doi.org/10.1016/j.ins.2014.12.040
13. Kondylakis, H., Plexousakis, D.: Ontology evolution without tears. J. Web Semant. **19**, 42–58 (2013)
14. Kozierkiewicz, A., Pietranik, M.: A formal framework for the ontology evolution. In: Nguyen, N.T., Gaol, F.L., Hong, T.-P., Trawiński, B. (eds.) ACIIDS 2019. LNCS (LNAI), vol. 11431, pp. 16–27. Springer, Cham (2019). https://doi.org/10.1007/978-3-030-14799-0_2
15. Qawasmeh, O., Lefrançois, M., Zimmermann, A., Maret, P.: Observing the impact and adaptation to the evolution of an imported ontology. In: IC3K 2019–11ht International Joint Conference on Knowledge Discovery, Knowledge Engineering and Knowledge Management, 11 pages (2019)
16. dos Reis, J.C.: Mapping refinement based on ontology evolution: a context-matching approach. In: ONTOBRAS, pp. 251–256 (2018)
17. Yamamoto, V.E., dos Reis, J.C.: Updating Ontology Alignments in Life Sciences based on New Concepts and their Context. Revista dos Trabalhos de Iniciacao Cientifica da UNICAMP, Campinas, SP, vol. 27 (2019)

# Theme-Based Summarization for RDF Datasets

Mohamad Rihany$^{(\boxtimes)}$, Zoubida Kedad, and Stéphane Lopes

DAVID Lab, University of Versailles Saint-Quentin-en-Yvelines, Versailles, France
mohammad.rihany92@gmail.com,{zoubida.kedad,Stephane.lopes}@uvsq.fr

**Abstract.** A growing number of RDF datasets are published on the web. These datasets can be viewed as graphs; querying, analyzing and visualizing such data graphs are critical challenges facing the applications willing to use them, especially when their size is important. Summarization can help addressing these challenges. In this paper, we present a summarization approach which exploits the underlying themes of an RDF graph, and builds a global summary from the theme summaries. To this end, we propose some node relevance metrics. We present some experiments to illustrate the effectiveness of our approach.

**Keywords:** RDF datasets · Summarization · Theme identification

## 1 Introduction

An increasing amount of interlinked datasets are published on the web, making a huge amount of knowledge available. They are described in languages proposed by the W3C, such as the Resource Description Framework (RDF) [9], in which the building block is a triple *(subject, predicate, object)*. An RDF dataset can be viewed as a labelled directed graph where nodes represent resources or literals and where labelled edges represent properties. Understanding these graphs is a critical challenge for their meaningful use.

RDF graph summarization has been the topic of several research works [3], aiming to provide the user with a concise representation containing the most relevant information in the graph in order to ease its exploitation.

In this paper, we describe a summarization approach for RDF datasets which provides a representation taking into account the various underlying topics or themes existing in the dataset. It ensures that the most relevant nodes of each theme are present and that their representativity is reflected in the summary.

We consider that a theme is represented in the graph by a dense area, where nodes are highly connected. The intuition behind this is that the more connected a set of nodes, the more likely these nodes are related to the same theme. To identify these dense areas, we rely on the algorithm proposed in [11]. In order

This work was partially funded by the National Council for Scientific Research of Lebanon (CNRS-L).

S. Hartmann et al. (Eds.): DEXA 2020, LNCS 12392, pp. 312–321, 2020.
https://doi.org/10.1007/978-3-030-59051-2_21

to build a summary for a given theme, we propose an extended definition of node centrality to select the most relevant nodes and we propose an aggregation method to find the best paths between them.

The rest of this paper is organized as follows. An overview of the approach is provided in Sect. 2. Section 3 presents the process of summarizing the extracted themes. The generation of the graph summary is presented in Sect. 4. Section 5 presents our experiments and Sect. 6 reviews the related works. Finally, we conclude the paper and present some future works in Sect. 7.

## 2   Overview of the Approach

The goal of our approach is to summarize an RDF graph not only by selecting the most relevant nodes, but also by maximizing the number of themes represented in the summary. This is done firstly by identifying the different themes described in the graph, secondly by building a summary for each of them, and finally building a global summary by aggregating the theme summaries. Our framework, presented in Fig. 1, comprises three main components: *theme identification*, *theme summarization* and *building the final summary*.

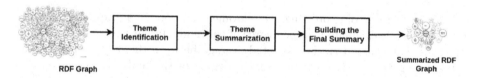

**Fig. 1.** Thematic-based summarization framework

*Theme Identification* consists in identifying the underlying themes in the input RDF graph. To this end, we use the approach proposed in [11], which identifies highly connected areas in the graph. The intuition is that the more connected a set of nodes in the graph, the more likely they are related to the same topic. The algorithm used in the approach is MCODE, a density-based clustering algorithm producing possibly overlapping clusters [1]. Each one represents a theme.

The second component, *theme summarization*, generates a summary for each theme. This summary contains the most relevant graph elements. In order to identify these elements, we propose two definitions of centrality to distinguish between schema-level nodes (classes) and instance-level nodes (entities). We also consider the number of instances of the classes when selecting the most relevant ones. The theme summary is built by connecting the set of most relevant nodes in the theme while minimizing the number of other nodes.

Finally, the third component of our framework builds the global summary from the theme summaries. This is done by selecting the most central node from each theme summary and aggregating these nodes, which we have formulated as a Steiner tree problem [4]. During this process, we ensure that all the themes are represented in the summary proportionally to their size.

The following sections are focused on the summarization process, details on theme identification can be found in [11].

## 3    Theme Summarization

In this section, we present the concepts and algorithms of our theme summarization approach. Consider the sub-graph $G_i$ representing a theme $t_i$ in the RDF data graph G. Summarizing $G_i$ consists in building a graph $S_{G_i}$ such that: (i) $S_{G_i}$ contains the top-k most relevant nodes of $G_i$, and (ii) the number of other nodes is minimal. This requires the computation of a relevance score for each node in $G_i$.

### 3.1    Evaluating Node Relevance

An RDF graph comprises schema-related information, such as classes or type definitions, and instance-related information, such as entities. In our work, we distinguish between these two types of nodes when computing the relevance score.

A common way of expressing the relevance of a node in a graph is node centrality, which is calculated as the degree of this node. In our work, the centrality of a class is evaluated as the weighted average between the degree of the node representing this class, and the average degree of the nodes representing the entities belonging to this class. We do not take into account the edges labelled "type", which only represent the link between a class and its instances.

**Definition 1.** *Class Centrality. Consider the class $C_i$ in the RDF graph G, and its set of entities $E_i$. The centrality score of $C_i$ is:*

$$CS(C_i) = \frac{\sum_{e \in E_i} degree(e)}{|E_i|} \times w + D(C_i) \times (1 - w)$$

*where $D(C_i)$ is the degree of the node corresponding to $C_i$ in the graph G excluding all the edges labelled "type" and w a value ranging from 0 to 1.*

The average degree of the entities in a class is calculated in order to reflect the importance of this class in the graph. If the score of the class was calculated according to the degree of the class only, then the node "Sci-Fi Movie" which has a degree of three in the graph of Fig. 2 would be more relevant than the node "Comedy Movie" which has a degree of two. Yet we can see that the node "Comedy Movie" has an instance with a degree of four while the two instances of the class "Sci-Fi Movie" both have a degree of one; this means that according to their instances, the class "Comedy Movie" has more links with the other nodes of the graph than the class "Sci-Fi Movie".

When computing the centrality score of a class, the weight $w$ expresses the level of confidence in the schema compared to the instances: a high value will give more weight to the average degree of the entities, which could represent the fact that the description of the classes in the graph is incomplete; a low value will give more weight to the class degree, which could express the fact that the description of the class is complete and accurate.

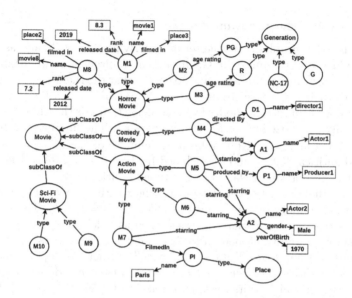

**Fig. 2.** Example of an RDF graph describing movies

The set of classes in the graph of Fig. 2 is C = {Horror Movie, Comedy Movie, Action Movie, Sci-Fi Movie, Generation, Place}. A centrality score will be calculated for each of these classes. For the "Horror Movie" class, the set of entities is $E_i = \{M1, M2, M3, M8\}$. The degree of each of these nodes is 1, 1, 4 and 4 respectively. The average degree of the entities in the class "Horror Movie" is therefore 2.5. The degree of the class "Horror Movie" is computed considering all the edges except those corresponding to the "type" property and is equal to 1. If we assume that $w$ is equal to 0.7, then $CS("HorrorMovie") = 2.05$.

Centrality is not the only score that reflects the relevance of a class. Indeed, a class could have a low centrality score, but a high number of instances. For example, the centrality score of the "Generation" class is equal to 0.35, which is one of the lowest in the example. However, this class has the highest number of instances. During the selection of the most relevant classes, we also take into account their representativity, defined as follows.

**Definition 2.** *Class Representativity. The representativity of a class $C_i$ in a graph G is the number of entities of this class, i.e., the number of nodes connected to $C_i$ by an edge labelled with the "type" property.*

Beside the nodes corresponding to classes, some nodes in the graph may correspond to entities for which no type declaration has been specified. The centrality of these nodes is defined hereafter.

**Definition 3.** *Untyped Entity Centrality. The centrality of an untyped entity in a graph G is the degree of the node corresponding to the entity in G, i.e., the number of incoming and outgoing edges for this node.*

## 3.2 Identifying the Relevant Nodes

Consider the RDF graph G and the set of underlying themes T, and assume we want to build a top-k summary for G. One of our requirements is to maximize the number of themes in the summary, and to ensure that the themes are represented according to their respective size in G.

The number $k_i$ of relevant nodes for each theme $t_i$ in T is therefore determined proportionally to its size in G as follows: $k_i = \frac{|t_i|}{|G|} \times k$. The summary of the theme $t_i$ is determined firstly by defining both the number of classes and the number of untyped entities that should be included in the summary. This is done so as to reflect the proportion of classes and untyped entities in $t_i$.

Let $p_i$ be the proportion of classes and their entities in $t_i$. The number of classes in the summary of $t_i$ is defined as follows: $nc_i = p_i \times k_i$. The number of untyped entities in the summary of $t_i$ is defined as follows: $ne_i = (1 - p_i) \times k_i$.

The relevance of an untyped entity is evaluated as its centrality, defined in the previous section. The selection process will extract the top-$ne_i$ most relevant untyped entities. The relevance of a class can be considered according to two facets, its centrality or its representativity. In the graph of Fig. 2 the top-3 most central classes are "Comedy Movie", "Action Movie" and "Horror Movie", while the top-3 most representative ones are "Horror Movie", "Action Movie" and "Generation". In our approach, a weight $w$ specifies the importance of centrality with respect to representativity. If $nc_i$ classes are to be selected, $w \times nc_i$ classes having the highest centrality and $(1 - w) \times nc_i$ classes having the highest representativity will be selected.

Algorithm 1 describes the identification of the relevant nodes in a given theme.

## 3.3 Building the Theme Summary

Building the theme summary $S_{Gi}$ from the set of top-$k_i$ most relevant nodes in a given theme $t_i$ can be stated as a Steiner tree problem [4]: given a graph $G_i$ (V, E), a subset $Ter \subseteq V$ of vertices called terminals, and a weighted function $d : E \to \mathbb{R}$ on the edges, the goal is to find a sub-graph S of minimal weight in $G_i$ containing all the terminals. Other nodes than the terminals, called Steiner nodes, can be added to S. S should be a tree, which means that from any pair of nodes $i$ and $j \in V$ there should exist exactly one path.

In our context, the set of relevant nodes represents the set of terminals. We have adapted the distance network heuristic(DNH) [10], which has an approximation ratio equal to $2 - \frac{2}{p}$ where p is the total number of the terminal nodes.

---

**Algorithm 1.** Identifying Relevant Nodes

---

**Input:** Graph $G_i$, set of classes C, set of untyped entities E, number of relevant nodes $k_i$, number of relevant classes $nc_i$ and weight $w$

**Output:** Set of relevant nodes $R_i$

1: **procedure** RELEVANT NODES($G_i$, C, E, $k_i$, $nc_i$, w)
2:     $R_i = \phi$ /*set of relevant nodes*/
3:     $CCS \leftarrow sort(Classes)$ /*sort classes according to the centrality score*/
4:     $CRS \leftarrow sort(Classes)$ /*sort classes according to the representativity score*/
5:     $ECS \leftarrow sort(entities)$ /*sort untyped entities according to the centrality score*/
6:     $l = w * nc_i$ /*number of classes according to centrality*/
7:     $m = (1 - w)nc_i$ /*number of classes according to representativity*/
8:     $n = (k_i - nc_i)$ /*number of relevant untyped entities*/
9:     $R_i \leftarrow top - l$ $classes$ $of$ $CCS \cup top - m$ $classes$ $of$ $CRS$ /*Identify relevant classes*/
10:    $R_i \leftarrow R_i \cup top - n$ $entities$ $of$ $ECS$ /*Identify relevant untyped entities*/
11:    **return** $R_i$
12: **end procedure**

---

Instead of randomly selecting a path when there are several paths between a given pair of terminals, we compute a centrality score for each path based on the centrality of its nodes. The centrality score of a path is calculated as follows. Let p $= [(v_1, e_1, v_2)(v_2, e_2, v_3)....(v_{n-1}, e_{n-1}, v_n)]$ be the path connecting the two terminal nodes $v_1$ and $v_n$; the centrality score of p $CS(p) = \frac{\sum_n^{i=1} deg(v_i)}{n}$ is the average degree of the nodes in the path. We can also limit the computation of $CS(p)$ to the top k nodes having the highest centrality.

The adapted distance network heuristic is described below. First the distance graph $DG_i$ is computed from $G_i$ by using the shortest path between all the relevant nodes. Let $i$ and $j$ be two relevant nodes, and $n_{ij}$ the number of edges in the shortest path connecting $i$ and $j$. We compute the minimum spanning tree of $DG_i$ using the Kruskal algorithm, which consists in creating a forest F (a set of trees) where each node in the graph is a separate tree, and creating a set S containing all the edges in the graph. The algorithm finds an edge with minimal weight connecting any pair of trees in the forest without forming a cycle. If several edges match these criteria, then one is chosen arbitrarily. The edge is added to the spanning tree, and this step is repeated until there are |V| -1 edges in the spanning tree (where |V| is the number of vertices).

We have adapted Kruskal's algorithm by modifying the weights of the edges and taking into account the centrality score of a path. If $n_{ij}$ in $DG_i$ represents the shortest path connecting i with j, the centrality score is calculated for each path. The paths are first sorted according to their length, and if two or more paths have the same length, we select the path having the highest centrality score.

Assume that R = { "Horror Movie", "Comedy Movie", "Generation", "A2"} is the set of relevant nodes of the graph in Fig. 2. The generated distance graph is

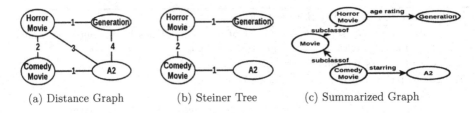

(a) Distance Graph          (b) Steiner Tree          (c) Summarized Graph

**Fig. 3.** From distance graph to summarized graph

shown in Fig. 3a. Figure 3b shows the Steiner tree constructed using the adapted Kruskal's algorithm. After replacing the edges in the Steiner tree by the paths from G, we obtain the sub-graph of Fig. 3c.

Finally, we enrich the resulting summary by adding some edges corresponding to properties that describe the selected nodes, such as "label" and "name". Figure 4 shows the final summary, where the edge labelled "name" has been added for the node "A2".

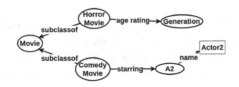

**Fig. 4.** Final summarized graph

## 4   Building the Final Summary

Let S = { $S_{G1}$, $S_{G2}$ ... $S_{Gn}$} be the set of summaries corresponding to the themes of a graph G such that each $S_{Gi}$ is the summary of a sub-graph corresponding to the theme $t_i$. The summary of G is generated by connecting all the elements of S. To this end, we select in each $S_{Gi}$ the most central node $s_i$ and we build the graph in which all the nodes $s_1$, $s_2$,..., $s_n$ are connected. Similarly to theme summarization, this can be stated as a Steiner tree problem where the set of terminal nodes is the set of all the nodes $s_i$ corresponding to each theme.

We also use the DNH algorithm to solve this problem; the default weight of each edge is equal to one. Let us consider the graph in Fig. 5a where two theme summaries are represented in red and blue respectively; consider "a1" and "b1", the most central nodes of the two themes respectively.

The shortest path connecting nodes "a1" and "b1" is P1 = [a1, c1, b1], but we can observe that choosing this path will introduce the node "c1", while the path P2 = [a1, a3, b2, b1] connects the two theme summaries without introducing any additional node. In our approach, we assign a weight to each edge. This

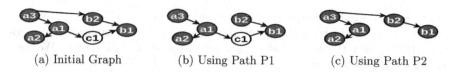

(a) Initial Graph          (b) Using Path P1          (c) Using Path P2

**Fig. 5.** Path selection for building the final summary

weight is equal to zero if the origin and destination nodes of the edge are in the same theme summary, and it is equal to one otherwise. Using such weights leads to the graph depicted in Fig. 5c. The path P2 will be selected as its weight is equal to one while the weight of P1 is equal to two.

## 5    Evaluation

In this section, we describe our experiments to validate the performances of our approach and to compare the resulting summary with the one of a baseline approach where summarization is performed solely on the basis of centrality, without considering the underlying themes. Our approach is implemented in Java, we have used the Jena API for the manipulation of RDF data. The Jung API is used for graph manipulation and visualization. All the experiments have been done on Intel Core i7 with 32 GB RAM.

We have used three datasets: AIFB, DBpedia and Olympics. These data sets consist of 29 233 triples, 30 793 triples and 1 781 625 triples respectively.

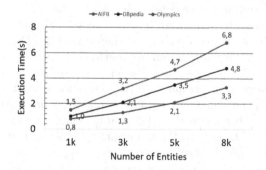

**Fig. 6.** Average execution time

Figure 6 shows the execution time with respect to the size of the dataset. It shows that the execution time increases when the number of triples increases for all the datasets. We can also see that the execution time for Olympics is greater than the execution time of AIFB and DBpedia.

To check the effectiveness of our approach we have used 8 simple graphs where the number of entities varies from 25 to 34, and asked five users to select the relevant elements in these graphs. We have then compared their answers to the results obtained by our summarization algorithms (TBA). We have also compared our results with the application of a baseline summarization algorithm (BA), where the top-k nodes are selected based on their degree. We have computed the precision at k as follows: $P@K = \frac{NumberOfrelevantelements}{K}$.

According to Table 1, we can see that P@K varies between 0.84 and 0.98 for our thematic-based approach. These results are better than the ones obtained using the baseline approach, where P@K varies between 0.62 and 0.88; on these graphs, the results achieved with the thematic-based approach were more accurate according to the users.

**Table 1.** Top-k precision

| Data | G1 | | G2 | | G3 | | G4 | | G5 | | G6 | | G7 | | G8 | |
|------|-----|-----|-----|-----|-----|-----|-----|-----|-----|-----|-----|-----|-----|-----|-----|-----|
| K | 5 | 10 | 5 | 10 | 5 | 10 | 5 | 10 | 5 | 10 | 5 | 10 | 5 | 10 | 5 | 10 |
| TBA | 0.88 | 0.84 | 0.84 | 0.88 | 0.88 | 0.98 | 0.84 | 0.86 | 0.96 | 0.94 | 0.92 | 0.94 | 0.92 | 0.94 | 0.96 | 0.96 |
| BA | 0.84 | 0.88 | 0.72 | 0.80 | 0.72 | 0.84 | 0.88 | 0.82 | 0.8 | 0.72 | 0.68 | 0.62 | 0.68 | 0.72 | 0.64 | 0.62 |

# 6   Related Works

Summarizing RDF graphs has been the topic of several research works [3]. In [6] the notion of quotient graph is introduced. It consists in defining an equivalence relation on the nodes and then summarizing the graph by replacing the equivalent nodes by one representative equivalence node. The approach is extended in [5], where summaries are built in an incremental way. [7] presents an approach for parallelizing the summarization of big RDF graphs. In [12] several centrality metrics are proposed. Similarly to our approach, the generation of the summary from the relevant nodes is stated as a Steiner tree problem. In [13], the summarized graph is created using graph patterns, identified using the binary matrix of the graph. In [2], the summarized graph is created by first extracting the type and predicate for each node in the RDF graph then grouping the nodes which share the same set of types into the same summary node. The nodes for which no type definition is provided are grouped according to their attributes. In [8], all the nodes that share the same type and properties are grouped together in a set and the summary is built from these sets. Some statistics such as the number of instances per class or per property are aggregated with this structural information.

Unlike the existing works, our approach relies on the identification of clusters containing semantically related nodes in the graph, then summarizes each partition before merging the partial summaries into a global summary. The relevance of the nodes takes into account the centrality but also the representativity of the classes.

# 7    Conclusion

In this paper we have provided an approach for summarizing RDF graphs based on their underlying themes. We have proposed relevance metrics for the nodes in the graph, as well as algorithms for building both the theme summaries and the global summary. We have also conducted some experiments to evaluate our approach. A possible extension would be to insert in the resulting summary properties or nodes which do not exist in the graph, but have some semantic relationships with existing nodes; for instance, this could lead to replacing a path in the summary by a single edge conveying the same meaning, or replacing two classes by their parent class if they are close in meaning. This could be done using external knowledge sources such as domain ontologies.

# References

1. Bader, G.D., Hogue, C.W.: An automated method for finding molecular complexes in large protein interaction networks. BMC Bioinform. **4**, 2 (2003). https://doi.org/10.1186/1471-2105-4-2
2. Campinas, S., Delbru, R., Tummarello, G.: Efficiency and precision trade-offs in graph summary algorithms. In: Proceedings of the 17th International Database Engineering & Applications Symposium, pp. 38–47. ACM (2013)
3. Čebirić, Š., et al.: Summarizing semantic graphs: a survey. VLDB J. **28**(3), 295–327 (2018). https://doi.org/10.1007/s00778-018-0528-3
4. Garey, M.R., Johnson, D.S.: The rectilinear Steiner tree problem is NP-complete. SIAM J. Appl. Math. **32**(4), 826–834 (1977)
5. Goasdoué, F., Guzewicz, P., Manolescu, I.: Incremental structural summarization of RDF graphs. In: EDBT 2019–22nd International Conference on Extending Database Technology (2019)
6. Guzewicz, P., Manolescu, I.: Quotient RDF summaries based on type hierarchies. In: 2018 IEEE 34th International Conference on Data Engineering Workshops (ICDEW), pp. 66–71. IEEE (2018)
7. Guzewicz, P., Manolescu, I.: Parallel quotient summarization of RDF graphs (2019)
8. Khatchadourian, S., Consens, M.P.: Exploring RDF usage and interlinking in the linked open data cloud using ExpLOD. In: LDOW (2010)
9. Klyne, G.: Resource description framework (RDF): Concepts and abstract syntax (2004). http://www.w3.org/TR/2004/REC-rdf-concepts-20040210/
10. Kou, L., Markowsky, G., Berman, L.: A fast algorithm for Steiner trees. Acta Inform. **15**(2), 141–145 (1981). https://doi.org/10.1007/BF00288961
11. Ouksili, H., Kedad, Z., Lopes, S.: Theme identification in RDF graphs. In: Ait Ameur, Y., Bellatreche, L., Papadopoulos, G.A. (eds.) MEDI 2014. LNCS, vol. 8748, pp. 321–329. Springer, Cham (2014). https://doi.org/10.1007/978-3-319-11587-0_30
12. Troullinou, G., Kondylakis, H., Stefanidis, K., Plexousakis, D.: RDFDigest+: a summary-driven system for KBs exploration. In: International Semantic Web Conference (P&D/ Industry/BlueSky) (2018)
13. Zneika, M., Lucchese, C., Vodislav, D., Kotzinos, D.: RDF graph summarization based on approximate patterns. In: Grant, E., Kotzinos, D., Laurent, D., Spyratos, N., Tanaka, Y. (eds.) ISIP 2015. CCIS, vol. 622, pp. 69–87. Springer, Cham (2016). https://doi.org/10.1007/978-3-319-43862-7_4

# Contextual Preferences to Personalise Semantic Data Lake Exploration

Devis Bianchini$^{(\boxtimes)}$, Valeria De Antonellis, Massimiliano Garda,
and Michele Melchiori

Department of Information Engineering, University of Brescia,
Via Branze 38, 25123 Brescia, Italy
{devis.bianchini,valeria.deantonellis,m.garda001,
michele.melchiori}@unibs.it

**Abstract.** In the latest years, the availability of data collected within Smart Cities is enabling citizens to take decisions about their daily life in an autonomous way. In this landscape, data aggregation according to different analysis dimensions may help users to take decisions, leveraging indicators as powerful tools for meaningful exploration. However, due to the volume and heterogeneity of Smart City data, data lakes have to be used as flexible repositories for enabling data storage and organisation. Despite they are usually based on centralisation of data storage, data lakes compel to consider pay-as-you-go or on-demand solutions, where integration is progressively carried out, to cope with the cumbersome nature of Big Data. Given the variety of interested users, their goals and preferences on available data, personalised data access, as well as representation and use of preferences, are required and need to be adapted to the unique characteristics of data lakes. In this paper, we describe an approach to model preferences on Smart City indicators built on top of a data lake. Preferences are used for personalised data exploration. Main contributions of this paper concern: (a) the definition of users' preferences and preference constructors over the semantic representation of indicators; (b) the definition of users' contexts and contextual preferences; (c) preference-based personalised exploration of Smart City data.

**Keywords:** Semantic Data Lake · Contextual preferences ·
Personalised data exploration · Smart City

## 1 Introduction

In modern Smart Cities, the ever growing availability of data has created the opportunity for different categories of users (such as citizens, service providers, building administrators and energy managers) to explore data autonomously, to take decisions on their daily life. In this landscape, data aggregation according to different analysis dimensions may help users to take decisions in a more efficient manner, leveraging indicators as powerful tools for meaningful data aggregation and exploration. However, data comes from heterogeneous infrastructures and

© Springer Nature Switzerland AG 2020
S. Hartmann et al. (Eds.): DEXA 2020, LNCS 12392, pp. 322–332, 2020.
https://doi.org/10.1007/978-3-030-59051-2_22

new generation devices, considering different formats, models and storage systems [2]. Recent initiatives suggest to adopt data lakes to store and share both structured and unstructured data, given their flexibility and schema-on-read nature. Despite they are usually based on the centralisation of data storage, data lakes compel to consider pay-as-you-go or on-demand solutions, where integration is progressively carried out, due to the cumbersome nature of Big Data [8]. Hence, the definition of indicators in the context of data lakes deserves more investigation. Based on these premises, in our previous work [1] we proposed a three-layered approach for personalised Smart City data exploration based on semantic representation of indicators, built on top of a data lake. Distinct layers reflect the separation of concerns and competencies within the informative model: (i) at the bottom, domain experts define proper Semantic Models, to semantically abstract data lake representation and ensure a unified view of data; (ii) in the middle, data analysts design indicators and analysis dimensions, in terms of concepts and relationships of the Semantic Models, weaving the Smart City Exploration Graph and (iii) at the top, users' profiles are exploited to personalise the exploration of indicators, over the Smart City Exploration Graph.

With reference to exploration personalisation, preferences, formulated as soft constraints, have proven to be effective in supporting users to focus on the most interesting data [3,7]. The variety of users and their goals determine a multiplicity of contexts under which data exploration is performed; therefore, preferences have to be contextualised according to users' roles and activities (*context-aware preferences*) and need to be adapted to the unique characteristics of data lakes. In this paper, we extend our previous work by describing an approach for modelling contextual preferences on Smart City indicators built on top of a data lake and to use them for enabling personalised data exploration. Main contributions of this paper concern: (a) the definition of users' preferences and preference constructors over the semantic representation of indicators; (b) the definition of users' contexts and contextual preferences; (c) preference-based personalised exploration of Smart City data.

The paper is organised as follows: cutting-edge features of the proposed approach with respect to the literature are discussed in Sect. 2; Sect. 3 explains the Smart City Exploration Graph; Sect. 4 presents users' profiles and contextual preferences formulation, while preference-based exploration of Smart City data is described in Sect. 5; Sect. 6 gives an excerpt of the implementation of the approach and preliminary validation; finally, Sect. 7 closes the paper, sketching future research directions.

## 2   Related Work

In the literature of the last decades, preferences have been studied under a two-fold perspective: *quantitative* [9], relying upon a scoring function exploited to determine a total order of results, and *qualitative* [5,6,10], expressed using binary relations to deliver a strict partial order of results. For instance, delving into the field of databases and data warehouses, authors in [6] discuss about qualitative preferences formulated over aggregation levels of facts. Conversely, the work presented in [5] suggests the adoption of conditional preferences, to extend the classic SQL query language with soft constraints specified as a set of if-then rules. In these approaches, despite multi-dimensional data is handled, preferences are defined over data marts content, without providing a seamless and unified view of underlying data. Concerning the Semantic Web research, in [9] preferences are formalised as constraints on resources, mixing both quantitative and qualitative aspects, whilst [10] devises an extension of SPARQL query language, apt to express qualitative preferences. Overall, these semantics-based approaches focus mainly on the extension made to the query language. Beyond the qualitative and quantitative aspects, a pivotal position has been fulfilled by *contextual preferences*, adapting their behaviour according to the different situations users are involved in. In particular, the work presented in [4] unravels the propagation of preferences between contexts, considering effects of context changes. Multi-dimensional data is not handled and personalisation aspects are limited to provide a formal definition of context, disregarding exploration issues.

Our approach, instead, deals with both multi-dimensional and Semantic Web aspects. In particular, Smart City indicators and related dimensions are semantically represented, starting from the semantic abstraction of data sources over the data lake, in order to ease data exploration experience of users. To increase the expressiveness of the approach, we adopt qualitative preferences, which are not expressed directly over data sources content, but involve the semantic conceptualisation of indicators and related dimensions. Moreover, with respect to the existing Semantic Web research efforts, our approach introduces novel preference constructors capitalising on the semantic relationships concerning indicators and the associated dimension hierarchies. Lastly, to further personalise exploration of indicators, contextual preferences are defined and associated with the users' profiles.

## 3    Smart City Exploration Graph

In our previous work [1], we proposed a three-layered approach to extract the Smart City Exploration Graph from a semantically enriched data lake, thus enabling personalised data exploration and analysis over heterogeneous Smart City data sources. Indeed, due to the complex structures and heterogeneity of data lake sources, the content of each data source in its raw format cannot be directly exploited by users in order for them to get fruitful insights. Therefore, domain experts, relying on their knowledge and expertise, weave an overlay of Semantic Models on the data lake, to ensure interoperability among heterogeneous sources and improve data access. In this respect, concepts and relationships contained in the Semantic Models are leveraged to design the Smart City Exploration Graph $\mathcal{G}$, conceived as a starting point for users to explore Smart City data in terms of indicators, explorable according to different perspectives (dimensions). Base concepts and relationships exploited for modelling indicators and dimensions are defined within the so-called Multi-Dimensional Ontology (MDO), whose design has been inspired by the existing literature concerning

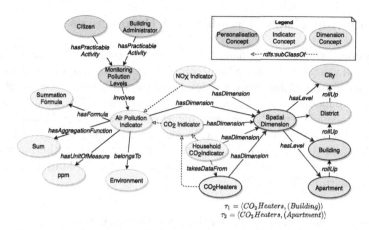

$$\tau_1 = \langle CO_2 Heaters, (Building) \rangle$$
$$\tau_2 = \langle CO_2 Heaters, (Apartment) \rangle$$

**Fig. 1.** Portion of the Smart City Exploration Graph $\mathcal{G}$ with two sample MDDs $\tau_1$ and $\tau_2$ highlighted.

OLAP paradigm constructs (please refer to [1] for a comprehensive explanation of the MDO design methodology). Starting from the base concepts and relationships of the MDO (i.e., personalisation concepts, indicators and dimensions, as shown in Fig. 1), the Smart City Exploration Graph $\mathcal{G}$ is designed by the data analyst (whose role and knowledge are clearly distinguished from those of domain experts) and contains the semantic representation of indicators, their dimensional characterisation and formulas to aggregate them. Figure 1 illustrates a portion of $\mathcal{G}$ containing the semantic definition of a set of environmental indicators, along with the associated spatial dimension hierarchy and related granularity levels. Personalisation concepts refer to the categories of users and their activities for which indicators have been designed (e.g., building administrators, citizens). Based on the Smart City Exploration Graph $\mathcal{G}$, each indicator can be formalised through the notion of *Multi-Dimensional Descriptor* (in brief, MDD), that resembles the association of the indicator to its dimensional coordinates and is defined as follows.

**Definition 1 (Multi-Dimensional Descriptor).** *A Multi-Dimensional Descriptor (in brief, MDD), defined over the Smart City Exploration Graph $\mathcal{G}$, is denoted as $\tau_i = \langle ind_i, L_i \rangle$, where: (i) $ind_i$ represents an indicator concept from $\mathcal{G}$ and (ii) $L_i$ is a vector of dimensional level concepts. Specifically, let:*

- *$D_i = \{\overline{d} \mid ind_i$ hasDimension $\overline{d}\}$ be the set of dimensions associated with $ind_i$ and $|D_i| = n$;*
- *$Levels(d) = \{\overline{l} \mid d$ hasLevel $\overline{l}\}$ be the set of dimensional levels associated with dimension $d$ in $\mathcal{G}$.*

*then the vector $L_i$ is represented as $L_i = (l_i^1, l_i^2, \ldots, l_i^n)$, $l_i^j \in Levels(dim_i^j)$, $dim_i^j \in D_i$. We denote with $\mathcal{T}$ the overall set of MDDs for the Smart City Exploration Graph $\mathcal{G}$. For an indicator $ind_i \in \mathcal{G}$, the maximum number of possible MDDs is equal to $\prod_{j=1}^{n} |Levels(dim_i^j)|$.*

# 4  Modelling Users' Profiles and Preferences

In this section, we provide a definition for user's profile, describing its pivotal elements and the role they cover for accomplishing personalised indicators exploration based on the Smart City Exploration Graph. Being $\mathcal{U}$ the set of Smart City users, we model the profile of a user $u \in \mathcal{U}$ as $p(u) = \langle \mathcal{P}_u, \mathcal{CT}_u \rangle$, composed of two sets, namely: (a) $\mathcal{P}_u$, representing *Preferences* or *Soft Constraints*, apt to declare the most interesting data for the user, in order to better focus the exploration and (b) $\mathcal{CT}_u$, representing the *User's Contexts*, gathering information characterising the situation under which the user explores indicators, in compliance with his/her needs.

## 4.1  User's Preferences and Preference Constructors

The set $\mathcal{P}_u$ contains *preferences*, formalised according to an algebra grounded on the foundation constructs presented in [7], but properly adapted to be applied to concepts of the Smart City Exploration Graph. Preferences have as target objects the Multi-Dimensional Descriptors (MDDs) formalised in Definition (1) and are defined as follows.

**Definition 2 (Preference).** *A preference* **P** *is a couple* $(<_{\mathbf{P}}, \cong_{\mathbf{P}})$ *where* $<_{\mathbf{P}} \subseteq \mathcal{T} \times \mathcal{T}$ *is binary relation expressing a strict partial order over the set of MDDs* $\mathcal{T}$ *and* $\cong_{\mathbf{P}} \subseteq \mathcal{T} \times \mathcal{T}$ *is a binary relation for expressing substitutability. Given* $\tau_1, \tau_2 \in \mathcal{T}$, *the semantics of* $\tau_1 <_{\mathbf{P}} \tau_2$ *is that* $\tau_2$ *is preferred to* $\tau_1$, *whereas the semantics of* $\tau_1 \cong_{\mathbf{P}} \tau_2$ *is that* $\tau_1$ *is equivalent (or substitutable) to* $\tau_2$.

Preferences on the space of MDDs are expressed through *preference constructors*, focusing on either indicators, dimensional hierarchies, granularity levels or indicators domains. Hereafter, the four main base constructors of our approach are briefly explained.

1. $\text{IND}(i)$ expression denotes that $\tau_2$ is preferred to $\tau_1$ ($\tau_1 <_{\text{IND}(i)} \tau_2$) iff the distance of indicator $ind_2$, contained in $\tau_2$, from indicator $i$ (in the hierarchy induced by the *is-a* relationships, denoted with $\sqsubseteq$) is less than the distance between $ind_1$, contained in $\tau_1$, and $i$. This constructor relies on the function $dist_I(i_1, i_2)$, computing the distance between indicators concepts $i_1$ and $i_2$ within the hierarchy induced by $\sqsubseteq$ relationships.
2. $\text{LEV}(dim, l)$ expression denotes that $\tau_2$ is preferred to $\tau_1$ ($\tau_1 <_{\text{LEV}(dim,l)} \tau_2$) iff the distance of level $l_2 \in Levels(dim)$, contained in $\tau_2$, from $l$ (in the hierarchy induced by $\texttt{rollUp}$ relationships) is less than the distance between $l_1 \in Levels(dim)$, contained in $\tau_1$, and $l$. This constructor relies on the function $dist_L(l_i, l_j)$, computing the distance between two levels $l_i$ and $l_j$ within the hierarchy $dim$, induced by $\texttt{rollUp}$ relationships.
3. $\text{DEP}(i)$ expression denotes that $\tau_2$ is preferred to $\tau_1$ ($\tau_1 <_{\text{DEP}(i)} \tau_2$) iff the formula of indicator $ind_2$, contained in $\tau_2$, is calculated starting from the indicator $i$ (modelled through the $\texttt{takesDataFrom}$ relationship in $\mathcal{G}$, see Fig. 1), while $ind_1$, contained in $\tau_1$, does not. This constructor exploits the boolean

function $tdf(i_a, i_b)$ returning *True* if $i_a$ takesDataFrom $i_b$ in $\mathcal{G}$, *False* otherwise.

4. DOM($dom$) expression denotes that $\tau_2$ is preferred to $\tau_1$ ($\tau_1 <_{\text{DOM}(dom)} \tau_2$) iff the indicator $ind_2$, contained in $\tau_2$, belongs to domain $dom$, while $ind_1$, contained in $\tau_1$, does not. This constructor exploits the boolean function $domain(i, dom)$ returning *True* if $i$ belongsTo $dom$ in $\mathcal{G}$, *False* otherwise.

*Example.* Given the two MDDs $\tau_a = \langle$CO2Indicator, (District)$\rangle$ and $\tau_b = \langle$HouseholdCO2Indicator, (Apartment)$\rangle$, with reference to Fig. 1 the constructor IND(AirPollutionIndicator) identifies as preferred $\tau_a$, since $\tau_b.ind_b \sqsubseteq \tau_a.ind_a \sqsubseteq$ AirPollutionIndicator. In fact, $dist_I($AirPollutionIndicator, $\tau_a.ind_a) = 1$ is less than $dist_I($AirPollutionIndicator, $\tau_b.ind_b) = 2$. For the same MDDs, the constructor LEV(SpatialDimension, City) identifies again as preferred $\tau_a$ because $dist_L($City, District$) = 1$ is less than $dist_L($City, Apartment$) = 3$.

## 4.2   User's Contexts and Contextual Preferences

The context under which the user explores indicators is modelled by considering the *category* of the user (e.g., citizen, building administrator) and the *activity* that the user is currently performing (e.g., monitoring electrical consumption, checking air pollution levels). For instance, the citizen Alice, living in the Smart City, while checking air pollution levels, may prefer to inspect each pollutant concentration in detail (e.g., $CO_2$, $SO_x$, $NO_x$), since she is also a connoisseur of environmental issues. Such exploration preference holds in the former specific context, but may be not applicable elsewhere (e.g., for a building administrator performing building monitoring activity). Remarkably, indicators can be explored by users according to the exploration perspectives available for them (for instance, a building administrator is allowed to inspect indicators for the administered buildings only).

Formally, a context $ctx^i$ belonging to $\mathcal{CT}_u$ is a triple $\langle cat^i, a^i, CP^i \rangle$ where: (i) $cat^i$ represents the category of the user $u$, (ii) $a^i$ is an activity performed by the user and (iii) $CP^i$ denotes the set of preferences valid for the context, so that $CP^i \subseteq \mathcal{P}_u$. Categories and activities delimit in $\mathcal{G}$ the set of indicators explorable by the user within the context. Noteworthy, $CP^i$ contains, amongst the others, the so-called *base preferences* for the user's category, designed by the data analyst. They ensure a first-level exploration support within a context, so that indicators exploration can be performed preventing the user from relying on a trial-and-error approach only, during the exploration sessions.

## 5   Personalised Exploration

**Context Selection and Request Formulation.** Let us consider again the user Alice introduced in Sect. 4.2. Once logged in to the exploration platform, she decides to take up an exploration session as a Citizen, to accomplish

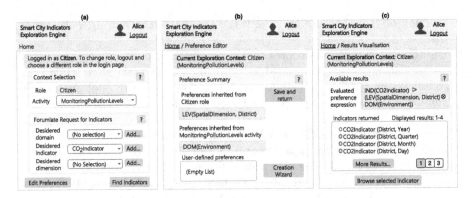

Fig. 2. GUI for personalised indicators exploration.

MonitoringPollutionLevels activity. Afterwards, to formulate a request, Alice exploits the form in Fig. 2(a) specifying the desired indicators, domains and dimensions through proper dropdown lists, whose content depends on the current context. Noteworthy, the request formulated by the user becomes a *short-term* preference, aimed at addressing an imminent exploration need. Back to the example, Alice chooses to express such a preference only for a desired indicator, thus IND(CO2Indicator). Beyond the former short-term preferences, the user, by clicking on the "Edit Preferences" button, is redirected to a page containing the summary of preferences stored in her profile and compatible with the current context, as shown in Fig. 2(b). User-defined preferences only can be manually edited, to be reused across exploration sessions, in the same context. For the selected exploration context, Fig. 2(b) shows that Alice has at her disposal two base preferences, whilst no user-defined preferences have been created yet.

**Evaluating Preferences.** Preference evaluation procedure starts with the construction of a compound preference, using two composition operators from the literature, namely the *Pareto composition* ($\otimes$), composing two preferences with equal priority, and the *prioritization* ($\triangleright$) [7]. Short-term preferences have *priority* over the others, as they address an imminent need, whereas the rationale is to combine the other preferences of the profile using the Pareto composition operator, since they all assume an equal importance for the user. Therefore the resulting expression is: IND(CO2Indicator) $\triangleright$ (LEV(SpatialDimension, District) $\otimes$ DOM(Environment)). The compound preference undergoes an evaluation process. In our approach, we foster the WeSt algorithm [6], properly adapted to work with our preference model. This algorithm partitions the space $\mathcal{T}$ of the MDDs in the so-called *S-classes* (that is, groups of MDDs fulfilling preferences in the same way), adopting a *preference graph* to guide the overall evaluation process, avoiding to search the entire space $\mathcal{T}$. In the graph, each node is associated with a boolean predicate, selecting a subset of MDDs from $\mathcal{T}$. Since we deal with an ontology-based structure ($\mathcal{G}$ has been deployed in OWL), the boolean predicate within each node of the preference graph had to be translated into a SPARQL query, progressively assembled with a procedure decomposing the predicate into its sub-predicates, separately processed depending on the preference constructor they refer to (e.g., DOM(Environment) is mapped to the SELECT query $\mathcal{T}_{env} = \{\tau_i \in \mathcal{T} \mid \tau_i.ind_i \text{ belongsTo Environment}\}$).

**Presenting Preference Evaluation Results.** By default, the output of WEST algorithm is the set of MDDs not worse than others (complying with the so-called *best match only* model), which are displayed in the first page of the output list in Fig. 2(c). Nonetheless, the user can further broaden the list, clicking on the "More Results" button, visualising MDDs which are differently ranked according to the evaluated preference (by virtue of the definition of *implicit* ranking [3]). This is achieved by descending the preference graph mentioned in the previous step, moving to its next levels, so that next best-matching MDDs can be retrieved and inspected. For the sake of clarity, referring to the example, the first page of the output list contains MDDs with CO2Indicator and District level (i.e., the first-best MDDs), then MDDs with CO2Indicator and City/Building level will follow (that is, the second-best MDDs) and so forth. Finally, the user can select one of the available indicators from the list and, clicking on the "Browse selected indicator" button in Fig. 2(c), the multi-dimensional query apt to retrieve indicator values will be issued over the underlying Semantic Data Lake and the result will be rendered relying upon external Business Intelligence tools.

# 6 Implementation and Preliminary Validation

In this section, we present the Personalisation Level of the architecture proposed in [1], implemented in Java and deployed under the Apache TomEE[1] application server. The Personalisation Level contains: (a) *End-User Application Service*, elaborating inputs and outputs of personalised exploration; (b) *Profile Management Service*, responsible for the maintenance of users' profiles and preferences and (c) *Personalisation Service*, implementing the preference evaluation procedure discussed in the previous section. Numbers in Fig. 3 give the order of the interaction flow between services and related modules. Firstly, the user specifies her short-term preferences, whereas the ones retained in her profile (long-term) are retrieved (1, 2). Then, the *Request Handler* module assembles the compound preference expression, which is sent to the *Preference Evaluator* module (3a) and the output MDDs are pushed back to the Request Handler (3b). Finally, the user selects a MDD from the result set (4) and the *Indicator Visualiser* module composes the multi-dimensional query involving the chosen indicator (5), which is sent to the *Semantic Data Lake Access Layer* to retrieve the values of the indicator (this process is not detailed here, as it is out of the scope of the paper). At this point, indicator values can be visualised and navigated invoking external Business Intelligence platforms services.

---

[1] http://tomee.apache.org/.

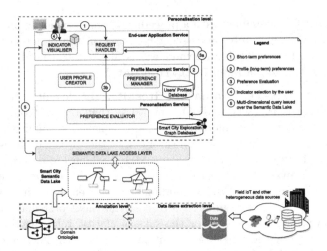

**Fig. 3.** Architecture overview (detail of Personalisation Level services).

**Preliminary Validation.** Preliminary evaluation of our approach is compliant with those performed in similar efforts [6], aimed at: (i) evaluating the degree of usability of our system and (ii) demonstrating the effectiveness of preferences in pruning the search results, improving the selectivity of best matches using preferences. Usability experiments were conducted on a group of 14 users from the Smart City domain (citizens, representatives of Public Administration, utility and energy providers) to investigate if they could easily understand how to express their preferences through the GUI. After an initial training, to let them create their profile and acquire basic notions on preferences working principles, they were asked to formulate 10 requests of increasing complexity (including from 1 to 5 base constructors), starting from the corresponding description in natural language. Collected statistics evidenced an average formulation time and formulation errors (e.g., due to a wrong interpretation of described requests) proportional to the number of base constructors involved. Concerning the effectiveness of our system, experiments have been conducted considering a Smart City Exploration Graph of around 40 indicators, linked (on average) to three dimensions of three granularity levels each. In particular, to assess the behaviour of the preference evaluation procedure, the *preference selectivity* has been computed, defined as the ratio between the size of the set of MDDs not worse than others and the cardinality of the whole set of MDDs $\mathcal{T}$. The workload consisted of a set of preference expressions prepared using different combinations and numbers (from 1 to 5) of base constructors, composed using both Pareto composition and prioritization operators. As an example, expressions containing only $\mathrm{DOM}(\cdot)$ constructors led to an average selectivity of 61%, expressions with only $\mathrm{LEV}(\cdot)$ to 29% whereas in the case of expressions with both $\mathrm{DOM}(\cdot)$ and $\mathrm{LEV}(\cdot)$ the average selectivity was 8%, thus demonstrating that, for instance, $\mathrm{DOM}(\cdot)$ is less selective when applied alone (as it determines large S-classes).

Ongoing experiments are devoted to evaluate response times of the proposed system using different preference expressions, with respect to the size of the Smart City Exploration Graph (number of indicators, depth and width of dimension hierarchies) and the complexity of preference expressions, adopting the experimentation methodology suggested in [6].

## 7    Conclusions and Future Work

In this paper, we described an approach to model contextual preferences on Smart City indicators built on top of a data lake, for personalising data exploration. Main contributions of this paper concerned: (a) the definition of users' preferences and preference constructors over the semantic representation of indicators; (b) the definition of users' contexts and contextual preferences; (c) preference-based personalised exploration of Smart City data. The approach has been discussed in the Smart City domain, given its variety both in terms of data and involved users' roles. To better address users' exploration goals, prone to endless changes, future research efforts will be devoted to enhance the preference model described in this paper, by considering issues related to both preference propagation amongst exploration contexts [4] and preference recommendation, to assure a higher level of automation. In this respect, also the automatic evolution of preferences, based on users' exploration habits and choices, deserves a more detailed examination, fostering adaptive evolution models for users' profiles, to achieve implicit preferences learning.

## References

1. Bagozi, A., Bianchini, D., De Antonellis, V., Garda, M., Melchiori, M.: Personalised exploration graphs on semantic data lakes. In: 27th International Conference on Cooperative Information Systems (CoopIS 2019), Rhodes, Greece, pp. 22–39 (2019)
2. Chauhan, S., Agarwal, N., Kar, A.: Addressing big data challenges in smart cities: a systematic literature review. Info **18**(4), 73–90 (2016)
3. Chomicki, J.: Preference formulas in relational queries. ACM Trans. Database Syst. (TODS) **28**(4), 427–466 (2003)
4. Ciaccia, P., Martinenghi, D., Torlone, R.: Foundations of context-aware preference propagation. J. ACM (JACM) **67**(1), 1–43 (2020)
5. de Amo, S., Ribeiro, M.R.: CPref-SQL: a query language supporting conditional preferences. In: 24th Annual ACM Symposium on Applied Computing (SAC 2009), Honolulu, Hawaii, USA, pp. 1573–1577 (2009)
6. Golfarelli, M., Rizzi, S., Biondi, P.: myOLAP: an approach to express and evaluate OLAP preferences. IEEE Trans. Knowl. Data Eng. **23**(7), 1050–1064 (2010)
7. Kießling, W.: Foundations of preferences in database systems. In: 28th International Conference on Very Large Databases (VLDB 2002), Hong Kong, China, pp. 311–322 (2002)
8. Nargesian, F., Zhu, E., Miller, R.J., Pu, K.Q., Arocena, P.C.: Data lake management: challenges and opportunities. Proc. VLDB Endow. **12**(12), 1986–1989 (2019)

9. Polo, L., Mínguez, I., Berrueta, D., Ruiz, C., Gómez, J.M.: User preferences in the web of data. Semant. Web **5**(1), 67–75 (2014)
10. Troumpoukis, A., Konstantopoulos, S., Charalambidis, A.: An extension of SPARQL for expressing qualitative preferences. In: d'Amato, C., et al. (eds.) ISWC 2017. LNCS, vol. 10587, pp. 711–727. Springer, Cham (2017). https://doi.org/10.1007/978-3-319-68288-4_42

# Stream Data Processing

# Feature Drift Detection in Evolving Data Streams

Di Zhao$^{(\boxtimes)}$ and Yun Sing Koh📷

School of Computer Science, The University of Auckland, Auckland, New Zealand
dzha866@aucklanduni.ac.nz, ykoh@cs.auckland.ac.nz

**Abstract.** Most current stream mining techniques can adapt to data distribution changes, known as concept drift. Common concept drift detectors focus on detecting and signaling drift when a model's prediction accuracy deteriorates. To allow us to evaluate a model's accuracy we need data with ground truth. We focus on feature drift that shifts the model's boundaries, and present a framework to detect feature drift without labels. The framework detects abrupt and gradual feature drift by two distance functions, Wasserstein and Energy, and discuss feature changes in the data stream. A less explored area is describing the changes in the data stream. Crucially, the ability to describe changes in the data stream would enable a better understanding of the changing dynamics in the relationships that take place over time. In particular, we seek to answer the following question: Whether the distribution changes of important features will also cause concept drift. Experimental results show that the proposed framework detects and describes the feature drift.

**Keywords:** Feature drift · Drift detection · Distance measures

## 1 Introduction

The field of data stream mining has been extensively researched in the last decade. In particular, robust models for incremental construction of predictive models and drift detection have been proposed. This research focuses on a specific type of concept drift, namely feature drift. In this research, Feature Drift is defined to the distribution of feature values changes affecting decision boundaries. We are interested in both detecting and interpreting the feature drift. While there is a range of techniques to detect features, including tree-based and ensemble-based algorithms, most of these techniques require datasets with class labels. Our proposed approach does not rely on the class labels of datasets.

Recently there have been researches that have started to explore the concept drift and its interpretations. Inspired by this new thread of work, we are interested in investigating feature drift, specifically detecting and describing the changes that occur within the data stream. We designed a feature drift detection and interpretation framework. Figure 1 shows the phases in our framework.

© Springer Nature Switzerland AG 2020
S. Hartmann et al. (Eds.): DEXA 2020, LNCS 12392, pp. 335–349, 2020.
https://doi.org/10.1007/978-3-030-59051-2_23

We assume the data arrives in batch and batch$_t$ becoming available at times-tamp $t$, then the batch is added to the sliding window. There are two sliding windows, the reference window, and the current window. The reference window represents the old data stream distribution, whereas the current window repre-sents the immediate data distribution. If the changes of distribution measured by the Wasserstein distance or Energy distance for features between two sliding windows exceed a predefined threshold, a feature drift is signaled. Once a drift is signaled, the framework evaluate feature drift based on different known fea-ture importance detection methods, namely, SHAP (SHapley Addictive Explana-tions) [10], LIME (Local Interpretable Model-agnostic Explanations) [13] and PI (Permutation Importance) [1]. Intuitively, we are detecting distribution changes of important features.

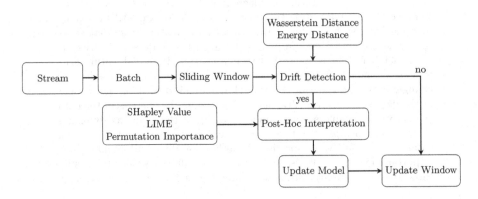

**Fig. 1.** Flow chart of feature drift detector

The advantage of using Wasserstein distance and Energy distance is that these distance measurement does not rely on the class labels and can be scaled up to higher dimensions easily. The Wasserstein distance can be scaled up to higher dimensions if the samples can be divided into a small number of clusters, and the computation of energy distance is linear with respect to the number of dimensions [9].

Our main contribution is developing a framework to detect abrupt and grad-ual feature drift and describe the distribution changes of the important features in the data stream. We investigate the effect of different distance measures for drift detection along with applying different feature importance ranking meth-ods. The rest of the paper is organized as follows. In Sect. 2, we review related works in the area of model and drift detection. Sect. 3 details our approach, detecting and interpreting feature drift. Our experimental results are presented in Sect. 4, and we conclude the paper in Sect. 5.

## 2    Related Work

In this section, we provide an overview of concept drift detection, abrupt and gradual drift, and feature drift techniques.

**Concept Drift Detection:** The concept drift detection problem has a statistical interpretation: given a sample of data, does this sample represent a single homogeneous distribution or is there some point in the data (*i.e.*, the concept change point) at which the data distribution has undergone a significant shift from a statistical of view? All concept change detection approaches in the literature formulate the problem from this viewpoint, but the models and the algorithms used to solve this problem differ significantly in their detail.

The problem of drift detection and analysis can be more formally framed. Let $P = (x_1, x_2, \ldots, x_m)$ and $Q = (x_{m+1}, \ldots, x_n)$ with $0 < m < n$ representing two samples of instances from a stream with population means $\mu_p$ and $\mu_q$ respectively. The drift detection problem can be expressed as testing the null hypothesis $H_0$ such that $\mu_p = \mu_q$, *i.e.* the two samples are drawn from the same distribution against the alternate hypothesis $H_1$ that they arrive from different distributions with $\mu_p \neq \mu_q$. In practice the underlying data distribution is unknown and a test statistic based on sample means needs to be constructed by the drift detector. If the null hypothesis is accepted incorrectly when a change has occurred then a false negative has taken place. On the other hand if the drift detector accepts $H_1$ when no change has occurred in the data distribution then a false positive has occurred. Since the population mean of the underlying distribution is unknown, sample means need to be used to perform the above hypothesis tests. The hypothesis tests can be restated as the following. We accept hypothesis $H_1$ whenever $Pr((|\hat{\mu}_p - \hat{\mu}_q|) \geq \epsilon) > \delta$, where $\delta$ lies in the interval $(0, 1)$ and is a parameter that controls the maximum allowable false positive rate, while $\epsilon$ is a function of $\delta$ and the test statistic used to model the difference between the sample means [6].

**Abrupt and Gradual Drift:** Webb et al. [19] defined the concept $a$ by $S_a$ and $E_a$ where $S_a$ indicates the start of the concept $a$ and $E_a$ indicates the end of the concept $a$. They also defined the magnitude of concept drift, which is the distance between the concepts at the start and the end of the period of drift. The magnitude of drift between times $t$ and $u$ is: $Magnitude_{t,u} = D(t, u)$. Abrupt drift occurs when a stream with the concept $a$ suddenly changes to the concept $a+1$. Given that $\delta$ is some nature number greater than 0 that defines the maximum duration over which abrupt drift can occur. The value of the constant $\delta$ depends on the context of data stream, and may be different for different streams. Abrupt drift between concepts $a$ and $a + 1$ can be defined as occurring when: $S_{a+1} - E_a \leq \delta$. Gradual drift may or may not be a steady progression from one concept towards another. Given that $\mu$ is a maximum allowed difference between concepts over the time period $v$ during a period of drift for the drift to be considered gradual, the gradual drift between concepts $a$ and $a + 1$ is defined as: $\forall_{t \in [E_a, S_{a+1-v}]} D(t, t + v) \leq \mu$.

**Feature Drift Detection**: Feature drift is defined to whenever a subset of features becomes relevant or irrelevant to the learning task [3,12]. Given a feature space $\mathcal{D}$ at a timestamp $t$, it is possible to select the ground-truth relevant subset $\mathcal{D}_t^* \subseteq \mathcal{D}$ such that $\forall \mathcal{D}_i \in \mathcal{D}_t^*$. A drift occurs if, at any two time instants $t_i$ and $t_j = t_i + \Delta$, $\mathcal{D}_{t_i}^* \neq \mathcal{D}_{t_j}^*$ holds. Most research for feature drift detection focus on feature selection [2,5,11,20].

## 3    Detecting and Describing Feature Drift

Here we use the notion of feature drift as a drift that occurs when the distribution of the important features changes affecting the decision boundary. Most of the feature drift detection research focus on detecting the distribution changes for a subset of feature space or the entire feature space. Moreover, most of these concept drift detection techniques do not consider the interpretation of the feature that has changed and is geared towards datasets that are fully labeled. Beyond that, the evaluation of the feature drift enable us to better understand the data stream. We proposed a framework that detects feature distribution changes independently by using distance measurements, which gives an in-depth understanding of how each feature affects the concept drift. The distance measurement allows the framework to detect the feature drift without class labels. In this section, we will first discuss the overview of our framework.

In Fig. 2, the data arrives in batches with a fixed batch size $b$, and $Batch_t$ indicates the batch of data at timestamp $t$. $Batch_\lambda$ indicates the last timestamp in which drift was detected and be initialized as 1. The reference window stores the data from $Batch_\lambda$ to $Batch_{t-1}$. The current window stores the most recent data in $Batch_t$. Distance measurements require the two windows with the same size. Therefore, we need to create a Sample Window $S$ from the reference window and compare it with the current window. The Sample Window randomly samples $b$ instances from the reference window where $b$ is the size for the current window since the current window stores one batch instances. If the distance of a feature $f$'s distribution between the sample window $S_f$ and the current window $C_f$ exceeds a predefined threshold $\delta$, Distance$(S_f, C_f) > \delta$, a feature drift is signaled. Once a feature drift is detected, the interpretation techniques are applied to investigate the features have changed. The interpretation technique interprets all

Data Stream

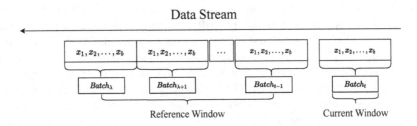

**Fig. 2.** Batch and sliding window

---

**Algorithm 1:** Feature Drift Detection and Interpretation

---

**Data:** Incoming examples in Batches $B$ with $k$ features, Reference Window $R$,
Sample Window $S$, Current Window $C$, distance threshold $\delta$, P-value $p$, Drift
Detected Flag $Drift\_Flag$;

$Drift\_Flag$ = False;

**for** $Batch\ B_i\ where\ i = 1,2,\ldots,t$ **do**

  **if** $B_1$ **then**

    extend($R, B_1$);

    train_model($R$);

  **else**

    extend($C, B_i$);

    $S$ = sample($R$);

    **for** $Feature\ f_j\ where\ j = 1, 2, \ldots, k$ **do**

      **if** $Distance(Sf_j, Cf_j) > \delta$ **then**

        Signal Drift in Feature $f_j$ at Batch $B_i$;

        $Drift\_Flag$ = True;

        **for** $Feature\ f_j\ where\ j = 1, 2, \ldots, k$ **do**

          $SI_j$ = Feature_Importance($S,\ f_j$);

          $CI_j$ = Feature_Importance($C,\ f_j$);

          **if** $Two\_Sample\_T\_test(SI_j, CI_j) < p$ **then**

            Signal Interpretation Shift in $f_j$;

    **if** $Drift\_Flag == True$ **then**

      clear($R$);

    extend($R,\ C$);

    clear($C$)

---

features for each data in the sample window and the current window. Then the mean and standard deviation of interpretations for each feature in both window is computed and compared by Two-Sample T-Test. $H_0$: $\forall f(\mu_{sf} = \mu_{cf})$, $H_1$: $\exists f(\mu_{sf} \neq \mu_{cf})$. Further details of distance measures are discussed in Sect. 3.1, and the interpretation of feature drift detection is discussed in Sect. 3.3. The pseudocode for our framework is described in Algorithm 1.

## 3.1 Distance Measurements

In this section we discuss the two different distance measurements used in our feature drift detection framework.

**Wasserstein Distance:** The Wasserstein distance is also known as Earth Mover's Distance (EMD). Intuitively, it calculates the minimum cost to transform from one distribution to another distribution. Rubner [15] defines EMD as follows. Assume you have two signatures $P$ and $Q$ where $P = \{(p_1, w_{p1}),$ $(p_2, w_{p2}),..., (p_n, w_{pm})\}$ and $Q = \{(q_1, w_{q1}), (q_2, w_{q2}),...,(q_n, w_{qn})\}$. The $p_i$ represent the mean of the $i_{th}$ cluster and $w_{pi} \in [0,1]$ is the weight of $i_{th}$ cluster. The sum weights of each cluster is 1 ($\sum_{i=1}^{n} w_{pi} = 1$).

Let $D = [d_{ij}]$ be the ground distance matrix where $d_{ij}$ is the ground distance between $p_i$ and $q_j$. We are trying to find a flow $F = [F_{ij}]$ which minimises the total cost: $COST(P, Q, F) = \sum_{i=1}^{m} \sum_{j=1}^{n} d_{ij} F_{ij}$ that satisfy the following constraints:

$$F_{ij} \geq 0, 1 \leq i \leq m, 1 \leq j \leq n$$

$$\sum_{j=1}^{n} F_{ij} \leq w_{p_i} 1 \leq i \leq m$$

$$\sum_{i=1}^{m} F_{ij} \leq w_{q_j} 1 \leq j \leq n \tag{1}$$

$$\sum_{i=1}^{m} \sum_{j=1}^{n} F_{ij} = \min \left( \sum_{i}^{m} w_{p_i}, \sum_{j}^{n} w_{q_j} \right)$$

The formula for computing EMD is:

$$EMD(P, Q) = \frac{\sum_{i=1}^{m} \sum_{j=1}^{n} d_{ij} F_{ij}}{\sum_{i=1}^{m} \sum_{j=1}^{n} F_{ij}} \tag{2}$$

**Energy Distance:** Energy Distance measures the distance between two objects. The intuition behind energy distance of two objects is inversely correlated to the distance between the two objects, thus, if the distance between two objects is 0, then the energy is 0. The proof provided by [17,18] shows that the Energy distance is zero only when the two distributions are identical. Assume we have batch of data $P$ and $Q$ with the size $n$ and $m$, respectively. The estimation of normalised energy distance is computed as following and $0 \leq H \leq 1$:

$$H(P, Q) = \frac{2A - B - C}{2A}$$

$$A = \frac{1}{mn} \sum_{i=1}^{n} \sum_{j=1}^{m} \|p_i - q_j\|$$

$$B = \frac{1}{n^2} \sum_{i=1}^{n} \sum_{j=1}^{n} \|p_i - p_j\| \tag{3}$$

$$C = \frac{1}{m^2} \sum_{i=1}^{m} \sum_{j=1}^{m} \|q_i - q_j\|$$

### 3.2   Examples for Distance Measures

Assume there is a reference window $R$ with two instances and a current window $C$ with two instances. The data only has one feature. Both $R$ and $C$ are divided into two equal weights batches, which means each batch contains only one instance. The computation of EMD distance based on Eq. 2 is shown in Example 1, and the computation of Energy Distance based on Eq. 3 is shown in Example 2.

*Example 1.*

$$\text{EMD}(R,C) = \frac{\sum_{i=1}^{2}\sum_{j=1}^{2} d_{ij}F_{ij}}{\sum_{i=1}^{2}\sum_{j=1}^{2} F_{ij}}$$

$$\text{EMD}(R,C) = \frac{d_{11}F_{11} + d_{12}F_{12} + d_{21}F_{21} + d_{22}F_{22}}{F_{11} + F_{12} + F_{21} + F_{22}}$$

$$F_{11} = F_{12} = F_{21} = F_{22}$$

$$\text{EMD}(R,C) = \frac{F_{11}(d_{11} + d_{12} + d_{21} + d_{22})}{4F_{11}}$$

$$\text{EMD}(R,C) = \frac{d_{11} + d_{12} + d_{21} + d_{22}}{4}$$

(4)

*Example 2.*

$$H(R,C) = \frac{2A - B - C}{2A}$$

$$A = \frac{1}{2^2}(||p_1 - q_1|| + ||p_1 - q_2|| + ||p_2 - q_1|| + ||p_2 - q_2||)$$

$$B = \frac{1}{2^2}(||p_1 - p_1|| + ||p_1 - p_2|| + ||p_2 - p_1|| + ||p_2 - p_2||)$$

$$C = \frac{1}{2^2}(||q_1 - q_1|| + ||q_1 - q_2|| + ||q_2 - q_1|| + ||q_2 - q_2||)$$

(5)

From Examples 1 and 2, alongside Eqs. 2 and 3, if the features' values are normalized, then $\forall i \forall j (0 \leq d_{ij} \leq 1, 0 \leq ||p_i - q_j|| \leq 1, 0 \leq ||p_i - p_j|| \leq 1, 0 \leq ||q_i - q_j|| \leq 1)$, therefore, the Wasserstein distance $\text{EMD}(R,C)$ and the Energy distance $H(P,Q)$ between two features are always between 0 and 1.

Figure 3 shows an example of how the distance changes when drift occurs. In the example, we inject drift in the middle of the data stream by swapping the values of two features specifically, Feature 29 and Feature 34 as noted in the figure. The threshold of Wasserstein distance is set to 0.05. Note that the distance of each feature remains below the threshold when a drift does not occur. At the drift point of at Batch9, we notice that the distance measure spikes above the 0.05 threshold which indicates a drift within the stream.

## 3.3    Interpretation of Feature Importance

Let $f$ be the original prediction model to be explained and $g$ the explanation model. Here, we focus on local methods designed to explain a prediction $f(x)$ based on a single input $x$, as proposed in LIME [13]. Explanation models often use simplified inputs $x'$ that map to the original inputs through a mapping function $x = h_x(x')$. Local methods try to ensure $g(z') \approx f(h_x(z'))$ whenever $z' \approx x'$.

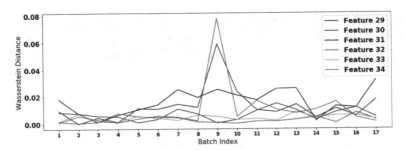

**Fig. 3.** Bank dataset example

**Additive Feature Attribution Methods** have an explanation model that is a linear function of binary variables:

$$g\left(z'\right) = \phi_0 + \sum_{i=1}^{M} \phi_i z_i' \tag{6}$$

where $z' \in \{0,1\}^M$, $M$ is the number of simplified input features, and $\phi_i \in R$.

**LIME:** The LIME method interprets individual model predictions based on locally approximating the model around a given prediction [13,14]. The local linear explanation model that LIME uses adheres to Eq. 6 exactly and is thus an additive feature attribution method. LIME refers to simplified inputs $x'$ as "interpretable inputs", and the mapping $x = h_x(x')$ converts a binary vector of interpretable inputs into the original input space. To find $\phi$, LIME minimizes the following objective function:

$$\xi = \underset{g \in \mathcal{G}}{\arg \min} L\left(f, g, \pi_{x'}\right) + \Omega(g) \tag{7}$$

Faithfulness of the explanation model $g(z')$ to the original model $f(h_x(z'))$ is enforced through the loss $L$ over a set of samples in the simplified input space weight by the local kernel $\pi_{x'}$. The $\Omega$ penalizes the complexity of $g$. Since in LIME $g$ follows Eq. 6 and $L$ is a squared loss, Eq. 7 can be solved using penalized linear regression.

**SHAP:** SHAP assigns each feature an importance value for a particular prediction [10,21]. It novel components include: (1) the identification of a new class of additive feature importance measures, and (2) theoretical results showing there is a unique solution in this class with a set of desirable properties.

**Permutation Importance:** The Permutation importance is a heuristic for correcting biased measures of feature importance [1]. The method normalizes the biased measure based on a permutation test and returns significant P-values for each feature. It can be used together with any learning method that assesses feature relevance, providing significance P-values for each predictor variable.

# 4   Experiments and Results

Five synthetic datasets with abrupt drift, five synthetic datasets with gradual drift, and two real datasets were evaluated, and the Hellinger distance was chosen as the benchmark since it has been widely used in concept drift detection in unlabelled data. The experiment results show that the Wasserstein distance and Energy distance outperform the Hellinger distance. Wasserstein distance and Energy distance has higher True Positive Rate and lower False Positive Rate than Hellinger distance. The evaluation results are shown in Table 3. Figure 5 shows that SHapley value outperform LIME and Permutation Importance. The evaluation results are shown in Table 4.

**Experimental Setting.** The classifier used in experiments is the random forest classifier with 20 base Hoeffding trees. The threshold for each dataset is shown in Table 2.

**Dataset Generators.** The synthetic datasets are used to test the sensitivity and robustness of the distance measurements, and the real datasets are used to observe how the framework performs in the real world. The synthetic datasets generated by the datasets chosen from UCI [7] and MOA [4]. The MOA synthetic datasets are generated by the Random RBF generator. The Random RBF generator generates a random radial basis function data.

For the UCI synthetic datasets, We followed the steps provide in [16] to manually inject drifts. The steps are as follows.

- **Task Conversion:** If the data has multi labels, it should be converted to binary labels since the interpretation of Additive Feature Attribution Methods only support binary labels.
- **One-Hot Encoding:** All categorical features with more than two values should be One-Hot encoded.
- **Normalization:** All numeric values should be normalized between $[0, 1]$ since it makes the range of distance threshold $\delta$ becomes $[0, 1]$.
- **Injecting Feature Drift:** The features are ranked by information gain metric firstly [8]. Then the values of top 25% important features are swapped since the top 25% important features has higher impact on the data distribution than the rest features [16]. The change point is set to the middle of data. Then instance from $X_0$ to $X_t$ stay unchanged and the instance from $X_{t+1}$ to $X_{last}$ with label 1 have their feature values swapped. For example: an instance has Features $F1$, $F2$, $F3$ with values $f1$, $f2$, $f3$ respectively. After swapping, $F1$ has value $f3$, $F2$ has value $f1$ and $F3$ has value $f2$.

Both abrupt and gradual drift is generated for each synthetic dataset. The width is set to 1 for abrupt drift, and 10% of total data for gradual drift where the width is the length of the window of the drift. There are two real datasets: Phishing dataset from UCI [7] and the ElecNorm from MOA [4]. The date feature of ElecNorm is removed. The details of data are shown in Table 1.

**Table 1.** Datasets details

| Dataset | # instances | # original feature | # converted feature | Batch size |
|---------|-------------|--------------------|--------------------|------------|
| Adult | 48842 | 14 | 65 | 2500 |
| Bank | 45211 | 16 | 48 | 2500 |
| Credit | 30000 | 24 | 26 | 1500 |
| Gamma | 19020 | 10 | 10 | 1000 |
| RBF | 100000 | 10 | 10 | 2500 |
| ElecNorm | 45312 | 8 | 8 | 1440 |
| Phishing | 11055 | 46 | 46 | 1000 |

**Table 2.** Threshold settings

| Dataset | Wasserstein | Energy | Hellinger |
|---------|-------------|--------|-----------|
| Adult_Abrupt | 0.04 | 0.05 | 25 |
| Adult_Gradual | 0.04 | 0.05 | 25 |
| Bank_Abrupt | 0.05 | 0.1 | 25 |
| Bank_Gradual | 0.05 | 0.1 | 25 |
| Credit_Abrupt | 0.045 | 0.1 | 19 |
| Credit_Gradual | 0.045 | 0.1 | 19 |
| Gamma_Abrupt | 0.04 | 0.06 | 7 |
| Gamma_Gradual | 0.04 | 0.06 | 7 |
| RBF_Abrupt | 0.05 | 0.1 | 12 |
| RBF_Gradual | 0.05 | 0.08 | 12 |
| ElecNorm | 0.24 | 0.24 | 21 |
| Phishing | 0.2 | 0.3 | 16 |

## 4.1 Feature Drift Detection Evaluation

To evaluate the performance of distance measurements on drift detection, we describe several metrics used in experiment results. For the synthetic datasets, average accuracy, True Positive Rate (TPR), False Positive Rate (FPR), and Delay are applied. For the real datasets, only average accuracy and the number of alarms are evaluated since we do not know the change point of real datasets.

Figure 4 and Table 3 show Wasserstein and Energy distances outperform Hellinger distance in the synthetic datasets. Table 3 shows Wasserstein and Energy distance has higher True Positive rates, lower False Positive rates, and lower delay than the Hellinger distance in the synthetic datasets. Table 3 also shows that all three distance measurements perform slightly worse in gradual drift datasets than abrupt drift datasets. The gradual drift is more difficult to detect than the abrupt drift since the gradual drift does not always be a steady progression from one concept towards another. However, the framework detects

**Table 3.** Evaluation of synthetic datasets

| Dataset | Distance | TPR | FPR | Delay |
|---|---|---|---|---|
| Adult_Abrupt | Wasserstein | 0.98 ± 0.14 | 0.08 ± 0.04 | 629.00 ± 353.55 |
| | Energy | 1.00 ± 0.00 | 0.13 ± 0.05 | 579.00 ± 0.00 |
| | Hellinger | 0.26 ± 0.44 | 0.51 ± 0.09 | 2429.00 ± 1107.72 |
| Adult_Gradual | Wasserstein | 0.78 ± 0.42 | 0.09 ± 0.04 | 1129.00 ± 1046.13 |
| | Energy | 1.00 ± 0.00 | 0.12 ± 0.05 | 579.00 ± 0.00 |
| | Hellinger | 0.20 ± 0.40 | 0.49 ± 0.11 | 2579.00 ± 1010.15 |
| Bank_Abrupt | Wasserstein | 1.00 ± 0.00 | 0.001 ± 0.01 | 2394.00 ± 0.00 |
| | Energy | 1.00 ± 0.00 | 0.00 ± 0.00 | 2394.00 ± 0.00 |
| | Hellinger | 0.60 ± 0.49 | 0.56 ± 0.14 | 3394.00 ± 1237.18 |
| Bank_Gradual | Wasserstein | 1.00 ± 0.00 | 0.00 ± 0.00 | 2394.00 ± 0.00 |
| | Energy | 1.00 ± 0.00 | 0.00 ± 0.00 | 2394.00 ± 0.00 |
| | Hellinger | 0.54 ± 0.50 | 0.57 ± 0.11 | 3544.00 ± 1258.64 |
| Credit_Abrupt | Wasserstein | 1.00 ± 0.00 | 0.05 ± 0.04 | 1499.00 ± 0.00 |
| | Energy | 1.00 ± 0.00 | 0.00 ± 0.00 | 1499.00 ± 0.00 |
| | Hellinger | 0.98 ± 0.14 | 1.00 ± 0.13 | 1529.00 ± 212.13 |
| Credit_Gradual | Wasserstein | 1.00 ± 0.00 | 0.05 ± 0.05 | 1499.00 ± 0.00 |
| | Energy | 1.00 ± 0.00 | 0.00 ± 0.00 | 1499.00 ± 0.00 |
| | Hellinger | 1.00 ± 0.00 | 0.99 ± 0.02 | 1499.00 ± 0.00 |
| Gamma_Abrupt | Wasserstein | 0.96 ± 0.20 | 0.06 ± 0.00 | 530.00 ± 197.95 |
| | Energy | 1.00 ± 0.00 | 0.06 ± 0.01 | 490.00 ± 0.00 |
| | Hellinger | 1.00 ± 0.00 | 0.92 ± 0.06 | 490.00 ± 0.00 |
| Gamma_Gradual | Wasserstein | 0.72 ± 0.45 | 0.06 ± 0.00 | 770.00 ± 453.56 |
| | Energy | 1.00 ± 0.00 | 0.06 ± 0.01 | 490.00 ± 0.00 |
| | Hellinger | 1.00 ± 0.00 | 0.92 ± 0.05 | 490.00 ± 0.00 |
| RBF_Abrupt | Wasserstein | 1.00 ± 0.00 | 0.00 ± 0.00 | 2499.00 ± 0.00 |
| | Energy | 1.00 ± 0.00 | 0.00 ± 0.00 | 2499.00 ± 0.00 |
| | Hellinger | 0.00 ± 0.00 | 0.25 ± 0.12 | 4849.06 ± 599.50 |
| RBF_Gradual | Wasserstein | 0.88 ± 0.33 | 0.00 ± 0.01 | 2549.10 ± 353.54 |
| | Energy | 0.98 ± 0.14 | 0.00 ± 0.00 | 2549.00 ± 353.56 |
| | Hellinger | 0.00 ± 0.00 | 0.23 ± 0.15 | 4849.06 ± 599.50 |

the gradual drift precisely by comparing the distribution changes batch by batch. The evaluation results show that a large batch size can handle the gradual drift more precisely but takes more computation cost and more storage space.

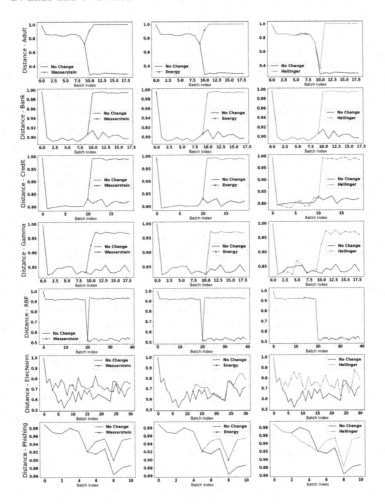

**Fig. 4.** Evaluation results: average accuracy vs batch size

## 4.2 Feature Drift Interpretation Evaluation

Since we have ground truth for the feature drifts in the synthetics datasets, we use the ground truth to verify whether the interpretation methods can correctly describe the features drift. Figure 5 shows that SHapley Value outperforms LIME and Permutation Importance. The x-axis represents all features in the datasets. For example, the Adult dataset has 65 features. The x-axis represents all features in the datasets. For example, the Adult dataset has 65 features. The cyan dots represent the features we manually swap its value to inject feature drift. The red dots represent the features with feature drift detected by Wasserstein distance. The yellow dots, green dots, and blue dots represent the features with feature importance shifts identified by PI, LIME, and SHapley values, respectively.

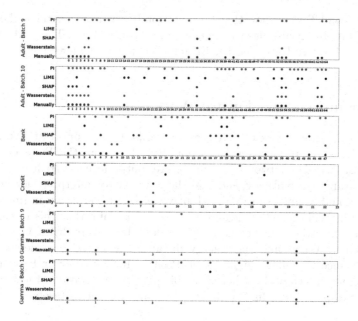

**Fig. 5.** Shifts in feature importance (Color figure online)

**Table 4.** Interpretation evaluation

| Datasets | Interpretation | TPR | FPR |
|---|---|---|---|
| Adult Batch$_9$ | PI | $0.900 \pm 0.044$ | $0.571 \pm 0.091$ |
| | LIME | $0.000 \pm 0.000$ | $0.023 \pm 0.029$ |
| | SHAP | $0.100 \pm 0.044$ | $0.000 \pm 0.000$ |
| Adult Batch$_{10}$ | PI | $0.900 \pm 0.044$ | $0.810 \pm 0.029$ |
| | LIME | $0.000 \pm 0.000$ | $0.379 \pm 0.029$ |
| | SHAP | $0.156 \pm 0.053$ | $0.004 \pm 0.009$ |
| Bank | PI | $0.833 \pm 0.056$ | $0.667 \pm 0.110$ |
| | LIME | $0.000 \pm 0.000$ | $0.143 \pm 0.017$ |
| | SHAP | $0.158 \pm 0.047$ | $0.000 \pm 0.000$ |
| Credit | PI | $0.857 \pm 0.269$ | $0.805 \pm 0.168$ |
| | LIME | $0.000 \pm 0.000$ | $0.222 \pm 0.204$ |
| | SHAP | $0.214 \pm 0.101$ | $0.000 \pm 0.000$ |
| Gamma Batch$_9$ | PI | $0.967 \pm 0.105$ | $0.443 \pm 0.081$ |
| | LIME | $0.000 \pm 0.000$ | $0.188 \pm 0.164$ |
| | SHAP | $0.333 \pm 0.000$ | $0.000 \pm 0.000$ |
| Gamma Batch$_{10}$ | PI | $0.967 \pm 0.105$ | $0.971 \pm 0.060$ |
| | LIME | $0.000 \pm 0.000$ | $0.410 \pm 0.036$ |
| | SHAP | $0.033 \pm 0.105$ | $0.043 \pm 0.136$ |

Figure 5 and Table 4 show that Permutation Importance has the highest True Positive Rate, but it also has the highest False Positive Rate. The LIME has the lowest True Positive Rate, and the SHapley Value has the lowest False Positive Rate.

## 5    Conclusions and Future Work

We proposed a framework to detect and describe the feature drift. We use two distance measures proposed for concept drift detection: the Wasserstein and Energy distance. These two distance measures outperforms current baseline distance measures on synthetic and real datasets. Three interpretation methods are applied and compared: LIME, SHapley value, and Permutation Importance. While the individual techniques are not novel, we claim that this research is a preliminary investigation into how we can describe feature drifts. Overall both Wasserstein distance, and the Energy distance was effective in detecting feature drift since they have higher True Positive Rate and lower False Positive Rate than the benchmark. When describing the interpretation of feature value changes, SHapley value had a slight advantage for the datasets used in our research.

As future work, we will focus on constructing a unified threshold for the distance measures which allows the threshold of distance measures independent to the dataset. Moreover, further investigation is needed to develop an interpretation system for feature drifts.

## References

1. Altmann, A., Toloşi, L., Sander, O., Lengauer, T.: Permutation importance: a corrected feature importance measure. Bioinformatics **26**(10), 1340–1347 (2010)
2. Barddal, J.P., Enembreck, F., Gomes, H.M., Bifet, A., Pfahringer, B.: Merit-guided dynamic feature selection filter for data streams. Expert Syst. Appl. **116**, 227–242 (2019)
3. Barddal, J.P., Gomes, H.M., Enembreck, F., Pfahringer, B.: A survey on feature drift adaptation: definition, benchmark, challenges and future directions. J. Syst. Softw. **127**, 278–294 (2017)
4. Bifet, A., Holmes, G., Kirkby, R., Pfahringer, B.: MOA: massive online analysis. J. Mach. Learn. Res. **11**(May), 1601–1604 (2010)
5. Cohen, A.M., Bhupatiraju, R.T., Hersh, W.R.: Feature generation, feature selection, classifiers, and conceptual drift for biomedical document triage. In: TREC (2004)
6. Ditzler, G., Roveri, M., Alippi, C., Polikar, R.: Learning in nonstationary environments: a survey. IEEE Comput. Intell. Mag. **10**(4), 12–25 (2015)
7. Dua, D., Graff, C.: UCI machine learning repository (2017). http://archive.ics.uci.edu/ml
8. Duch, W., Wieczorek, T., Biesiada, J., Blachnik, M.: Comparison of feature ranking methods based on information entropy. In: 2004 IEEE International Joint Conference on Neural Networks (IEEE Cat. No. 04CH37541), vol. 2, pp. 1415–1419. IEEE (2004)

9. Goldenberg, I., Webb, G.I.: Survey of distance measures for quantifying concept drift and shift in numeric data. Knowl. Inf. Syst. **60**, 591–615 (2018)

10. Lundberg, S.M., Lee, S.I.: A unified approach to interpreting model predictions. In: Advances in Neural Information Processing Systems, pp. 4765–4774 (2017)

11. Méndez, J.R., Fdez-Riverola, F., Iglesias, E.L., Díaz, F., Corchado, J.M.: Tracking concept drift at feature selection stage in SpamHunting: an anti-spam instance-based reasoning system. In: Roth-Berghofer, T.R., Göker, M.H., Güvenir, H.A. (eds.) ECCBR 2006. LNCS (LNAI), vol. 4106, pp. 504–518. Springer, Heidelberg (2006). https://doi.org/10.1007/11805816_37

12. Nguyen, H.-L., Woon, Y.-K., Ng, W.-K., Wan, L.: Heterogeneous ensemble for feature drifts in data streams. In: Tan, P.-N., Chawla, S., Ho, C.K., Bailey, J. (eds.) PAKDD 2012. LNCS (LNAI), vol. 7302, pp. 1–12. Springer, Heidelberg (2012). https://doi.org/10.1007/978-3-642-30220-6_1

13. Pedersen, T.L., Benesty, M.: Lime: local interpretable model-agnostic explanations. R Package version 0.4 1 (2018)

14. Ribeiro, M.T., Singh, S., Guestrin, C.: Model-agnostic interpretability of machine learning. arXiv preprint arXiv:1606.05386 (2016)

15. Rubner, Y., Tomasi, C., Guibas, L.J.: The earth mover's distance as a metric for image retrieval. Int. J. Comput. Vis. **40**(2), 99–121 (2000)

16. Sethi, T.S., Kantardzic, M.: On the reliable detection of concept drift from streaming unlabeled data. Expert Syst. Appl. **82**, 77–99 (2017)

17. Székely, G.J.: E-statistics: the energy of statistical samples. Bowling Green State University, Department of Mathematics and Statistics Technical report, vol. 3, no. 05, pp. 1–18 (2003)

18. Székely, G.J., Rizzo, M.L.: A new test for multivariate normality. J. Multivar. Anal. **93**(1), 58–80 (2005)

19. Webb, G.I., Hyde, R., Cao, H., Nguyen, H.L., Petitjean, F.: Characterizing concept drift. Data Min. Knowl. Disc. **30**(4), 964–994 (2016). https://doi.org/10.1007/s10618-015-0448-4

20. Yuan, L., Pfahringer, B., Barddal, J.P.: Addressing feature drift in data streams using iterative subset selection. ACM SIGAPP Appl. Comput. Rev. **19**(1), 20–33 (2019)

21. Zheng, S., van der Zon, S.B., Pechenizkiy, M., de Campos, C.P., van Ipenburg, W., de Harder, H.: Labelless concept drift detection and explanation. In: NeurIPS 2019 Workshop on Robust AI in Financial Services: Data, Fairness, Explainability, Trustworthiness, and Privacy (2019)

# An Efficient Segmentation Method: Perceptually Important Point with Binary Tree

Qizhou Sun[ID] and Yain-Whar Si[✉][ID]

Department of Computer and Information Science, University of Macau,
Avenida da Universidade, Macau, Taipa, China
{yb87460,fstasp}@umac.mo

**Abstract.** Segmentation is an important preprocessing step for pattern classification in financial time series. In this paper, we propose a novel segmentation method called Perceptually Important Point with Binary tree (PIP-Btree) for efficient preprocessing of financial time series for classifying chart patterns. PIP-Btree takes advantage of a standard binary tree for improving the efficiency without compromising the accuracy of original Perceptually Important Point (PIP) method. Besides, attribute parameters of PIP-Btree support self-updating when a new data point arrives in a streaming time series. In the experiments, efficiency of PIP-Btree is compared to other segmentation methods. Accuracy of PIP-Btree method is also evaluated for financial chart pattern classification after it is integrated with a rule-based pattern matching approach. Experiment results reveal that PIP-Btree achieves best score in both efficiency and accuracy performance.

**Keywords:** Financial time series · Segmentation · Perceptually Important Point · Binary tree · Chart patterns

## 1 Introduction

Majority of the economic data are often represented in the form of time series. Stock time series is one of the most well-known representations in Financial community. One of the major applications of time series analysis is for predicting the future value based on the historical data. Historical stock data is often used to predict the future price through various methods including identification of chart patterns [1,15].

Due to the randomness and stochastic factors of the stock time series, researchers have proposed various approaches to extract important points which maintain the characteristics of the original time series. Piecewise Linear Approximation (PLA) [5,7], Perceptually Important Points (PIP) [2], Turning Points (TP) [11] methods are designed to extract important points from the original time series. Piecewise Aggregate Approximation (PAA) [6] uses the points with

S. Hartmann et al. (Eds.): DEXA 2020, LNCS 12392, pp. 350–365, 2020.
https://doi.org/10.1007/978-3-030-59051-2_24

the average value of each segment as the important points. However, they all have their own drawbacks. PLA, PIP and TP have low efficiency due to their time complexity $O(n^2)$. In addition, according to [13], PAA has lower classification accuracy when it is used to pre-process the time series. Besides, in streaming time series generated by stock markets, the number of data points can increase exponentially.

During the segmentation process, a sliding window is commonly used to extract the important points. An example of a sliding window shifting backward or forward, expansion or shrinking on the time axis is illustrated in Fig. 1.

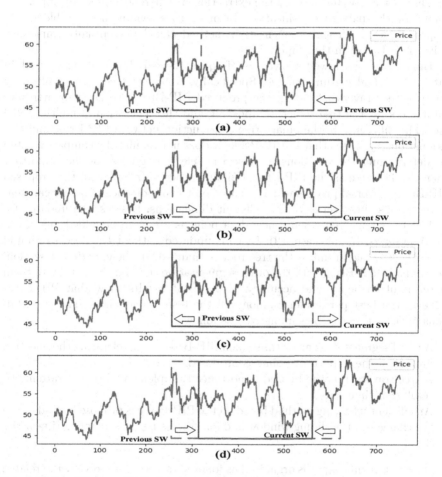

**Fig. 1.** Sliding Window (SW): (a) shifting backward (b) shifting forward (c) expansion (d) shrinking

In PLA, PIP and TP methods, any changes in the sliding window (shifting, expansion and shrinking) during the segmentation process force these algorithms to redo the whole procedure again for the new situation. In this case, majority of previously computed points are recalculated again in these algorithms. Such calculations are often unnecessary and time consuming. The overhead incurred in such recalculation is even more serious when streaming time series are required to be segmented for pattern matching in real-time situations. Besides, according to [14], each chart pattern requires different number of important points during the classification process. For example, the minimum number of the important points for classifying 'Head and Shoulder' pattern is 7. However, the minimum number of the important points for classifying 'Triangles, Symmetrical' is 6. Therefore, a new round of segmentation is triggered not only by the changes of the sliding window but also by the extraction of different number of important points for the pattern classification. In order to alleviate these problems, we propose a novel efficient segmentation method called Perceptually Important Point (PIP-Btree) in this paper.

Binary tree is a well-known and frequently-used data structure used in database and algorithm studies. Specifically, a binary tree can significantly decrease the traversal time. In the proposed PIP-Btree, we use four attribute parameters for the tree structure updating and the generation of multiple PIP lists. The difference in the binary tree construction between PIP-Btree and traditional binary tree is that in PIP-Btree, it does not require the comparison step for determining the positions because the selection process of the important points is the same as the PIP-Btree building process. In addition, the proposed PIP-Btree traversal process which is based on the importance of points can generate various times series of the different dimensions. Besides, the update of a PIP-Btree is triggered when a new data point is about to be added in streaming time series. In this case, PIP-Btree is updated without the need to rebuild the entire tree again unless the tree root is changed. In the experiment section, we compare PIP-Btree with the other segmentation methods and evaluate them in terms of efficiency and accuracy. Experiment results show that PIP-Btree achieves the best performance among all the tested algorithms. The contributions of the paper can be summarized as follows:

1. A novel segmentation algorithm called PIP-Btree for accelerating the selection process of the important points is proposed.
2. A tree traversal method based importance is implemented to generate multi-resolution time series.
3. An efficient updating method based on the PIP-Btree is proposed for adapting the changes of the sliding window and for processing the streaming time series data.

The rest of this paper is organized as follows. In Sect. 2, we review the related work. In Sect. 3, we briefly introduce the preliminaries regarding the original PIP and the relationship between PIP and Binary tree. The proposed PIP-Btree algorithm is detailed in Sect. 4. In Sect. 5, experiment results are given. Conclusion is drawn in Sect. 6.

# 2  Related Work

In 2012, Rakthanmanon et al. proposed an approach called UCR-suite [10] for finding a particular motif pattern from 20,145,000 ECG data points in 35 s. UCR-suite is based on Dynamic time warping (DTW) and achieves fast performance on processing large dataset. In the context of financial chart pattern matching, one of the main problems is about classification of the patterns in a highly efficient way from the streaming time series data. Until now, majority of the efficient online chart pattern classification methods are designed based on the combination of segmentation pre-processing and rule-based pattern matching algorithms. Classification methods based on machine leaning such as Neural Network (NN) and Support Vector Machine (SVM) models impose a significant training step and they are often only effective for fixed-length patterns [12].

Segmentation methods are widely used in financial time series analysis. Well-known segmentation methods include PIP [2], PLA [7], and TP [11]. In [13], Wan and Si have analyzed the accuracy comparison of chart pattern classification methods based on segmentation and concluded that PIP achieved better performance than other segmentation methods such as PLA, PAA and TP. PAA has lower computation cost, but has lowest classification accuracy. The original PLA, PIP and TP all face the efficiency problem. And they were improved from the following researches.

PLA was proposed by Keogh et al. as an online segmentation method in [7] in 2001. PLA includes two approximating line methods: linear interpolation and linear regression. Linear interpolation tends to give a smooth shape of the segmented time series by aligning the endpoints during segmentation. Linear regression will produce a disjointed result which has the minimum residual error due to the least square function. In 2018, Hu et al. [5] proposed an approach called multi-resolution piecewise linear representation-important data point (MPLR-IDP). MPLR utilized the tree structure to accelerate the speed of PLA.

In [11], Si et al. proposed an approach called optimal binary search tree (OBST) by assigning distance value to the turning points (TP) and storing them into a heap according to their distance. OBST exploits the advantages of the heap data structure for storing and retrieving TP. In 2008, Fu et al. proposed an approach called specialized binary tree of PIP (PIP-SBT) [4]. PIP-SBT adopts the end point of the original time series as the root of the tree and the starting point as the left child of the tree root. PIP-SBT ensures that the resulting tree is not a balanced tree from the beginning of the segmentation process. However, the tree construction process involves comparisons of the distance values such as Vertical Distance (VD), Euclidean Distance (ED), and Perpendicular Distance (PD) from the root to leaf nodes in downward direction in each round. The updating process of PIP-SBT also includes locating the affected tree nodes. These comparison and identification steps could effect the overall efficiency of the method. In order to overcome these short-comings, we adopt a standard binary tree in the proposed algorithm.

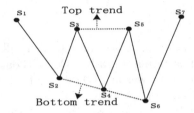

**Fig. 2.** "Broadening Formation, right-angled and descending" pattern. The dash lines are the top trend line and bottom trend line.

Segmented time series are commonly used for chart pattern classification in technical analysis. Comprehensive studies of the known chart patterns in financial domain were reported in [1]. In addition, definitions of the known chart patterns from [1] are translated into the first-order logic representation from the natural language in [14]. Such formal representation allows researchers and practitioners to easily adopt them for further analysis and application in real world systems. For example, Fig. 2 shows a "Broadening Formation, right-angled and descending" (BFRAD) pattern after the segmentation. Some of the distinguishable characteristics of BFRAD are the horizontal top trend line and the slope-down bottom trend line. According to [14], the first-order logic representation of BFRAD pattern can be defined as:

$$number\ of\ Points(Time\ series) >= 6\ \wedge$$
$$number\ of\ Points(Top\ trend) >= 2\ \wedge$$
$$number\ of\ Points(Bottom\ trend) >= 2\ \wedge$$
$$horizontal(Top\ trend) \wedge slopedown(Bottom\ trend)$$

Next, the first-order logic representation of each pattern can be transformed into rules for the pattern matching. In this paper, a rule-based (RB) pattern matching algorithm is implemented in Sect. 5 for evaluating the accuracy of the proposed approach.

## 3    Preliminary

PIP algorithm was first introduced by Chung [2] in 2001. In [3], Fu et al. have compared the residual errors between the original sequence and the segmented sequences generated by PIP-ED (PIP with Euclidean Distance), and PIP-VD (PIP with Vertical Distance), PIP-PD (PIP with Perpendicular Distance). Their analysis showed PIP with VD outperformed PIP with ED and PIP with PD.

In this section, we describe the definitions of a time series ($P$), PIP lists ($S$), and the index translation function from $S$ to $P$. We also provide a detailed analysis of PIP algorithm and uncover the relationship between PIP and the binary tree. The observed relationship is illustrated in Fig. 3.

**Definition 1.** *A Time Series is defined as* $P = \{(t_1, p_1), ..., (t_i, p_i), ..., (t_m, p_m)\}$, *where* $t_i \in T$, $T$ *is a set of discrete time points,* $1 \leq i \leq m$, $2 \leq m$.

**Definition 2.** *A PIP list* $S$ *of* $P$ *can be defined as* $S = \{s_1, s_2, ..., s_j, ..., s_n\}$, *where* $1 \leq j \leq n, 2 \leq n, t_1 = s_1.t, t_m = s_n.t$.

**Function 1.** $Tindex(s)$ *returns the corresponding data point index in* $T$ *for the salient point* $s$ *of* $S$.

The main idea of PIP process is to select a salient point from $P$, after which PIP algorithm continues to select the salient points one by one from the segments which are divided by the previous salient points. Figure 3 shows the process of extracting 8 PIPs. Initially, PIP list only contains the first point $(t_1, p_1)$ and last point $(t_m, p_m)$ of $P$ which depicts in blue. The initial PIP list can be donated as $\{s_1, s_2\}$, where $s_1 = (t_1, p_1)$, $s_2 = (t_m, p_m)$. Next, the first salient point $s_{x1}$ (except the $s_1$ and $s_2$), which has the maximum VD ($d_1$ is the distance between $s_{x1}$ and $s'_{x1}$) to the dash line $(t_1, t_m)$ in Fig. 3(1), is selected and inserted into the PIP list as $\{s_1, s_2, s_3\}$, where $s_1 = (t_1, p_1)$, $s_2 = s_{x1}$ and $s_3 = (t_m, p_m)$. The index $x_1$ of $s_{x1}$ represents the sequential order of this salient point in the final result of $S$. It will be determined after producing all the salient points. $s_{x1}$ represents both the first selected salient point in $S$ and the first salient node in binary tree which is used for discussion in this paper.

In the second round, $s_{x2}$ is chosen as the second salient point since it has the maximum VD ($d_2$) in the segment $(Tindex(s_{x1}), t_m)$ of $T$ comparing to the points in their segments. Next, $s_{x2}$ is inserted to the right of $s_{x1}$, that is $\{s_1, s_{x1}, s_{x2}, s_4\}$, where $s_4 = (t_m, p_m)$. In third round, $s_{x3}$ with maximum VD ($d_3$) is inserted to the left of $s_{x3}$. Similarly, $s_{x4}$ with $d_4$, $s_{x5}$ with $d_5$, $s_{x6}$ with $d_6$ are selected and inserted into the PIP list as $\{s_1, s_{x5}, s_{x3}, s_{x1}, s_{x6}, s_{x2}, s_{x4}, s_6\}$ where $s_6 = (t_m, p_m)$. Finally, 6 salient points are selected and the list of 8 PIPs is produced. At the same time, these salient points from Fig. 3(1) to Fig. 3(6) form an hierarchical structure. When these salient points are connected with respect to the hierarchical structure, a binary tree is generated.

**Table 1.** Attributes of a node in a PIP-Btree

| Attributes | Description |
|---|---|
| *Time* | It stores the time stamp of a salient point |
| *Value* | It stores the price (closing price) for financial time series |
| *Dist* | Distance, such as VD, ED or PD |
| *SPR* | Salient point rank. The initial value is 0 |
| *SSP* | Segment start point |
| *SEP* | Segment end point |
| *l_child* | The left child |
| *r_child* | The right child |

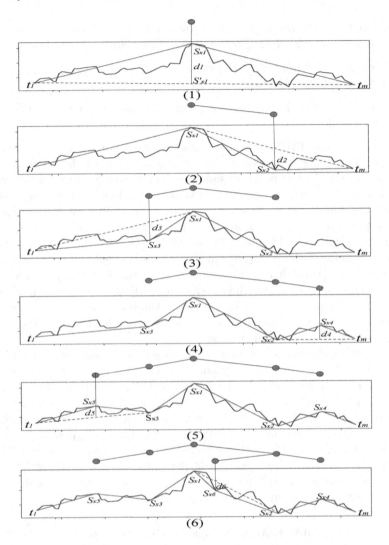

**Fig. 3.** Relationship between PIP and binary tree. The green circles are the tree nodes. The blue line is the raw time series. The red line is the PIP sequence. (Color figure online)

## 4    PIP-Btree

From the above analysis, we also found that the $d_3$ of $s_{x3}$ in Fig. 3(3) was calculated in the second round of Fig. 3(2). If $s_{x3}$ and $d_3$ are kept in the second round, $s_{x3}$ will be selected directly without repeating the computation in the third round. Based on this observation, we propose an approach called PIP-Btree which is designed to avoid repetitive computation. PIP-Btree is also able to generate multiple PIP lists and extend the time series without reconstructing a new tree. The construction, generation of the multi-resolution PIP lists, and updating of a PIP-Btree are detailed in the following subsections.

The attributes of the node used in a PIP-Btree are first defined in Table 1. *Time* and *Value* are used to store key information of an important point. *Dist* represents the distance values such as VD, ED or PD. *SPR* is the rank of a salient point and indicates the level of importance. *SPR* with value 0 means no rank. *SPR* with value 1 means the current node is the most important node in the tree. The larger the *SPR*, the less important the node. *SSP* and *SEP* store the start and end points of the segment in which the node is selected from. *l_child* and *r_child* refer to the left and right child of the node.

## 4.1   Construction

The process of PIP-Btree construction is depicted in Fig. 4. In the first round, $s_{x1}$ with maximum VD is selected as the root of the tree. The *SPR* of $s_{x1}$ is set to 1 (depicted in light-red circle of Fig. 4(1)). Next, its two children are created by selecting the data points with maximum VD from the two segments which are resulted from selecting $s_{x1}$. *SPRs* of two children are 0 initially. In the second round, the right child of $s_{x1}$ is selected as $s_{x2}$ since *Dist* of the right child is larger than the left one. In Fig. 4(2), *SPR* of $s_{x2}$ is set to 2 and two children are created. In the third round, the left child of $s_{x1}$ is selected as $s_{x3}$ due to its largest *Dist* in comparison to other nodes with *SPR* 0. Similarly, $s_{x4}$, $s_{x5}$ and $s_{x6}$ are selected and their children are created in Fig. 4.

During the construction, the attribute *SPR*, *SSP*, *SEP* of the important nodes are set in each round (see Table 2). At the same time, *Dist* is set to the maximum VD of the point in a *T*-segment (*SSP*, *SEP*) through Eq. 1, 2. The pseudo code of PIP-Btree construction is given in Algorithm 1. According to Algorithm 1, the time complexity of PIP-Btree construction is $O(m)$ for the best case and $O(mn)$ for the worst case, where $m$ is the length of $P$ or $T$ and $n$ is the length of $S$.

$$VD((t_c, p_c)) = \|SSP.p - p_c + (SEP.p - SSP.p)\frac{t_c - SSP.t}{SEP.t - SSP.t}\|, \tag{1}$$

$$SSP.t < c < SEP.t.$$

$$Dist = \max(\{..., VD((t_c, p_c)), ...\}) \tag{2}$$

## 4.2   Generation of Multi-resolution PIP Lists

Depending of the size and shape of each pattern, different number of important points are required for the pattern classification task. Therefore, to be able to classify patterns of various types, it is essential for generating multi-resolution PIP lists from a given time series.

After the construction of PIP-Btree, the tree stores important nodes which are identified by *SPR*. Important nodes can be retrieved from the PIP-Btree by simple traversal and sorted in ascending order of *SPR* as shown in Fig. 5(a).

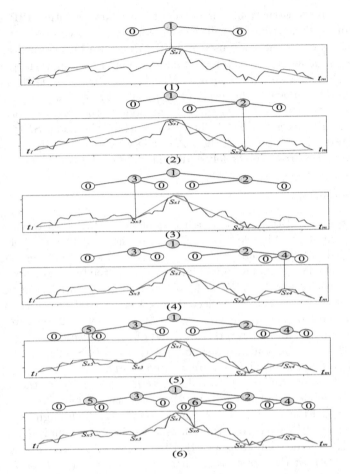

**Fig. 4.** An example of PIP-Btree construction. $SPR$ is depicted in the circle. The light-red circles indicates the important points. The white circles are candidates for the next round of important point selection. (Color figure online)

**Table 2.** $SPR, SSP, SEP$ of salient nodes in PIP-Btree

| Round | Sailent node | $SPR$ | $SSP$ | $SEP$ |
|---|---|---|---|---|
| 1 | $s_{x1}$ | 1 | $(t_1, p_1)$ | $(t_m, p_m)$ |
| 2 | $s_{x2}$ | 2 | $s_{x1}$ | $(t_m, p_m)$ |
| 3 | $s_{x3}$ | 3 | $(t_1, p_1)$ | $s_{x1}$ |
| 4 | $s_{x4}$ | 4 | $s_{x2}$ | $(t_m, p_m)$ |
| 5 | $s_{x5}$ | 5 | $s_{x1}$ | $s_{x3}$ |
| 6 | $s_{x6}$ | 6 | $s_{x1}$ | $s_{x2}$ |

---

**Algorithm 1.** Construction pseudo code of PIP-Btree

---

**Input:** $P$, lengh $n$ of $S$
**Output:** PIP-Btree

1  Create Root from the point with maximum VD in $P$;
2  Set Root.$SPR$ to 1, Root.$SSP$ to start point of $P$, Root.$SEP$ to end point of $P$;
3  Create Root.$l\_child$ from the point with maximum VD in the $T$-segment (Root.$SSP$, Root) ;
4  Create Root.$r\_child$ from the point with maximum VD in the $T$-segment (Root, Root.$SEP$) ;
5  **for** $spr \in [2, n-2]$ **do**
6     Select a node with maximum $Dist$ as Important Node(SN) among nodes with $SPR$ 0;
7     Set SN.$SPR$ to spr;
8     Create SN.$l\_child$ from the point with maximum VD in the $T$-segment (SN.$SSP$, SN);
9     Create SN.$r\_child$ from the point with maximum VD in the $T$-segment (SN, SN.$SEP$);
10  **end**

---

Multi-resolution PIP lists can be obtained by selecting the first $k$ ($k \leq n-2$) tree nodes for multiple times. The selected $k$ tree nodes are then sorted in the ascending order of $Time$. The sorted result is assembled into a segment using $SSP$ and $SEP$ for generating a PIP list. Figure 5 depicts the first 4 and 6 tree nodes are selected and assembled using $SSP$ and $SEP$ to form two PIP lists.

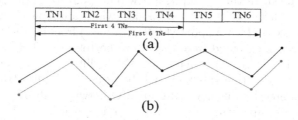

**Fig. 5.** Generating multi-resolution PIP lists. (a) Tree nodes ($TN_i$, $i = SPR$) in the ascending order of $SPR$. (b) Left most 6 TNs are selected for the first PIP list shown in black color. Left most 4 TNs are selected for the second PIP list shown in red color. (Color figure online)

### 4.3   Updating of PIP-Btree

Updating of PIP-Btree is triggered when a new data point arrives or a node from the tree is deleted. PIP-Btree updating can realize the shifting, expansion and shrinking of the sliding window as depicted in Fig. 1. The process of PIP-Btree updating has four steps: (a) Affecting (b) Pruning, (c) Resetting $SPR$ to 0, and (d) Re-ranking and Regenerating.

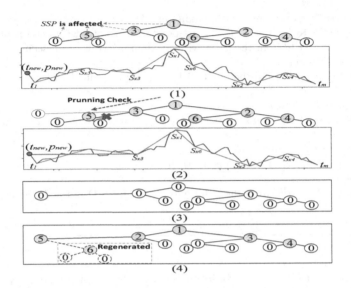

**Fig. 6.** A new data point $(t_{new}, p_{new})$ triggers the PIP-Btree update (1) Affecting (2) Pruning (3) Resetting $SPR$ to 0 (4) Re-ranking and Regenerating

In Fig. 6, we illustrate the updating process of a PIP-Btree when the sliding window is expanded. In this example, a new data point $(t_{new}, p_{new})$ on the far left of $P$ directly affects $SSP$ of several nodes (which are shown in green arrows in Fig. 6(1)). These $SSP$s are updated to $(t_{new}, p_{new})$ from $(t_1, p_1)$. In Fig. 6(2), the pruning check is executed from the top to bottom left. If the time $t$ of the newly selected data point with maximum VD is different from $Time$ value of this node, the subtree from this node will be pruned and all attributes of this node will be updated by this new data point. Otherwise, only $Dist$ of this node is updated. For example, the subtree from $s_{x5}$ is replaced by the green node. The purpose of setting $SPR$ to 0 is for re-ranking the important nodes. If an important node has no child during re-ranking, its children are generated. If the number of PIPs is not enough, new subtrees are regenerated. The pseudo code of re-ranking and regenerating is given in Algorithm 2.

## 5    Experiment and Performance

In this section, we design two experiments to evaluate the efficiency and effectiveness of the PIP-Btree. These experiments are both conducted on the computer (Intel (R) Core (TM) i5-6200U CPU @ 2.30 GHz 2.40 GHz, Memory 8G and Windows 10. The algorithms are written in Python and run on Python 3.7.3 environments.

---

**Algorithm 2.** Re-ranking and Regenerating

---

    **Input:** Root, Number of PIPs($n$)
    **Output:** PIP-Btree
1  Function findnodes_spriszero(node):
2    node_list={};
3    **If** node.$SPR$ is zero **then**
4      node_list.append(node)
5    **else**
6      node_list.append(findnodes_spriszero(node.$l\_child$))
7      node_list.append(findnodes_spriszero(node.$r\_child$))
8    **return** node_list
9  **End** Function
10 Function reranking_regenerating(Root,$n$):
11    **for** $spr \in [1, n-2]$ **do**
12      Set spr to $SPR$ of the node with maximum $Dist$ in findnodes_spriszero(Root)
13      **If** This node has no child **then**
14        Create children for this node
15    **end for**
16 **End** Function

---

## 5.1 Efficiency of PIP-Btree

PIP-Btree is a multi-functional segmentation method which supports the shifting, expansion and shrinking of the sliding window and allows the generation of multi-resolution PIP lists for the classification of different chart patterns with varying number of PIPs. In this section, we examine the efficiency of 3 different usages of PIP-Btree. They are *PIP-Btree construction*, the multi-resolution PIPs generation (*PIP-Btree Once Construction*), the sliding window expansion (*PIP-Btree Expansion*). According to Fig. 1, expansion of the sliding window requires more computation compared to shifting and shrinking in most cases. PIP-Btree Construction means that PIP-Btree will be reconstructed again when the required number of PIPs is changed. PIP-Btree Once Construction means that construction of PIP-Btree is performed only once for generating the multi-resolution PIP lists (see Sect. 4.2). PIP-Btree Expansion means that PIP-Btree construction process begins with the initial (minimum) number of PIPs required for each pattern and increases the number of PIPs gradually by expanding the sliding window (see Sect. 4.3). We have also implemented original PIP, PIP-SBT, TP-OBST, and MPLR for comparison. We collected a dataset of Apple Inc. (AAPL) stock price from 1996-1-1 to 2019-1-1, which contains 5,747 close prices. The experiment results are listed in Table 3. $NoP$ from Table 3 is number of PIPs which are extracted from the raw time series for evaluating the efficiency of each approach. PIP-Btree Once Construction uses 5,747 close prices for building a tree with 260 PIPs. PIP-Btree Expansion begins with 10 close prices for

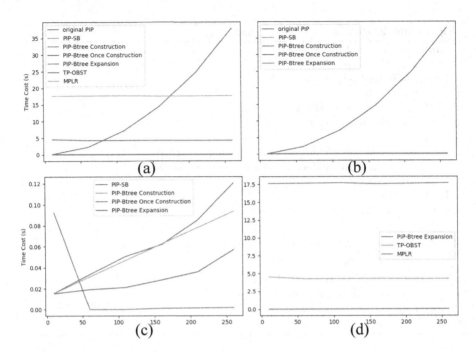

**Fig. 7.** (a) Comparison among all listed methods (b) comparison among all PIP methods (c) comparison among the tree-structure PIP methods (d) comparison among PIP-Btree Expansion, TP-OBST and MPLR

building a tree with 10 PIPs and the tree is gradually added with more close prices until all 5,747 prices are added.

**Table 3.** Efficiency comparison among segmentation algorithms (in seconds)

| NoP | Original PIP | PIP-SBT | PIP-Btree construction | PIP-Btree once construction | PIP-Btree expansion | TP-OBST | MPLR |
|---|---|---|---|---|---|---|---|
| 10 | 0.0936 | 0.0157 | 0.0156 | 0.0923 | 0.0152 | 4.5396 | 17.6124 |
| 60 | 2.2308 | 0.0334 | 0.0312 | 0 | 0.019 | 4.2744 | 17.6436 |
| 110 | 7.1448 | 0.0506 | 0.0468 | 0 | 0.021 | 4.283 | 17.6904 |
| 160 | 14.6796 | 0.0618 | 0.0624 | 0.001 | 0.0281 | 4.29 | 17.55 |
| 210 | 24.8508 | 0.0853 | 0.078 | 0.0013 | 0.0358 | 4.2588 | 17.6124 |
| 260 | 38.0017 | 0.1204 | 0.0936 | 0.0016 | 0.057 | 4.2748 | 17.6904 |

From Table 3 and Fig. 7, we can observe that the time cost of the original PIP when it is compared to other methods. We also find that the time cost of PIP-SBT and PIP-Btree Construction are quite similar and PIP-Btree Once Construction and PIP-Btree Translation outperform other methods. Figure 7(d) shows that PIP-Btree is also superior in term of time cost.

## 5.2    Classification Effectiveness of PIP-Btree

In [13], Wan and Si have evaluated the segmentation methods (PIP, PLA, PAA, TP) with respect to several pattern classification methods (Template-based approach [3], Rule-based approach [3], hybrid approach [16], decision tree [13]). They concluded that PIP achieves the best performance than other segmentation methods especially for the rule-based approach and the hybird approach and PLA also works well on the rule-based approach. Note that PIP-Btree, MPLR and TP-OBST have the same segmentation results as PIP, PLA and TP. Therefore, in this paper, instead of conducting similar experiments as stated in [13], we rather focus on the effectiveness of PIP-Btree on the rule-based pattern classification approaches. Besides, to achieve a fair evaluation, PIP-Btree will not be compared with machine learning approaches such as neural networks and support vector machines because these model-based classification methods rely on training datasets. In this experiment, we select 10 chart patterns (see Table 4) and generate 10 templates with the different lengths for each of them. The length of the template varies 1 month to 10 months. Then by using Eq. 3, Gaussian white noise is added into the 1000 templates for obtaining the synthetic time series.

$$Y_{new} = Y_{old} + \varepsilon \tag{3}$$

$Y_{new}$ is the new time series. $Y_{old}$ is the old time series. $\varepsilon$ is the noise series with $\mu = 0$ and $\sigma = 1$. These 1000 generated templates are then filtered according to the definitions given in [1,8,9]. Finally, we obtain 474 true patterns and 526 false patterns.

At last, we integrate PIP-Btree, MPLR, and TP-OBST with the rule-based pattern classification method given in [14] for classifying the 1000 templates generated in the previous step. The confusion matrix and accuracy (Acc) of the classification results are shown in Table 4. The comparison of the classification results of PIP-Btree, MPLR and TP-OBST is depicted in Fig. 8. From Table 4 and Fig. 8, we can observe that PIP-Btree is superior than MPLR and TP-OBST on the financial chart pattern classification based on the rule-based approach.

# 6    Conclusion

This paper proposes an efficient segmentation method entitled Perceptually Important Point with Binary Tree (PIP-Btree). PIP-Btree utilizes the characteristics of important point selection and the structure of a binary tree for building the tree without downward comparison process. Four attributes of a node such as distance ($Dist$), salient point rank ($SPR$), segment start point ($SSP$) and segment end point ($SEP$) are used to guide the traversal and updating of PIP-Btree. $SPR$ enables the tree traversal process for generating multiple PIP series after sorting. By using $SPR$, $SSP$ and $SEP$, PIP-Btree can realize shifting, expansion and shrinking of the sliding window in order to process streaming time series data. Experiment results reveal that PIP-Btree is more efficient than

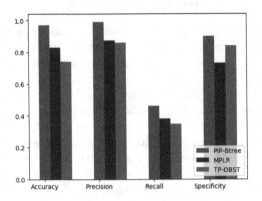

**Fig. 8.** The effectiveness of PIP-Btree

**Table 4.** Classification results of PIP-Btree, MPLR and TP-OBST

| Confusion matrix | | PIP-Btree | | | MPLR | | | TP-OBST | | |
|---|---|---|---|---|---|---|---|---|---|---|
| | | P | N | Acc | P | N | Acc | P | N | Acc |
| Broadening Formations, Right-angled and Descending | T | 27 | 5 | 95% | 19 | 13 | 81% | 10 | 22 | 69% |
| | F | 0 | 68 | | 6 | 62 | | 5 | 63 | |
| Three Falling Peaks | T | 73 | 3 | 95% | 53 | 13 | 84% | 56 | 20 | 75% |
| | F | 2 | 22 | | 3 | 31 | | 5 | 19 | |
| Triangles, Symmetrical | T | 74 | 0 | 100% | 56 | 18 | 81% | 50 | 24 | 75% |
| | F | 0 | 26 | | 1 | 25 | | 1 | 25 | |
| Broadening Formations, Right-Angled and Ascending | T | 25 | 3 | 96% | 20 | 8 | 92% | 7 | 21 | 75% |
| | F | 1 | 71 | | 4 | 68 | | 4 | 68 | |
| Broadening Wedges, Ascending | T | 44 | 4 | 96% | 40 | 8 | 85% | 25 | 23 | 72% |
| | F | 0 | 52 | | 7 | 45 | | 5 | 47 | |
| Triangles, Descending | T | 33 | 0 | 100% | 20 | 13 | 79% | 11 | 22 | 71% |
| | F | 0 | 67 | | 8 | 59 | | 7 | 60 | |
| Head and Shoulders, Bottoms | T | 37 | 6 | 94% | 29 | 14 | 84% | 20 | 23 | 75% |
| | F | 0 | 57 | | 2 | 55 | | 2 | 55 | |
| Wedges, Rising | T | 42 | 0 | 100% | 26 | 16 | 76% | 13 | 29 | 64% |
| | F | 0 | 58 | | 8 | 50 | | 7 | 51 | |
| Head and Shoulders, Tops | T | 42 | 6 | 94% | 32 | 16 | 83% | 31 | 17 | 83% |
| | F | 0 | 52 | | 1 | 51 | | 0 | 52 | |
| Broadening Tops | T | 50 | 1 | 99% | 44 | 7 | 87% | 34 | 17 | 77% |
| | F | 0 | 49 | | 6 | 43 | | 6 | 43 | |

the original PIP, PIP-SBT, MPLR and TP-OBST. The construction, once construction, and expansion of PIP-Btree also achieve the best performance. As for the effectiveness, we perform experiments similar to [13] and show that PIP-Btree is more effective than MPLR and TP-OBST.

**Acknowledgment.** This research was funded by University of Macau (File No. MYRG2019-00136-FST).

# References

1. Bulkowski, T.N.: Encyclopedia of Chart Patterns, vol. 225. Wiley, Hoboken (2011)
2. Chung, F.L., Fu, T.C., Luk, R., Ng, V.: Flexible time series pattern matching based on perceptually important points (2001)
3. Fu, T.C., Chung, F.L., Luk, R., Ng, C.M.: Stock time series pattern matching: template-based vs. rule-based approaches. Eng. Appl. Artif. Intell. **20**(3), 347–364 (2007)
4. Fu, T.C., Chung, F.L., Luk, R., Ng, C.M.: Representing financial time series based on data point importance. Eng. Appl. Artif. Intell. **21**(2), 277–300 (2008)
5. Hu, Y., Jiang, Z., Zhan, P., Zhang, Q., Ding, Y., Li, X.: A novel multi-resolution representation for streaming time series. Procedia Comput. Sci. **129**, 178–184 (2018)
6. Keogh, E., Chakrabarti, K., Pazzani, M., Mehrotra, S.: Dimensionality reduction for fast similarity search in large time series databases. Knowl. Inf. Syst. **3**(3), 263–286 (2001)
7. Keogh, E., Chu, S., Hart, D., Pazzani, M.: An online algorithm for segmenting time series. In: Proceedings 2001 IEEE International Conference on Data Mining, pp. 289–296. IEEE (2001)
8. Kirkpatrick II, C.D., Dahlquist, J.A.: Technical Analysis: The Complete Resource for Financial Market Technicians. FT Press, Upper Saddle River (2010)
9. Meyers, T.: The Technical Analysis Course: Learn How to Forecast and Time the Market. McGraw Hill Professional, London (2011)
10. Rakthanmanon, T., et al.: Searching and mining trillions of time series subsequences under dynamic time warping. In: Proceedings of the 18th ACM SIGKDD International Conference on Knowledge Discovery and Data Mining, pp. 262–270. ACM (2012)
11. Si, Y.W., Yin, J.: OBST-based segmentation approach to financial time series. Eng. Appl. Artif. Intell. **26**(10), 2581–2596 (2013)
12. Velay, M., Daniel, F.: Stock chart pattern recognition with deep learning. arXiv preprint arXiv:1808.00418 (2018)
13. Wan, Y., Gong, X., Si, Y.W.: Effect of segmentation on financial time series pattern matching. Appl. Soft Comput. **38**, 346–359 (2016)
14. Wan, Y., Si, Y.W.: A formal approach to chart patterns classification in financial time series. Inf. Sci. **411**, 151–175 (2017)
15. Wu, K.P., Wu, Y.P., Lee, H.M.: Stock trend prediction by using k-means and aprioriall algorithm for sequential chart pattern mining. J. Inf. Sci. Eng. **30**(3), 669–686 (2014)
16. Zhang, Z., Jiang, J., Liu, X., Lau, R., Wang, H., Zhang, R.: A real time hybrid pattern matching scheme for stock time series. In: Proceedings of the Twenty-First Australasian Conference on Database Technologies, vol. 104, pp. 161–170. Australian Computer Society, Inc. (2010)

# SCODIS: Job Advert-Derived Time Series for High-Demand Skillset Discovery and Prediction

Elisa Margareth Sibarani[1,2(✉)] and Simon Scerri[1]

[1] Fraunhofer IAIS, Sankt Augustin, Germany
{elisa.margareth.sibarani,simon.scerri}@iais.fraunhofer.de
[2] University of Bonn, Bonn, Germany

**Abstract.** In this paper, we consider a dataset compiled from online job adverts for consecutive fixed periods, to identify whether repeated and automated observation of skills requested in the job market can be used to predict the relevance of skillsets and the predominance of skills in the near future. The data, consisting of co-occurring skills observed in job adverts, is used to generate a skills graph whose nodes are skills and whose edges denote the co-occurrence appearance. To better observe and interpret the evolution of this graph over a period of time, we investigate two clustering methods that can reduce the complexity of the graph. The best performing method, evaluated according to its modularity value (0.72 for the best method followed by 0.41), is then used as a basis for the SCODIS framework, which enables the discovery of in-demand skillsets based on the observation of skills clusters in a time series. The framework is used to conduct a time series forecasting experiment, resulting in the F-measures observed at 72%, which confirms that to an extent, and with enough previous observations, it is indeed possible to identify which skillsets will dominate demand for a specific sector in the short-term.

**Keywords:** Graph mining · Network clustering · Time-evolving network · Graph-based time-series prediction

## 1 Introduction

The employment sector is overwhelmed by the rapid changes witnessed in areas of expertise such as big data, data science, and artificial intelligence. A shortage of prospective candidates with relevant skillsets required for on-demand and highly-specific job positions is widely reported at the national, regional, and international levels. In 2016, the gap between supply and demand in careers contributing to the European 'Data Economy' was estimated at 420,000, with

---

Co-funded by the European Union's Horizon 2020 research and innovation programme under the QualiChain Project, Grant Agreement No. 822404.

S. Hartmann et al. (Eds.): DEXA 2020, LNCS 12392, pp. 366–381, 2020.
https://doi.org/10.1007/978-3-030-59051-2_25

a possibility to reach 2 million by 2020[1]. To counteract this phenomenon, there is an urgent need to up-skill and re-skill the combined talent pool to meet the needs of fast-growing employment sectors. To do so effectively, training bodies, corporations, and prospective candidates alike require a dynamic overview of high-demand skillsets being requested in the job market. Early and some recent research for monitoring the labor market and analyzing skills in-demand from job adverts has concentrated on the individual skills based on its frequencies from the static point of view [2,15,16], neglecting both the interactions between skills and its dynamic character in nature. Complementary to these efforts, there were also few studies for tracking skills of Information Technology (IT)-related job profiles from job adverts that have been considering the trends and changes of the skills requirements over time [1,14,17,18]; unfortunately, all relied solely on the 'flat' representation and information like individual skills and still do not exploit the relational behavior and network structure naturally encoded within this labor market domain. In this paper, we aim at overcoming these issues utilizing a graph-based method to represent skills in-demand in the labor market. By representing skills that are mentioned in job adverts as complex interaction (or, is called co-occurrence) networks where nodes represent skills and edges mimic the co-occurrences among them. Prior to this reported effort, we have conducted earlier studies both from the static point of view and the dynamic nature of job market networks [5,13]. The results of this previous research were applied here in this reported study, particularly its information extraction (IE) pipeline for compiling dataset and the generic model for mining evolving networks. In summary, the contributions of this work are as follows:

- We propose a combination of graph-based labor demand model as co-occurrence networks to identify clusters of nodes that are naturally formed, with a task to characterizing each cluster's importance and evolution and use it to identify the skillset which represents the real demand, and finally tracking the changes and predicting the future skillset's importance.
- We present the Louvain algorithm [4], a state-of-the-art large network partitioning approach based on modularity optimization, since identifying clusters is paramount in this approach. In contrast to the Coulter algorithm, which we considered in previous work [13], Louvain is, by definition, better suited for the nature of our task.
- We propose several cluster categorizations: *isolated*, *secondary*, or *principal* (see Sect. 4.2 for detail), based on its association to other clusters. Also, we propose an additional distinction whose aim is to highlight crucial clusters with a strong ability to structure the skills network, which is called *crossroads*.
- We present a variety of indices, *centrality* and *density*, in order to equip each cluster with a presentation of their respective position and their relative stability. The *centrality* characteristic is used to see the cluster's position by the bundle of links uniting it to other clusters in the network, while *density* is to see a cluster, that is made up of words linked with each other, to define a more or less dense group, and more or less coherent and robust. We need this

---

[1] https://ec.europa.eu/digital-single-market/en/news/final-results-european-datam arket-study-measuring-size-and-trends-eu-data-economy.

double analytical perspective, which allows us to give a synthetic and simplified presentation of the network, and provide a stepping-stone for dynamic analysis.

- We provide the skills demand tracking attributes by using each cluster's *centrality* and *density* values and plot it into a strategical diagram. This diagram is obtained by ordering clusters horizontally (along the x-axis), and vertically (along the y-axis), that allows us to classify all aggregates into four general categories, which correspond to the four quadrants of the graph (see Sect. 3), that is adapted from [8].

As a result of this, we are able to provide a comprehensive Skills Cluster Observation and Discovery (SCODIS) framework that presents a quadrant-based categorization of the important and emerging skillsets based on its density and centrality plotted into a strategic diagram. The novelty aspect lies in the whole integrated approach comprises the most fitted clustering algorithm and generates a forecasting model as well as its application to extract the labor market demand as skillsets. We begin by converting an evolving graph into static snapshot graphs at different time points and obtain clusters at each of these snapshots independently. Next, we characterize each cluster and generate a series of clusters to see its transformations, detect the stable clusters, and tracking the changes. Using this framework, we then carry out an experiment to detail a forecasting model for evolving skills networks and demonstrate their application for the task of predicting future high-demand and emerging skillsets.

## 2 Related Work

There has been enough interest in finding skills requirements for IT as well as non-IT related occupations from job advertisements. However, the majority of these studies [2,15,16] have focused on mining individual skills that are highly in-demand from a static point of view (i.e., only a snapshot of the demand from one point in time) based on its frequency of occurrence. Recently, tracking the skills requirements and its trends and changes over several snapshots of different periods have attracted the interest of several groups. Smith and Ali [1] conducted a study to assess the employable skills in programming to guide curriculum decisions. This study concluded that skill is still in demand based on the trend line that remains strong and continues to grow relative to the total number of jobs. Surakka [17] conducted a trend analysis from 1990 to 2004 to identify technical skills for the software developer by merely calculating the frequencies of different phrases. Todd et al. [14] compiled skill requirements over time by manually collecting, classifying, counting, and building an index of keywords from a pilot sample of 200 job adverts. Kennan et al. [18] studied 400 information systems (IS) job adverts over two months-period to get the skills and competencies demanded of early-career IS graduates in Australia. Although they analyze the trends over different time, the result is simply a list of skills which are ranked based on its frequency of occurrence, ignoring associations between them and unable depicting the holistic overview of the structure of the labor market. Most importantly, none of the above works address the skills cluster (or skillset) evolution tracking problem.

The problem of finding appropriate network clustering methods has been studied for some decades in many fields. Girvan-Newman [6] introduced a network clustering algorithm that implements divisive hierarchical clustering based on the breadth-first search. Xu et al. [7] proposed a structural clustering algorithm for networks (SCAN) that is based on the hierarchical agglomerative method. However, despite the high effectiveness of discovering community structure in networks, unfortunately, both algorithms [6,7] were purposely designed to process an unweighted graph. In contrast, an agglomerative hierarchical clustering algorithm specifically built for a weighted graph with breadth-first search described in [3] (henceforth referred to as the Coulter algorithm) utilizes the number of co-occurrence between pairs of vertices (such as words or noun phrases) in a corpus of text, in order to give a weight to each link. Because it always considers the strongest link every time it visits a vertex according to the breadth-first search order, unfortunately, it tends to find only the cores of clusters and disregard peripheral nodes with no substantial similarity, leading to incomplete observed network structures. Blondel et al. [4] introduced the Louvain algorithm: a heuristic method that is designed for a weighted graph and is based on modularity optimization that unfolds a complete hierarchical community structure for a network. The accuracy is excellent for the top-level hierarchy and extremely fast for networks of unknown sizes, as confirmed by a survey paper [12] that emphasizes its excellent performance and low computational complexity. Because it always visits and includes all links and vertices into the sub-graphs, therefore, the depiction of a complete structure of the network is evident. Recently, mining from dynamic graphs for tracking the clusters' evolution and its dynamic behavior has attracted the interest of several groups [9,10,19,20]. However, none of these reported efforts were applied or evaluated for analyzing labor market demand as well as its changes. Lee et al. [9] proposed an incremental cluster evolution algorithm where its main contribution lies in generating the skeletal graph to summarize the information in the dynamic network in order to find the changes and track its composite evolution behaviors (i.e., merging and splitting). Another effort is by Kim et al. [20], which first clusters individual snapshots into quasi-cliques and then maps them over time using the density of bipartite graphs between quasi-cliques in adjacent snapshots. They mainly focused on handling the evolution behaviors such as birth, growth, decay, death of clusters. Asur et al. [19] focused on identifying a set of critical events that occur and influence the behavior of clusters in real dynamic networks. According to those events, communities can newly form or dissolve (i.e., start or stop) at any time as well as continue with some change (i.e., evolve). Hopcroft et al. [10] introduced a tracking evolving community approach that utilizes the centroid-based agglomerative clustering algorithm based on cosine similarity. Based on the growth rate of the cluster size (i.e., number of nodes), their goal is to detect the emerging clusters and track temporal changes in the underlying structure of a dynamic network. Although the outcomes of these works are cluster evolution behaviors in dynamic networks, unfortunately, none of those can guide users' decisions for determining the significance level of a cluster in the network where it is organized.

## 3    Problem Definition

Before describing our quadrant-based skills cluster observation framework in detail, we introduce the necessary notations used throughout the paper. As we explained earlier, this study focuses on the evolution of graphs, in particular, to identify the importance and significance of each cluster in the graph over time. In order to fully understand the temporal evolution of graphs, it is critical to identify all clusters together with its density and centrality attributes and track the importance transformations undergone by the graph at different point of time along the way. In this regard, we use temporal snapshots to examine static versions of the evolving graph at different time points. Definition 1 describes a Skill Graph (or Network), more formally.

**Definition 1.** *A skill graph is said to be evolving if its associations vary over time. For each country $a$, let $G_a = (V, E)$ denote a temporally varying skill graph where $V$ represents the total unique skills and $E$ the total co-occurrence associations that exist among the skills. We define a temporal snapshot $S_{i,a} = (V_{i,a}, E_{i,a})$ of $G_a$ to be a graph representing only skills and co-occurrences active in a particular time interval $[T_i, T_x]$, called the snapshot interval. In other words, a dynamic skill network of a particular country $a$, $G_a$, is a sequence of networks $S_{i,a}(V_{i,a}, E_{i,a})$, i.e., $G_a = \{S_{1,a}, ..., S_{t,a}, ...\}$.*

To build a dynamic skill network $G_a$ as a weighted graph, each edge has a weight attached to it, which is what we called strength of association. A simple count of co-occurrence frequency cannot sufficiently measure the strength of association between co-skills, as it favors high-frequency pairs over those with low frequency. Therefore, normalization is necessary. In SCODIS, we apply the equivalent coefficient (based on the systematic study of all such possible indices reported in [8]) as the value of strength, as follows:

$$E_{ij} = \frac{(C_{ij})^2}{(C_i \cdot C_j)}, 0 \leq s_{ij} \leq 1 \tag{1}$$

where, $C_{ij}$ is the number of job adverts in which both skill $i$ and $j$ appear; $C_i$ is the number of job adverts in which skill $i$ appears; $C_j$ is the number of job adverts in which skill $j$ appears.

To investigate the graph's evolution, we need to represent its structure, by identifying clusters from each snapshot of the graph, which is explained in Definition 2. For the clustering task, we have examined the nature of several clustering algorithms thoroughly, which can be found in Sect. 2. We decided on the Louvain algorithm [4], to be the fittest unsupervised clustering algorithm to obtain all clusters at different timestamps. However, the alternative Coulter algorithm, which we have investigated in the same context (i.e., skills clustering) in the earlier related work [13], is considered to be the state-of-the-art and thus be used as a benchmark for the experiment of this work, see Sect. 5 for a detail report. The Louvain algorithm does not require a parameter specifying the number of clusters, nor a threshold to forcibly terminate its process. Consequently, the number of clusters for each snapshot may vary depending on the co-occurrence interactions in that time interval.

**Definition 2.** *For each snapshot from the collection of all $i$ temporal snapshots $S_{i,a}$ of graph $G_a$ of country $a$, we partitioned it into $k$ communities or clusters indicated by $C_i = \{C_i^1, C_i^2, ..., C_i^k\}$. The $l^{th}$ cluster of $S_{i,a}$, $C_i^l$ is also a graph indicated by $(V_i^l, E_i^l)$ where $V_i^l \in V_{i,a}$, and $V_{i,a}$ is a set of nodes in $S_{i,a}$. Hence, for each snapshot $S_{i,a} = (V_{i,a}, E_{i,a})$, $V_i^1 \cup V_i^2 \cup ... \cup V_i^k = V_{i,a}$. Finally, $E_i^l$ indicates the edges between nodes in $V_i^l$.*

There are several well-known quality measures for graph clustering, such as min-max cut, normalized cut, a measure of agreement (e.g., adjusted rand index (ARI)), and modularity. Among them, min-max cut and normalized cut can only measure the quality of a binary clustering, while ARI is only possible when we have the expected clustering ahead of time. Modularity, which is introduced in [6], however, can measure the quality of clustering with multiple clusters, and when no prior clusters are known, thus, it is more suitable for our problem. The modularity $Q$ is defined as follows:

$$Q = \sum_{s=1}^{k} \left[ \frac{l_s}{L} - \left( \frac{d_s}{2L} \right)^2 \right] \tag{2}$$

where, $k$ is the number of clusters, $L$ is the total strength between all pairs of nodes in the graph, $l_s$ is the total strength of a pair of nodes within a cluster $s$, $d_s$ is the total strength between a node in the cluster $s$ and any node in the graph. To enrich the information about a cluster, we identify its *density* and *centrality*, which are described in Definition 3.

**Definition 3.** *For a cluster $s$, a **density** index is a measure of the cluster's internal strength (local context), is calculated by: $s_{den} = \frac{l_s}{x}$, where, $x$ is the number of edges that associate all pairs of nodes within a cluster $s$ and $l_s$ is the total strength of those edges. Whilst a **centrality** value measures the strength of a cluster's interaction with other clusters (global context), that is calculated with: $s_{cen} = \sqrt{\sum_1^n d_n^2}$, where, $n$ is the number of edges that associate all pairs of nodes in cluster $s$ with other nodes outside cluster $s$, and $d_n$ is the strength value of those edges.*

Both indices are very critical in this study. For centrality, the higher is its value, the more a cluster designates a set of skills considered crucial by the employer. On the other hand, the higher a density value is, the more the skills corresponding to the cluster constitute a coherent and integrated whole. Thus, a centrality occupies a strategical position, and density provides a good representation of the cluster's capacity to maintain itself and to develop over time. A cluster also has an additional attribute: a label or a descriptor; that is, a particular skill represents this cluster. The label's candidates must be from all nodes within but also is possible from outside that cluster; however, it is associated with the node inside that cluster. As the members of each cluster are also normalized weighted elements, thus, for each node within cluster $Cl$ or any node in the graph that is connected with the node within cluster $Cl$, we calculate its weight $w$ which was adopted from [3] as follows:

$$w_{Cl}(a) = \frac{K_{Cl}(a)}{n_{Clin} + n_{Clex}}, 0 < K_{Cl}(a) \le n_{Clin} + n_{Clex}, 0 < w_{Cl}(a) \le 1 \tag{3}$$

where, $n_{Clin}$ is the total strength of a pair of nodes within cluster $Cl$; $n_{Clex}$ is the total strength between a node in the cluster $Cl$ and any node in the graph; and $K_{Cl}(a)$, if node $a$ is within-cluster $Cl$, then it is the total strength between node $a$ and nodes that are within and outside cluster $Cl$, else if node $a$ is not within-cluster $Cl$ but is connected to a node in the cluster $Cl$, then it is the total strength between node $a$ and nodes within cluster $Cl$. Hence the keyword with most weight serves to label cluster $Cl$.

Once all clusters have their density and centrality, we further trace and classify each cluster, plotted it in the form of a strategical diagram. As an example, we randomly chose the graph of country code 'at' $G_{at}$ and depicted its temporal snapshot for period T1 (Jan–Mar 2016) $S_{T1,at}$ in Fig. 1a. In total of 145 nodes and 1029 edges, notice that this figure shows the network after the clustering process, where each color represents one cluster, and the Louvain algorithm identifies 39 clusters.

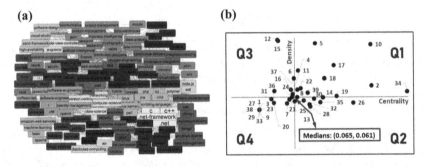

**Fig. 1.** (a) Temporal snapshot period T1 (Jan–Mar 2016) in country code 'at', and (b) temporal snapshot period T1 country code 'at' centrality and density.

Plots of centrality and density for all clusters $\{C_{T1}^1, C_{T1}^2, ..., C_{T1}^{39}\}$ being identified from $S_{T1,at}$ is shown in Fig. 1b. The origin of this diagram is the median of the represent axis values (the horizontal axis represents centrality; the vertical axis represents density). This operation allows us to classify all aggregates into four general groups, which correspond to the four quadrants of the diagram: (1) *Q1*, a quadrant that contains the *core* of the network because it strongly connects to other clusters and has strong internal links; (2) *Q2*, comprises *emerging* clusters that have already shown its importance and their growing maturity although the density of their internal links is relatively low; (3) *Q3* contains *marginalized* clusters with strong internal links yet gradually (if not progressively) become less and less interest due to its low centrality index; and (4) *Q4* comprises clusters that are both peripheral and little developed, represents the margins of the network and might gradually *disappear* from the network.

The most interesting clusters are the ones with both strong density and strong centrality (which is classified as *Q1*) and clusters with strong centrality but less density (is classified as *Q2*). From Fig. 1b, 12 clusters are categorized in Q1, and among them are Clusters-10, -17, and -18, which has a slightly balance high values of both indices, together with 8 clusters which are classified in Q2. On the contrary, Cluster-12 and -15 show strong density but weak centrality values. These attributes play a vital role if and only if it is combined with the cluster classifications (described in Sect. 4.2, depicted in Fig. 2). Together it will provide the basis for tracking importance (or demand) evolution of clusters from different snapshots under study. Note that when two clusters are close to one another in the strategic diagram, it does not mean that they are closely linked to each other. The only conclusion that can be drawn from this observation is that their indices of centrality and density have neighboring values.

Algorithm 1 shows the outline of the SCODIS framework we propose. We design a quadrant-based strategy to mine each snapshot of the graph by identifying clusters and its density and centrality transformations over time, referred to as quadrant locations, along with its cluster classifications. These quadrants are then used to generate a forecasting model, to predict high-demanded skillsets.

---

**Algorithm 1.** Mine-demands $(G_a, T)$

---

**Input:** Skills graph $G_a = (V, E)$ and $T$, the number of intervals
    Convert graph $G_a = (V, E)$ into $T$ temporal snapshots $S = \{S_1, S_2, ..., S_T\}$.
    **for** $i = 1$ to $T$ **do**
        $C_i \leftarrow$ Cluster $(S_i)$                                             # $C_i = \{C_i^1, C_i^2, ..., C_i^k\}$
        **for** each cluster $C_i^k$ in $C_i$ **do**
            $C_i^k \leftarrow$ Define-DensityCentralityLabel $(C_i^k)$
        **end for**
        **for** each cluster $C_i^k$ in $C_i$ **do**
            $C_i^k \leftarrow$ Set-quadrant $(C_i^k,$ MEDIAN (centrality,density) $C_i)$
            $C_i^k \leftarrow$ Set-classification $(C_i^k)$ [Section 4.2]
        **end for**
    **end for**
    **for** $i = 1$ to $T - 1$ **do**
        $Series =$ Generate-series $(S_i, S_{i+1})$ [Section 4.3]
    **end for**
    Implement-forecasting $(Series)$ [Section 5]

---

## 4    The SCODIS Framework

This section contains a comprehensive overview of the Skills Cluster Observation and Discovery (SCODIS) framework, in which we introduce and afford a formal definition to certain cluster classifications and the skillset series that is being identified in evolving graphs. The cluster classifications described in this section are inspired by a similar notion described by Callon et al. [8], in the context of tracking and visualizing clusters.

### 4.1    Dataset

The dataset used by SCODIS is compiled using a method we have implemented in earlier work [5], but covers a longer period of time (2016–2017) [13]. It is derived from 620,760 job adverts collected from various online job portals, namely Adzuna[2], Indeed[3], and Trovit[4]. The adverts are mostly related to 'Data Science' and 'Data Analysis' and hail from 17 European countries, and were indexed by a total of 1,287,994 skill keywords (a mean of 2.07 per job advert). The OBIE skills extraction process (fully described in [5]) is guided by the *Skills and Recruitment Ontology*[5] (SARO) [11], and utilizes a de-duplication framework to remove multiple entries (posted on multiple portals) and to minimize noise and improve accuracy. The resulting data is transformed to comply with the Resource Description Framework (RDF) W3C standard for data representation[6]. All job ad-related information (including skills) are stored as RDF instances, and consist primarily of `JobPosting` and `Skill` instances adhering to the descriptions in the SARO ontology.

We then used the resulted RDF-based job adverts descriptions to generate a skill co-occurrence graph, where each skill is represented as a node, and an edge between two skills corresponds to a joint demand of these two skills mentioned by the employer. Given a graph spanning for two years (2016–2017), we chose the snapshot interval to be a three-months, resulting in 8 consecutive snapshot graphs, for each 17 EU countries. Although this should generate 136 snapshot graphs, due to missing data for some countries, the result is 92 graphs having a total of 14,017 nodes and 122,697 edges. These graphs are then clustered and analyzed to identify the importance, quadrants, and transformations.

### 4.2   Cluster Classification

The primary benefit of the classification task that we implement for each cluster of a particular snapshot is to group them accordingly to distinguish their importance. Moreover, this task is considered to be crucial in the SCODIS framework to prevent the cluster map resulted from the Louvain algorithm being too cluttered for large networks. Thus, the aim is to trim out all clusters with low-degree node and low connectivity from the evolution tracking and prediction task. It also ensures that only potential clusters are included for the next crucial task, i.e., establish the series of the cluster of skills. For initial distinction, each cluster can be grouped based on three categories, which are as follows:

**1. Isolated:** a cluster $C_i^k$ of snapshot $S_i$ is *isolated* if all nodes within cluster $C_i^k$ does not associate with any nodes from other clusters of snapshot $S_i$.

**2. Principal and Secondary:** a cluster $C_i^k$ of snapshot $S_i$ is a *principal*, if cluster $C_i^k$ has nodes within it which are connected with the node of another cluster $C_i^x$ of snapshot $S_i$ with the strength values higher than the minimum strength of all pair of nodes within cluster $C_i^k$. This condition also makes cluster

---

[2] https://www.adzuna.com/.
[3] https://de.indeed.com/?r=us.
[4] https://de.trovit.com/.
[5] https://elisasibarani.github.io/SARO/.
[6] https://www.w3.org/RDF/.

$C_i^x$ as a *secondary* to a *principal* cluster $C_i^k$. The number of connections that satisfy this condition must be at least $\delta$, a predefined link threshold.

Thus based on the above distinct categories, we can confidently decide to omit all *isolated clusters* and instead focus more on the *principal clusters*, which in some sense are the core of a given network, and so are the best candidates to be put under consideration for further processes. To simplify the analysis of characterizing cluster's content and following their evolution, we propose an additional distinction, aiming to highlight clusters with a strong ability to structure the general network, which is called a *crossroads cluster*. As a result, any analysis must therefore start with the *principal clusters* – and in particular *crossroads clusters*.

**3. Crossroads:** is a *principal cluster* which has $y$ secondary clusters, where $y$ has a value greater than $\alpha$, a predefined secondary threshold. By their power to connect, *crossroads clusters* play an essential role in the transformation of a network. Setting the threshold at $\delta = 3$ qualifying links for *principal cluster*

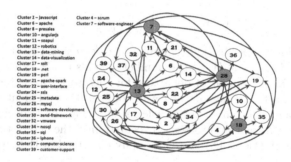

**Fig. 2.** Cluster map after classification step for snapshot T1 country code 'at'.

classification, the classification result for snapshot $S_{T1,at}$ of graph $G_{at}$ depicted in Fig. 1a is shown in Fig. 2, with each node, represents a skills cluster and the legend provides the cluster's label. Some observations include: Cluster-13 is a principal cluster and a secondary cluster relative to Cluster-6, -7, -26, -28, and -37; and Cluster-4, -10, and -35 are a secondary cluster of principal cluster Cluster-18. Putting this in context of the graphs' contents, we might conclude that:

1. In general, four clusters play a vital role in shaping the skills in-demand for data science, labeled with software-engineer (Cluster-7), data-mining (Cluster-13), .net (Cluster-18), and software-development (Cluster-28).
2. Referring to the strategic diagram in Fig. 1b and this network map, we can conclude that Cluster-18 comprises four skills (.net, .net-framework, c++, c), and is the dominant software development environment during T1 period.

## 4.3   Skillsets Series Analysis

Once each cluster in all snapshots of a graph has been classified, then SCODIS
will continue to establish the series of skills clusters. These generated series
represent the evolution of demand given the labor market network, and further
can be observed to see the transformation and trends based on the quadrant-
strategical position of each cluster. Later, this observation will be the foundation
for extracting interesting trends in the labor market that afforded insight into
the demand evolution as well as the demand forecasting for the near future. Let
$S_i$ and $S_{i+1}$ be snapshots of S at two consecutive time intervals with $C_i$ and $C_{i+1}$
denoting the set of clusters respectively. The framework computes the similarities
between clusters in $C_i$ and $C_{i+1}$, which shows the overlap of two consecutive
clusters through a comparative analysis process. The similarity is calculated by
using the *Similarity Index* (SI), derived from Callon's Dissimilarity Index [8],
to measure the intersection of skills in two given clusters. Although it does not
include the corresponding links in clusters directly, all skills in a cluster are
nonetheless indirectly linked. Thus the metric can sufficiently capture a portion
of cluster similarity. Given two clusters $i$ and $j$, their similarity is defined as
follows:

$$SI(W_i, W_j, W_{ij}) = 2 \times \left( \frac{W_{ij}}{W_i + W_j} \right), 0 < SI \leq 1 \tag{4}$$

where, $W_{ij}$ is the number of skills common to cluster $i$ and cluster $j$; $W_i$ is the
number of skills in $C_i$; and $W_j$ is the number of skills in $C_j$. When generating a
skills cluster series, the point of departure for each series must be a *crossroads
cluster*. The choice for *crossroads*, however, confirms that all identified series are
representative and play an essential role in depicting the transformation of a
network. The series creation process is executed until the end of the whole time
interval under evaluation (T8) and is formally described in Definition 4.

**Definition 4.** *For each graph $G_a$, SCODIS framework generates a number of
$i$ series $Ser_a$ in a time-ordered manner, by implementing comparative analysis
between two consecutive clusters $C_i^k \in C_i$ from snapshot $S_i$ and $C_{i+1}^x \in C_{i+1}$
from snapshot $S_{i+1}$, measured by a similarity index. To ensure notable intersection between two clusters, we set a minimum similarity threshold $\theta$ that defines
the least number of similar skills should be matched between two compared clusters.*

## 4.4   Trends-Based Analysis

The final process in our proposed framework is to projecting the *centrality* and
*density* indices of each cluster in each series that has been established for each
graph $G_a$. We use these primary indices to identify more complex trends and
behavior, to monitor their stability and strategic position in order to decide
which skillset is still on high-demand or currently emerging. In particular, we
are interested in capturing the strategical tendencies of skills that contribute to
the evolution of the graph. We define these strategical patterns to strengthen
further our finding of the stability of the resulted series of clusters. By manually

investigating the evolving clusters identified in our experiment, we found four categories of transition curves described as follows:

1. **Category A** represents clusters exhibiting a stable and steady quadrant location.
2. **Category B** displays a gradual shift from being central (Q1) to a less developed but possibly the emerging one (Q2), eventually reaching the margin of the whole network (Q4); or the other way around.
3. **Category C** embodies series that have clusters exhibit a steady and stable quadrant location during period T1 until T3, but then suddenly change track to other quadrants in T4.
4. **Category D** shows cluster series with less consistent trends that reflect a complex curve, with frequent and repeated changes in the clusters' quadrant location.

## 5 Results and Evaluation

The objectives of our study are two-fold: i) to evaluate the adequacy of the Louvain algorithm in finding the relevant clusters aiming to complement or substitute Coulter algorithm, and ii) to identify whether the proposed method has the potential to correctly track established high-demanded skills clusters and the emergence of new skills clusters identified by their 'quadrant' location. Taking each of the above into consideration will enable us to determine whether we can confirm our hypothesis – that skill demand can be uncovered and predicted based on a sufficient amount of earlier observations. The first exercise returns the quantitative scientific result by determining the modularity gain of each clustering algorithm result, while the second returns the F-measure of the prediction result on each cluster's quadrant location (see Sect. 3).

**Clustering Results and Evaluation.** We implemented SCODIS framework in Java environment utilizing *Java Universal Network/Graph*[7] JUNG library for graph implementation. We determined the clusters for every 92 graphs constructed from the dataset of job adverts published in 17 countries from the year 2016–17 that were split into eight periods, resulting in a total of 3,310 clusters. We evaluated the clustering results of both the Louvain and Coulter algorithms by calculating each graphs' clustering modularity ($Q$) using Formula 2 and compared the two results. Our first observation is that on all clustering results for all 92 graphs by the Louvain algorithm, the modularity is consistently higher (average of 0.72) than the results of the Coulter algorithm (average of 0.41). Thus these results suggest that the Louvain clustering can identify a more robust community structure, with a modularity value approaching the maximum of $Q = 1$. In practice, values for such networks typically fall in the range of about 0.3 to 0.7, with higher values considered rare [6]. Second, unlike the Louvain algorithm, the Coulter algorithm generates clusters that only include part of the total nodes and edges from the source graph because it only considers the most 'important' nodes relative to the strongest weights of the edge.

---

[7] http://jrtom.github.io/jung/javadoc/.

**Series Generation Results and Evaluation.** By taking the clustering result of the Louvain algorithm as an input, we establish the list of the series of skills clusters for each country. The series parameters $(\delta, \theta, \alpha)$ described in Sect. 4.2 and 4.3, control the construction of the resulted series. Two variables will help us decide which parameter values give a satisfying series: the amount of series, and the similarity index (SI).

**(a)**

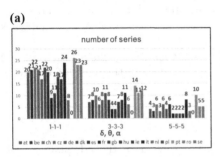

**(b)**

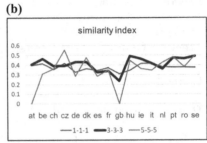

**Fig. 3.** The trends of: (a) the number of series, and (b) the similarity index, when increasing $\delta, \theta, \alpha$.

Figure 3a shows the number of established series as $\delta, \theta, \alpha$ increases from 1 to 5 respectively. We can see that the amount of series decreases when the parameters increase because it is difficult to find a very exact similar cluster. For the data and period (T1–T8) in consideration, there are 302, 135, and 71 established series when the framework uses series parameters 1-1-1, 3-3-3, and 5-5-5, respectively. From those established series, we got only 190, 30, and 10 complete series that have consistently one cluster in four quarterly periods (T1–T4), and 21, 1, and 1 series has a cluster in every eight periods (T1–T8). However, results obtained from calculating the *similarity index* (SI) that is shown in Fig. 3b confirm that the average of SI is higher (see the red line) when the parameters are based on the median value of the internal links, internal nodes, and secondary amount of each cluster (i.e., 3-3-3), are being used. Therefore, we set $\delta = 3$, $\theta = 3$, $\alpha = 3$ as a trade-off between the number of series one hand and the *similarity index* on the other, proven better outcomes based on manual observation on the resulted series.

**Prediction Results and Evaluation.** To return a quantitative scientific result for the discovery and prediction of skills in high demand, the second objective of our evaluation considers the forecast quadrant location of established clusters against their next known location. An F-measure is thus computed to determine the forecast accuracy. Repeating the exercise based on the method presented in [13], we implement three traditional statistical prediction models that can sufficiently handle the trend characteristic and variability of our dataset, namely Naive, Simple Exponential Smoothing (SES), and Holt's linear trend method. A walk-forward validation is carried out to compare forecast values with actual values observed later than the period in question. Forecast calculations were conducted based on the 30 series (each containing four versions of an evolving skills cluster), starting by using two data points (T1-T2) to train the model, then the previous two to forecast the third, and the previous three to forecast the fourth. Thus, a total of 60 cluster forecasts are generated. We then compare the quadrant location of each cluster based on the forecast result to its actual quadrant in order to assess the performance of the forecasting method. By extension, the evaluation also determines whether our proposed framework is feasible to identify and monitor labor market skills demand. The precision, recall, and F-measure were computed for 60 clusters between the two datasets and is summarized in Table 1.

**Table 1.** Prediction F-measure for each quadrant location in **all** investigated series.

| Quadrant label | Precision | | | Recall | | | F-measure | | |
|---|---|---|---|---|---|---|---|---|---|
| | Naive | SES | Holt | Naive | SES | Holt | Naive | SES | Holt |
| Q1 | 69% | 60% | 57% | 76% | 86% | 83% | **72%** | 70% | 68% |
| Q2 | 71% | 72% | 71% | 67% | 43% | 40% | **69%** | 54% | 51% |

Overall the F-measure ranges between 68%–72% for Q1 and 51%–69% for Q2, with no prediction potential observed for Q3 (0 clusters) and Q4 (1 cluster). Since both Q1 and Q2 are the focus of this study in which in-demand and emerging skills clusters are contained; thus, these strategical clusters are considered to be the answer to the question of this study. The F-measure result for the Q1 label is quite reliable and seems to indicate that the proposed framework is suitable for predicting highly-demanded and emerging skills. Interestingly, for both Q1 and Q2 label, the highest F-measure (69% precision, 76% recall) is achieved by the Naive method, and so confirming that the series of clusters are indeed stable and steady over time.

Furthermore, we fix our investigation on the series, which shows a stable transformation during the period being studied. By using four categories (**A, B, C,** and **D**) we have identified in Sect. 4.4, we re-calculate the F-measure based on these groupings. The results as shown in Table 2, focus on the predictions for Q1 and Q2, i.e., we only consider the category-based predictions to determine which of the skill clusters are expected to shift to Q1 (most stable and developed) and Q2 (emerging and become more important). The second column shows the distribution (ratio) of clusters within the series assigned to these four categories. The results indicate that forecasts are most accurate when the clusters are already within a stable and steady series (category A), followed by those in category B, and finally D, however no established series found in category C.

By looking at the ratio, the proposed framework can generate a stable and steady series of clusters more than half of the whole generated series. In summary, demand trends ascribed to category A (which represents 63.3% of the total), the

**Table 2.** F-measure results organized by identified category (and percentage of series belonging to each category).

| Category | Percentage | Quadrant | Precision | | | Recall | | | F-measure | | |
|---|---|---|---|---|---|---|---|---|---|---|---|
| | | | Naive | SES | Holt | Naive | SES | Holt | Naive | SES | Holt |
| A | 63.3% | Q1 | 63% | 48% | 46% | 86% | 93% | 86% | **73%** | 63% | 60% |
| | | Q2 | 89% | 91% | 91% | 71% | 42% | 42% | **79%** | 57% | 57% |
| B | 16.7% | Q1 | 100% | 100% | 100% | 60% | 80% | 80% | 75% | **89%** | **89%** |
| | | Q2 | 0 | 0 | 0 | – | – | – | – | – | – |
| D | 20% | Q1 | 57% | 57% | 50% | 80% | 80% | 80% | **67%** | **67%** | 62% |
| | | Q2 | 60% | 60% | 50% | 50% | 50% | 33% | **55%** | **55%** | 40% |

highest F-measure is achieved by the Naive method at 73% and 79% for Q1 and Q2 respectively. Therefore it is clear that the proposed framework can generate a series of steady and stable clusters and to use it further to track established high-demanded skills clusters and the emergence of new skills clusters identified by their 'quadrant' location.

## 6    Conclusion

In this paper, we have proposed SCODIS: a quadrant-based framework that can observe the evolution of dynamic skills graphs and forecast which skillsets that are being requested. In fast-evolving sectors impacted by technology, their labor market tends to be large and, at the same time, very dynamic with rapid changes in the skills being referenced, thus a scalable and realistic evolutionary demand tracking method is required. The framework is based on the use of certain critical indices (*centrality, density*) and cluster categories (*isolated, secondary, principal, crossroads*), that facilitate our ability to compute and reason about novel strategical-oriented observations, which can offer new and interesting insights for the identification of dynamic requirement of such labor market domain. We have presented a dense and central-based model for constructing a series of skills clusters and have shown the use of quadrant location patterns for demand prediction. We have demonstrated the efficacy of our framework in predicting our job adverts dataset. The application of the quadrant-based method we obtained to a skillset demand prediction (emerging or re-inforced skillsets) scenario provided favorable results with sufficient prior observations. A critical next step for us is to extend our framework for comparing clusters over time. In this context, we would like to improve the performance of series generation by leveraging the knowledge graph for semantic similarity, which currently relies only on the contents from each cluster. We would also like to extend our framework to predict in-demand skillsets using other types of the learning-based forecasting model.

## References

1. Smith, D., Ali, A.: Analyzing computer programming job trend using web data mining. Issues Inf. Sci. Inf. Technol. **11**, 203–214 (2014)

2. Sodhi, M.S., Son, B.-G.: Content analysis of OR job advertisements to infer required skills. J. Oper. Res. Soc. **61**, 1315–1327 (2010)
3. Coulter, N., Monarch, I., Konda, S.: Software engineering as seen through its research literature: a study in co-word analysis. J. Am. Soc. Inf. Sci. **49**, 1206–1223 (1998)
4. Blondel, V., Guillaume, J.-L., Lambiotte, R., Lefebvre, E.: Fast unfolding of communities in large networks. J. Stat. Mech: Theory Exp. **2008**, P10008 (2008)
5. Sibarani, E.M., Scerri, S., Morales, C., Auer, S., Collarana, D.: Ontology-guided job market demand analysis: a cross-sectional study for the data science field. In: SEMANTiCS, pp. 25–32 (2017)
6. Newman, M.E.J., Girvan, M.: Finding and evaluating community structure in networks. Phys. Rev. E **69**, 026113 (2004)
7. Xu, X., Yuruk, N., Feng, Z., Schweiger, T.A.J.: SCAN: a structural clustering algorithm for networks. In: KDD, pp. 824–833 (2007)
8. Callon, M., Courtial, J.P., Laville, F.: Co-word analysis as a tool for describing the network of interactions between basic and technological research: the case of polymer chemistry. Scientometrics **22**(1), 155–205 (1991)
9. Lee, P., Lakshmanan, L.V.S., Milios, E.E.: Incremental cluster evolution tracking from highly dynamic network data. In: ICDE, pp. 3–14 (2014)
10. Hopcroft, J., Khan, O., Kulis, B., Selman, B.: Tracking evolving communities in large linked networks. Proc. Nat. Acad. Sci. **101**(suppl 1), 5249–5253 (2004)
11. Dadzie, A.-S., Sibarani, E., Novalija, I., Scerri, S.: Structuring visual exploratory analysis of skill demand. Web Semant. **49**, 51–70 (2018)
12. Fortunato, S., Lancichinetti, A.: Community detection algorithms: a comparative analysis. In: The Fourth International ICST Conference on Performance Evaluation Methodologies and Tools, pp. 27:1–27:2 (2009)
13. Sibarani, E.M., Scerri, S.: Generating an evolving skills network from job adverts for high-demand skillset discovery. In: Cheng, R., Mamoulis, N., Sun, Y., Huang, X. (eds.) WISE 2020. LNCS, vol. 11881, pp. 441–457. Springer, Cham (2019). https://doi.org/10.1007/978-3-030-34223-4_28
14. Todd, P.A., McKeen, J.D., Gallupe, R.B.: The evolution of IS job skills: a content analysis of IS job advertisements from 1970 to 1990. MIS Q. **19**(1), 1–27 (1995)
15. Wowczko, I.A.: Skills and vacancy analysis with data mining techniques. Informatics **2**, 31–49 (2015)
16. Aken, A., Litecky, C., Ahmad, A., Nelson, J.: Mining for computing jobs. IEEE Softw. **27**(1), 78–85 (2010)
17. Surakka, S.: Analysis of technical skills in job advertisements targeted at software developers. Inform. Educ. **4**(1), 101–122 (2005)
18. Kennan, M.A., Willard, P., Cecez-Kecmanovic, D., Wilson, C.S.: A content analysis of Australian IS early career job advertisements. Austr. J. Inf. Syst. **15**(2), 169–190 (2009)
19. Asur, S., Parthasarathy, S., Ucar, D.: An event-based framework for characterizing the evolutionary behavior of interaction graphs. In: KDD, pp. 913–921 (2007)
20. Kim, M.-S., Han, J.: A particle-and-density based evolutionary clustering method for dynamic networks. In: VLDB, pp. 622–633, August 2009

# Temporal, Spatial, and High Dimensional Databases

# Temporal Aggregation of Spanning Event Stream: A General Framework

Aurélie Suzanne[1,2]($\boxtimes$) (ID), Guillaume Raschia[1] (ID), and José Martinez[1]

[1] Université de Nantes, 1 Quai de Tourville, 44035 Nantes, France
{aurelie.suzanne,guillaume.raschia,jose.martinez}@ls2n.fr
[2] Expandium, 15 Boulevard Marcel Paul, 44800 Saint-Herblain, France

**Abstract.** The Big Data era requires new processing architectures among which the stream systems that have become very popular. Those systems are able to summarize infinite data streams with aggregates on the most recent data. However, up to now, only point events have been considered and spanning events, which come with a duration, have been let aside, restricted to the persistent databases world only. In this paper, we propose an unified framework to deal with such stream mechanisms on spanning events. To this end, we formally define a spanning event stream with new stream semantics and events properties. We then review and extend usual stream windows to meet the new spanning events requirements. Eventually, we validate the soundness of our new framework with a set of experiments, based on a straightforward implementation, showing that aggregation of spanning event stream is providing as much new insights on the data as effectiveness in several use cases.

**Keywords:** Data stream · Spanning events · Temporal aggregates · Temporal database · Window query

## 1 Introduction

Data stream processing has been widely studied in recent years [6,12], and many industrial systems are now using it [13] with applications such as monitoring systems for networks, marketing, transportation, manufacturing or IoT systems. Retrieving useful insights from this continuously produced data has hence become a key issue. A common way to process those streams is to aggregate events with respect to the instant they occurred or arrived in the system.

Time is a first class dimension for stream events as it determines how they will be aggregated, but up to now a strong assumption was made about point events, leaving aside spanning events. Let us consider a network monitoring system where we want to evaluate the load of an antenna, with spanning transactions, e.g., phone calls, happening continuously. In a classical streaming system, the load would be based either on the start or on the end time of the event. With a spanning event stream the full event duration would be interpreted.

Figure 1 models a series of calls: events $a_i$ show their reception time, while $b_i$'s show the full call duration. We want to analyse the load of the antenna every

© Springer Nature Switzerland AG 2020
S. Hartmann et al. (Eds.): DEXA 2020, LNCS 12392, pp. 385–395, 2020.
https://doi.org/10.1007/978-3-030-59051-2_26

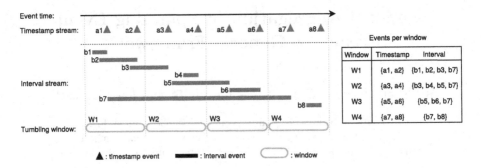

**Fig. 1.** End time vs. full-time events aggregation in a window-based stream system.

5 min showed by $W_i$'s. With spanning events, stream intervals $b_3$ and $b_5$ span over resp. windows $\{w_1, w_2\}$ and $\{w_2, w_3\}$, while their matching timestamps $a_3$ and $a_5$ are uniquely assigned to resp. $w_2$ and $w_3$. Natively modeling event duration also allows to detect events which have no bounds in the window, like event $b_7$ crossing window $w_2$ and $w_3$. Spanning event stream (SES) hence allows to get not only information about (dis)connections, but also to the full connection information, providing more accurate results than point event stream (PES).

To be able to provide such results, interval comparison predicates, inspired by Allen's algebra [1], need to be properly set up to allow assignment of spanning events to windows. Furthermore, as a side-effect of spanning events, past windows can be affected by fresh new events and this without any delay in the stream: on Fig. 1, $b_7$ not only impacts $w_4$, but also past windows $w_1$, $w_2$, and $w_3$. Those simple observations motivate the need for a close review of the many windows types, and a new window-based aggregation framework to address SES.

The rest of this paper is organized as follows: Sect. 2 focuses on the formal requirements to elaborate the framework. Section 3 discusses adaptation of common windows to spanning events. We propose, in Sect. 4, a straightforward implementation of the framework. Finally, we review prior works in Sect. 5, before the general conclusion in Sect. 6.

## 2   Problem Analysis and Definitions

### 2.1   About Time

Following the dominant view point, one defines the time domain as an infinite, totally ordered, discrete set $(\mathbb{T}, \prec_{\mathbb{T}})$, with units coined *chronons* [2]. As in temporal databases [2], two dimensions can be used: *valid time* and *transaction time*, being resp. the lifespan of an event in the real world and in the system. In streaming systems, transaction time is usually reduced to the start time point.

In this bi-temporal model, intervals are required to represent valid time. As an adjacent series of time points in $\mathbb{T}$, they are entirely defined by their lower and upper bounds, as pairs $(\ell, u) \in \mathbb{T} \times \mathbb{T}$ with $\ell \prec_{\mathbb{T}} u$. We denote by $\mathbb{I} \subset \mathbb{T} \times \mathbb{T}$ the set of time intervals. For any $t$ in $\mathbb{I}$, $\ell(t) \in t$ and $u(t) \notin t$ are resp. the lower and upper bounds of interval $t$, such that a chronon can be represented by $[c, c+1)$. Two intervals can be compared with the 13 Allen's predicates [1] which define precisely how the bounds compare to each other.

## 2.2 Spanning Event Stream

In this article, we extend the stream concept to incorporate spanning events with a lifespan as valid time. As a consequence, it is required to distinguish valid time (interval) and transaction time (point) of an event. Since data stream applications require near-real-time computation, valid and transaction time should be ideally connected, e.g., a phone call is recorded as soon as it ends.

We define a SES in Eq. 1, where $\Omega$ is any composite type that brings the content of each event $e \in S$; $t$ is the valid time interval; and $\tau$ is the transaction timestamp of $e$. We denote by $t(e)$ and $\tau(e)$ those values for an event $e$.

$$S = (e_i)_{i \in \mathbb{N}} \quad \text{with } e_i = (x, t, \tau) \in \Omega \times \mathbb{I} \times \mathbb{T} \tag{1}$$

We assume that we record events when they finish, in a no-delay stream setting, such that $u(t(e)) = \tau(e)$. In the following, we denote by $S(.)$ any projection of stream $S$ on one or more of its 3 components. For instance, $S(t)$ refers to the sequence of valid time intervals of all the events. We also denote $\tau_i$ the transaction time of the $i$th event in $S$, ie., $\tau_i = \tau(e_i)$.

## 2.3 Temporal Windowing

In data stream processing, data are transient and queries persistent, meaning that queries are continuously re-evaluated as data arrive. Blocking operators, such as aggregates, are a challenge for those *continuous queries* as they require to scan all the data set before producing the first answer [6]. A popular way to bypass this issue is to operate on a bounded sub-stream given by a *window* [6]. Indeed, closing a window unblocks the operation and an answer can be given with respect to the events "inside" the window. A common practice is to define infinite family of windows such like "each hour". Many window flavors exist [4,6,12], and in this paper, we propose a new categorisation for those windows.

**Measures.** Measures are useful to set up the shape and/or frequency of windows in a window family. Each measure is time related, to provide finite window bounds. We denote the measure set by $\mathcal{M} = \{\mathbb{T}\} \cup \{S(t)\} \cup \{S(\tau)\} \cup \{S(x, \tau)\}$ and consider it as extensive. Measures are the following:

- **Stream independent with a wall-clock time** $\mathbb{T}$: a system clock is used. Opening and/or closing windows are then independent from the stream.
- **Stream dependent with a stream projection:**
  - **Valid-time** $S(t)$ uses the valid time of events, e.g., sessions where bounds depend on the traffic flow.

- **Shape** $S(\tau)$ uses the transaction time of the events, more common example is *Count-based* which counts events arriving in the stream.
- **Data** $S(x,\tau)$ uses the data of events and orders them with their transaction time. Most common types are: *Delta-based* which uses a data part [6], e.g., a transaction counter; *Punctuation-based*: where window bounds are sent in the stream as special events [6].

**Formal Definition.** Formally, a temporal window is a regular time interval. A family of windows is $W = (w_k)_{k \in \mathbb{N}}$, $w_k \in \mathbb{I}$ ordered by increasing $\ell(w_k)$ or $u(w_k)$. $W \in \mathcal{W}$ is the set of windows families. We propose a couple $(F_{bounds}, P_{insert})$ to compute any window family.

$F_{bounds}^n : \mathcal{M}^n \rightarrow \mathcal{W}$, defines the bounds of the window. It uses one or several measures, and outputs a set of intervals (as shown in Table 1).

**Table 1.** Examples of windows created with $F_{bounds}$ and $F_{bounds}^2$

| $\mathcal{M}$ | Example of $\mathcal{W}$ | $\mathcal{M}^2$ | Example of $\mathcal{W}$ |
|---|---|---|---|
| $\mathbb{T}$ | 15 min each 5 min | $\mathbb{T}, S(\tau))$ | 5 min each 100 events |
| $S(t)$ | session | $\mathbb{T}, S(x,\tau))$ | 50 transactions each 5 min |
| $S(\tau)$ | 100 events each 10 events | $S(t), S(\tau)$ | 100 events each session |
| $S(x,\tau)$ | 50 transactions each punctuation | $S(t), S(x,\tau)$ | Session each punctuation |

$P_{insert} : \mathbb{I}^2 \rightarrow \mathcal{B}$, determines event belonging to a window with an Allen-like predicate. It takes as input two intervals: window bounds and event valid time, and outputs a Boolean. We define two specializations:

- $P_\triangle$: True if $\ell(w) \leq u(e) < u(w)$ else False, deals with point events by using a chronon and a window interval
- $P_\cap$: True if $u(e) > \ell(w) \wedge \ell(e) < u(w)$ else False, asserts if an event has at least one chronon in a window or not

**Common Windows Review.** We will now review window types which are often used in the literature [6,12]. Within PES, event valid time is a point, and hence for all those window types, $P_{insert} = P_\triangle$.

*Sliding window*: window is defined by $\omega$ the range or size of the window, and $\beta$ the step which sets up the delay between two successive windows. Most common measures are wall-clock time: $F_{bounds}(\mathbb{T}) = ((i, i+\omega) : i \mod \beta = 0)_{i \in \mathbb{T}}$ and count-based: $F_{bounds}(S(\tau)) = ((\tau_i, \tau_{i+\omega}) : i \mod \beta = 0)_{i \in \mathbb{N}}$

*Tumbling window*: tumbling windows can be seen as a specialization of sliding windows where $\omega = \beta$, meaning that only one window is open at a time.

*Session window*: a session is defined as a period of activity followed by a period of inactivity. Parameter $\varepsilon$ gives the inactivity threshold time range. Session window family is $F_{bounds}(S(t)) = ((\tau_i, \tau_j) : i < j \wedge (\tau_i - \tau_{i-1} \geq \varepsilon \vee \tau_i = \tau_0) \wedge \tau_j + \varepsilon \leq \tau_{j+1} \wedge \forall p \in [\![1, j-i]\!], \tau_{i+p} - \tau_{i+p-1} < \varepsilon)_{i,j \in \mathbb{N}}$.

# 3    Assigning Spanning Events to Temporal Windows

## 3.1    Adaptation of Point Events Windows to Spanning Events

We now detail modifications that using SES implies to the previously defined windows. This impact depends on the measures used and we will review of all common windows in a mono-measure context, $F_{bounds} : \mathcal{M} \rightarrow \mathcal{W}$.

Within PES, we saw that we can always use $P_{\triangle}$ for $P_{insert}$. With SES we have to take into account intervals. $P_{insert}$ depends on the window bounds definition $F_{bounds}$. If it used only transaction time we can keep $P_{\triangle}$, otherwise it needs to be modified to use an Allen's predicate (see Table 2).

**Table 2.** Most common spanning event window definition with $(F_{bounds}, P_{insert})$

| Window type | Point | | Spanning | |
|---|---|---|---|---|
| | $\mathcal{M}$ | $P_{insert}$ | $\mathcal{M}$ | $P_{insert}$ |
| Time sliding/tumbling | $\mathbb{T}$ | $P_{\triangle}$ | $\mathbb{T}$ | $P_{Allen}$ |
| Stream shape sliding/tumbling | $S(\tau)$ | $P_{\triangle}$ | $S(\tau)$ | $P_{\triangle}$ |
| Stream data sliding/tumbling | $S(x, \tau)$ | $P_{\triangle}$ | $S(x, \tau)$ | $P_{\triangle}$ |
| Session | $S(t) = S(\tau)$ | $P_{\triangle}$ | $\mathbf{S(t)}$ | $P_{Allen}$ |

Hence, we claim that among the most popular windows, stream shape and data sliding/tumbling windows can be used straightforwardly with SES. Time-based sliding/tumbling and session windows must conversely be extended.

## 3.2    Time-to-Postpone

When dealing with a SES where events are released only once ended, lifespan yields two problems: (1) the system should be able to wait for (expected) event completion before closing any window; and (2) long-standing events may be assigned to multiple windows. Figure 1 shows an example of such duration constraint with, for instance, the event $b_7$ released in window $w_4$, but assigned to $w_1$, $w_2$, and $w_3$ as well. Therefore, working on a valid time interval requires to postpone the release date of the aggregates in order to accept events that started in, or before the window. We call this waiting time the Time-To-Postpone (TTP).

Of course, TTP is a patch to overcome limitations of exact aggregate computation for temporal windows, and can lead to approximate results since events may be ignored and aggregates already released. Advanced TTP techniques deserve to be explored in future works to leverage those approximations.

## 3.3    Time-Based Sliding and Tumbling Windows

Event assignment to a window depends on how their lifespan compare: an event $e$ is assigned to window $w$ if $P_{Allen}(t(e), w)$, with $P_{Allen}$ any Allen-like predicate searching for event in the window. Release time of the window $\tau_R$ depends on the TTP parameter $\delta$, satisfying $\tau_R \geq u(w) + \delta$. A full overview of the needed changes to deal with SES is presented in Table 3.

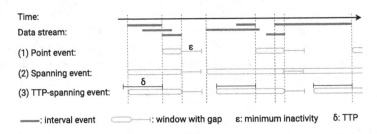

**Fig. 2.** 3 strategies for session windows.

**Table 3.** Comparison between PES and SES for sliding and session windows

| Properties | Time sliding/tumbling | | Session | |
|---|---|---|---|---|
| | Point | Spanning | Point | Spanning |
| Parameters | $(\omega, \beta)$ | $(\omega, \beta, P_{Allen}, \delta)$ | $(\varepsilon, \omega_{max})$ | $(\varepsilon, \omega_{max}, \delta)$ |
| New window | $i \mod \beta = 0, i \in \mathbb{T}$ | | $\nexists i \in \mathbb{N}, P_\triangle(\tau_i, \varepsilon(w))$ | $\nexists i \in \mathbb{N}, P_\cap(\Lambda(e_i), \varepsilon(w))$ |
| $\ell(w)$ | $i, i \in \mathbb{T}$ | | $(\min\{\tau(e)\})_{e \in S_w}$ | $(\min\{\lambda(e)\})_{e \in S_w}$ |
| $u(w)$ | $i + \omega, i \in \mathbb{T}$ | | $(\max\{\tau(e)\} + 1)_{e \in S_w}$ | $(\max\{u(t(e))\})_{e \in S_w}$ |
| $P_{insert}$ | $P_\triangle$ | $P_{Allen}$ | $P_\triangle$ | $P_\cap$ |
| Release time | $u(w)$ | $u(w) + \delta$ | $u(w) + \varepsilon + 1$ | $u(w) + \varepsilon + \delta$ |

### 3.4   Session Windows

In session windows, each received event either enters in the current window, or creates a new one. The upper bound of a session depends only on the end of the assigned events $u(w) = \max\{u(t(e))\}_{e \in S(w)}$. As for the lower bound, it must be chosen carefully. With PES, one can definitely decide the start of a session window as a fresh new event arrives. With SES, we must live-adjust this lower bound, since it requires to define an instant from a set of spanning events.

We model this problem as $\ell(w) = \min\{\lambda(e)\}_{e \in S_w}$, where $\lambda : S \to \mathbb{T}$ is a choice function that gives a reference point for an event. For a PES, this function is written as $\lambda(e) = u(t(e)) - 1$, using only the end bound as shown in strategy (1) in Fig. 2. When considering the lifespan of the event, a first estimate of the lower bound is $\lambda(e) = \ell(t(e))$, such that the event is starting in (and covers) the session. However, this strategy can lead to problems, as illustrated with strategy (2) where the last event leads to either re-opening or creating a session, causing impossible situations since the aggregate has already been released, and session overlaps are not allowed. To overcome this problem, we apply the TTP parameter $\delta$ to restrict back-propagation of the update; with $\lambda(e) = \max(\ell(t(e)), \tau(e) - \delta)$ as with strategy (3) which makes the long-standing event problem disappear.

Release of the session depends on the minimum inactivity period $\varepsilon$ and the TTP $\delta$, satisfying $\tau_R \geq u(w) + \varepsilon + \delta$. Several sessions can be active at the same time, and long events can yield to merge sessions. Table 3 details the adaptation between PES and SES sessions. We use $\varepsilon(w) = (u(w), u(w) + \varepsilon)$ as the inactivity interval, and $\Lambda(e) = (\lambda(e), u(t(e)))$ as the re-considered event with the TTP.

# 4    Experiments

**Experimental Setup.** In this series of experiments, data is not received at specific instant based on machine clock, but better "as fast as possible."

**Data Set.** We use 2 kinds of data sets: *Generated data set* allows fine-grained synthesis of SES with configurable parameters: event size, session duration, and inactivity. For each chronon, an event is created, which can be canceled with session creation. Each event size is generated by a normal distribution ($\mu = 100$, $\sigma = 10$) around the event size parameter. The generated set is 200K events. *SS7 data set* replays real-world-like data coming from a telephony network, assembling 1 min of communication with 3.2M events.

**Aggregates.** Aggregation in all experiments is a multi-measure of three aggregate functions: count, sum, and max.

**Setup.** All experiments were executed with an Intel(R) Core(TM) i7-8650U CPU @ 1.90 GHz with 16 GB RAM running under Linux Debian 10. Implementation is done in modern C++, using a single core.

**Implementation.** Implementation uses an event-at-a-time execution. For PES windows an unique FIFO queue is used, with new events added and old ones removed each time the window is released. With SES, such an implementation is not possible. Instead, events pointers are stored in a *bucket* per window.

**Results.** All the scenarios chosen in this series of experiments have been motivated by industrial requirements, especially in the field of telecommunication.

**Time-Based Windows.** The predicate used for event assignment is $P_\cap$. As expected the error rate between PES and SES increases with the event size, and decreases with window range (see Fig. 3a). This validates the soundness of using PES but also the urge to choose wisely the window range. When using an Oracle, which knows all the stream, we can validate the need for a TTP within SES, which should be chosen accordingly to the event size (see Fig. 3b). Concerning the throughput, TTP has a restricted impact when the window range evolves (see Fig. 3c), which is not the case for increasing events size. SES yield to many duplicates among the windows, which comes with a cost in throughput. For increasing duplications, the throughput goes down, but it stays roughly the same

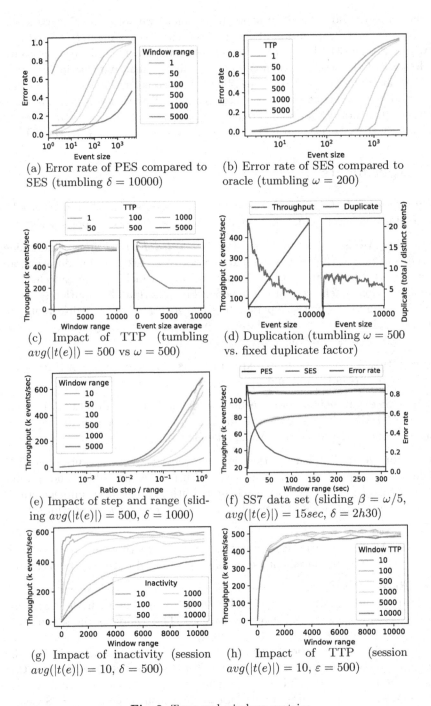

(a) Error rate of PES compared to SES (tumbling $\delta = 10000$)

(b) Error rate of SES compared to oracle (tumbling $\omega = 200$)

(c) Impact of TTP (tumbling $avg(|t(e)|) = 500$ vs $\omega = 500$)

(d) Duplication (tumbling $\omega = 500$ vs. fixed duplicate factor)

(e) Impact of step and range (sliding $avg(|t(e)|) = 500$, $\delta = 1000$)

(f) SS7 data set (sliding $\beta = \omega/5$, $avg(|t(e)|) = 15sec$, $\delta = 2h30$)

(g) Impact of inactivity (session $avg(|t(e)|) = 10$, $\delta = 500$)

(h) Impact of TTP (session $avg(|t(e)|) = 10$, $\varepsilon = 500$)

**Fig. 3.** Temporal windows metrics

for increasing window range with the same duplication rate (see Fig. 3d). Dupli-cation induced by overlapping in sliding windows also has a strong impact on throughput (see Fig. 3e). Concerning real-world applications, on a naïve imple-mentation the throughput of SES is only 30% slower than PES, and the error rate is still around 20% for 1 min windows, as we can see on Fig. 3f.

**Session Windows.** As shown on Figs. 3g and 3h, reducing window size as a negative impact on throughput. This is in accordance with time-based windows and refers to how often aggregates should be computed. Figure 3g highlights the impact of inactivity duration on throughput which is quite high. When main-taining inactivity periods at a same level, we can observe that TTP has a small influence on the throughput (see Fig. 3h).

**Summary.** This series of experiments shows that our framework is consistent with the all required assumptions for window-based aggregation on SES, and in particular the TTP. It then deserves to be pushed further in order to gain efficiency and completely meet the industrial requirements.

## 5    Related Work

The work done in this paper elaborates on previous work on data stream pro-cessing and temporal databases.

**Window Aggregation in Data Stream Processing.** Windowing is a com-mon technique and a common categorization of window characteristics is given in [4,12] with CF, FCF and FCA classes. Depending on those characteristics, several optimisations techniques have been proposed, such as sub-aggregating the input stream, and using aggregate tree indexes [4,12]. Studied in the context of PES, we believe that the extension of such methods would be of great interest to fasten window-based aggregation of SES. Nevertheless, the window approach has been criticized for its inability to take into account delayed or out-of-order streams. Some methods have been proposed to fix the delay issue, among which an allowed waiting time (TTP in this paper), the use of punctuation in the stream [7], or even the generation of heartbeats [11].

**Temporal Databases.** Queries in a temporal database can be of various forms [2]. Among those, sequenced queries, where the query spans over a time range, are close to our temporal window-based aggregates. However, pure sequenced query is resource demanding and barely evaluated with a one-pass algorithm [8]. Several methods were proposed to evaluate sequenced queries, mainly with graph or indexes [2,9], but as an open issue, only few market databases implement them [3,5,10]. Temporal aggregates on spanning events have not been widely studied. In [14], the authors combine windows and full history with a fine-to-coarse grain along the timeline, using SB-tree structure to index events and evaluate the queries. However, the approach is out-dated w.r.t. recent advances in window-based stream processing and temporal databases.

# 6    Conclusion and Future Work

This paper aimed to introduce a brand new consideration in stream data with the integration of spanning events. To do so, we first introduced notions common to temporal databases with a valid time range and a transaction time point for events in a bi-temporal model. Then, a common solution to overcome the infinite stream problem with blocking operators is to use windowing. To that extent, we conducted a careful review that yielded to a new categorization of usual measures and the definition of a pattern (function, predicate) to define every popular window family as well as the forthcoming ones. We showed that, among the real-life window families, only time-based sliding/tumbling and session windows need to be adapted to handle spanning events.

Among those changes, we introduced pairwise interval comparison, as for Allen's algebra, for event assignment to windows. We also had to define a Time-To-Postpone parameter that allows for long-standing events to be properly assigned to past windows. In the experiments, we showed that spanning events can be processed by a stream system. We demonstrated that our framework is effective for fixing PES errors. We also pointed out some behaviors, like assignment duplication of events, which is a great challenge for real-life applications.

As future work, we anticipate that the implementation should use more advanced techniques to share parts of computations among windows. Delay also should be studied in more details. Finally, extension of the aggregation, such as new operations like grouping or filtering, should also be considered with the ultimate goal of making the system fully operational in real-world conditions.

# References

1. Allen, J.F.: Maintaining knowledge about temporal intervals. Commun. ACM **26**(11), 832–843 (1983). https://doi.org/10.1145/182.358434
2. Böhlen, M.H., Dignös, A., Gamper, J., Jensen, C.S.: Temporal data management – an overview. In: Zimányi, E. (ed.) eBISS 2017. LNBIP, vol. 324, pp. 51–83. Springer, Cham (2018). https://doi.org/10.1007/978-3-319-96655-7_3
3. Böhlen, M.H., Dignös, A., Gamper, J., Jensen, C.S.: Database Technology for Processing Temporal Data. In: 25th International Symposium on Temporal Representation and Reasoning, TIME 2018 (2018, Invited Paper). https://doi.org/10.4230/lipics.time.2018.2
4. Carbone, P., Traub, J., Katsifodimos, A., Haridi, S., Markl, V.: Cutty: aggregate sharing for user-defined windows. In: CIKM 2016, pp. 1201–1210 (2016). https://doi.org/10.1145/2983323.2983807
5. Dignos, A., Glavic, B., Niu, X., Bohlen, M., Gamper, J.: Snapshot semantics for temporal multiset relations. Proc. VLDB Endow. **12**(6), 639–652 (2019). https://doi.org/10.14778/3311880.3311882
6. Gedik, B.: Generic windowing support for extensible stream processing systems. Softw. - Pract. Exp. **44**(9), 1105–1128 (2014). https://doi.org/10.1002/spe.2194
7. Kim, H.G., Kim, M.H.: A review of window query processing for data streams. J. Comput. Sci. Eng. **7**(4), 220–230 (2013). https://doi.org/10.5626/JCSE.2013.7.4.220

8. Moon, B., Lopez, I.F.V., Immanuel, V.: Efficient algorithms for large-scale temporal aggregation. IEEE Trans. Knowl. Data Eng. **15**(3), 744–759 (2003). https://doi.org/10.1109/TKDE.2003.1198403
9. Piatov, D., Helmer, S.: Sweeping-based temporal aggregation. In: Gertz, M., et al. (eds.) SSTD 2017. LNCS, vol. 10411, pp. 125–144. Springer, Cham (2017). https://doi.org/10.1007/978-3-319-64367-0_7
10. Snodgrass, R.T.: A case study of temporal data. Teradata Corporation (2010)
11. Srivastava, U., Widom, J.: Flexible time management in data stream systems. In: PODS 2004, pp. 263–274 (2004). https://doi.org/10.1145/1055558.1055596
12. Traub, J., et al.: Efficient window aggregation with general stream slicing. In: EDBT 2019, pp. 97–108. OpenProceedings (2019). https://doi.org/10.5441/002/edbt.2019.10
13. Yang, P., Thiagarajan, S., Lin, J.: Robust, scalable, real-time event time series aggregation at Twitter. In: SIGMOD 2018, pp. 595–599 (2018). https://doi.org/10.1145/3183713.3190663
14. Zhang, D., Gunopulos, D., Tsotras, V.J., Seeger, B.: Temporal aggregation over data streams using multiple granularities. In: Jensen, C., et al. (eds.) EDBT 2002. LNCS, vol. 2287, pp. 646–663. Springer, Heidelberg (2002). https://doi.org/10.1007/3-540-45876-X_40

# Leveraging 3D-Raster-Images and DeepCNN with Multi-source Urban Sensing Data for Traffic Congestion Prediction

Ngoc-Thanh Nguyen[1,2], Minh-Son Dao[3(✉)], and Koji Zettsu[3]

[1] Vietnam National University, Ho Chi Minh City, Vietnam
[2] University of Information Technology, Ho Chi Minh City, Vietnam
thanhnn.13@grad.uit.edu.vn
[3] National Institute of Information and Communications Technology, Tokyo, Japan
{dao,zettsu}@nict.go.jp

**Abstract.** Nowadays, heavy traffic congestion has become an emerging challenge in major cities, which should be tackled urgently. Building an effective traffic congestion predictive system would alleviate its impacts. Since the transit of vehicles heavily depends on its spatial-temporal correlations and effects of exogenous factors such as rain and accidents, they should be simultaneously considered. This study proposes a deep learning approach based on 3D-CNN to utilize many urban sensing data sources wrapped into 3D-Raster-Images. Armed with this, the spatial and temporal dependencies of the data can be entirely preserved. Furthermore, traffic congestion status of different geographical scales at various time horizons can be fully explored and analyzed. We also propose data fusion techniques to (1) fuse many environmental factors that affect vehicles' movements, and (2) incorporate social networking data to improve predictive performance further. The experiments are performed using a dataset containing four sources of urban sensing data collected in Kobe City, Japan, from 2014–2015. The results show that the predictive accuracy of our models improves significantly when using multiple urban sensing data sources. Finally, to encourage further research, we publish the source code of this study at https://github.com/thanhnn-uit-13/Fusion-3DCNN-Traffic-congestion.

**Keywords:** Traffic congestion · Deep learning · Spatio-temporal data

## 1 Introduction

In [1], Lana et al. argued that traffic congestion prediction, which is a part of the urban computing domain, is a crucial and urgent topic. The authors also introduced many exciting new topics such as concept drift in data-driven models, big data, and architecture implementation. Finally, they pointed out that

S. Hartmann et al. (Eds.): DEXA 2020, LNCS 12392, pp. 396–406, 2020.
https://doi.org/10.1007/978-3-030-59051-2_27

increasing the prediction time horizon (i.e., longer than 1 h) and incorporating exogenous factors to models are two of the major challenges.

In [2], the authors introduced a fascinating overview of fusing urban big data based on deep learning. This research emphasized the critical role of urban big data fusion based on deep learning to get more values for urban computing.

In [3], the vital role of traffic data visualization was seriously concerned. The authors compiled several techniques designed for visualizing data based on time, location, spatial-temporal information, and other properties in traffic data. They aim to satisfy the need to have good traffic data visualization to analyze as well as reveal hidden patterns, distributions, and structure of the data.

In light of the above discussions, we introduce a new deep learning and multi-source data fusion approach to predict the average traffic congestion length from different road segments concurrently appearing in the predefined area by using multiple urban sensing data sources with variable time-horizons. To do this, we first wrap spatial-temporal information of multi-source data into 2D/3D raster-images by using the method introduced in [4]. Then, we utilize recent advances in convolutional neural networks to build a deep learning model for traffic congestion prediction. The inputs of the model are (1) environmental sensing data that are fed via separate channels and (2) social networking data which is incorporated with the channels based on the content of the posts via a weighted function. The techniques significantly enhance the prediction performance of our models. The contribution of this research is summarized as follows:

- **Multi-scale Time Horizons:** [1] argued that short-term traffic (60 min or below) had been a big interest of the research community, but the investigations on medium- and long-term are neglected. Some more recently published works like [5–7], and [8] also did not address it. We explore all the time windows mentioned. The predictive time ranges from 1.5 h to 1 day with multiple immediate steps varying from 30 min to 4 h.
- **Multi-source/Multimodal Big Data Fusion:** This study proposes data-fusion techniques to simultaneously extract and feed to models various urban sensing data types including (1) traffic congestion, (2) precipitation, (3) vehicle collisions, and (4) social networking data. While the first three categories were included in many highly influential works such as [5,6], the last one has attracted less interest and even required more research to conclude its impact [9]. Our proposed models success them by considerably improving the predictive performance compared to models using single-source data.
- **Spatial-temporal Wrapping:** The raster-image-based wrapping technique utilized in this study can reserve the correlations between spatial and temporal information, which are crucial in urban-related domains. It also allows us to look back and give predictions over variable time horizons with multiple steps unlike [5,10] which give predictions only on single time steps. Last but not least, by using this data format, we can leverage sophisticated 2D-CNN and 3D-CNN layers to build learning models.

The paper is organized as follows: Sect. 2 describes procedures to pre-process multi-source data. Section 3 explains fusion techniques to merge many sources of sensing data. Section 4 discusses contributions of the proposed method. Finally, Sect. 5 concludes the work and suggests possible future research directions.

# 2   Data Preprocessing

This section discusses the procedure to prepare data that is fed to models.

## 2.1   Data Storage

This study uses four different urban data sensing sources:

- Traffic congestion: offered by Japan Road Traffic Information Center, JAR-TIC (www.jartic.or.jp),
- Precipitation: received from Data Integration and Analysis System Program, DIAS (www.diasjp.net),
- Vehicle collisions: got from Institute for Traffic Accident Research and Analysis, ITARDA (www.itarda.or.jp), and
- Social networking data: bought from TWITTER (www.twitter.com). Twitter provided us with posts that contain certain keywords relating to bad traffic status, heavy rain, and vehicle collisions in the requested/examined area. Each post contains *username, location, timestamp, concerned_keywords* information. In this study, this data is referred to as "SNS".

The mentioned types of sensing data are stored in a data warehouse system called Event Warehouse (EvWH) introduced in [4]. The table *analysis.raw_transation* stores four different urban sensing values at each local area identified by *meshcode* per timestamp $t$. Each location has a size of 250 m × 250 m.

## 2.2   Converting Urban Sensing Data to Raster-Images

This section presents the procedure to convert spatiotemporal urban sensing data stored in time-series format to 2D multi-layer raster-images. We store the four urban sensing sources to a single 2D multi-layer raster-image per timestamp $t$. Using this data format, we can simultaneously analyze their effects at a specific location at any time $t$ given. For example, Fig. 1 illustrates how (1) traffic congestion (Blue channel), (2) rain (Green channel), and (3) traffic accidents (Red channel) affected Osaka Bay, Kobe City, Japan at 10:00:00 AM July $17^{th}$, 2015. Their effects are well-interpreted via a single RGB raster-image. Since this problem deals with multiple time steps, we place many raster-images consecutively to form a 3D raster-image. Thus, it helps reserve the temporal correlations of the data.

In this study, we examine the area of 20 km × 50 km, which is represented on a raster-image sized 80 × 200 ($W \times H$). It is illustrated in Fig. 2. Each pixel on a raster-image pixel equals to a local area sized 250 m × 250m. Green points denote locations that have traffic congestion in the ground truth data. We use four different sensing data types ($L$), so each produced raster-image has a size

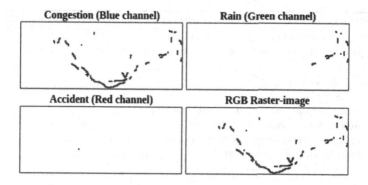

**Fig. 1.** Leveraging raster-images to visualize urban data

of $80 \times 200 \times 4$ $(W \times H \times L)$. Each sensing event is triggered and recorded every 5 min, so 288 raster-images are generated per day. For traffic prediction, we divide the examined area further into three smaller regions as denoted in red squares in Fig. 2. Since the other areas are waters or have very little traffic information, ignoring them will significantly alleviate the impact of sparse data.

**Fig. 2.** Region for traffic congestion prediction

Next, we explain the detailed procedure to convert time-series sensing data to 3D multi-layer raster-images in Algorithm 1. Following is the configurations of 3D multi-layer raster-images:

- Global map's size: $H \times W$ (Height × Width);
- Number of sensing data types: $L$ (Layer);
- Top-left coordinate of the map: $base\_lat$, $base\_lon$;
- Delta between 2 adjacent geographical locations: $d\_lat$, $d\_lon$;
- Number of time steps considered: $S$.

---

**Algorithm 1** Convert multi-source sensing data to a 3D multi-layer raster-image

    **Input:** Time-series sensing data: *data*
    **Output:** A 3D multi-layer raster-image $R$, sized $S \times H \times W \times L$.

  Initialize a raster-image $R$
  **while** $s \in S$ **do**
    Read sensing data at time $t + s$ into $data\_ts$
    **while** $l \in data\_ts$ **do**
      //Extract relative position on the raster-image
      $loc\_x \leftarrow (loc\_lat - base\_lat)/l\_lat$
      $loc\_y \leftarrow (loc\_lon - base\_lon)/l\_lon$

      //Assign sensing data to the desired location
      $R[s, loc\_x, loc\_y, l] \leftarrow l\_sensingdata$
    **end while**
  **end while**

---

## 3    Learning Model

### 3.1    Leveraging Computer Vision's Breakthroughs

After completing the procedures explained in Sect. 2, the data is stored in 3D multi-layer raster-images. They are similar to a video of conventional RGB pictures, so we can utilize research breakthroughs of the Computer Vision domain. [2] indicated that spatial and temporal relationships should be simultaneously considered when analyzing traffic movements. In [11], Tran et al. proposed 3D Convolutional Neural Network (3D-CNN) which can fully reserve such dependencies. Therefore, we decide to utilize this network in our learning models.

To better explain the complete solution of the study, we illustrate the whole workflow in Fig. 3. Firstly, the urban sensing data sources are converted to 3D Multi-layer Raster-Images by Algorithm 1 in Sect. 2. Then, different layers of the 3D multi-layer raster-images are either fed to our proposed learning models (called **Fusion-3DCNN**) in separate channels or integrated with others. We will explain them shortly. The input and output of Fusion-3DCNN are 3D Single-layer Raster-Images. The models look back $k$ ($k \geq 1$) historical milestones of multiple environmental factors to predict $m$ periods ($m \geq 1$) of traffic congestion. Depending on the values of $k$ and $m$, we will flexibly stack the number of 2D multi-layer raster-images to form a 3D multi-layer raster-image accordingly. This technique allows us to give predictions for various immediate time steps in different time horizons.

### 3.2    Data Fusion Techniques

This section reveals fusion techniques to feed multiple data sources to Fusion-3DCNN. The traffic may be affected by two types of sensing data: environmental factors and social networking data. The former contains traffic congestion,

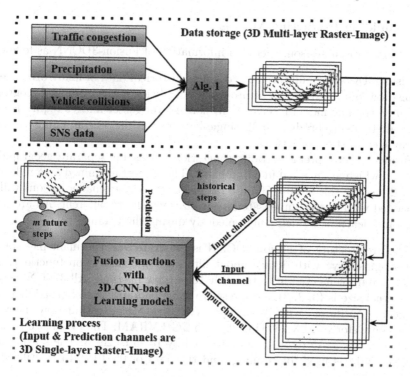

**Fig. 3.** Complete workflow of the study

precipitation, and vehicle collisions. They are called "environmental factors". They are learnable features and directly supplied to Fusion-3DCNN as follows: $W \times x = \sum_{i=1}^{N} W_{f_i} \times x_{f_i}$; where $f_i$ denotes individually learnable factors. The second sensing data group includes Tweets of Twitter's users complaining/warning about the bad surrounding environment. There are three types of criticism: (1) heavy traffic congestion, (2) heavy rain, and (3) vehicle collisions. As a matter of fact, if one is warned about bad environmental conditions at a location in his commuting route, he would avoid reaching it. Therefore, the negative effects of these factors on traffic congestion would be mitigated. The following formula denotes how large each explicit factor changes based on the community's online activities: $x_{(i,j)}^f = \begin{cases} (1-p) \times x_{(i,j)}^f, & \text{if } y_{(i,j)}^f = 1 \\ x_{(i,j)}^f, & \text{otherwise} \end{cases}$ where $x_{(i,j)}^f$ is the normalized value of the affected factor type $f$ at the location $(i,j)$ on raster-images; $y_{(i,j)}^f$ is a binary value indicating whether a warning related to that factor is detected (1: yes/0: no); and $p$ is a hyperparameter defining impact level of the SNS data on environmental factors.

### 3.3 Building Predicting Models

This section discusses some detailed information of Fusion-3DCNN as follows:

1. Fusion-3DCNN simultaneously considers historical data of many urban sensing data sources to predict future traffic congestion in the examined geographical areas. The model receives $k$ 3D single-layer raster-images (equivalent to $k$ milestones) to produce $m$ 3D single-layer raster-images containing $m$ future traffic congestion situations.
2. Fusion-3DCNN predicts traffic congestion status for separate geographical areas, which are marked in red in Fig. 2. The biggest examined area indicated by 2 has the size of $60 \times 80$ on a raster-image, so the smaller regions will be padded with 0s to compensate for the sizes which are different. All areas are fed into learning models simultaneously during the training process.

The architecture of Fusion-3DCNN is illustrated on the main page of the work's repository[1] with details about filter size and activation functions. The models are implemented in Keras with Tensorflow backend. All 3D-CNN layers have kernel size is (3, 3, 3) and are set to same padding, one-step striding. The models are optimized with Adam optimizer with MSE loss function. They are trained on Geforce GTX 750Ti GPU with 2 GB VRAM. The batch size is 1, the learning rate is 3e−5 to 5e−5, and decayed by 1e−5 to 2e−5. The sensing data of 05–10/2014 is used for training, and the data of 05–10/2015 is utilized for testing.

## 4 Empirical Evaluation

### 4.1 Evaluative Preparation

This section discusses settings prepared to evaluate models. Firstly, we prepare three distinct datasets with different looking back and predicting time horizons, as in Table 1. Next, we use four baselines namely (1) Historical Average - HA, (2) Sequence-to-Sequence Long Short-Term Memory - Seq2Seq LSTM, (3) 2D Convolutional Neural Network - 2D-CNN, and (4) 3D Convolutional Neural Network - 3D-CNN. They only use traffic congestion data. The first three neglect either spatial or temporal dependencies, while the last one reserves both. Therefore, 3D-CNN-based models are expected to be the best baseline. Beside, to show the effectiveness of using multi-source data, Fusion-3DCNN models are expected to be better than all the baselines. To evaluate the models' performance, we use Mean absolute error (MAE). Subsequently is the baselines' information:

- **HA:** traffic congestion value at each geographical area is calculated by: $C_{i,j}^{f} = \frac{1}{N} \sum_{h=1}^{N} C_{i,j}^{h}$; where $C_{i,j}^{h}$ and $C_{i,j}^{f}$ represents traffic congestion value at the location $(i, j)$ on raster-images at historical milestones $h$ and future stages $f$, respectively. This model only reserves temporal information.

---

[1] https://github.com/thanhnn-uit-13/Fusion-3DCNN-Traffic-congestion.

**Table 1.** Experimental datasets

| Data type description | Dataset type (-term) | | |
|---|---|---|---|
| | Short | Medium | Long |
| No. looking back frames | 6 | 6 | 6 |
| Delta between looking back frames | 30 min | 1 h | 4 h |
| Total looking back time | 3 h | 6 h | 24 h |
| No. prediction frames | 3 | 3 | 6 |
| Delta between prediction frames | 30 min | 1 h | 4 h |
| Total prediction time | 1.5 h | 3 h | 24 h |

– **Seq2Seq LSTM:** this network has achieved great successes on different tasks relating to sequential data[12]. However, it only learns temporal dependencies. We use five ($[400 \times 5]$) hidden LSTM layers for both encoder and decoder components, and train in 300 epochs.
– **2D-CNN:** this network is very efficient in learning and predicting data that requires the reservation of spatial information [13]. However, it totally neglects the temporal dimension. We use eight hidden 2D-CNN layers $[[128 \times 2] - [256 \times 4] - [128 \times 1] - [64 \times 1]]$ in this baseline, and train in 1 epoch.
– **3D-CNN:** this network can reserve both spatial and temporal dimensions [11]. We use eight hidden layers $[[128 \times 2] - [256 \times 4] - [128 \times 1] - [64 \times 1]]$, and train in 1 epoch.

Next, the models that use multi-source data are prepared as follows:

1. One Fusion-3DCNN model that uses (1) traffic congestion, (2) precipitation, and (3) vehicle collisions; and
2. Two Fusion-3DCNN models that gather the above three factors and social networking posts (SNS data). In the experiment, we perform a grid search to see how much the SNS data affects the other factors. Two impact levels are evaluated in this study: $[25\%, 75\%]$.

### 4.2 Evaluative Results Discussion

This section discusses conclusions extracted from empirical evaluation. Models' performance are presented in Table 3. Better models have lower values. Insights concluded from the results are also discussed subsequently. Table 2 identifies models presented in Table 3 with shorter names to reduce the space consumed.

Firstly, by considering the results of baselines which only use traffic congestion data, some conclusions can be drawn as follows:

– The predictive performance of Seq2Seq models in all datasets are quite bad compared to the others. That is because there is too much sparse data in our dataset, which accounts for 95%. Alvin et al. indicated that LSTM-based

**Table 2.** Shortened model names

| Model | Shortened name |
|---|---|
| Historical average | HA |
| Seq2Seq LSTM | Seq2Seq |
| 2D-CNN | 2D-CNN |
| 3D-CNN using congestion | 3D-CNN with C |
| Fusion-3DCNN using congestion & Precipitation & accidents | fusion-3DCNN CPA |
| Fusion-3DCNN using congestion & precipitation & Accidents & reduce their 25% impacts by SNS data | fusion-3DCNN CPA*$^1/_4$SNS |
| Fusion-3DCNN using congestion precipitation & Accidents & reduce their 75% impacts by SNS data | Fusion-3DCNN CPA*$^3/_4$SNS |

**Table 3.** Experimental results (measured in MAE)

| Dataset (-term) | Model | | | | | | |
|---|---|---|---|---|---|---|---|
| | HA | Seq2Seq | 2D-CNN | 3D-CNN with C | Fusion-3DCNN CPA | Fusion-3DCNN CPA*$^1/_4$SNS | Fusion-3DCNN CPA*$^3/_4$SNS |
| Short | 5.51 | 54.47 | 5.37 | 5.19 | 5.04 | **4.90** | 4.98 |
| Medium | 6.60 | 52.05 | 6.45 | 6.31 | 6.12 | 5.70 | **5.64** |
| Long | 7.05 | 56.33 | 6.85 | 6.63 | 6.16 | 5.86 | **5.72** |

models are data-hungry models [14], so this argument explains our problem. Because of that, we will ignore this baseline and only analyze the other baselines. Furthermore, deep learning models that use 2D-CNN and 3D-CNN show acceptable performance (better than the naive baseline - HA). It indicates that convolutions of those networks can effectively tackle data sparsity problems.

- The predictive performance of 2D-CNN models are slightly better than HA in all datasets. Besides, the predictive accuracy of 3D-CNN models are higher than the other baselines. It indicates that spatial and temporal dependencies are crucial. Therefore, leveraging 3D-CNN layers in building Fusion-3DCNN models is a correct choice in this study.

Next, some observations can be extracted when comparing the predictive performance 3D-CNN and Fusion-3DCNN models as follows:

- Generally, the predictive performance of Fusion-3DCNN models considering multiple sources of data are better than that of 3D-CNN models that use only traffic congestion data in all datasets. The predictive accuracy of Fusion-3DCNN CPA models are higher than 3D-CNN models by 3% to 7%. The gaps even extend when compare Fusion-3DCNN CPA-SNS with 3D-CNN. Particularly, the predictive accuracy of Fusion-3DCNN CPA-SNS models are higher than 3D-CNN by 4% to 14%. It indicates that all external environmental factors considered in this study really affect traffic congestion. It also proves that our proposed data fusion functions and strategies to leverage many urban sensing data sources simultaneously is effective in enhancing the predictive performance. Moreover, incorporating SNS data helps further increase forecasting accuracy. Therefore, this study has opened a promising research direction to utilize this abundant and easy-to-collect source of information in the studies relating to traffic congestion prediction.

– Considering the short-term dataset, once leveraging rainfall and traffic accident data, the predictive performance of Fusion-3DCNN CPA increases by 3% compared to 3D-CNN. Incorporating SNS data to the environmental factors makes the predictive performance of Fusion-3DCNN CPA*1/4SNS higher than Fusion-3DCNN CPA by 3%. Besides, since the predictive accuracy of Fusion-3DCNN CPA*3/4SNS is lower than Fusion-3DCNN CPA*1/4SNS, it indicates that aggressively reducing environmental impacts via SNS data could make models perform poorly. Therefore, it could be concluded that a minority of the population (about 25%) tend to update the latest environmental status during their short-distance commute.

– Moving to the longer-term datasets (medium-term and long-term), external factors have a more significant impact on traffic congestion compared to the short-term time window. In more detail, Fusion-3DCNN CPA models are better than 3D-CNN by 5% and 8% for medium-term and long-term datasets, respectively. It shows that when heavy rain or a traffic accident occurs, they are likely to affect traffic flows on longer time horizons (3 h–24 h versus 1 h 30 min). In addition, the best models in these two datasets are Fusion-3DCNN CPA*3/4SNS. The predictive performance of these models are better than Fusion-3DCNN CPA by 8% in both the datasets. The results reveal that a majority of people try to catch up with the latest happenings when planning their travel in these longer forecasting time horizons.

## 5   Conclusions

This study proposes a deep learning approach to predict the traffic congestion status of each geographical area in the examined regions by reserving both spatial and temporal information. It also simultaneously considers external factors that can affect the flows of vehicles. We successfully show the positive impacts of using environmental factors such as rain, traffic accident, and social networking contents in enhancing predictive performance. Owing to the advances of modern-day technology, collecting a variety of urban sensing data and social networking information is feasible. Thus, this work has opened a promising research direction to utilize various additional factors besides traffic congestion data. Finally, our raster-image-based wrapping solution could be utilized to integrate more urban data types perfectly. They play a vital role in building an ideal learning model to predict traffic congestion.

# References

1. Lana, I., Del Ser, J., Velez, M., Vlahogianni, E.I.: Road traffic forecasting: recent advances and new challenges. IEEE Intell. Transp. Syst. Mag. **10**(2), 93–109 (2018)
2. Liu, J., Li, T., Xie, P., Du, S., Teng, F., Yang, X.: Urban big data fusion based on deep learning: an overview. Inf. Fusion **53**, 123–133 (2020)
3. Chen, W., Guo, F., Wang, F.: A survey of traffic data visualization. IEEE Trans. Intell. Transp. Syst. **16**(6), 2970–2984 (2015)
4. Dao, M., Zettsu, K.: Complex event analysis of urban environmental data based on deep-CNN of spatiotemporal raster images. In: 2018 IEEE International Conference on Big Data (Big Data), pp. 2160–2169. IEEE (2018)
5. Yuan, Z., Zhou, X., Yang, T.: Hetero-convlstm: a deep learning approach to traffic accident prediction on heterogeneous spatio-temporal data. In: Proceedings of the 24th ACM SIGKDD International Conference on Knowledge Discovery & Data Mining, KDD 2018, pp. 984–992 (2018)
6. Tseng, F.H., Hsueh, J.H., Tseng, C.W., Yang, Y.T., Chao, H.C., Chou, L.D.: Congestion prediction with big data for real-time highway traffic. IEEE Access **6**, 57311–57323 (2018)
7. Chen, M., Yu, X., Liu, Y.: PCNN: deep convolutional networks for short-term traffic congestion prediction. IEEE Trans. Intell. Transp. Syst. **19**(11), 3550–3559 (2018)
8. Pan, Z., Liang, Y., Wang, W., Yu, Y., Zheng, Y., Zhang, J.: Urban traffic prediction from spatio-temporal data using deep meta learning. In: Proceedings of the 25th ACM SIGKDD International Conference on Knowledge Discovery & Data Mining, KDD 2019, pp. 1720–1730 (2019)
9. Pourebrahim, N., Sultana, S., Thill, J.C., Mohanty, S.: Enhancing trip distribution prediction with twitter data: comparison of neural network and gravity models. In: Proceedings of the 2nd ACM SIGSPATIAL International Workshop on AI for Geographic Knowledge Discovery, GeoAI 2018, pp. 5–8 (2018)
10. Di, X., Xiao, Y., Zhu, C., Deng, Y., Zhao, Q., Rao, W.: Traffic congestion prediction by spatiotemporal propagation patterns. In: 2019 20th IEEE International Conference on Mobile Data Management (MDM), pp. 298–303, June 2019
11. Tran, D., Bourdev, L., Fergus, R., Torresani, L., Paluri, M.: Learning spatiotemporal features with 3D convolutional networks. In: Proceedings of the 2015 IEEE International Conference on Computer Vision (ICCV), ICCV 2015, pp. 4489–4497, Washington, DC, USA, 2015. IEEE Computer Society (2015)
12. Shi, T., Keneshloo, Y., Ramakrishnan, N., Reddy, C.: Neural abstractive text summarization with sequence-to-sequence models. arXiv preprint arXiv:1812.02303 (2018)
13. Liu, W., Wang, Z., Liu, X., Zeng, N., Liu, Y., Alsaadi, F.E.: A survey of deep neural network architectures and their applications. Neurocomputing **234**, 11–26 (2017)
14. Kennardi, A., Plested, J.: Evaluation on neural network models for video-based stress recognition. In: Gedeon, T., Wong, K.W., Lee, M. (eds.) ICONIP 2019. CCIS, vol. 1143, pp. 440–447. Springer, Cham (2019). https://doi.org/10.1007/978-3-030-36802-9_47

# Privacy-Preserving Spatio-Temporal Patient Data Publishing

Anifat M. Olawoyin, Carson K. Leung(✉), and Ratna Choudhury

University of Manitoba, Winnipeg, MB, Canada
kleung@cs.umanitoba.ca

**Abstract.** As more data become available to the public, the value of information seems to be diminishing with concern over what constitute privacy of individual. Despite benefit to data publishing, preserving privacy of individuals remains a major concern because linking of data from heterogeneous source become easier due to the vast availability of artificial intelligence tools. In this paper, we focus on preserving privacy of spatio-temporal data publishing. Specifically, we present a framework consisting of (i) a 5-level temporal hierarchy to protect the temporal attributes and (ii) temporal representative point (TRP) differential privacy to protect the spatial attributes. Evaluation results on big datasets show that our framework keeps a good balance of utility and privacy. To a further extent, our solution is expected be extendable for privacy-preserving data publishing for the spatio-temporal data of coronavirus disease 2019 (COVID-19) patients.

**Keywords:** Database · Database system application · Privacy · Spatio-temporal data · Data publishing · Differential privacy

## 1 Introduction

In recent years, there are increasing pressure on various levels of government to embrace open data [6, 14]. Besides promoting transparency, there are also scientific, economic and social development benefits of open data initiatives. When publishing as open data, attributes—such as names and social insurance number (SIN)—that may uniquely identify individuals within the published records are usually de-identified. However, this de-identification method may fail to account for the possibility of uniquely identifying individuals by combining their spatio-temporal attributes from heterogeneous external datasets. Temporal attributes, when fully published, provide detailed-level information. Combining this information with spatial attributes may increase the risk of re-identification by an adversary having background knowledge of individuals.

Over the past decades, several models and techniques have been introduced to prevent re-identification of individual in data publishing. They can be classified into two broad categories: (a) syntactic models [16, 21, 23] and (b) *differential privacy* [2, 3, 30]. The latter applies some random (e.g., Laplace) noise such that

© Springer Nature Switzerland AG 2020
S. Hartmann et al. (Eds.): DEXA 2020, LNCS 12392, pp. 407–416, 2020.
https://doi.org/10.1007/978-3-030-59051-2_28

inserting a single record into (or deleting a single record from) the dataset would not have significant effects on the outcome of data analysis.

In this paper, we focus on privacy-preserving data publishing of spatio-temporal dataset. In particular, we exploit the notion of *temporal hierarchy* (which has been shown to improve accuracy of forecasting particularly under increased modelling uncertainty) [5] and time series aggregation [26]. In addition, we extend the notion of *representative point* [27] in the geographical information system (GIS) to temporal correlated dataset. To elaborate, in the GIS, near objects are considered more relevant when compared to far objects. So, we extend this concept from capturing only spatial relationships to capturing also temporal relationships. Specifically, we first group all spatial coordinates within the same temporal hierarchy into a single point called *temporal representative point*. We then apply differential privacy mechanism to temporal representative points to further improve the privacy of the released dataset. To maintain a good balance between anonymization and data utility, the count of all other attributes within the dataset remain the same. *Our key contributions of this paper* include (a) the introduction of the temporal representative point, as well as (b) a non-trivial integration of temporal hierarchy, time series aggregation, and these temporal representative points.

The remainder of this paper is organized as follows. The next section provides background and related works. Section 3 presents our privacy-preserving framework for spatio-temporal patient data publishing. Evaluation results are shown in Sect. 4. Finally, conclusions are drawn in Sect. 5.

## 2   Background and Related Works

Over the years, privacy-preserving techniques have been applied to various tasks (e.g., data mining [15,17,19,20,28], information retrieval [4]). These result in privacy-preserving tasks like privacy-preserving data mining (PPDM) [18,22,24, 31], privacy-preserving keyword data search [29], and *privacy-preserving data publishing* (*PPDP*) [11–13]. The latter is the focus of this paper.

The concept of anonymization in data publishing has been widely studied, which include $k$-anonymity [16], $l$-diversity [23], and $t$-closeness [21]. However, most of these models were designed and applicable mainly to relational databases. Due to high dimensionality and sparseness nature of spatio-temporal data [25], the computation cost of these models can be exponential.

Abul et al. [1] investigated the problem of anonymity preserving in moving objects databases, and designed a $(k, \delta)$-anonymity algorithm based on clustering. However, not all spatio-temporal dataset contains moving objects [25], and $(k, \delta)$-anonymity may also distort information.

Mohammed et al. [25] proposed an LKC-privacy algorithm for anonymizing high-dimensional, sparse and sequential trajectory data. It preserves the information utility for trajectory data. As a preview, our temporal hierarchy can be considered as its generalization, in which temporal items at the leaf level are suppressed and their associated identifiers are aggregated without losing information.

Several recent studies on privacy-preserving data publishing focus on *differential privacy* [2,7,9,30], which describes the pattern of a group within the dataset while keeping information private for individuals. In other words, it is a constraint on the algorithms for publishing aggregate information about a statistical database while limiting the disclosure of individual's sensitive information in the database. An algorithm is differentially private when an adversary seeing the output of the database cannot differentiate from original database, and thus, cannot tell if a particular individual's information was used in the computation. Hence, differential privacy prevents identification and re-identification attacks [10].

## 3   Our Privacy-Preserving Framework

Our privacy-preserving framework for spatio-temporal patient data publishing consists of three key components: temporal hierarchy, temporal representative point and Laplace mechanism for differential privacy. With this framework, given a spatio-temporal dataset, we first generate temporal hierarchy, then compute $\delta$-location sets. Afterward, we generate temporal representative point, and apply Laplace mechanism for differential privacy. Consequently, the published spatio-temporal dataset would preserve privacy.

### 3.1   Temporal Hierarchy

A *temporal hierarchy* is a non-overlapping temporal aggregated frequency designed over predefined periodic levels. It is similar to a bottom-up generalization, but with a condition that the parent is exactly an aggregate of its descendants. To elaborate, the smallest unit of time (e.g., hour) is at the leaf, and the aggregation is constructed bottom-up to the root. Each temporal hierarchy level has business implication. For example, operational level management may be interested in detail transaction by the hour for daily decision, staff scheduling, or resource allocation. Monthly, quarterly and annual aggregations are useful for both tactical and strategic levels of management for budgeting and several other business decisions.

Hence, we design a *5-level temporal hierarchy* with (a) hours at the bottom, (b) days, months and quarters at intermediate levels, and (c) the year at the top (i.e., root) of the hierarchy. See Fig. 1. Generally, each level is an aggregate of its intermediate level below. For example, the year is an aggregate of 4 quarters:

$$Y = \sum_{i=1}^{4} Q_i \tag{1}$$

Each of the four quarters is an aggregate of the corresponding 3 months. For example, Quarter 1 is an aggregate of the first 3 months in a year (i.e., January, February and March):

$$Q_j = \sum_{i=1}^{3} M_i \tag{2}$$

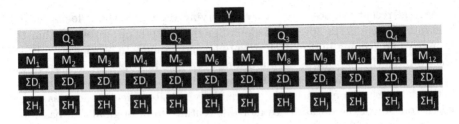

**Fig. 1.** Our 5-level temporal hierarchy

Similarly, each of three months within a quarter is an aggregate of the corresponding number of days (e.g., ranging from 28 to 31) within the month. For example, Month 1 (i.e., January) is an aggregate of its 31 days:

$$M_j = \sum_{i=1}^{d} D_i \qquad (3)$$

where $d$ is the maximum number of days within the Month $j$. For example, $d = 31$ for January. Finally, each day is an aggregate of its 24 h:

$$D_j = \sum_{i=1}^{24} H_i \qquad (4)$$

### 3.2    Temporal Representative Points

Recall from Sect. 1, a (spatial) representative point is a point that optimally represents a geographical area, a block-face, a multi-point, a line, or a polygon in the GIS. Here, we extend this notion of spatial representative point to a *temporal representative point*. In spatio-temporal context, a temporal representative point is a point representing all points (a line, multi-points or polygon) within a specific level of temporal hierarchy. The temporal representative point is guaranteed to be within the set of coordinate matrices constructed for any level of the temporal hierarchy.

Note that this temporal representative point is not necessary a centroid. For example, consider a user who visited multiple locations in the states of Missouri, New York, Oklahoma, and Texas. There may be six centroids for his trajectory: One centroid in Missouri, in New York, in Oklahoma, and three centroids in Texas. However, only four his temporal representative points are needed for his trajectory because three of his centroids are in the same state of Texas. This demonstrates a benefit of using the notion of temporal representative points—namely, a reduction in the number of representatives.

The pseudocode in Algorithm 1 shows the key steps for computing the temporal representative point. Here, we first generate the temporal hierarchy and create a location set (in the form of a matrix of coordinates) within the same temporal level. We then compute the geometry, which consists of multiple points in most cases (but just a single points in a few extreme cases). Then, we compute representative points for the selected temporal level. The temporal representative point is the same point for single-point cases. However, it is not necessary to isolate such cases because our last step eventually adds Laplace noise to preserve the privacy.

---

**Algorithm 1.** Temporal representative point

---
1: INPUT: Dataset $D$
2: OUTPUT: Differential private aggregated dataset with temporal point $D'$
3: $T \leftarrow$ temporal hierarchy
4: $M \leftarrow$ coordinate matrix
5: $G \leftarrow$ geometry
6: $D \leftarrow$ dissolve geometry by group ID
7: $R \leftarrow$ representative point from geometry
8: **return** differential private aggregated dataset with temporal point $D'$

---

### 3.3  Differential Privacy via Laplace Mechanism

Our framework guarantees that the actual timestamp and the true location are both protected. Specifically:

- the actual timestamp is protected within temporal hierarchy, and
- the true location is protected with differential privacy mechanism.

Generally, at any temporal hierarchy level $t$, a randomized mechanism $A$ satisfies $\epsilon$-differential privacy on temporal representative point if, for any location output from spatio-temporal database and a neighboring spatio-temporal database obtained by adding or removing a record is not significantly different:

$$\frac{Pr\left(A(x_t = z_t)\right)}{Pr\left(A(x_t^* = z_t)\right)} \leq e^\epsilon \tag{5}$$

At any temporal hierarchy level, the released dataset will not help an adversary or any other external stakeholders to learn more than their prior knowledge. A strong adversary may have accurate true knowledge but cannot identify the true location or the actual time using the published dataset.

The pseudocode in Algorithm 2 shows our Laplace mechanism. The algorithm takes as input the output of the temporal representative point module and applies Laplace noise to the temporal representative point. The probability density function for the Laplace mechanism is computed as:

$$Lap(x|\mu, \epsilon) = \frac{1}{2\epsilon} e^{-\left|\frac{x-\mu}{\epsilon}\right|} \tag{6}$$

---

**Algorithm 2.** Laplace mechanism

---
1: INPUT: Dataset $D$
2: OUTPUT: Differential private dataset $D'$
3: **for each** temporal geometry in $D'$ **do**
4:     generate temporal hierarchy with level = month
5:     $(x, y)$.add(Laplace noise)
6: **return** differential private dataset $D'$

---

## 4   Evaluation

To evaluation our presented framework, we conducted experiments on a 2.6 GHz Intel(R) core i7 64-bit operating system laptop with 8 GB installed memory. The algorithm is implemented in Python on a Spyder scientific development environment. We applied two real-world datasets for our evaluations:

- GeoLife version 1.3 [32], which is a dataset from Microsoft Research Asia Lab capturing outdoor activities of 182 users over a 5-year period from April 2007 to August 2012. There are 24,876,978 points within the 1.49 GB file. Among them, we extracted the data within the city of Beijing, China, with latitude between 32.0°N and 48.0°N as well as longitude between 114.0°E and 120.0°E. This resulted in 21,593,600 points (i.e., tuples). Each tuple contains a user ID, timestamp, longitude, latitude, and altitude.
- Gowalla [8], which is a check-in location dataset from Stanford Network Analysis Platform (SNAP) capturing 6,442,890 check-in records of 196,586 users collected from February 2009 to October 2010. By removing some erroneous tuples, we focused on the resulting 6,442,857 check-in records. Each record contains a user ID, timestamp, latitude, longitude, and location ID.

We implemented the baseline differential privacy using Laplace mechanism and compared it with our temporal hierarchy framework. For a fair comparison, in the baseline Laplace solution, the timestamp is generalized to the day level without aggregation, and Laplace noise is applied to each spatial record.

In terms of metrics, we measured the *utility* of our temporal representative point differential privacy by computing the *haversine distance* $d$ between the released location $(lat_1, lon_1)$ and true location $(lat_2, lon_2)$:

$$d = 2r \sin^{-1} \left( \sqrt{\sin^2 \left( \frac{lat_2 - lat_1}{2} \right) + \cos(lat_1) \cos(lat_2) \sin^2 \left( \frac{lon_2 - lon_1}{2} \right)} \right) \quad (7)$$

where (a) $r$ is the approximate radius of the Earth (e.g., Equatorial radius of about 6,378 km, Polar radius of about 6,356 km), (b) $lat_1$ and $lon_1$ are respectively the latitude and longitude of the released location, and (c) $lat_2$ and $lon_2$ are respectively the latitude and longitude of the true location. Note that, when measuring distance between two global positioning system (GPS) locations expressed in terms of $(latitude, longitude)$, it is common to use haversine distance.

To evaluate the *impact of parameters*, we varied the value of $\epsilon$ from 0.001 to 0.01 for both the baseline Laplace solution and our temporal representative point. We measured the spatial utility, and haversine distance $d$ for each $\epsilon$ value. See Table 1. Results show that:

- For Gowalla dataset, there is no significant difference between the baseline Laplace solution and our temporal representative point differential privacy solution. Our result is consistent with that of related work [30], which stated that the impact of $\epsilon$ is negligible (as check-ins are not frequent).
- For GeoLife, the optimal privacy is observed when $\epsilon$ is between 0.007 and 0.008.

**Table 1.** Average haversine distance $d$

|                                  | GeoLife | Gowalla |
|----------------------------------|---------|---------|
| Baseline laplace                 | 0.7708  | 0.7612  |
| Our temporal representative point | 0.7613  | 0.7572  |

Then, we evaluated the *spatial difference over time* for the entire dataset and for a user in GeoLife. See Table 2. Results show that our temporal point representative gave maximum difference 11 km while the baseline Laplace mechanism gave 23 km. This demonstrates that our framework is capable to provide differential private protection in a group setting while the Laplace mechanism provides differential privacy to each individual point in the dataset.

**Table 2.** Spatial difference over time on GeoLife dataset

|                              | 2007  | 2008  | 2009  | 2010  | 2011  | 2012  |
|------------------------------|-------|-------|-------|-------|-------|-------|
| Baseline laplace             | 16.36 | 21.50 | 23.30 | 20.72 | 20.56 | 18.87 |
| Our temporal representative point | 8.83  | 8.37  | 11.00 | 4.50  | 4.79  | 5.37  |

Generally, aggregation reduces the number of records in a dataset and thus reduces the file size. Moreover, our temporal representative point also reduces the number of spatial information released for the dataset. Hence, to evaluate the *compression ratio*, we measured (a) *file size compression ratio (FCR)* and (b) *spatial compression ratio (SCR)*. The FCR is defined as the ratio of the anonymized dataset $D'$ to the original dataset $D$:

$$FCR(D, D') = \frac{\text{size}(D')}{\text{size}(D)} \tag{8}$$

See Table 3.

Similarly, the SCR is defined as a ratio of distinct spatial points in anonymized dataset $D'$ to the spatial points in original dataset $D$:

$$SCR(D, D') = \frac{\text{count}(P')}{\text{count}(P)} \tag{9}$$

where $P$ and $P'$ are the numbers of points in datasets $D$ and $D'$, respectively. See Table 3. Results show that file size for our temporal representative framework was substantially lower than the original file size. Similarly, the number of points also reduced drastically. We reduced 157,646 points in the original map to 8 temporal representative points while retaining the count of location visited by the user. In addition, the prior knowledge that all user activities take place within the city of Beijing remain the same, none of the new points is outside the city.

**Table 3.** File and spatial compression ratios

|  | File compression ratio (FCR) | | Spatial compression ratio (SCR) | |
|---|---|---|---|---|
|  | GeoLife | Gowalla | GeoLife | Gowalla |
| Baseline laplace | 100% | 100% | 100% | 100% |
| Our temporal hierarchy | 0.0048% | 8.71% | 0.0039% | 5.87% |

## 5    Conclusions

In this paper, we presented a framework for privacy-preserving spatio-temporal patient data publishing. The framework consists of three key components. The first key component is the 5-level temporal hierarchy, which protects temporal attributes. The second component is temporal representative points, which are computed for any set of spatial points within a specific temporal hierarchy. The temporal representative point differential privacy protects spatial attributes in spatio-temporal dataset publishing. The third key component is the differential privacy via Laplace mechanism, which improves the location protection. With these key components, we were able to publish spatio-temporal dataset with privacy preserved. Moreover, we managed to extend the traditional spatial representative points to temporal representative points for privacy-preserving spatio-temporal data publishing. We also managed to non-trivially integrate temporal hierarchy and time series aggregation with these temporal representative points. To evaluate our framework, we conducted experiments on two big datasets: (a) GeoLife dataset, which captures GPS trajectories; and (b) Gowalla dataset, which captures check-in data for a located-based social networking site. The results show that our temporal hierarchy framework provides additional privacy protection through temporal correlation and addition of Laplace noise to ensure the actual location of individuals within a group is preserved (even when there is only one point in the spatio-temporal grouping). The year-by-year temporal distance variation shows the performance of our framework over the baseline Laplace mechanism. Furthermore, our framework achieves high compression of spatial points and input file size to output file size ratio. As ongoing and future work, we consider conducting more exhaustive evaluation and/or comparing with other related works. Moreover, we are extending our solution to privacy-preserving data publishing of coronavirus disease 2019 (COVID-19) patients.

**Acknowledgements.** This work is partially supported by NSERC (Canada) and University of Manitoba.

# References

1. Abul, O., Bonchi, F., Nanni, M.: Never walk alone: uncertainty for anonymity in moving objects databases. In: IEEE ICDE 2008, pp. 376–385 (2018)
2. Acs, G., Castelluccia, C.: A case study: privacy preserving release of spatio-temporal density in Paris. In: ACM KDD 2014, pp. 1679–1688 (2014)
3. Andrés, M.E., Bordenabe, N.E., Chatzikokolakis, K., Palamidessi, C.: Geo-indistinguishability: differential privacy for location-based systems. In: ACM CCS 2013, pp. 901–914 (2013)
4. Arora, N.R., Lee, W., Leung, C.K.-S., Kim, J., Kumar, H.: Efficient fuzzy ranking for keyword search on graphs. In: Liddle, S.W., Schewe, K.-D., Tjoa, A.M., Zhou, X. (eds.) DEXA 2012, Part I. LNCS, vol. 7446, pp. 502–510. Springer, Heidelberg (2012). https://doi.org/10.1007/978-3-642-32600-4_38
5. Athanasopoulos, G., Hyndman, R.J., Kourentzes, N., Petropoulos, F.: Forecasting with temporal hierarchies. EJOR **262**(1), 60–74 (2017)
6. Audu, A.-R.A., Cuzzocrea, A., Leung, C.K., MacLeod, K.A., Ohin, N.I., Pulgar-Vidal, N.C.: An intelligent predictive analytics system for transportation analytics on open data towards the development of a smart city. In: Barolli, L., Hussain, F.K., Ikeda, M. (eds.) CISIS 2019. AISC, vol. 993, pp. 224–236. Springer, Cham (2020). https://doi.org/10.1007/978-3-030-22354-0_21
7. Cao, Y., Yoshikawa, M., Xiao, Y., Xiong, L.: Quantifying differential privacy under temporal correlations. In: IEEE ICDE 2017, pp. 821–832 (2017)
8. Cho, E., Myers, S.A., Leskovec, J.: Friendship and mobility: user movement in location-based social networks. In: ACM KDD 2011, pp. 1082–1090 (2011)
9. Dandekar, A., Basu, D., Bressan, S.: Differential privacy for regularised linear regression. In: Hartmann, S., Ma, H., Hameurlain, A., Pernul, G., Wagner, R.R. (eds.) DEXA 2018, Part II. LNCS, vol. 11030, pp. 483–491. Springer, Cham (2018). https://doi.org/10.1007/978-3-319-98812-2_44
10. Dwork, C.: Differential privacy. In: Bugliesi, M., Preneel, B., Sassone, V., Wegener, I. (eds.) ICALP 2006, Part II. LNCS, vol. 4052, pp. 1–12. Springer, Heidelberg (2006). https://doi.org/10.1007/11787006_1
11. Eom, C.S., Lee, C.C., Lee, W., Leung, C.K.: Effective privacy preserving data publishing by vectorization. Inf. Sci. **527**, 311–328 (2020)
12. Eom, C.S., Lee, W., Leung, C.K.: STDP: secure privacy-preserving trajectory data publishing. In: IEEE Cybermatics 2018, pp. 892–899 (2018)
13. Fung, B.C., Wang, K., Chen, R., Yu, P.S.: Privacy-preserving data publishing: a survey of recent developments. ACM CSur **42**(4), 14:1–14:53 (2010)

14. Khouri, S., Lanasri, D., Saidoune, R., Boudoukha, K., Bellatreche, L.: LogLInc: LoG queries of linked open data investigator for cube design. In: Hartmann, S., Küng, J., Chakravarthy, S., Anderst-Kotsis, G., Tjoa, A.M., Khalil, I. (eds.) DEXA 2019, Part II. LNCS, vol. 11706, pp. 352–367. Springer, Cham (2019). https://doi.org/10.1007/978-3-030-27615-7_27

15. Lakshmanan, L.V.S., Leung, C.K., Ng, R.T.: The segment support map: scalable mining of frequent itemsets. ACM SIGKDD Explor. **2**(2), 21–27 (2000)

16. LeFevre, K., DeWitt, D.J., Ramakrishnan, R.: Incognito: efficient full-domain $k$-anonymity. In: ACM SIGMOD 2005, pp. 49–60 (2005)

17. Leung, C.K.: Frequent itemset mining with constraints. In: Encyclopedia of Database Systems, 2nd edn., pp. 1531–1536 (2018)

18. Leung, C.K., Hoi, C.S.H., Pazdor, A.G.M., Wodi, B.H., Cuzzocrea, A.: Privacy-preserving frequent pattern mining from big uncertain data. In: IEEE BigData 2018, pp. 5101–5110 (2018)

19. Leung, C.K.-S., Tanbeer, S.K., Cameron, J.J.: Interactive discovery of influential friends from social networks. Soc. Netw. Anal. Min. **4**(1), 154:1–154:13 (2014). https://doi.org/10.1007/s13278-014-0154-z

20. Leung, C.K., Zhang, H., Souza, J., Lee, W.: Scalable vertical mining for big data analytics of frequent itemsets. In: Hartmann, S., Ma, H., Hameurlain, A., Pernul, G., Wagner, R.R. (eds.) DEXA 2018, Part I. LNCS, vol. 11029, pp. 3–17. Springer, Cham (2018). https://doi.org/10.1007/978-3-319-98809-2_1

21. Li, N., Li, T., Venkatasubramanian, S.: $t$-closeness: privacy beyond $k$-anonymity and $l$-diversity. In: IEEE ICDE 2007, pp. 106–115 (2007)

22. Lin, J.C.-W., Zhang, Y., Fournier-Viger, P., Djenouri, Y., Zhang, J.: A metaheuristic algorithm for hiding sensitive itemsets. In: Hartmann, S., Ma, H., Hameurlain, A., Pernul, G., Wagner, R.R. (eds.) DEXA 2018, Part II. LNCS, vol. 11030, pp. 492–498. Springer, Cham (2018). https://doi.org/10.1007/978-3-319-98812-2_45

23. Machanavajjhala, A., Kifer, D., Gehrke, J., Venkitasubramaniam, M.: $l$-diversity: privacy beyond $k$-anonymity. ACM TKDD **1**(1), 3:1–3:52 (2007)

24. Mendes, R., Vilela, J.P.: Privacy-preserving data mining: methods, metrics, and applications. IEEE Access **5**, 10562–10582 (2017)

25. Mohammed, N., Fung, B.C.M., Debbabi, M.: Walking in the crowd: anonymizing trajectory data for pattern analysis. In: ACM CIKM 2009, pp. 1441–1444 (2009)

26. Olawoyin, A.M., Chen, Y.: Predicting the future with artificial neural network. Procedia Comput. Sci. **140**, 383–392 (2018)

27. Statistics Canada: Representative point. Illustrated Glossary, Census Year 2011 (2011). https://www150.statcan.gc.ca/n1/pub/92-195-x/2011001/other-autre/point/point-eng.htm

28. Tanbeer, S.K., Leung, C.K., Cameron, J.J.: Interactive mining of strong friends from social networks and its applications in e-commerce. JOCEC **24**(2–3), 157–173 (2014)

29. Wodi, B.H., Leung, C.K., Cuzzocrea, A., Sourav, S.: Fast privacy-preserving keyword search on encrypted outsourced data. In: IEEE BigData 2019, pp. 6266–6275 (2019). https://doi.org/10.1109/BigData47090.2019.9046058

30. Xiao, Y., Xiong, L.: Protecting locations with differential privacy under temporal correlations. In: ACM CCS 2015, pp. 1298–1309 (2015)

31. Yin, Y., Kaku, I., Tang, J., Zhu, J.: Privacy-preserving data mining. In: Data Mining. Decision Engineering, pp. 101–119. Springer, London (2011). https://doi.org/10.1007/978-1-84996-338-1_6

32. Zheng, Y., Xie, X., Ma, W.: GeoLife: a collaborative social networking service among user, location and trajectory. IEEE Data Eng. Bull. **33**(2), 32–39 (2010)

# Correction to: Database and Expert Systems Applications

Sven Hartmann, Josef Küng, Gabriele Kotsis, A Min Tjoa,
and Ismail Khalil

## Correction to:
**S. Hartmann et al. (Eds.): *Database and Expert Systems Applications*, LNCS 12392,
https://doi.org/10.1007/978-3-030-59051-2**

The paper "Bounded Pattern Matching Using Views" was accidentally published twice in the book "Database and Expert Systems Applications, Part II", and the paper "View Selection for Graph Pattern Matching" was omitted. This has been corrected. Each paper is now included once.

The updated version of the book can be found at
https://doi.org/10.1007/978-3-030-59051-2

# Author Index

Printed in the United States
by Baker & Taylor Publisher Services

Printed in the United States
by Baker & Taylor Publisher Services